VOLUME TWO HUNDRED AND THIRTY

Advances in
IMAGING AND
ELECTRON PHYSICS

Coulomb Interactions in Particle Beams

EDITOR-IN-CHIEF

Martin Hÿtch
CEMES-CNRS
Toulouse, France

ASSOCIATE EDITOR

Peter W. Hawkes
CEMES-CNRS
Toulouse, France

VOLUME TWO HUNDRED AND THIRTY

Advances in
IMAGING AND ELECTRON PHYSICS
Coulomb Interactions in Particle Beams

GUUS JANSEN

Edited by

MARTIN HŸTCH
CEMES-CNRS Toulouse, France

PETER W. HAWKES
CEMES-CNRS Toulouse, France

Cover photo credit: "Monte Carlo simulation of the electron displacements caused by Coulomb interactions" (Ref. Chapter 14, Figure 14.6)

Academic Press is an imprint of Elsevier
125 London Wall, London EC2Y 5AS, United Kingdom
525 B Street, Suite 1650, San Diego, CA 92101, United States
50 Hampshire Street, 5th Floor, Cambridge, MA 02139, United States

Copyright © 2024 Elsevier Inc. All rights are reserved, including those for text and data mining, AI training, and similar technologies.

Publisher's note: Elsevier takes a neutral position with respect to territorial disputes or jurisdictional claims in its published content, including in maps and institutional affiliations.

For accessibility purposes, images in this book are accompanied by alt text descriptions provided by Elsevier.

MATLAB® is a trademark of The MathWorks, Inc. and is used with permission.
The MathWorks does not warrant the accuracy of the text or exercises in this book.
This book's use or discussion of MATLAB® software or related products does not constitute endorsement or sponsorship by The MathWorks of a particular pedagogical approach or particular use of the MATLAB® software.

No part of this publication may be reproduced or transmitted in any form or by any means, electronic or mechanical, including photocopying, recording, or any information storage and retrieval system, without permission in writing from the publisher. Details on how to seek permission, further information about the Publisher's permissions policies and our arrangements with organizations such as the Copyright Clearance Center and the Copyright Licensing Agency, can be found at our website: www.elsevier.com/permissions.

This book and the individual contributions contained in it are protected under copyright by the Publisher (other than as may be noted herein).

Notices
Knowledge and best practice in this field are constantly changing. As new research and experience broaden our understanding, changes in research methods, professional practices, or medical treatment may become necessary.

Practitioners and researchers must always rely on their own experience and knowledge in evaluating and using any information, methods, compounds, or experiments described herein. In using such information or methods they should be mindful of their own safety and the safety of others, including parties for whom they have a professional responsibility.

To the fullest extent of the law, neither the Publisher nor the authors, contributors, or editors, assume any liability for any injury and/or damage to persons or property as a matter of products liability, negligence or otherwise, or from any use or operation of any methods, products, instructions, or ideas contained in the material herein.

ISBN: 978-0-443-29784-7
ISSN: 1076-5670

For information on all Academic Press publications
visit our website at https://www.elsevier.com/books-and-journals

Publisher: Zoe Kruze
Acquisitions Editor: Jason Mitchell
Editorial Project Manager: Palash Sharma
Production Project Manager: James Selvam
Cover Designer: Gopalakrishnan Venkatraman

Typeset by STRAIVE, India

Contents

Preface	*xi*

1. Introduction — 1

    1.1 Introduction — 1
    1.2 Focused particle beam systems — 2
    1.3 Classification of interaction phenomena — 3
    1.4 Organization of the chapters — 6

2. Historical notes — 9

    2.1 Introduction — 9
    2.2 Particle optics and particle interactions — 9
    2.3 Boersch effect — 11
    2.4 Trajectory displacement effect — 19
    2.5 Space charge effect in low density particle beams — 23
    2.6 Monte Carlo simulation — 25

3. General beam properties — 27

    3.1 Introduction — 27
    3.2 Beam parameters — 28
    3.3 Classification of beams — 31
    3.4 Hamilton's formalism and Liouville's theorem — 32
    3.5 Boltzmann equation — 35
    3.6 Conservation of beam emittance and brightness — 37
    3.7 Beam temperature — 41
    3.8 Thermodynamic limits for particle interaction effects — 49
    3.9 Debeye screening — 51
    3.10 Potential energy relaxation — 55

4. The many body problem of particles interacting through an Inverse Square force law — 59

    4.1 Introduction — 59
    4.2 Vlasov equation — 60
    4.3 Fokker–Planck equation — 63

4.4	Aspects of the diffusion approximation	67
4.5	Calculation of coefficients of dynamical friction and diffusion	69
4.6	Coulomb logarithm	75
4.7	Discussion of the Fokker–Planck approach	79
4.8	Validity of the Fokker–Planck approach for particle beams	81
4.9	Holtsmark distribution	86
4.10	Conclusions	89

5. Concepts of an analytical model for statistical interactions in particle beams   91

5.1	Introduction	91
5.2	General formulation of the problem	92
5.3	Reduction of the N-particle problem	96
5.4	Calculation of the displacement distribution	106
5.5	Moments and cumulants of the displacement distribution	109
5.6	On- and off-axis reference trajectories	112
5.7	Models in one, two, and three dimensions	113
5.8	Distribution of the interaction force in cylindrical beams	115
5.9	Representation in the k-domain	120
5.10	Addition of the effects generated in individual beam segments	127
5.11	Slice method	131

6. Two-particle dynamics   133

6.1	Introduction	133
6.2	Basic properties	134
6.3	Coordinate representation in the orbital plane	138
6.4	Dynamics of a complete collision	141
6.5	General analysis of the two-particle dynamical problem	144
6.6	Numerical approach to the dynamical problem	147
6.7	Dynamics of a nearly complete collision	148
6.8	First order perturbation dynamics	153
6.9	Collisions with zero initial relative velocity	157
6.10	Expressions for the longitudinal velocity shift Δv_z	160
6.11	Expressions for the transverse velocity shift $\Delta v_\perp$	162
6.12	Expressions for the spatial shift Δr	163
6.13	Coulomb scaling	166
	Appendix 6.A. Mathematics of nearly complete collision dynamics	167

7. Boersch effect — 173

7.1 Introduction — 173
7.2 General aspects — 174
7.3 Beam segment with a narrow crossover — 177
7.4 Homocentric cylindrical beam segment — 189
7.5 Beam segment with a crossover of arbitrary dimensions — 198
7.6 Results for Gaussian angular and spatial distributions — 213
7.7 Thermodynamic limits — 218

8. Statistical angular deflections — 221

8.1 Introduction — 221
8.2 General aspects — 222
8.3 Beam segment with a narrow crossover — 225
8.4 Homocentric cylindrical beam segment — 231
8.5 Beam segment with a crossover of arbitrary dimensions — 240
8.6 Application of the slice method — 261
8.7 Results for Gaussian angular and spatial distributions — 262

9. Trajectory displacement effect — 267

9.1 Introduction — 267
9.2 General aspects — 268
9.3 Homocentric beam segment with a crossover — 272
9.4 Homocentric cylindrical beam segment — 286
9.5 Beam segment with a crossover of arbitrary dimensions — 288
9.6 Trajectory displacement and angular deflection distribution — 293
9.7 Results for Gaussian angular and spatial distributions — 296

10. Further investigations on statistical interactions — 299

10.1 Introduction — 299
10.2 Exact approach for off-axis reference trajectories — 300
10.3 Approximating approach for off-axis reference trajectories — 307
10.4 Nonmonochromatic beams — 312
10.5 Beams in an external uniform axial electrostatic field — 320
10.6 Relativistic beams — 324
Appendix 10.A. Distributions for the average reference trajectory — 327

11. Space charge effect in low density particle beams — 329

- 11.1 Introduction — 329
- 11.2 General aspects — 330
- 11.3 Beams with laminar flow — 338
- 11.4 First order perturbation theory — 341
- 11.5 First order optical properties of the space charge lens — 343
- 11.6 Third order geometrical aberrations of the space charge lens — 346
- 11.7 Beam segment with a narrow crossover — 350
- 11.8 Homocentric cylindrical beam segment — 352
- 11.9 Addition of the effects generated in individual beam segments — 355

12. Calculation of different spot- and edge-width measures — 357

- 12.1 Introduction — 357
- 12.2 Spot-width obtained by knife-edge scans — 358
- 12.3 Edge-width of a shaped spot — 361
- 12.4 Trajectory displacement effect — 365
- 12.5 Chromatic aberration — 370
- 12.6 Spherical aberration — 384
- 12.7 Space charge effect — 388
- 12.8 Results for a truncated Gaussian angular distribution — 390

13. Monte Carlo simulation of particle beams — 393

- 13.1 Introduction — 393
- 13.2 Source routine — 395
- 13.3 Optical elements — 398
- 13.4 Numerical ray tracing — 401
- 13.5 Analytical ray tracing — 406
- 13.6 Simulation of large currents near the source — 412
- 13.7 Correction of finite-size effects — 413
- 13.8 Data analysis — 416
- 13.9 Accuracy limitations of the MC program — 421
- 13.10 MC simulation vs analytical modeling — 423
- 13.11 Program organization and examples — 424
- Appendix 13. A. Random number routine — 430
- Appendix 13. B. Polynomial fit algorithm — 433

14. Comparison of analytical theory with Monte Carlo simulations — 437

- 14.1 Introduction — 437
- 14.2 General aspects — 438

14.3	Voltage and current dependencies for a fixed geometry	441
14.4	Geometry and current dependencies for a fixed beam voltage	445
14.5	Discussion of the results	450

15. Comparison of recent theories on statistical interactions — 455

15.1	Introduction	455
15.2	Boersch effect	457
15.3	Statistical angular deflections	467
15.4	Trajectory displacement effect	476
15.5	Conclusions	482

16. Summary for the one-minute designer — 487

16.1	Introduction	487
16.2	Physical aspects	488
16.3	Parameter dependencies	499
16.4	Equations for the Boersch effect	507
16.5	Equations for the trajectory displacement effect	513
16.6	Equations for the space charge effect	516
16.7	Addition of the effects generated in individual beam segments	518

References — *521*
Index — *529*

Preface

In this volume of Advances in Electronic and Electron Physics, we are pleased to make available again the work of G.H. Jansen that first appeared as a supplement of Advances in Electronics and Electron Physics. All the regular volumes of AEEP are available on ScienceDirect but the supplements were never digitized. Some of these are extremely important and, like the present work, are routinely cited as monographs. Elsevier have generously agreed to reprint a selection of these supplements as volumes of these Advances, beginning with Saxton (AIEP Volume 214), H.J.A.M. Heijmans (AIEP Volume 216) and Ximen (AIEP Volume 226). We are very grateful to the author of the current volume for continuing this endeavour.

Martin Hÿtch and Peter W. Hawkes

This book is concerned with the theory of Coulomb interactions between charged particles in low- and medium-density nonrelativistic beams used in electron and ion beam lithography instruments, scanning microscopes, and other similar devices. The different phenomena of Coulomb interactions—Boersch effect, trajectory displacement effect, and space charge defocussing and blurring—have been the subject of a great deal of experimental and theoretical study during the past 30 years and also the subject of a certain amount of controversy. One objective of this book is to review and compare the various contributions to the field. An effort has been made to indicate the relation with the approaches taken in plasma physics and stellar dynamics to handle similar many-body problems. The main objective of this book, however, is to present, in some detail, an analytical model that yields accurate results for all phenomena involved over a wide range of operating conditions. The accuracy of this model has been verified by comparing its predictions to the outcomes of a large number of numerical Monte Carlo simulations. The principles of this technique are described in detail.

The book's chapters are organized in such a way that, in principle, each chapter can be read independently of the others. The last chapter, entitled "Summary for the one-minute designer," is especially written for those who do not wish to go through all the detailed calculations but do wish to understand the final equations and the underlying physics.

The emphasis is on fundamentals rather than applications. Nevertheless, the book's purpose is to provide the designer of particle beam systems with practical and accurate tools to predict the impact of Coulomb interactions on the system's resolution. At present, such tools are lacking in the standard aberration theory of particle optics. In the course of the research prior to the writing of this book, two computer programs have been developed to assist the designer of particle beam systems. One, called INTERAC, is based on the equations resulting from the analytical model. It can best be compared with a dedicated calculator for Coulomb interactions. The other, called MONTEC, is a fast Monte Carlo program. It is based on a direct ray-tracing of the particles through a user-defined system. It involves fewer physical assumptions than the analytical model and is more computation-intensive. INTERAC and MONTEC are complementary in many respects; their combination was found to be very instrumental in the design process.

The physics on which the programs are based is described in this book. The actual source code and executable modules are made available by the Delft Particle Optics Foundation. Further information on the programs and conditions of sale can be obtained through the research group Particle Optics, Faculty of Applied Physics, Delft University of Technology, PO Box 5046, 2600 GA Delft, The Netherlands.

This work resulted from a Ph.D. project that was carried out partly at Delft University of Technology and partly at the electron beam lithography group of IBM, General Technology Division, East Fishkill, New York. I am indebted to Prof. Dr. J.M.J. van Leeuwen, Prof. Dr. ir. J.B. Le Poole, and Dr. J. Barth for introducing me to the subject. Important contributions and valuable criticism concerning the fundamentals of the theory as well as the preliminary draft of this book were made by Prof. Dr. J.M.J. van Leeuwen. I thank Prof. Dr. ir. K.D. van der Mast for creating a suitable position at Delft University in which I could complete this work; for the stimulating discussions on the applications of the theory and further improvements of the Monte Carlo program in its final stage; and for his valuable comments on the preliminary draft of this book. I am most grateful to my managers at IBM, Dr. W. Shekel and Dr. H.C. Pfeiffer, for giving me the opportunity to experience the industrial environment of an electron beam lithography development laboratory and for pointing out the real problems and needs in the design of electron beam systems. I hope that this work will provide them with the "screwdriver" they asked for. I would also like to acknowledge the valuable discussions with my colleagues at IBM, Dr. T.R. Groves and Dr. W.D. Meisburger, especially with respect to the technique of Monte Carlo simulation. I am grateful to the management of my present affiliation, the Royal Dutch/Shell Laboratories, Amsterdam, the Netherlands, for providing facilities in the final stage of this project. Finally, I would like to thank my family and all my friends for the support and diversion provided during the writing of this book.

G.H. JANSEN
December 1989

CHAPTER ONE

Introduction

Contents

1.1. Introduction	1
1.2. Focused particle beam systems	2
1.3. Classification of interaction phenomena	3
1.4. Organization of the chapters	6

1.1. Introduction

The standard theory of charged particle optics is mainly concerned with the calculation of the imaging properties of electrostatic and magnetic fields utilized in electron microscopes and similar devices. In this type of instrument, it is usually sufficient to study the trajectory of an individual particle in an external field, neglecting the effect of particle interactions. This leads to a description that closely resembles the geometrical optics developed for light optical systems. However, the analogy between charged particle optics and photon optics breaks down for beams in which the interaction between the particles cannot be ignored. Such is, for instance, the case in the intense beams used for electron beam lithography and the low-voltage beams used for inspection and testing of electronic circuits and chip packaging. The resolution of these systems is not just determined by geometrical and chromatic aberrations, but also by the Coulomb interactions between the particles in the beam. With increasing particle density, their impact becomes dominant among the factors limiting the system performance.

One manifestation of the particle interactions is the energy broadening generated in the drift length of a column. This phenomenon was first systematically investigated by Boersch (1954) and therefore bears his name. It affects the system resolution via the chromatic aberration of lenses and deflectors. Another manifestation is the generation of stochastic lateral shifts in the particle positions and velocities, called the trajectory displacement effect. It directly causes a deterioration of the system resolution. Both phenomena stem from the discrete nature of the charges and should, in that respect, be distinguished from the space charge effect, which depends on the average charge density in the beam. The space charge effect causes (in first order approximation) only a defocusing of the spot and has no impact on the best obtainable system resolution.

This monograph is devoted to the theory of particle interactions in low and medium density, nonrelativistic, and time-independent beams of identical particles, in probe-forming instruments. The qualifications "low and medium density" and "nonrelativistic" refer to beam currents typically much smaller than 1 mA and beam voltages typically between 1 and 100 keV. "Time-independent" implies that the flow of particles is assumed to be constant. The theory is developed for electron beam lithography pattern generators and scanning electron microscopes. It should be applicable to focussed ion beam tools as well. The analytical models used are based on the ideas of van Leeuwen and Jansen (1983). This work is extended and confronted with other theories and the results of numerical Monte Carlo simulation. The major part of this monograph is concerned with the impact of particle interactions on beams in drift space. Some preliminary results are presented to extend the model to beams located in an external uniform acceleration field and to relativistic beams

1.2. Focused particle beam systems

The research into the different phenomena of Coulomb interactions in particle beams is strongly stimulated by the developments in the field of electron beam lithography. The fabrication of Very Large Scale Integrated (VLSI) circuits requires a technology that is able to produce patterns with a minimum feature size below 1 µm. With the trend towards further reduction of the desired feature size, the usefulness of the conventional optical lithographic process becomes restricted by fundamental limits related to the wavelength of light. In order to overcome these limits, the technology of electron beam lithography was developed during the late 1960s and 1970s. It has been demonstrated that electron beam tools can produce patterns of features as small as tens of nanometers. It became clear, however, that this technology suffers from some fundamental limits of its own. The main drawback of electron beam lithography is its limited machine throughput, due to the sequential nature of the exposure process. Therefore, at present, electron beams are primarily used for the production of masks, whereas direct slice writing is limited to special applications. The main objective of system designers is to decrease the feature size while maintaining or increasing the throughput. As the exposure time is inversely proportional to the current in the probe, machine throughput can be improved by increasing the beam current. However, the deteriorating impact of both the Boersch effect and the trajectory displacement effect becomes more severe for larger currents, imposing an upper limit to the beam current at a certain resolution. The system resolution is not affected by the space charge effect, provided that its defocusing action can be compensated by proper lens adjustment. In this respect, it should be noted that difficulties arise in variable shaped spot lithography systems, in which the beam current (and thus the space charge defocusing) changes on a spot-by-spot basis. An adequate theoretical description of the different phenomena involved is indispensable in the optimization of the design of high throughput electron beam lithography machines.

Scanning electron microscopes generally operate at smaller beam currents than electron beam lithography machines. Consequently, the effects of particle interactions are less dominant. However, they can not be ignored in systems with high-brightness guns operating at low beam voltages. In a scanning system, the brightness of the beam determines the obtainable signal to noise ratio (at a certain integration time per pixel). High brightness is thus desirable. Operation at a low beam voltage is necessary in some applications to prevent the specimen from electrical charging. Therefore, the systems employed for testing and inspection of electronic circuits and chips packaging, as well as the analytical systems for the research of isolators, typically operate around 1 kV. In the design of those systems, it should be anticipated that particle interactions may impact the performance of the beam considerably.

Focused ion beam systems have gained interest during the last decade and are now widely studied in research laboratories. Their main application is mask-repair, while focused ion beam systems for lithography and direct maskless implantation are under development. Presently, the highest brightness is produced by liquid metal ion sources. A severe disadvantage of the liquid metal ion source is its large energy spread. Focused ion beam systems employing this type of gun are therefore usually limited by chromatic aberration. The effect of virtual source growth is observed as well, but in general, it does not limit the system performance. It is not clear whether these phenomena should be entirely attributed to the Coulomb interactions between the particles in the beam. The observed current dependency indicates that they are at least partly responsible.

1.3. Classification of interaction phenomena

In order to classify the different manifestations of the Coulomb interactions it is necessary to introduce some theoretical concepts first. Consider a set of *test particles* running, successively, along a specific trajectory in the beam, called the *reference trajectory*. Assume that they all have the same velocity and are well separated in space. Each of the test particles experiences the Coulomb repulsion of the particles by which it is surrounded, referred to as *field particles*. As a result, it will experience a displacement from its unperturbed path (which is the reference trajectory) both in position and velocity. This displacement is entirely determined by the initial coordinates of all field particles relative to the test particle (with "initial coordinates" we denote both the positions and the velocities at the start of the interaction). Clearly, different configurations of initial coordinates lead to different displacements. Accordingly, the displacements of the test particles, successively arriving at the target, will follow a certain distribution, which is related to the statistics of the beam.

The three-dimensional distributions of displacements in position and velocity experienced by the test particles represent the effect of Coulomb interactions. In general, a distribution can be characterized by its mean value and a measure for its width, e.g., the root mean square (rms) of the displacements from the mean value. The mean values of the distributions investigated here correspond to the average shift in position and velocity of all test particles. These average shifts can be related to the smoothed out distribution of charge in the beam and are therefore classified as manifestations of the *space charge effect*. The space charge effect is strictly deterministic. In other words, it does not depend on the statistics of the beam. On the other hand, the widths of the distribution of displacements of the test particles are directly related to the stochastic fluctuations in the charge density. These fluctuations occur as a consequence of the discrete (particle) nature of the beam. The distribution of displacements in position and velocity, relative to the average shift of the test particles, represent the so-called *statistical effects* In summary, space charge (or deterministic) effects are related to the smoothed out charge density in the beam. Statistical effects are related to the stochastic fluctuations in the charge density due to the discrete nature of the beam.

The magnitude of the different manifestations of Coulomb interactions depends, to some extent, on the choice of the reference trajectory. The trajectory along the beam axis of a rotational symmetric beam represents a case that is of special interest. The strength of the electrostatic force, generated by the smoothed out distribution of charge, is zero on this trajectory due to the symmetry of the beam. The space charge effect is therefore absent and the resulting displacements are purely statistical. Particles moving along other trajectories in the beam will be subject to the combined action of space charge and statistical effects.

Space charge effect and statistical effects are sometimes presented as the result of collective interaction and collisions between individual particles respectively. This interpretation is instructive for beams of very high density in which every test particle is surrounded by a large number of field particles. The total force acting on an individual test particle is, indeed, produced by the simultaneous action of many field particles. Accordingly, it appears to be reasonable to distinguish between the influence of the system as a whole and the influence of the local neighborhood. The former is associated with the space charge force, which is a smoothly varying function of position and time. The latter gives rise to rapid fluctuations in the interaction force. The corresponding displacements of the test particle can be envisioned as the result of collisions with individual field particles. On the other hand, for low density beams the distinction between collective effects and individual particle effects becomes inadequate. In such beams, a test particle is surrounded by just a limited number of field particles. Accordingly, it is meaningless to separate the force acting on the test particle into a space charge component and a fluctuating component. For such conditions, space charge effect and statistical effects can only be identified as the mean and the width of the displacement distribution respectively, which follows by observing a large number of test particles. This viewpoint was described above.

Two manifestations of statistical effects are distinguished, as was mentioned previously. The *Boersch effect* corresponds to a broadening of the longitudinal velocities of the test particles, which can directly be related to a broadening of the normal energy distribution. The *trajectory displacement effect* corresponds to a spatial broadening in a certain reference plane. It is the result of stochastic shifts in both the lateral position and the lateral velocity of the test particles. These quantities are combined to a single spatial shift in the reference plane by extrapolating the final position of a test particle to a position in the reference plane by a straight line along its final velocity. In order to relate the trajectory displacement effect to a broadening of the final probe, it is essential to consider a reference plane that is imaged on the target.

From a theoretical point of view, it can be convenient to study the shifts in lateral position and lateral velocity separately. The random shifts in lateral velocity correspond to the so-called effect of *statistical angular deflections*. We emphasize that this concept is only of theoretical significance. In experiment, one is only interested in the contribution of the statistical interactions to the blurring of the final probe. In our terminology, this blurring is entirely the result of the trajectory displacement effect. Finally, it should be noted that different terms are used in literature to describe the impact of statistical Coulomb interactions on the lateral properties of the beam. Some of these terms are reproduced in Chapter 2.

The lateral component of the space charge force, acting on a particle within a rotational symmetric beam, is proportional to its radial distance to the axis, provided that the distribution of charge is uniform in the cross section of the beam. Accordingly, the space charge effect can be described as the action of a perfect negative lens, causing a defocusing of the final probe as well as some (de)magnification. These effects can be compensated by proper lens adjustment, leaving the system resolution unaffected. For nonuniform charge density distributions, the space charge effect corresponds to the action of a nonideal lens, causing a blurring of the probe that can not entirely be refocused. This nonrefocusable space charge broadening, due to nonuniformity of the smoothed out distribution of charge, should be well distinguished from the trajectory displacement effect.

1.4. Organization of the chapters

The chapters are organized such that one can, in principle, read each chapter separate from the others. Explicit references are therefore made when using the results presented elsewhere in this monograph. Some of the chapters start with a summary of the relevant results presented up to that point. Clearly, this causes some overlap in the text, but we hope it improves the readability of this work, especially for those who do not wish to go through all the material.

Chapter 2 reviews the various contributions to the subject of Coulomb interactions in particle beams and gives reference to some related fields of physics as well. The history of the Boersch effect, the trajectory displacement effect, the space charge effect and the technique of numerical Monte Carlo simulation are successively described, stressing the physical aspects involved. Chapter 3 introduces some general beam properties that are used throughout this monograph. Liouville's theorem, well known from statistical mechanics, is employed to derive the principle of emittance and brightness invariance. A thermodynamical description of the beam is considered, concentrating on the notion of beam temperature. The thermodynamical description of the beam is exploited to determine upper limits for the effect of statistical interactions, utilizing the concepts of relaxation of kinetic energy and relaxation of potential energy. Chapter 4 studies the N-body problem of particles interacting through a Coulomb-type of interaction force, from a general perspective. It reviews the Fokker-Planck approach, which is used in plasma physics and stellar dynamics. The applicability of this approach to the problem of Coulomb interactions in particle beams is examined. The Holtsmark distribution is introduced to describe the effect of statistical interactions in a cylindrical beam of moderate particle density. Chapters 2, 3, and 4, which approach the subject of Coulomb interactions from a general viewpoint, are intended to clarify the background of the analytical models presented in the remaining chapters and to elucidate the fundamental aspects involved.

Chapter 5 discusses the basic aspects of an analytical model for the statistical Coulomb interactions between the particles of a beam. Various methods of reducing this type of N-body problem are examined. Chosen is a model based on two-particle interactions, which is accurate as long as the test particle is not involved in two or more *strong* interactions, simultaneously or successively, in the same beam segment. The statistical aspects of this so-called *extended two-particle model* are described in detail. The Holtsmark distribution is derived within the model, as well as some related distributions, pertaining to a beam with small lateral dimensions relative to the average axial separation of the particles. Chapter 6 studies the dynamical problem of two particles interacting through the Coulomb force. The classic solution of the Kepler problem pertains to a collision that is complete, which means that the particles effectively come from infinity and recede to infinity. However, the extended two-particle model requires knowledge of collisions that are possibly incomplete. It is shown that the problem can be solved analytically for nearly complete collisions as well as for weak collisions. A numerical method to treat strong incomplete collisions is also presented. Chapter 7 utilizes the material of Chapters 5 and 6 to calculate the Boersch effect in a beam segment in drift space. Chapter 8 is concerned with the effect of statistical angular deflections, while Chapter 9 discusses the trajectory displacement effect. Chapter 10 is dedicated to some further investigations on statistical interactions to cover off-axis reference trajectories, initially nonmonochromatic beams, as well as uniform acceleration fields. The case of relativistic beams, in which the particles travel with a velocity comparable to the speed of light, is also investigated. An analytical model for the space charge effect in low density particle beams is described in Chapter 11. The first and third order geometrical properties of the associated "space charge lens" are calculated. Chapter 12 translates the results from the previous chapters in terms of some experimental measures for the spot width in source imaging (Gaussian beam) systems and the edge width in aperture imaging (shaped beam) systems. Chapters 5–12 cover the main results of the analytical investigations.

Chapter 13 describes the principles of numerical Monte Carlo simulation of charged particle beams. An *analytical* ray tracing method is described, which provides an alternative to the usual third order *numerical* ray tracing procedure. It is based on a decomposition of the full interaction into a sum of two-particle effects, quite similar to our analytical model for statistical interactions. While preserving the typical advantages of the Monte Carlo approach, it improves the speed of the program with one to two orders of magnitude, without loss of accuracy, for practical operating conditions. This approach is referred to as *fast* Monte Carlo simulation of charged particle beams. Several other features to improve the performance of the MC-program and to extend its applicability to high density beams are discussed.

Chapter 14 compares the analytically obtained expressions for the Boersch effect, the trajectory displacement effect, and the space charge effect with the results of Monte Carlo simulations (using numerical ray tracing). This material serves to verify the accuracy of the analytical models.

Chapter 15 confronts the extended two-particle approach with some other recent theories on statistical interactions. It focusses on a quantitative comparison of the different theories. It should be recalled that a qualitative discussion of the different approaches is incorporated in Chapters 2, 4, and 5.

Chapter 16 summarizes the results presented in this monograph which are relevant for the design of probe-forming instruments. A review is given of the physical aspects of Coulomb interactions in particle beams and the basic theoretical parameters involved. The dependency on the experimental parameters, for the various effects, is explained on the basis of some elementary physical considerations. Finally, an overview is given of the resulting equations for the calculation of the Boersch effect, the trajectory displacement effect, and the space charge effect.

CHAPTER TWO

Historical notes

Contents

2.1.	Introduction	9
2.2.	Particle optics and particle interactions	9
2.3.	Boersch effect	11
2.4.	Trajectory displacement effect	19
2.5.	Space charge effect in low density particle beams	23
2.6.	Monte Carlo simulation	25

2.1. Introduction

This chapter is concerned with the history of the subject of particle interactions in particle beams. Previous reviews, presented by Zimmermann (1970) and Rose and Spehr (1983), were rather short on the early theoretical work as well as the relation with activities in adjacent fields. Therefore, a somewhat more detailed reference is presented here. The subjects of the Boersch effect, the trajectory displacement effect, the space charge effect and the technique of numerical Monte Carlo simulation are treated separately. The various contributions to each of these subjects are arranged chronologically. The descriptions are only qualitative, emphasizing the physical aspects involved. A systematic discussion of the different approximations underlying the various analytical models is included in Chapter 5 (statistical interactions) and Chapter 11 (space charge effect), while the reader is referred to Chapter 15 for a quantitative comparison of the results of the different models for statistical interactions.

2.2. Particle optics and particle interactions

The branch of physics studying the properties of focussed electron or ion beams is called particle optics. It originates from the mid-1920s, when the focussing action of a magnetic field on an electron beam was first studied by Busch (1926). Particle optics, as it has developed ever since, is mainly concerned with the calculation of the trajectories of charged particles in an external electrostatic or magnetic field. The particles are

Advances in Imaging and Electron Physics, Volume 230
ISSN 1076-5670
https://doi.org/10.1016/B978-0-443-29784-7.00002-9

essentially treated as independent, neglecting the effect of particle interactions. In this respect, it is completely analogous to the geometrical optics used for light optical systems. The theory is centered around the concept of a paraxial beam. In such a beam the particles flow near some axis on a trajectory that makes only a small angle with that axis. The ray equation, which follows from Maxwell's field equations using classical mechanics, is a second order differential equation, specifying the lateral ray coordinates as functions of the axial coordinate. For a paraxial beam, it can be approximated by a power expansion in the angle and the distance to the axis. Truncation of the expansion after the first order terms (in this angle and distance) constitutes the so-called paraxial ray equation, which gives rise to Gaussian optics. It serves to relate the position of an object plane to the position of an image plane and to determine the magnification from the first to the second. The geometrical aberrations follow by taking the higher order terms of the ray equation into account, employing perturbation theory. In this approach it is sufficient to study the total deviation experienced by particles running along a few principal (first order) rays that correspond to the independent solutions of the paraxial ray equation. (One requires two principal rays for a rotational symmetric beam and four principal rays to treat the general case.)

The performance of a probe forming system, as a whole, follows from the optical properties of its components, usually specified in terms of the first order imaging quantities and the aberration constants. The quality of the beam in the image plane can be derived from the total geometrical and chromatic aberrations, found by addition of the contribution of the individual components. This approach is extensively described in the various textbooks on electron optics; for instance, see Glaser (1952), Klemperer (1953), Zvorykin et al. (1961), Grivet (1965), Septier (1967), Szilagyi (1988) or for high energy beams Steffen (1965) and Lawson (1977).

The analogy with light optics breaks down for high intensity particle beams in which the interaction between the particles cannot be neglected. The significance of collective interaction, referred to as the space charge effect, was already recognized in the early stages of electron optics. It was understood that the space charge effect can dominate the emission properties in thermionic guns and may limit the minimum obtainable spot size in oscilloscope tubes. In the last device, defocussing and beam spreading due to the space charge effect were calculated by assuming laminar flow. This denotes that the beam is visualized as a fluid in which particle trajectories do not cross each other. In case of laminar flow, it is sufficient to consider the trajectory of a particular running along the edge of the beam. The space charge force acting on this particle can directly be expressed in terms of the macroscopic properties of the beam, and the radial equation of motion can be solved by straightforward integration.

The physics of space charge flow is discussed by Pierce (1954), Nagy and Szilagyi (1974), and Kirstein et al. (1967). A comprehensive discussion of beam spreading and defocussing is given by Schwartz (1957) and van den Broek (1984). An introduction to the problem can as well be found in the standard textbooks on electron optics; for instance, see Glaser (1952), Klemperer (1953), Hutter (1967), El-Kareh and El-Kareh (1970), or Szilagyi (1988). Despite the considerable attention given to space charge dominated beams during the past 60 years, the design of electron guns is still an important subject in the industrial research of picture tubes; see van den Broek (1986b). However, the fundamental issues involved have not changed since the first notice of these effects.

The importance of the interaction between the individual beam particles was not appreciated until the early 1950s. Hines (1951) and Mott-Smith (1953) analyzed the effect of encounters between individual electrons on the velocity distribution in a traveling wave tube. The issue was initiated by Parzen and Goldstein (1951), who showed that the characteristics of such a device are related to the velocity distribution of the electrons.

At that time, the distinctions between collective and individual Coulomb interactions was well established in the field of plasma physics, due to the work of Langmuir, Cowling, Landau, Chandrasekhar, Spitzer, Pines, Bohm and others. However, application of their results to particle beams is not straightforward, due to the distinct differences between the physics of neutralized plasma's near equilibrium and nonneutralized beams having a short flight time relative to the relaxation time of the interaction phenomena involved. The correspondence with the field of plasma physics is further discussed in Chapter 4. An introduction into the problem of Coulomb interactions in plasmas is given by Bohm and Pines (1951), Pines and Bohm (1952), Ash and Gabor (1954), Trubnikov (1965), Sivukhin (1966), and Chapman and Cowling (1970). A general introduction to the principles of plasma physics is given by Spitzer (1962) and Ichimaru (1973).

2.3. Boersch effect

The subject of particle interactions in particle beams is strongly influenced by Boersch (1954). He carried out a systematic experimental investigation to the energy distribution of electrons in a high intensity focussed beam produced by thermionic emission. Boersch measured the energy distribution as function of the beam current for a fixed beam potential of 30 kV. The (half) Maxwellian energy distribution expected on theoretical grounds was only measured for very low current densities. With increasing current, a transition was observed to a broadened more symmetric distribution. This anomalous energy broadening is nowadays called the Boersch effect. Boersch concluded that the relevant beam parameter is the current density at the most dense point in the beam. He did not attribute the effect to the interaction between the individual electrons but assumed that it was caused by collective longitudinal space charge oscillations generated by the noise fed into the beam from the gun. Fack (1955) developed a model along these lines but did not publish any quantitative results.

The results obtained by Boersch were verified by Dietrich (1958), Miller and Dow (1961), Hartwig and Ulmer (1963a,b), Ulmer and Zimmermann (1964), and Simpson and Kuyatt (1966). These investigations are summarized by Simpson and Kuyatt (1966). All authors agreed that the effect exists and increases with the maximum current density in the beam. Ichinokawa (1968, 1969), however, suggested that the Boersch effect is not a genuine phenomenon but an artifact of the retarding field analyzer, related to its response to beam divergence. Ditchfield and Whelan (1977) repeated his experiment and concluded that his statement is incorrect, although the analyzer response can lead to an apparent energy broadening in extreme cases. See also the discussion with the reviewers in the paper by Pfeiffer (1972). Many other investigations of the energy distribution in electron beams from thermionic guns followed, confirming the Boersch effect: Andersen (1967), Andersen and Mol (1968), Speidel and Gaukler (1968), Hanszen and Lauer (1969), Pfeiffer (1971), Degenhardt and Koops (1982), and Troyon (1987). The effect of energy broadening is also extensively observed in beams from field emission guns: Gadzuk and Plummer (1973), Swanson (1975), Gaukler et al. (1975), Heinrich et al. (1977), Troyon (1976), Bell and Swanson (1979), Speidel et al. (1979), Groves (1981b), Speidel et al. (1985), Takaoka and Ura (1986), Zinzindohoue (1986), Troyon and Zinzindohoue (1986), and Troyon (1988). Similar experiments were carried out for different types of liquid metal ion sources: Krohn and Ringo (1975), Swanson et al. (1979), Swanson et al. (1980), Mayer (1985), Mair and Mulvey (1985), Umemura et al. (1986), and Ishitani et al. (1987b,c,d). It is not clear, whether the large energy spreads observed in these devices is entirely due to the Boersch effect.

In contrast to the experimental activity, the effort to explain the Boersch effect theoretically was rather limited during the decade following its discovery. It is interesting to note that no link was established to the work of Hines and Mott-Smith, mentioned previously, despite its close relation. Veith (1955) presented a qualitative analysis in case of space charge limited thermionic emission, based on the assumption that the Boersch effect is related to the dynamic behavior of the space charge barrier in front of the cathode, causing an amplification of shot noise.

Lenz (1958) showed that a calculation, based on the transition of momentum through collisions between free electrons and weakly bounded electrons in a crossover region (thus assuming the presence of neutralizing ions), leads to the right order of magnitude in comparison to the observed energy broadening. Lenz considered the presence of weakly bounded electrons to be essential, based on the erroneous idea that collisions between free identical particles do not alter the energy distribution. He argued that colliding free identical particles exchange energy and momentum, leaving the macroscopic properties of the beam unchanged. However, this is only true for central collisions, which are very rare in practical beams. Schiske (1962) advocated an approach starting from the Holtsmark distribution. This is the probability distribution of the fluctuating component of the electrostatic force acting on a point charge surrounded by a random distributed cloud of identical point charges. It was first derived by Holtsmark (1919) in connection with his work on the Stark broadening of spectral lines. Schiske's discussions are based on a method described by Chandrasekhar (1943) to calculate the statistics of the gravitational field in stellar dynamics. Miller and Dow (1961) investigated the energy exchange between electrons in a (magnetically confined) cross-field stream analyzer and suggested that the Boersch effect, in such a device, is associated with multiloop trajectories in the cathode region. It should be noted that the significance of the impact of the local interactions between the individual particles on the velocity distribution in the beam was, at that time, not generally recognized; for instance, see Lindsay (1960).

Considerable progress to the understanding of the Boersch effect was made by Zimmermann (1968, 1969, 1970). Proceeding on the ideas of Hartwig and Ulmer (1963a,b) and Ulmer and Zimmermann (1964), he developed the concept of relaxation of internal energy. The internal energy of a particle is defined as its kinetic energy in the frame of reference moving with the beam. The essential observation is that the axial internal energy spread of the particles is significantly reduced during acceleration. This is a consequence of the quadratic relation between kinetic energy and momentum. (Notice that the energy spread in the laboratory system remains equal during acceleration since every particle gains the same amount of energy.) While the axial internal energy spread is reduced, the lateral internal energy spread is not affected and a nonequilibrium situation is generated. Due to the Coulomb interaction between the particles, relaxation will occur toward a more isotropic distribution of internal energy. Consequently, the longitudinal energy spread (in the laboratory system) increases at a rate that can be estimated from the average gain in axial momentum spread of a single particle, due to the interaction with the other particles in the beam. Zimmermann found that the root mean square (rms) energy spread increases with the square root of the current density.

The process of relaxation of internal energy in accelerated beams was already qualitatively described by Hines (1951), but his paper was apparently not known to Zimmermann, Hartwig, or Ulmer. The result derived by Zimmermann for a cylindrical beam is identical to that of Mott-Smith (1953), which is, however, obtained by different reasoning. Zimmermann's approach is similar to the one used in plasma physics to describe Coulomb interactions between the individual particles; see Trubnikov (1965) and Sivukhin (1966). In order to prevent the results from diverging, it is necessary to assume a maximum interaction radius, usually expressed in terms of an upper limit for the impact parameter. This is the closest distance of approach between two particles in the imaginary case that they do not interact. The choice of this upper limit seems somewhat arbitrary. However, the result is not sensitive to its exact value, since it depends only logarithmically on it.

The method employed by Zimmermann is less suited for beams that are not cylindrical, thus diverging beams, converging beams, or beams containing a crossover. In such beams, the current density and the beam temperature associated with the transverse motion of the particles varies with the axial position in the beam (see Section 3.7). This can be taken into account by averaging the result obtain for a cylindrical slice over the axial coordinate, but the physical ground for such a procedure seems questionable.

Loeffler (1969) calculated the Boersch effect in a beam segment in drift space with a narrow crossover by considering the effect of weak binary encounters on a reference electron moving along the beam axis. He used first order perturbation theory to calculate the dynamics of two interacting particles, whereas Zimmermann (following the concepts of plasma physics) exploited the well-known solution of the full Kepler problem. It should be noted that these approaches lead to identical expressions in the limit of weak complete collisions (see Chapter 6). Loeffler considered the encounters between particles (in the geometry considered) to be a result of the crossover motion, which should be distinguished from the thermal motion. Consequently, he questioned the validity of the relaxation concept on the ground that the Boersch effect exists in beams with a zero transverse temperature, such as beams with a stigmatic (point) crossover.

Loeffler's results diverge for zero beam angle and therefore do not apply to cylindrical beams. In this respect, his results are complementary to those obtained by Zimmermann, which are most suited for cylindrical beams. This was recognized by Zimmermann (1970), who embodied Loeffler's method for the crossover case. He argued that it basically is in agreement with his own calculation, since both results demonstrate the same parameter dependency. He considered the distinction between thermal and crossover motion to be artificial. It should be added that

Zimmermann's and Loeffler's results only agree (qualitatively) in the high-current limit. In the low-current limit, Loeffler predicts a linear dependency on the beam current, whereas Zimmermann finds a square root dependency on this parameter for all experimental conditions. In this context, it is important to notice that Loeffler expresses his result in terms of the median of the energy distribution, while Zimmermann, like most others, evaluates a root mean square (rms) value.

Ever since the contributions of Ulmer, Zimmermann, and Loeffler, the Boersch effect is generally believed to be generated by the Coulomb interactions between the individual beam particles. All subsequent theories described in this section are based on this assumption. Approaches, however deviating from the mainstream, were taken by Fischer (1970) and Beck (1973). Fischer proceeded on the idea, first formulated by Boersch (1954), that the phenomenon is caused by longitudinal space charge oscillations generated by shot noise fed into the beam from the electron gun. In his approach, it is essential that the beam is neutralized. Thus for an electron beam, as considered by Fischer, the presence of a background of positive ions is assumed. This theory therefore does not explain the presence of the Boersch effect in beams at ultra high vacuum conditions, such as beams from field emission guns. Beck (1973) considered a nonneutralized electron beam and applied the theory for Brownian motion in an external field, described by Chandrasekhar (1943). In Beck's version of this model, the external field is produced by the (long-range) space charge force. The short-range effects, caused by strong binary collisions, are expressed in terms of dynamical friction and diffusion (see Section 4.3). Beck concluded that the initial velocity distribution can strongly be perturbed by the (long-range) space charge effect but that short range effects can be neglected in beams with practical current densities. Beck's results are confusing and inadequate since he represents the transverse space charge velocity broadening as a contribution to the energy spread, ignoring the fact that this motion is not random but systematic and can largely be compensated by refocussing.

Knauer (1979a,b) extended the work of Zimmermann for noncylindrical beams. He considered the transverse particle velocities to be a consequence of both thermal and crossover motion. This way, he obtained expressions that do not diverge in the limit of zero beam angle, as in Loeffler's approach. However, the physical basis for this procedure seems somewhat obscure. Knauer (1981) also discussed the interaction phenomena in a gun with a spherical cathode, introducing the concept of potential energy relaxation in collisionless three-dimensional beams. "Three-dimensional" refers to beams in which the average axial separation of the particles is small compared to the lateral beam dimensions. "Collisionless" means that the beam trajectories do not cross, which is a consequence of the strong radial field near the cathode. Knauer treats the influence of near and distant neighbors separately, applying different statistical procedures. The physical basis for this approach appears to be questionable, but it can be demonstrated that the resulting dependency on the experimental parameters is independent of the statistical procedure followed. Knauer's process of potential energy relaxation leads to a two-third power dependency on the beam current.

Crewe (1978a) considered a beam in drift space with a narrow crossover. He calculated the energy change of a central reference electron due to a single (not necessarily weak) collision with another electron. For simplicity, he assumed that the total effect can be represented by a single collision, choosing adequate characteristic collision parameters. Crewe's expression for the energy change is inversely proportional to the axial separation of the particles and therefore increases linearly with the beam current.

A similar, but statistically more sophisticated model was developed by Rose and Spehr (1980). They stated that the main contribution stems from nearest neighbors, that is pairs of electrons having the smallest impact parameters relative to each other. They determined the effect on a (not necessarily central) reference particle, taking the probability distribution of impact parameters of nearest neighbor colliding particles into account. The improvement on Crewe's model is established by considering the entire nearest neighbor probability distribution, instead of picking just one characteristic value. The procedure of Rose and Spehr also solves some of the disadvantages of the model of Zimmermann and Knauer. This model includes an integration over all impact parameters up to a somewhat arbitrary maximum value, as was mentioned previously. In the approach of Rose and Spehr, the contribution of large impact parameters is suppressed by a factor, which depends on the nearest neighbor probability distribution. In other words, colliding particles with a large impact parameter are unlikely to be the nearest neighbor of the reference particle, and their contribution therefore vanishes in this approach. The accuracy of the model of Rose and Sephr depends on the validity of the assumption that the contribution of particles other than nearest neighbors can be neglected. It might be anticipated that their result underestimates the Boersch effect, in particular for high current densities. For practical current densities, the model of Rose and Spehr predicts a square root dependency on the beam current. In addition, it shows that the Boersch effect strongly depends on the size of the crossover. Consequently, the effect can be reduced by forming beams with an astigmatic crossover.

De Chambost and Hennion (1979), Massey et al. (1981), and Sasaki (1984, 1986) presented models that are based on the calculation of the mean square of the fluctuations in the electrostatic force acting on a particle surrounded by randomly distributed point charges. This approach resembles the idea of Schiske (1962) and is, unlike the other models discussed so far, not based on two-particle collisions. A detailed description of a similar model in stellar dynamics is given by Chandrasekhar (1941, 1943). The basic assumption is that the time development of the distribution of field (or colliding) particles surrounding a test (or reference) particle can be described as succession of independent states. The electrostatic force acting on the test particle is separated into a systematic (space charge) component, corresponding to the smoothed-out distribution of charge, and stochastic fluctuating component, depending on the instantaneous distribution of field particles. The former will be zero for a test particle that is surrounded by a spherical symmetric distribution of field particles. The latter varies randomly, and its instantaneous value during two successive independent states is assumed to show no correlation. The total effect of the fluctuations on the velocity distribution of the test particle is found by adding statistically the deviations experienced during the subsequent independent states.

The models of this type encounter two inherent problems. Firstly, one is forced to introduce a lower limit for the distance between the test particle and its surrounding field particles; otherwise the mean square of the fluctuations in the electrostatic force acting on the test particle would diverge. The outcome depends critically on the value of this lower limit, but its choice seems arbitrary. Secondly, additional assumptions are required to estimate the speed of the fluctuations or, in other words, the size of the time intervals between the successive independent states. The approximations used to solve these problems vary for the three models mentioned, leading to significant differences in the results. The models of Massey et al. and Sasaki predict a square root dependency on the beam current, whereas the result of De Chambost and Hennion shows a 3/8 power dependency. Massey et al. stated that the Boersch effect can be generated by potential energy relaxation, for which they used the model described above, as well as by velocity (internal kinetic energy) relaxation. For the latter they followed Knauer (1979a).

van Leeuwen and Jansen (1983) presented a model for statistical interactions in a beam with a crossover, based on a reduction of the full N-particle problem to a sum of two-particle effects. Their model predicts the detailed energy distribution, whereas most earlier models evaluate only the (rms or median) width of the energy distribution, presupposing that its shape is Gaussian. According to van Leeuwen and Jansen, the energy distribution indeed is Gaussian for a significant range of operation, but it takes on a quasi-Lorentzian form (Lorentzian core with exponential tails) in the limit of low current densities. The Lorentzian distribution falls off quadratically, which implies that the tails of the distribution contain a substantially larger fraction of the particles than expected if a Gaussian shape is assumed.

This is in agreement with the results obtained by Rose and Spehr (1980), who explicitly predicted such a non-Gaussian behavior for low current densities. Groves et al. (1979) noted non-Gaussian features in their numerically obtained results as well. According to van Leeuwen and Jansen, the quasi-Lorentzian energy distribution is most adequately described by its Full Width at Half Maximum (FWHM). The rms and FWHM show different dependencies on the experimental parameters. The rms value is proportional to the square root of the beam current, for all conditions. The FWHM value shows an identical dependency in the Gaussian regime (FWHM = 2.35 rms) but follows a linear dependency in the quasi-Lorentzian regime. For this regime, the outcome coincides with the result given by Loeffler (1969) for low current densities, which is, however, larger by a factor $\sqrt{2}$.

Gesley and Swanson (1984) analyzed the energy shift and energy broadening generated in the vicinity of a spherical emitter with single file emission. In this model, all particles are uniformly accelerated along a line. The energy broadening is related to the distribution of initial distances between nearest neighbors, which is governed by shot noise. Gesley and Swanson exploited a semiempirical equation to calculate the energy change of a particle involved in a binary interaction during acceleration. This equation still contains one unknown parameter. Some kind of unification of different theories is established by varying this parameter as well as another geometry dependent quantity.

Jansen, Groves and Stickel (1985) presented an equation that covers the results of a number of theories on the Boersch effect, all pertaining to the case of a beam segment with a crossover in drift space. It says that the relative rms energy spread ($=\langle \Delta E^2 \rangle^{1/2}/E$) is directly proportional to the square root of the perveance (perveance $= I/V^{3/2}$), as well as a function of three dimensionless parameters, which differs for each of the theories. The theories of van Leeuwen and Jansen (1983) and Rose and Spehr (1980) were found to be in close agreement with each other and with the results obtained by numerical Monte Carlo simulation.

Furukawa (1986) determined an upper limit for the Boersch effect by assuming that it is caused by the conversion of potential energy into (internal) kinetic energy. Furukawa does not separate the generation of systematic velocity components (space charge effect) and the generation of random velocity components (statistical effects). It is erroneous to assume that the former contributes to the energy broadening. His expressions are therefore not very practical, although the actual energy spread will, indeed, always be much smaller than the value predicted by this upper limit.

Tang (1987) presented a model based on numerical integration of the so-called Fokker-Planck equation. His method resembles the ideas developed in plasma physics and stellar dynamics (see Section 4.3). The time evolution of the single-particle distribution function in six-dimensional phase space is studied under the influence of both the average space charge force as well as the fluctuating component of the interaction force. The impact of the latter is described as a diffusion process in velocity space, characterized by the coefficient of dynamical friction β and the diffusion coefficient D. This approximation presupposes that the generated velocity distribution is Gaussian. The expressions used to calculate β and D, as functions of the macroscopic parameters of the beam, are based on weak binary collisions. Accordingly, this approach does not improve on those of Zimmermann (1970) and Knauer (1979a), as far as the fundamental issues involved. However, Tang's discussion is useful in the sense that it clarifies the relation with some of the leading ideas in the field of plasma physics and stellar dynamics.

2.4. Trajectory displacement effect

Loeffler (1964) was probably the first to mention that the encounters between the individual beam particles do not only affect the longitudinal velocity distribution but also lead to stochastic displacements in the transverse direction. He observed that the demagnified image of a carbon microgrid became blurred when the beam current was increased. The major part of this blurring could be refocussed, but a significant part remained even after refocussing. Loeffler identified these contributions respectively as the space charge effect and the transverse counterpart of the Boersch effect, here called the trajectory displacement effect. Loeffler's experimental results and early theoretical analysis are summarized by Hamish et al. (1964).

Contrary to the Boersch effect, the published quantitative experimental data on the trajectory displacement effect is rather scarce. Most data pertains to high throughput electron beam lithography systems. Stickel and Pfeiffer (1978) measured the edge width of the shaped spot in the IBM EL3 variable shaped beam column as a function of the target current and suggested a two-third power dependency on this parameter. Similar measurements of the edge width of a shaped spot for different beams currents are presented by Saitou et al. (1981), Veneklasen (1985), and Morita et al. (1985). Experimental data of the current dependency of the spot size in a Gaussian beam system is given by Kelly et al. (1981). The current density profile of a focussed ion beam spot has been measured by Cummings et al. (1986) and Ward et al. (1987). They demonstrated the existence of long tails, which are probably due to the trajectory displacement effect. See Melngailis (1987) for a review on the subject of focussed ion beams.

A number of theoretical discussions concern the impact of the statistical interactions on the lateral properties of the beam. The distinction between the (deterministic) space charge effect and the (statistical) trajectory displacement effect is recognized in most contributions, but the terminology employed varies. In this section, the original terms are reproduced. Both the space charge effect and the trajectory displacement effect influence the lateral properties of the beam and are, in that respect, related. References to the literature dealing with the space charge effect in low density particle beams will be presented in Section 2.5. Many of the theoretical approaches to the trajectory displacement effect form an extension to models developed for the Boersch effect. Accordingly, some of the physical principles were already described in the previous section. The discussion here will be restricted to the new aspects.

Loeffler and Hudgin (1970) proceeded on an earlier approach to the Boersch effect in a beam segment with a narrow crossover, presented by Loeffler (1969). Three different phenomena of statistical interactions in a beam with a narrow crossover are distinguished, corresponding to each of the three components of the Coulomb force exerted by a colliding electron on a reference electron, which runs along the beam axis. The Boersch effect, which corresponds to an energy change, is associated with the longitudinal component. The lateral force is separated in a component in the direction of the transverse velocity of the colliding electron ($F_{\perp 1}$) and a component perpendicular to this direction ($F_{\perp 2}$), see Fig. 2.1. The latter

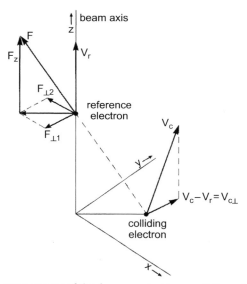

Fig. 2.1 The three components of the force exerted by a colliding electron on a central reference electron, according to Loeffler (1969).

causes an "angular change" of the reference particle. The former does not produce an angular deflection, because its sign changes when the particles pass the point of closest approach, which will roughly coincide with the passage through the crossover. The angular deflection, experienced before reaching the crossover, is thus compensated in the traject following the crossover. This force component causes, therefore only a lateral shift called "trajectory displacement." The median width measures of the distributions of energy changes, angular changes, and trajectory displacements are obtained by numerical integration over all possible trajectories of the colliding particles relative to the reference particle, employing a Monte Carlo method. Loeffler and Hudgin presented normalized curves from which the magnitude of the different phenomena can be determined. All three curves clearly indicate the existence of two separate regimes, which show a different dependency on the experimental parameters. For high electron densities, all effects ultimately increase with approximately the square root of the beam current, but for low electron densities, different powers (varying between 1 and 3) are found for the separate effects.

The discussion presented by Massey et al. (1981) includes a calculation of the effect of statistical interactions on the lateral properties of the beam. Lateral displacement and angular deflection are expressed in terms of a single quantity called "trajectory aberration," which refers to the virtual broadening of the source. De Chambost (1982) presented a similar calculation, using the term "electron-electron effect." The central quantity in both models is the mean square of the fluctuations in the electrostatic interaction force acting on a particle. The approach is essentially the same as utilized for the Boersch effect. Accordingly, the same dependency on the beam current is obtained, which is, however, different for both models. The fundamental aspects were discussed in the previous section.

Knauer (1981), applied his model of potential energy relaxation in three dimensional collisionless beams, mentioned in the previous section, to estimate the "virtual source growth" in a gun with a spherical emitter. As for the Boersch effect, the result shows a two-third dependency on the beam current.

In the papers by Weidenhausen et al. (1985) and Spehr (1985a,b), two phenomena associated with the effect of statistical interactions on the lateral properties of a beam with a crossover are distinguished. Weidenhausen et al. presented a discussion of "stochastic ray deflections," which corresponds to the angular spread in the electron trajectories generated near a crossover. This effect is equivalent to the broadening of the distribution of transverse velocities and can be treated according to the closest encounter model developed by Rose and Spehr (1980) for the Boersch effect. Stochastic ray deflections result in a spatial blurring in planes that are removed from the crossover. It can be neglected in source (crossover) imaging systems but is possibly of significance in shaped beam columns employing Koehler illumination. For practical current densities, the rms angular spread caused by stochastic ray deflections increases with the square root of the beam current.

Spehr (1985a) presented a calculation of the "stochastic probe broadening" effect, employing a thermodynamical description of the beam. Contrary to the effect of stochastic ray deflections, stochastic probe broadening is generated in the dilute parts of the beam. In these parts, the longitudinal beam temperature (which is possibly increased by the Boersch effect) exceeds the transverse beam temperature, which decreases with the inverse square of the beam radius (see Section 3.7). Consequently, relaxation of internal energy, from the longitudinal to the lateral degrees of freedom, occurs as a result of the weak collisions between the beam particles. This process is treated in a similar way as in the model of Zimmermann (1970) and Knauer (1979a) for the Boersch effect, which corresponds to an energy transfer in the opposite direction. Spehr utilizes the concept of beam entropy to relate the energy (or heat) transfer, into the lateral degrees of freedom, to a broadening of the virtual crossover. The concept of "stochastic probe broadening" explains, according to Sephr, the blurring of the final probe observed in systems, in which the source (or source crossover) is imaged on the target. It should be emphasized that Spehr's model presupposes that the beam is nonmonochromatic, since the probe broadening effect predicted by this model vanishes for a zero initial energy spread.

van Leeuwen and Jansen (1983) presented a detailed calculation of the full trajectory displacement distribution for a beam with a crossover. The trajectory displacement experienced by a reference particle is expressed as a virtual displacement in the crossover plane, obtained by extrapolation of its final coordinates back to this plane. The general behavior of the trajectory displacement distribution is similar to the behavior of the energy distribution produced by the Boersch effect. For high current densities, it is Gaussian, and its width increases with the square root of the beam current. For low and moderate densities, it shows Lorentzian characteristics, leading to a two-third power dependency on the beam current. The last result is given by Jansen et al. (1983) and forms a modification of the original work by van Leeuwen and Jansen (1983). Contrary to the Boersch effect, the Lorentzian type of regime prevails for practical operating conditions. Van Leeuwen and Jansen emphasized that the Boersch effect and the trajectory displacement effect stem from different types of collisions, leading to a different dependency on the experimental parameters. The Boersch effect results, predominantly, from collisions between particles having a large relative velocity. These collisions are complete, which means that the particles approach one another from a large separation and recede to a large separation. The Boersch effect is generated in the crossover area, where such collisions reach their point of closest approach. The effect, therefore, strongly depends on the size and shape of the crossover but is insensitive to the length of the beam. The trajectory displacement effect, on the other hand, is usually dominated by weak, incomplete collisions between particles having small relative velocities. It is not specifically generated in the vicinity of the crossover but rather in the entire beam. Consequently, it depends on the beam length but is insensitive to the size and shape of the crossover.

Venables and Cox (1987) reported on the development of a computer model to predict the performance of field emission gun scanning electron microscope columns, taking the trajectory displacement effect into account. For that they used the results of van Leeuwen and Jansen (1983) and Jansen et al. (1983). The modeling indicates that the unexpected large spot sizes observed by Venables and Janssen (1980) in the high resolution mode can be attributed to the impact of the trajectory displacement effect.

Jansen (1988a) presented a set of equations to predict the impact of the different manifestations of statistical interactions on the spot size in Gaussian beam systems and the edge width in shaped beam columns. The theoretical predictions for the trajectory displacement effect were found to be in good agreement with the edge width measurements reported by Stickel and Pfeiffer (1978) and Veneklasen (1985). Jansen also presented a general form for the different types of distributions produced by the statistical interactions. According to this model, the distribution can be Gaussian, Holtsmarkian, Lorentzian, or of the so-called "pencil beam type." This type of distribution occurs in beams in which the axial separation of the particles is large compared to the lateral dimensions of the beam. Ward et al. (1988) independently obtained a similar general form by generalizing the integral expression for the Holtsmark distribution.

2.5. Space charge effect in low density particle beams

The impact of the space charge effect on the properties of charged particles beams has been extensively studied during the past 60 years. However, most publications deal with the beam spreading in high density beams, such as found in oscilloscope tubes and picture tubes, or describe the space charge effect occurring in the vicinity of the cathode of a thermionic gun. Relatively little attention is given to the space charge effect in low and medium density beams, such as found in electron microscopes and electron beam lithography systems. However, space charge effects may affect the performance of such systems. For instance, in a variable-shaped spot lithography column, the space charge effect causes a current-dependent defocussing that is difficult to compensate on a spot-by-spot basis. It, therefore, limits the edge resolution of these systems during normal operation.

The space-charge effect in low density electron beams is discussed by Meyer (1958), Crewe (1978b), Massey et al. (1981), De Chambost (1982), Sasaki (1982) and van den Broek (1986a). Crewe calculated the average radial force acting on a beam particle as function of its position and concluded that the radial equation of motion is identical to the one found for high density beams. This result is not surprising. The radial equation of motion of a particle subject to the (average) space charge force can straightforwardly be obtained from Gauss's theorem for the electrostatic field produced by a configuration of charges (see Chapter 11). This law applies for arbitrary densities.

Although the same equation of motion applies to both high and low density beams, the analytical approach to solve this equation should be different for the distinguished cases. In general, the approximations used for high density beams do not apply to low density beams. In the former, it is valid to assume laminar flow, and the problem can be solved by studying the trajectory of a particle running along the edge of the beam, as was mentioned previously. In the latter, the particle trajectories can cross each other as well as the beam axis, and the laminar flow condition is not generally fulfilled. Therefore, a different approach is necessary. The problem can be solved analytically by exploiting the fact that the deviations from the unperturbed trajectories (trajectories in absence of interaction) are small in such beams. This was recognized by the authors mentioned above. Using a first order perturbation approximation, they evaluated the trajectory of a test particle by considering the space charge repulsion that would occur when all particles would follow their unperturbed trajectories. The space charge perturbation in low density beams can be determined in a similar manner as the geometrical aberrations, starting from the paraxial ray equation; see for instance Meyer (1958) or Zvorykin et al. (1961).

A straightforward solution can be obtained for a beam in field free drift space, with an uniform current density distribution in every cross section of the beam. In that case, the space charge effect corresponds to the action of an ideal negative lens and can thus be expressed in terms of a defocussing and a magnification effect. For nonuniform current density distributions, the space charge lens will be nonideal. Consequently, the space charge effect cannot entirely be refocussed. The nonrefocusable blurring (blurring produced in the plane of best focus) can be described in terms of the geometrical aberrations of the space charge lens. van den Broek (1984) studied the effect of nonuniform current density distribution on the spot growth numerically. He also obtained an analytical solution for the particular case of a parabolic space charge density distribution, employing first order perturbation theory; van den Broek (1986a). As most space charge phenomena, the nonrefocusable space charge blurring is directly proportional to the beam current and is expected to dominate the trajectory displacement effect for high (nonuniform) current densities. The subject of the space charge effect in low density beams is investigated in Chapter 11.

2.6. Monte Carlo simulation

A novel approach to the calculation of interaction effects in particle beams was introduced by Groves et al. (1979), Sasaki (1979, 1982), and El-Kareh and Smither (1979). They independently developed a numerical technique in which the trajectories of an ensemble (sample/bunch) of particles with randomly chosen initial conditions are traced through a system. The trajectories are determined by updating the position and velocity of every particle at regular time intervals, with each particle experiencing the Coulomb repulsion of every other particle in the ensemble. Lenses and other optical elements can easily be incorporated in thin-lens approximation, which means that their action is represented by a deflection of the particles in a single plane. In order to obtain acceptable statistics, the tracing procedure is repeated for a number of bunches, each bunch starting with a different "seed" of initial conditions. The final coordinates, accumulated from all seeds, are processed in order to reduce the information to some characteristic quantities, as the width of the energy distribution, the defocussing distance, and the spatial broadening in the plane of best focus.

This numerical approach is called Monte Carlo simulation, referring to the random initiation of the particle coordinates. The name is slightly misleading, since the ray tracing itself is strictly deterministic, unlike the Monte Carlo models used to simulate the scattering of particles, bombarding a solid-state material. Such models are used in electron beam lithography to determine the proximity effect; see, for instance, Murata and Kyser (1987). The Monte Carlo approach described above, should also be distinguished from the Monte Carlo calculation reported by Loeffler and Hudgin (1970). They employed a numerical Monte Carlo technique to approximate the integral over the coordinates of the colliding particles, within their binary encounter model. They did not attempt to solve the N-body problem by straightforward numerical integration.

Monte Carlo simulations of electron beams in drift space were also reported by Dayan and Jones (1981), Jones et al. (1983, 1985) and Tang (1983). Munro (1987) reported on a program written by Meisburger, which also covers uniform acceleration fields. Yau et al. (1983) presented Monte Carlo simulations of different types of ion beams emitted in a small cone from a spherical cathode under influence of a spherical acceleration field. Ward (1984) and Ward et al. (1987, 1988) employed Monte Carlo simulations to estimate the virtual source size growth of a liquid metal ion source.

Narum and Pease (1986) utilized the Monte Carlo technique to simulate the interaction between the ions in a focussed ion beam in the presence of a postlens retarding field. Allee et al. (1988) used the Monte Carlo approach to model the acceleration zone in a laser irradiated thermionic and photoemissive electron source. Similar integration methods are extensively used in stellar dynamics; see for instance Aarseth (1972). Jansen (1987) introduced the concept of Fast Monte Carlo simulation, which is based on a reduction of the N-body problem to binary interactions, for which analytical approximations are exploited. Some extensions to this model were reported by van der Mast and Jansen (1987). The principles of Monte Carlo simulation of particle beams will be discussed in Chapter 13.

The advantage of Monte Carlo simulation is its accuracy relative to the analytical models, which is a consequence of the minimum number of underlying physical assumptions. Obviously, this accuracy is only obtained when the numerical procedures are carried out with sufficient caution. This is not a trivial task and the quality of the simulation is often hard to judge from the presented results. A disadvantage of the Monte Carlo approach is that it does not give a direct appreciation of the physical mechanisms involved and does not lead to explicit equations, showing the dependencies on the experimental parameters. Consequently, the result of a single simulation only applies to the considered condition and every other condition requires a new simulation. Clearly, the analytical and the Monte Carlo approach are complementary in many respects, and a more firm theoretical foundation can be obtained by their combination. This was recognized by Jansen et al. (1985) and Jansen and Stickel (1984), who compared the different analytical theories on particle interactions with the results of Monte Carlo simulations for a simple beam geometry with a single crossover.

CHAPTER THREE

General beam properties

Contents

3.1. Introduction	27
3.2. Beam parameters	28
3.3. Classification of beams	31
3.4. Hamilton's formalism and Liouville's theorem	32
3.5. Boltzmann equation	35
3.6. Conservation of beam emittance and brightness	37
3.7. Beam temperature	41
3.8. Thermodynamic limits for particle interaction effects	49
3.9. Debeye screening	51
3.10. Potential energy relaxation	55

3.1. Introduction

This chapter discusses a number of well-known concepts used to model particle beams. It starts by surveying the experimental parameters, specifying the macroscopic properties of the beam. These parameters are then used to classify the different types of beams considered throughout this monograph.

After these preliminary sections we focus on some fundamental issues known from the field of statistical mechanics. Liouville's theorem is derived from Hamilton's equations of motion. This theorem covers the dynamics of an ensemble of systems in 6N-dimensional phase space, N being the number of particles in each system. It states that a small volume dV_{6N}, containing some dN points (each representing a single system), may arbitrarily change its shape in the course of time, but maintains its volume. For a system of identical noninteracting particles, a similar equation can be obtained covering the time development of a single system in six-dimensional phase space. This equation, which is known as the collisionless Boltzmann equation, shows the same properties as Liouville's theorem in 6N-dimensional phase space. The phase space to which Liouville's theorem applies can be further reduced when the different degrees of freedom of the beam particles are uncoupled. This leads to the concept of conservation of emittance and conservation of brightness, which are of great importance to the theory of particle optics.

Next, the concept of beam temperature is introduced, which is an essential quantity in the thermodynamical description of the beam. In a rotational symmetric beam, one has to distinguish two beam temperatures, $T_\parallel$ and $T_\perp$, pertaining to the longitudinal and lateral degrees of freedom of the beam respectively. Nonequilibrium conditions may occur as a consequence of the compression and expansion of the beam volume. The expansion of the beam in axial direction during acceleration causes a decrease of the longitudinal beam temperature $T_\parallel$. The transverse beam temperature $T_\perp$ changes with the lateral dimensions of the beam and is therefore affected by the focussing action of the lenses. The temperatures $T_\parallel$ and $T_\perp$ will be related to the geometrical properties of the beam and the conditions in the source.

Finally, the particular aspects of a beam of interacting particles are considered. In the thermodynamical description of the beam, Coulomb interactions cause a relaxation toward thermodynamic equilibrium. In this model, one can represent the Boersch effect as a relaxation of kinetic energy from the lateral degrees of freedom to the longitudinal degrees of freedom. This physical picture seems adequate for beams with a narrow crossover, but it is unable to explain the Boersch effect generated in a cylindrical beam with parallel rays. In order to complete the thermodynamical description of the beam, we will present a model for the relaxation of potential energy. The concept of Debeye screening will be utilized within this model and is therefore discussed in some detail.

Complete thermodynamic equilibrium is usually not reached in practical beams, due to their relatively low particle density and short time of flight. Thermodynamical considerations are, therefore, merely useful to provide insight in the physical mechanism involved and to determine upper limits for the impact of statistical interactions on the properties of the beam.

3.2. Beam parameters

In most analytical models for particle interactions, the beam is schematized as a succession of beams segments separated by thin optical components. We will follow this approach and study the dynamics of a single beam segment. This section is concerned with the experimental parameters specifying the macroscopic condition of such a segment. It also introduces some additional physical parameters that will be used throughout this monograph.

Fig. 3.1 shows a rotational symmetric beam segment in drift space of length L with a crossover. The particles, with mass m and charge e, enter

General beam properties

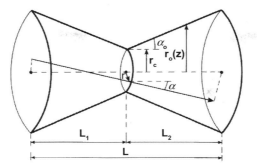

Fig. 3.1 Definition of the experimental parameters for a beam segment with a crossover in drift space. The beam dimensions are specified by the crossover radius r_c, the beam semi-angle α_0 and beam length L. The unperturbed trajectory of a particle is defined by the angle α and the radius r in the crossover, as well as two azimuthal angles (which are not drawn in the figure).

the segment with an average energy eV. V is called the beam potential. The energy distribution on entrance is $f_E(\Delta E)$. The width of this distribution is represented by ΔE_0. The total beam current is I. The unperturbed trajectory of a particle is a straight line, which can be specified by four cylindrical coordinates: the angle α with the axial direction, the radial distance r between the point of intersection with the crossover plane and the beam axis, and two azimuthal angles. As the beam is rotationally symmetric, the azimuthal angles will be uniformly distributed over 2π. The angular distribution is denoted as $f_\alpha(\alpha)$ and the spatial distribution as $f_r(r)$. The half-width of these distributions are represented by the crossover radius r_c and beam semi-angle α_0 respectively. The beam radius at an arbitrary axial position z is denoted as $r_0(z)$. When $\alpha_0 = 0$, the beam is cylindrical and $r_0(z) = r_c$. The crossover position parameter S_c and the image plane position parameter S_i are defined as

$$S_c = \frac{L_1}{L}, \qquad S_i = \frac{L_i}{L}, \qquad (3.2.1)$$

where L_1 is the distance from the entrance plane to the crossover plane and L_i is the distance from the entrance plane to a plane that is optically conjugated to the target plane. This plane, which is referred to as the image or the reference plane, can be located inside or outside the beam segment. The (virtual) broadening observed in the image plane can directly be related to a broadening of the final probe, since it is optically conjugated to the target plane. In source (crossover) imaging systems the image plane coincides with the crossover plane, but this is not true for all segments of the beam in a shaped beam column, which employs Koehler illumination.

Knowledge of S_i is immaterial for the Boersch effect but essential to determine the impact of the trajectory displacement effect on the final probe. The parameters I, V, ΔE_0, L, r_c, α_0, S_c and S_i are referred to as the experimental parameters. In most calculations, we will assume that the initial energy spread does not significantly influence the effect of the interactions between the particles in the beam. Accordingly, we will usually consider the theoretical case of a monochromatic beam, which corresponds to $\Delta E_0 = 0$.

We will now introduce a number of physical quantities that are frequently used and express them in terms of the experimental parameters. All presented relations are nonrelativistic. The average axial velocity of the particles $\langle v_\| \rangle$ is uniquely related to the beam potential as

$$\langle v_\| \rangle = \sqrt{\frac{2eV}{m}}. \qquad (3.2.2)$$

Since we will usually consider monochromatic beams, we will often denote the average axial velocity of the particles simply as $v_\|$ (or v_z) instead of $\langle v_\| \rangle$. The average time of flight T_f is equal to

$$T_f = \frac{L}{\langle v_\| \rangle}. \qquad (3.2.3)$$

It is here denoted as T_f, since the symbol T will, in this chapter, be used to denote temperature. In the remaining chapters we will usually omit the subscript and denote the time of flight as T instead of T_f. The linear particle density λ is defined as

$$\lambda = \frac{I}{e \langle v_\| \rangle} = I\sqrt{\frac{m}{2e^3 V}}. \qquad (3.2.4)$$

It is independent of the axial coordinate z, since the particle flow is assumed to be constant. The three-dimensional particle density n, however, depends on z through the beam radius $r_0(z)$:

$$n(z) = \frac{\lambda}{\pi r_0(z)^2} = \frac{I}{\pi r_0(z)^2}\sqrt{\frac{m}{2e^3 V}}. \qquad (3.2.5)$$

When the current density distribution in the cross section of the beam at position z is nonuniform, the local density n will depend both on z and on the radial coordinate r, thus $n = n(z, r)$. In that case, Eq. (3.2.5) merely provides a measure for the average density at position z. The average axial separation of the particles s and the average distance between the particles d follow from λ and n respectively, using

$$s = \lambda^{-1}, \quad d(z) = n(z)^{-1/3} \qquad (3.2.6)$$

Finally, the total number of particles N simultaneously present in the beam segment is equal to

$$N = \lambda L = IL\sqrt{\frac{m}{2e^3 V}}. \qquad (3.2.7)$$

As a typical case, let us consider an electron beam with a crossover, with a uniform angular and spatial distribution and the following values for the experimental parameters: $I = 1\,\mu\text{A}$, $V = 10\,\text{kV}$, $L = 0.1\,\text{m}$, $S_c = 0.5$, $r_c = 1\,\mu\text{m}$, and $\alpha_0 = 10\,\text{mR}$. By utilizing Eqs. (3.2.2)–(3.2.7) one obtains

$\langle v_{\parallel} \rangle = 5.9 \times 10^7 \text{m/s}$, $T_f = 1.7 \times 10^{-9}\,\text{s}$, $\lambda = 1.1 \times 10^6\,\text{m}^{-1}$
$n_c = 3.3 \times 10^{17}\,\text{m}^{-3}$, $n_1 = 1.3 \times 10^{12}\,\text{m}^{-3}$, $s = .95\,\mu\text{m}$
$d_c = 1.4\,\mu\text{m}$, $d_1 = 91\,\mu\text{m}$, $N = 1.1 \times 10^5$

The subscript c refers to the condition in the crossover, and the subscript 1 refers to the condition at the entrance plane. The values chosen for the experimental parameters are representative for the conditions encountered in the final beam sections of a state of the art, high throughput, electron beam lithography machine. Substantially higher beam currents can occur in the gun area.

3.3. Classification of beams

Having specified the experimental parameters, we are now able to classify different beams. Clearly, this can be done in number of ways, depending on what aspects of the beam are considered. This section just explains the terminology used in this monograph and is not intended to review the classifications used in literature. We will distinguish beams by their geometry and by the particle density within the beam.

Two types of beam geometries are considered: *cylindrical beam segments* and *beam segments with a crossover*. In the former, the current density is constant along the beam axis. In the latter, the current density varies with the axial position and reaches a maximum in the crossover. Divergent beams and convergent beams are included in this representation as limiting cases of a beam with a crossover. They correspond to the situation that the crossover coincides with the entrance plane ($S_c = 0$) and the exit plane ($S_c = 1$) respectively. Both cylindrical beam segments and beam segments with a crossover are assumed to be rotational symmetric with respect to the beam axis, unless specified otherwise.

Homocentric beams are of special interest to the theory, since they represent a simple limiting case. A beam with a crossover is called homocentric when all trajectories intersect each other in the crossover. The crossover is then stigmatic ($r_c = 0$). A cylindrical beam is called homocentric when all trajectories are parallel ($\alpha_0 = 0$). One could say that the trajectories intersect in infinity. Clearly, homocentric beams cannot exist in reality, since they require infinite brightness (the concept of beam brightness will be discussed in Section 3.6).

Another classification of beams is established on the basis of the particle density within the beam. In this classification one compares the average distance between the beam particles to the geometrical dimensions of the beam. A *pencil beam* is defined as a beam with small lateral dimensions relative to the axial separation of the particles s. Other beams are called *extended beams*. Related expressions used in literature are "single file regime" and "three dimensional regime," referring to the conditions encountered in pencil beams and extended beams respectively.

Throughout this monograph, it is assumed that the particle density is low enough to neglect quantum mechanical effects. Accordingly, we will treat the beam as a *classic system* and use Maxwell-Boltzmann statistics to describe its microscopic properties. This approach is justified as long as the de Broglie wavelength λ_B associated with a particle is short compared to the average separation of the particles d. In the frame of reference moving with the beam, the particle velocities are of the order $\alpha_0 \langle v_\parallel \rangle$, where the average axial velocity $\langle v_\parallel \rangle$ is given by Eq. (3.2.2). Accordingly, the appropriate de Broglie wavelength λ_B is of the order

$$\lambda_B \approx \frac{h}{m\alpha_0 \langle v_\parallel \rangle} = \frac{h}{(2me)^{1/2}} \frac{1}{\alpha_0 V^{1/2}}, \qquad (3.3.1)$$

where h is the Planck constant. For the example considered in the previous section (an electron beam with $\alpha_0 = 10\,\text{mR}$ and $V = 10\,\text{kV}$), one finds $\lambda_B \approx 1.2\,\text{nm}$, which is indeed much smaller than the average separation of the particles d_c at the most dense point in the beam.

Finally, relativistic effects are disregarded in most calculations, which is justified for an electron beam for voltages below approximately $100\,\text{kV}$. Accordingly, we ignore magnetic interactions and consider the effect of the mutual Coulomb repulsion only. The possibility to include relativistic corrections will be investigated in Chapter 10.

3.4. Hamilton's formalism and Liouville's theorem

Consider a nonrelativistic system of N identical particles, interacting through the Coulomb force. Let us denote the position vector of particle i as $\mathbf{r}_i$ and its velocity vector as $\mathbf{v}_i$. Both vectors have three components, corresponding to the three degrees of freedom of the particle. The microstate $\Gamma(t)$ of the system at time t is specified by the positions and velocities of all particles at time t (constituting $6N$ coordinates):

$$\Gamma(t) = (\mathbf{r}_1(t), \mathbf{r}_2(t), \ldots, \mathbf{r}_N(t); \mathbf{v}_1(t), \mathbf{v}_2(t), \ldots, \mathbf{v}_N(t)). \quad (3.4.1)$$

The macrostate of the system is represented by a point in 6N-dimensional phase space (Γ_{6N} space). Let $\Gamma(t_0)$ be the initial micro-state on initial time t_0. The development of the microstate in the course of time corresponds to a trajectory $\Gamma(t)$ in Γ_{6N} space, which starts at $\Gamma(t_0)$. This trajectory can, at least in principle, be obtained by integration of the equations of motion. This problem is the subject of classical mechanics.

We will start from the equations of motion in the Hamilton form, which can be derived from the Newtonian equations of motion; see for instance Goldstein (1980). Hamilton's equations are given by

$$\frac{\partial H}{\partial \mathbf{p}_i} = \dot{\mathbf{r}}_i, \quad \frac{\partial H}{\partial \mathbf{r}_i} = -\dot{\mathbf{p}}_i \quad (i = 1, 2, \ldots, N), \quad (3.4.2)$$

where H is the Hamilton function, representing the total energy of the system, and $\mathbf{p}_i$ is the conjugate momentum vector of particle i. Notice that the $2N$ equations given by (3.4.2) are expressed in vector form, each vector corresponding to three coordinates. Accordingly, the Hamilton equations constitute a set of 6N first order differential equations in the phase space coordinates.

For the nonrelativistic system of particles interacting through the Coulomb force considered here, the conjugate momentum of particle i is simply given by $\mathbf{p}_i = m\mathbf{v}_i$, and the Hamilton function H can be expressed as

$$H = \sum_{i=1}^{N} \left(\frac{\mathbf{p}_i^2}{2m} + V(\mathbf{r}_i) \right) + \frac{1}{2} \sum_{\substack{i,j=1 \\ i \neq j}}^{N} \frac{C_0}{r_{ij}}, \quad (3.4.3)$$

where $V(\mathbf{r}_i)$ is the potential energy of particle i in an external field, $C_0 = e^2/4\pi\varepsilon_0$ (the Coulomb potential strength), and $r_{ij} = |\mathbf{r}_i - \mathbf{r}_j|$ is the distance between particle i and particle j.

The dynamics of the system with initial state $\Gamma(t_0)$ is fully specified by Eqs. (3.4.2) and (3.4.3). The representative trajectory in phase is uniquely determined by the initial microstate of the system. Different initial states lead to different trajectories in Γ_{6N} space, which do not intersect. However, different microstates may correspond to the same macroscopic condition of the system. This condition is defined by the value of the experimentally observable parameters. Of special interest is the set of microstates representing the same macroscopic condition. They can be considered as independent realizations of the same system. Such a set is called an ensemble of realizations or also an ensemble of systems (the reader should be alert to the different usage of the word system in the various contexts). Instead of studying the properties of a single system over a period of time, it is mathematically more convenient to consider an ensemble of systems at one time and represent the macroscopic quantities of the system as an average over the members of the ensemble. The assumption that ensemble averages are equivalent to time averages constitutes the basis of statistical mechanics.

Let us examine the properties of an ensemble more closely. In Γ_{6N} space the members of the ensemble are, at a given instant of time, represented by a set of points. Let $\rho(\Gamma, t)$ describe the distribution of these points in phase space as function of time. This function is called the ensemble density function. The total number of systems in an ensemble does not change with time. Therefore, the ensemble density function must satisfy the equation of continuity

$$\frac{\partial \rho}{\partial t} + \sum_{i=1}^{N}\left(\frac{\partial}{\partial \mathbf{r}_i}(\dot{\mathbf{r}}_i\rho) + \frac{\partial}{\partial \mathbf{v}_i}(\dot{\mathbf{v}}_i\rho)\right) = 0, \qquad (3.4.4)$$

which expressed that what flows into an element of volume either comes out again or builds up in the volume element. This equation can be simplified using the fact that the velocity of a representative point in phase space is divergence free,

$$\sum_{i=1}^{N}\left(\frac{\partial \dot{\mathbf{r}}_i}{\partial \mathbf{r}_i} + \frac{\partial \dot{\mathbf{v}}}{\partial \mathbf{v}_i}\right) = 0, \qquad (3.4.5)$$

which is a direct consequence of the Hamilton Eq. (3.4.2). Substitution of Eq. (3.4.5) into the equation of continuity (3.4.4) gives the celebrated theorem of Liouville,

$$\frac{\partial \rho}{\partial t} + \sum_{i=1}^{N}\left(\dot{\mathbf{r}}_i \cdot \frac{\partial \rho}{\partial \mathbf{r}_i} + \dot{\mathbf{v}}_i \cdot \frac{\partial \rho}{\partial \mathbf{v}_i}\right) = \frac{d\rho}{dt} = 0, \qquad (3.4.6)$$

where $d\rho/dt$ is the total (or hydrodynamic) derivative. The Liouville theorem states that the density of representative points ρ is constant along a flow line. A small volume dV_{6N} in phase space, containing dN points, may arbitrarily change its shape in the course of time but maintains its volume, since the density of points $\rho = dN/dV_{6N}$ is constant. Thus $dV_{6N}/dt = 0$. Notice that points within the volume cannot cross the volume boundary, because phase space trajectories do not intersect.

For the particular case of a system of N particles interacting through the Coulomb force, Liouville's theorem can be written as

$$\frac{\partial \rho}{\partial t} + \sum_{i=1}^{N} \mathbf{v}_i \cdot \frac{\partial \rho}{\partial \mathbf{r}_i} + \frac{1}{m}\sum_{i=1}^{N}\left(\sum_{j \neq i}^{N} \frac{C_0(\mathbf{r}_i - \mathbf{r}_j)}{|\mathbf{r}_i - \mathbf{r}_j|^3} + \mathbf{F}_i\right) \cdot \frac{\partial \rho}{\partial \mathbf{v}_i} = 0, \qquad (3.4.7)$$

where $\mathbf{F}_i$ is the (conservative) external force acting on particle i.

Eq. (3.4.7) applies to the general case of a system of N identical particles interacting through the Coulomb force. Now, consider the particular case of a particle beam in a probe-forming optical instrument. Particles are continuously emitted from the source, producing a constant flow, and are "absorbed" at the target. The composition of the beam changes continuously, while the total number of particles present at one time will approximately be constant. However, Eq. (3.4.7), as it was derived here, refers to a fixed system of N particles. In order to solve this conceptual problem it is useful to visualize the beam as a succession of bunches and follow the motion of each bunch separately in the frame of reference moving with the beam. The bunches are assumed to be short compared to the total beam length but long enough to represent the macroscopic condition of the beam. Thus, the bunches reaching the target in the course of time can be considered as different realizations of the same beam. In this view the properties of the beam observed at the target represent an average over the bunches reaching the target during the period of observation. Clearly, this physical interpretation directly reflects the mathematical procedure of calculating a macroscopic property as an average over the members of an ensemble (assuming that beam bunches and ensemble members can be considered as equivalent).

Eq. (3.4.7) is here presented as the fundamental equation covering the dynamic behavior of a nonrelativistic system of particles interacting through the Coulomb force. It should be noted that the same equation arises in the theory of stellar dynamics; for instance see Chandrasekhar (1942) and Gilbert (1972). The only mutation required is the replacement of the Coulomb constant C_0 by its gravitational counterpart. This constant depends on the mass, which may vary per "particle."

3.5. Boltzmann equation

Liouville's theorem in Γ_{6N}, as given by (3.4.7), is not practical for the actual evaluation of the beam properties, due to the fact that N is, in general, a large number. It would be more convenient to have an equation for the single-particle distribution function $f(\mathbf{r}, \mathbf{v}, t)$, corresponding to the coordinates of the beam particles in the six-dimensional phase space (Γ_6-space or μ-space). Formally, this function can be obtained from the ensemble density $\rho(\Gamma, t)$ by integration over the coordinates of all particles but one,

$$f(\mathbf{r}_1, \mathbf{v}_1, t) = \int \rho(\Gamma, t)\, d\mathbf{r}_2\, d\mathbf{r}_3 \cdots d\mathbf{r}_N\, d\mathbf{v}_2\, d\mathbf{v}_3 \cdots d\mathbf{v}_N, \qquad (3.5.1)$$

which can be envisioned as a projection of the coordinates in Γ_{6N} space on the subspace Γ_6. Accordingly, the fundamental equation covering the dynamical behavior of $f(\mathbf{r}, \mathbf{v}, t)$ can be found from Liouville's theorem (3.4.7) by a similar integration.

Let us examine the result of such an integration for the different terms in Eq. (3.4.7) separately. The first term, $\partial \rho/\partial t$, transform directly to $\partial f/\partial t$. Next, consider the terms included within a single summation (over index i):

$$\sum_{i=1}^{N} \mathbf{v}_i \cdot \frac{\partial \rho}{\partial \mathbf{r}_i} \quad \text{and} \quad \frac{1}{m} \sum_{i=1}^{N} \mathbf{F}_i \cdot \frac{\partial \rho}{\partial \mathbf{v}_i}.$$

All terms $i \neq 1$ vanish, since it is allowed to assume, without loss of generality, that ρ is zero at the integration boundaries. The other terms (corresponding to $i = 1$) can directly be expressed into the derivatives of the single particle function f. Finally, consider the terms included within a double summation (over indices i and j), which correspond to the Coulomb interaction between the particles. These terms depend on the position of a particle i, relative to the other particles j in the system. Consequently, the results obtained after integration cannot be expressed in the single particle function alone. Evaluation of these terms requires the knowledge of the two-particle correlation function, which specifies the conditional probability that a particle j ($\neq 1$) is located at $\mathbf{r}_j$, if particle 1 is located at $\mathbf{r}_1$. Formally, this function can be found by integrating Liouville's theorem (3.4.7) over the coordinates of all particles but two. However, as it turns out, this requires the knowledge of the three-particle correlation function, and a straightforward explicit solution is still not obtained. In general, the equation for the n-particle distribution involves the $(n+1)$-particle distribution. The resulting set of coupled equations is referred to as the BBGKY hierarchy (for Bogolioubov, Born, Green, Kirkwood, and Yvon). Formally speaking, the problem of determining the impact of particle interactions on the time evolution of the single-particle function $f(\mathbf{r}, \mathbf{v}, t)$ is to find a method of truncating the BBGKY hierarchy. A discussion of the approach to this problem taken in plasma physics and stellar dynamics is postponed to the next chapter, and the interaction term is here left unspecified as $[\partial f/\partial t]_i$.

Hence, the resulting equation for the single-particle distribution function $f(\mathbf{r}, \mathbf{v}, t)$ can be expressed as

$$\frac{df}{dt} = \frac{\partial f}{\partial t} + \mathbf{v} \cdot \frac{\partial f}{\partial \mathbf{r}} + \frac{\mathbf{F}}{m} \cdot \frac{\partial f}{\partial \mathbf{v}} = \left[\frac{\partial f}{\partial t}\right]_i. \tag{3.5.2}$$

This is the Boltzmann transport equation, sometimes also referred to as Liouville's theorem in Γ_6 space (especially when $[\partial f/\partial t]_i = 0$) or as the kinetic equation. The quantity df/dt is the total (or hydrodynamic) derivative. Boltzmann's equation states that the rate of change of $f (= df/dt)$ along a free (unperturbed) particle trajectory is entirely the result of the interaction between the particles.

3.6. Conservation of beam emittance and brightness

Now we will focus on the properties of a system of N identical particles that do not interact. Thus, we take $[\partial f/\partial t]_i = 0$. The resulting (collisionless) Boltzmann Eq. (3.5.2) shows the same features in Γ_6 space as the Liouville Eq. (3.4.6) in Γ_{6N} space (which is, however, also valid for interacting particles). Thus, the time invariance of the phase space volume enclosing a group of representative points applies to Γ_6 space as well: $dV_6/dt = 0$. A further reduction of the phase space to which Liouville's theorem applies is achieved when the three degrees of freedom of a particle are uncoupled. Liouville's theorem then pertains to the three subspaces Γ_2 in x, y, and z direction separately.

In a beam of noninteracting particles, the two lateral degrees of freedom will in general be independent of the longitudinal degree of freedom. Consequently, Liouville's theorem applies to the subspaces Γ_4 and Γ_2 of lateral and longitudinal coordinates respectively: $dV_\parallel/dt = 0$ and $dV_\perp/dt = 0$. The two lateral degrees of freedom may be independent, too, but this is not required in the further analysis.

First, let us investigate the consequences of Liouville's theorem in the subspace of axial coordinates Γ_2. Consider a small volume ΔV_2, occupied by N particles. Let $\Delta r_\parallel$ be the axial spatial length occupied by the particles and $\Delta v_\parallel$ the corresponding length in axial velocity space. Liouville's theorem implies that $\Delta V_2 = \Delta \mathbf{r}_\parallel \Delta v_\parallel$ is conserved along the particle trajectories. This statement can be expressed in terms of the experimental parameters of the beam. For N large enough, one may write

$$\Delta r_\parallel \approx N s_v(\Delta v_\parallel) \sim s \sim \langle v_\parallel \rangle / I,$$

in which $s_v(\Delta v_\parallel)$ denotes the average axial separation of the particles within the velocity range $\Delta v_\parallel$, with s the average axial separation of all particles and I the beam current. The last step follows with Eqs. (3.2.6) and (3.2.4). The velocity range $\Delta v_\parallel$ can be expressed in the corresponding range in normal kinetic energy: $\Delta v_\parallel = \Delta E_\parallel / m v_\parallel$. Hence, $\Delta v_\parallel \Delta r_\parallel \sim \Delta E_\parallel / I$. Accordingly, Liouville's theorem applied to the subspace of axial coordinates gives

$$\Delta V_2 = \Delta v_\parallel \Delta r_\parallel = \text{constant} \quad \Rightarrow \quad \Delta E_\parallel / I = \text{constant}. \tag{3.6.1}$$

The second equation expresses that the normal energy spread of the group of N particles is conserved during the flight as long as the beam current associated with these particles is constant (and vice versa).

Next, consider a small volume ΔV_4 in the subspace of lateral coordinates Γ_4, occupied by N particles. The trajectories of these particles constitute a bundle of rays within the beam. Let $\Delta r_\perp^2$ be the surface area occupied by the cross section of the bundle in an arbitrary reference plane, perpendicular to the beam axis. Let $\Delta v_\perp^2$ be the corresponding surface area in lateral velocity space. Liouville's theorem implies that the volume $\Delta V_4 = \Delta r_\perp^2 \Delta v_\perp^2$ is independent of the choice of the reference plane. In particle optics, this principle is known as invariance of emittance. As an illustration of this principle, consider the single lens system of Fig. 3.2. A rotational symmetric paraxial beam emitted from the object at $z = z_0$ is imaged by a thin ideal (that is aberration free) lens located at $z = z_1$ to the image plane at $z = z_i$. Assume that all particles drift with the same axial velocity $v_\parallel$. Due to the rotational symmetry it is sufficient to study the rays located in a meridian plane, which is a plane through the beam axis. A meridian ray is fully specified by a radial distance r, at some axial position z and its derivative with respect to z, denoted as r'. Thus the appropriate phase space coordinates are r and r'. Notice that for a paraxial beam in drift space $v_\perp = r'v_\parallel = \tan(\alpha)v_\parallel \approx \alpha v_\parallel$. Now consider the four rays emitted from the object plane with values for r and r' respectively given by $(r = -r_0, r' = r_0')$, $(r = r_0, r' = r_0')$, $(r = r_0, r' = -r_0')$, and $(r = -r_0, r' = -r_0')$; see Fig. 3.2. The phase space volume enclosed by these four rays is depicted by the hatched areas in the diagrams of Fig. 3.3. Such plots are known as phase space diagrams or emittance plots. Fig. 3.3A refers to the object plane ($z = z_0$); Fig. 3.3B and C refer to the object and image principal planes of the lens respectively, which are both located at $z = z_1$, since the lens is assumed to be thin; and Fig. 3.3D refers to the image plane ($z = z_i$). Invariance of emittance implies that the surface of the hatched areas is identical in all diagrams.

Although the principle of emittance invariance can be applied to arbitrary planes, it is particularly useful to obtain relations between the

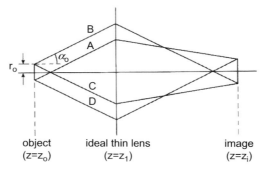

Fig. 3.2 Schematic view of a section of a rotational symmetric paraxial beam with a thin ideal lens. The lateral coordinates of the rays A, B, C and D enclose a volume of phase space, which is independent of z (see also Fig. 3.3).

General beam properties

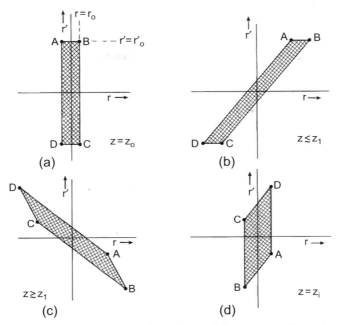

Fig. 3.3 Emittance plots representing the volume of phase space of lateral coordinates enclosed by the rays A, B, C and D (shown in Fig. 3.2) at four different locations in the beam. The surface of the hatched area is the same for all plots.

properties of the beam in planes that are optically conjugated. Such planes are related as image and object planes by the optical system. Let us investigate this more closely. Consider the object and image plane of a system containing an arbitrary number of optical, not necessarily thin, components. Assume, as previously, that the beam is paraxial. However, the beam can be nonrotational symmetric and is not necessarily entirely located in drift space. (Thus the beam potential may differ in image and object plane.) Consider a small bundle of rays. Let us denote $dS = dr_\perp^2$, which is the surface occupied by the bundle in a plane perpendicular to the beam axis. The corresponding volume in lateral velocity space can be expressed in terms of the solid angle $d\Omega$ formed by the bundle in that plane: $d\Omega = \Delta \mathbf{v}_\perp^2 / \pi v_\parallel^2$. As the analysis here is restricted to the object and image plane, the width of the angular distribution can be considered as independent of the radial position r, as can, for instance, be seen from Fig. 3.3A and D. Consequently, the emittance in these planes is simply equal to the product of $d\Omega$ and dS. Thus, for the object and image plane, one finds

$$\Delta V_4 = \Delta v_\perp^2 \Delta r_\perp^2 = \pi v_\parallel^2 \, d\Omega \, dS = (2\pi e/m) \, V \, d\Omega \, dS = \text{constant}, \qquad (3.6.2)$$

utilizing the nonrelativistic relation $v_\parallel^2 = 2\,eV/m$. The values of V, $d\Omega$ and dS may vary for different conjugated planes, but invariance of emittance implies that their product is the same. From Eq. (3.6.2) it follows that the spatial magnification from object to image plane $M_r (= \sqrt{dS_i}/\sqrt{dS_0})$ is related to the corresponding angular magnification $M_\alpha (= \sqrt{d\Omega_i}/\sqrt{d\Omega_0})$ as

$$M_r M_\alpha = \sqrt{\frac{V_0}{V_i}}, \qquad (3.6.3)$$

which is known as the law of Helmholtz-Lagrange.

An alternative way to express the invariance of emittance is obtained by introducing the concept of beam brightness. The normalized differential brightness of a bundle of rays carrying a current dI is defined as

$$\beta_n = \frac{dI}{V \, d\Omega \, dS}. \qquad (3.6.4)$$

It is called *differential* brightness, because it is a function of the infinitesimal quantities dI, $d\Omega$ and dS, whereas the average (or total) brightness is a function of I, Ω, and S. It is referred to as *normalized* in order to distinguish it from the usual differential brightness $\beta = (dI/d\Omega \, dS)$, which differs by a factor V. As the current dI remains constant during the flight, conservation of emittance, expressed by Eq. (3.6.2), implies conservation of normalized differential brightness β_n. The reader is referred to Lejeune and Aubert (1980) for a more detailed discussion on the subject of emittance and brightness invariance.

Brightness is an important concept in the description of particle beams in probe-forming instruments. It is used to characterize the gun performance. Given the brightness of the beam near the gun, a lower limit for the probe size in the target plane can be obtained as function of the beam current and the beam angle at this location. In order to explain this statement, consider a beam in the vicinity of the target plane, with a beam semi-angle α_p, beam current I, and beam potential V. Let β_n be the normalized brightness of the gun. Using Eq. (3.6.4), with $d\Omega = \Omega = \pi \alpha_p^2$ and $dS = S = \pi r_p^2$, one finds that the probe radius r_p should satisfy the following relation:

$$r_p \geq \frac{1}{\pi \alpha_p} \sqrt{\frac{I}{\beta_n V}}. \qquad (3.6.5)$$

The actual probe radius r_p will be larger than the value given by the right-hand side of Eq. (3.6.5), due to the aberrations of the optical components of the system and the effect of particle interactions. It should be noted that the aberrations of the optical components do not affect the differential brightness, which is invariant even in the presence of external (conservative) fields. But they do cause a reduction of the average brightness, which is the relevant parameter in Eq. (3.6.5). On the other hand, the presence of particle interactions renders Liouville's theorem inapplicable in Γ_6 space. Accordingly, the differential brightness is not invariant but deteriorates during the flight of the particles through the system.

3.7. Beam temperature

The velocity distribution of the particles within the beam depends in general on the position **r** within the beam, as is formally expressed by the single particle distribution function $f(\mathbf{r}, \mathbf{v}, t)$, given by Eq. (3.5.1). The properties of the local velocity distribution are often specified in terms of a beam temperature. In this section we will follow this approach and utilize some principles originating from the kinetic theory of gases.

For an isolated gas in thermodynamic equilibrium, it is well known that the particle velocities are distributed according to the Maxwell-Boltzmann distribution

$$f(\mathbf{v}) = \left(\frac{m}{2\pi k_B T}\right)^{3/2} e^{-mv^2/2k_B T}, \qquad (3.7.1)$$

where k_B is the Boltzmann constant. This is a three dimensional time-independent Gaussian distribution with zero average and a mean square value $\langle v^2 \rangle = 3k_B T/m$. The zero average value indicates that the gas is at rest, from a macroscopic point of view. According to Eq. (3.7.1), the average kinetic energy per particle is related to the temperature T as

$$\frac{1}{2}m\langle v^2 \rangle = \frac{1}{2}m(\langle v_x^2 \rangle + \langle v_y^2 \rangle + \langle v_z^2 \rangle) = \frac{3}{2}k_B T \qquad (3.7.2)$$

This equation expressed that a so-called thermal energy of $k_B T/2$ is associated with every degree of freedom. In absence of external forces the velocity distribution will show no correlation with the spatial coordinates.

Now, let us return to the case of a rotational symmetric particle beam. Clearly, relation (3.7.2) cannot straightforwardly be applied. First of all, the average axial velocity $\langle v_\| \rangle$ is not zero, since there is a net flow of particles down the beam axis. One can rule out this problem by considering the velocity distribution in the frame of reference moving at velocity $\langle v_\| \rangle$ with the beam. However, even in this system the average lateral velocity $\langle v_\perp \rangle$ is not necessarily zero at off axis locations. This should be taken into account, too. In addition, it should be noted that the distribution of lateral velocities may still depend on the position within the beam. Finally, it will be clear that the velocity distribution is not necessarily spherical symmetric, and one has to distinguish between the distribution of longitudinal velocities and the distribution of lateral velocities. From these arguments it appears to be reasonable to introduce two local beam temperatures, a longitudinal beam temperature $T_\parallel$ and a transverse beam temperature $T_\perp$, defined as

$$T_\parallel = \frac{m}{k_B}\left(\langle v_\parallel^2\rangle - \langle v_\parallel\rangle^2\right), \quad T_\perp = \frac{m}{2k_B}\left(\langle v_\perp^2\rangle - \langle v_\perp\rangle^2\right). \qquad (3.7.3)$$

The factor 2 in the denominator of the expression for $T_\perp$ stems from the fact that the transverse beam temperature corresponds to two degrees of freedom. Note that the beam temperatures defined by Eq. (3.7.3) are not related to the average absolute kinetic energy per particle, as in the case of a gas, but rather to the average kinetic energy of the particles in the local system of rest. This energy is sometimes referred to as *internal* kinetic energy. It should be emphasized that the temperatures are really local; that is, $T_\parallel$ and $T_\perp$ may depend on the spatial coordinates r and z. Thus, instead of the isotropic uniform temperature T, which was sufficient to describe the velocity distribution in an isolated gas in thermodynamic equilibrium, one needs, in general, an anisotropic local temperature to describe the velocity distribution in a particle beam.

The second equation of (3.7.3) relates the transverse beam temperature to the local spread in the transverse component of the particle velocities, irrespective of the origin of this motion. One sometimes separates the transverse motion of the particles into a component originating from the thermal motion near the cathode surface and a component corresponding to the crossover motion generated by the focussing action of the lenses. For instance, see Knauer (1979a). Indeed, one can distinguish locally a random component in the transverse velocities ($\sim (\langle v_\perp^2\rangle - \langle v_\perp\rangle^2)^{1/2}$) and a systematic component ($\sim \langle v_\perp\rangle$), as was discussed above. However, one should realize that, in general, both of these components originate from the initial thermal motion near the cathode and are both influenced by the focussing action of the optical components thereafter. Thus, one cannot attribute the random component of motion solely to the thermal motion near the cathode or the systematic component of motion solely to the focussing action of the optical components. Accordingly, it should be emphasized that Eq. (3.7.3) just express the random components of motion in terms of the effective beam temperatures. These equations do not imply any direct physical relation with the thermal motion near the cathode. In fact it would be less confusing to refer only to "random" motion and "systematic" motion and avoid the concept of beam temperature entirely. One does not really need the concept of beam temperature in the actual calculation of the impact of Coulomb interactions on the properties of the beam, and the main reason for introducing it is to establish a relation with the models developed in the fields of plasma physics and stellar dynamics. These models will be discussed in Chapter 4.

In order to illustrate the interrelation of crossover motion and thermal motion, let us consider the simple case of a rotational symmetric beam segment with a narrow crossover. Assume that the crossover is an image of a thermionic source (or source crossover). Let the angular distribution be uniform, with a beam semi-angle α_0. According to the second equation of (3.7.3), the transverse beam temperature in the crossover can now be expressed as

$$T_{c\perp} = \frac{m v_\parallel^2}{2 k_B} \langle \alpha^2 \rangle = \frac{eV}{2 k_B} \alpha_0^2, \qquad (3.7.4)$$

using $\langle \alpha^2 \rangle = \alpha_0^2/2$ and Eq. (3.2.2). For a Gaussian angular distribution, one obtains the same result by defining $\alpha_0 = 2\sigma_\alpha$ (with $\langle \alpha^2 \rangle = 2\alpha_\alpha^2 = \alpha_0^2/2$). The crossover temperature $T_{c\perp}$ can be regarded as the result of the initial thermal motion near the source but might as well be expressed in terms of the crossover motion related to the focussing action of the preceding lenses.

The definition of beam temperature allows us to describe the properties of the beam in terms of thermodynamical quantities, which obey thermodynamical relations. (By thermodynamical quantities we mean temperature, volume, particle density, entropy, etc.) We will not pursue this approach in detail, but will just note that such a description is equivalent to the description in terms of geometrical properties, as well as to the description in terms of statistical mechanical properties, such as the single-particle distribution function $f(\mathbf{r}, \mathbf{v}, t)$. As an illustration of the equivalence between the thermodynamical description of the beam and the statistical mechanical approach, we will show that the implications of Liouville's theorem discussed in Section 3.6 can also be understood from thermodynamical considerations. The application of Liouville's theorem to a rotational symmetric beam of identical noninteracting particles lead to the conclusion that the phase space volume $\Delta V_\parallel$ and $\Delta V_\perp$ are invariant. In order to obtain this result within the thermodynamical description, one should visualize the beam as an isolated ideal atomair gas of volume V and temperature T. As it flows through the system it changes its volume V. The corresponding change in the temperature T can be found from

$$T V^{2/f} = \text{constant}, \qquad (3.7.5)$$

which is the adiabatic equation for an ideal gas with f degrees of freedom. Since the longitudinal and the lateral motions of the particles are independent, Eq. (3.7.5) should be applied separately to the longitudinal degree of freedom ($f_\parallel = 1$) and the lateral degrees of freedom ($f_\perp = 2$). By substitution of Eq. (3.7.3), $V_\parallel \sim \Delta r_\parallel$ and $V_\perp \sim \Delta r_\perp^2$ into Eq. (3.7.5) one finds

$$\langle \Delta v_\parallel^2 \rangle \Delta r_\parallel^2 = \text{constant}, \qquad \langle \Delta v_\perp^2 \rangle \Delta r_\perp^2 = \text{constant}, \qquad (3.7.6)$$

which is equivalent to Eqs. (3.6.1) and (3.6.2) respectively, obtained from Liouville's theorem, q.e.d.

We like to investigate the dependency of the beam temperatures $T_\parallel$ and $T_\perp$ on the experimental parameters. We will first focus on the behavior of the longitudinal beam temperature $T_\parallel$ and next consider the transverse beam temperature $T_\perp$. In general, these temperatures are determined by the emission process, the acceleration of the beam, and the focal action of the optical components of the system. Let us examine each of these elements for the specific case of an electron beam produced by a thermionic source. The velocity distribution of the electrons emitted from the cathode, with temperature T_{ca}, is given by the so-called *half* Maxwell-Boltzmann distribution,

$$f_i(\mathbf{v}_i) = \frac{1}{2\pi}\left(\frac{m}{k_B T_{ca}}\right)^2 e^{-mv_i^2/2k_B T_{ca}} v_{i\parallel}\, \theta(v_{i\parallel}), \qquad (3.7.7)$$

in which the subscript i refers to the initial situation before acceleration. The step function $\theta(x)$ is defined as

$$\theta(x) = \begin{cases} 1 & \text{for } x \geq 0. \\ 0 & \text{for } x < 0. \end{cases} \qquad (3.7.8)$$

Eq. (3.7.7) is also used to describe the velocity distribution of the particles of a gas escaping from a container into vacuum through a small hole when the velocities of the particles inside the container obey the Maxwell-Boltzmann distribution given by Eq. (3.7.1). The step function accounts for the fact that only those electrons which have a positive axial velocity will emerge from the cathode. The factor $v_{i\parallel}$ stems from the fact that the electrons that leave the cathode surface area dS between times t and $t+dt$ with an axial velocity $v_{i\parallel}$ come from a depth $dz = v_{i\parallel}\, dt$. This group of electrons fills an element of volume of coordinate space $d\mathbf{r}$, which is equal to

$$d\mathbf{r} = v_{i\parallel}\, dS\, dt.$$

The number of electrons passing through dS in time dt with an axial velocity between $v_{i\parallel}$ and $v_{i\parallel}+dv_{i\parallel}$ follows by multiplication with the density inside the metal of particles within this velocity range. Consequently, the velocity distributions inside and outside the metal differ by a factor $v_{i\parallel}$. Clearly, the normalization constant appearing in the velocity distributions differs too. As a result of the additional factor $v_{i\parallel}$, Eq. (3.7.7) fulfills Lambert's law (or cosine law): the angular current density emitted in any direction is proportional to the cosine of the angle between this direction and the normal. Finally, we note that Eq. (3.7.7) presupposes that the workfunction is constant along the emitting part of cathode surface and disregards the effect of energy dependent reflections of the electrons at the inner surface of the cathode. For a detailed description of the theory of thermionic emission, the reader is referred to Nottingham (1956), Lindsay (1960), Franzen and Porter (1975), and Lauer (1982).

By applying the definitions of beam temperature, given by Eq. (3.7.3), to the velocity distribution of Eq. (3.7.7), one finds

$$T_{i\parallel} = 2(1 - \pi/4)T_{ca} \approx 0.429\, T_{ca}, \quad T_{i\perp} = T_{ca}. \tag{3.7.9}$$

Thus, the transverse beam temperature of the electrons leaving the cathode is equal to the cathode temperature, but a smaller value is found for the longitudinal beam temperature.

After emission, the electrons are accelerated in the axial direction by a potential V. The final axial velocity of an electron $v_{f\parallel}$ is related to its initial axial velocity $v_{i\parallel}$ as

$$v_{f\parallel}^2 = v_{i\parallel}^2 + 2eV/m. \tag{3.7.10}$$

The transverse velocity of the electrons remains unchanged, disregarding the focussing properties of the gun. Conservation of the number of particles implies

$$f_f(\mathbf{v}_f)\, d\mathbf{v}_f = f_i(\mathbf{v}_i)\, d\mathbf{v}_i. \tag{3.7.11}$$

By substituting Eqs. (3.7.7) and (3.7.10) in the right-hand side of Eq. (3.7.11), one finds for the velocity distribution after acceleration

$$f_f(\mathbf{v}_f) = \frac{1}{2\pi}\left(\frac{m}{k_B T_{ca}}\right)^2 e^{-(mv_f^2 - 2eV)/2k_B T_{ca}}\, v_{f\parallel}\, \theta(v_{f\parallel}^2 - 2eV/m). \tag{3.7.12}$$

As far as the lateral component of velocity is concerned, this distribution is identical to the initial distribution described by Eq. (3.7.7). However, it is significantly contracted in the axial direction. Utilizing Eqs. (3.7.10) and (3.7.11) one may write

$$\langle v_{f\parallel}^2 \rangle = \langle v_{i\parallel}^2 + 2eV/m \rangle = \langle v_{i\parallel}^2 \rangle + 2eV/m$$

$$\langle v_{f\parallel} \rangle^2 = \left\langle (v_{i\parallel}^2 + 2eV/m)^{1/2} \right\rangle^2$$

$$= (2eV/m)\left\{1 + (m/4eV)\langle v_{i\parallel}^2 \rangle - (m/4eV)^2 \langle v_{i\parallel}^4 \rangle/2 + \cdots \right\}^2$$

$$= \langle v_{i\parallel}^2 \rangle + 2eV/m - (m/8eV)\left(\langle v_{i\parallel}^4 \rangle - \langle v_{i\parallel}^2 \rangle^2\right) + \cdots.$$

Hence, it follows from Eqs. (3.7.3) and (3.7.7) that

$$T_{f\parallel} \approx \frac{m^2}{8k_B eV}\left(\langle v_{i\parallel}^4 \rangle - \langle v_{i\parallel}^4 \rangle^2\right) = \frac{k_B T_{ca}}{2eV}T_{ca} \tag{3.7.13}$$

assuming that $eV \gg m\langle v_{i\parallel}^2 \rangle$. Comparison with Eq. (3.7.9) shows that the acceleration leads to a decrease of the longitudinal beam temperature by factor $k_B T_{ca}/(4-\pi)\text{eV} \approx 1.16 k_B T_{ca}/\text{eV}$. In practical beams this factor is of the order 10^{-5}. Thus, in absence of interactions, the electrons leaving the thermionic gun, show a negligible axial velocity spread compared to the lateral velocity spread. The latter corresponds to the temperature T_{ca} of the cathode surface.

The reduction of axial velocity spread, due to the acceleration, is a phenomenon that is of crucial importance to the theory of statistical Coulomb interactions between the beam particles. It can be derived in different ways. It is sometimes presented as a consequence of the quadratic relation between kinetic energy and velocity. This interpretation might be clear from the derivation given above, but it can also directly be understood from the following argument. Consider two particles moving in the axial direction with velocities v and $v + \Delta v$ respectively. The difference in the kinetic energy of the particles ΔE is given by

$$\Delta E = \frac{1}{2}m\left[(v + \Delta v)^2 - v^2\right] \approx mv\,\Delta v \qquad (3.7.14)$$

assuming that $v \gg \Delta v$. The acceleration of the particles does not affect the value of ΔE, since both particles gain the same amount of energy. However, the velocity v is (by definition) increased during acceleration. Consequently, Δv must decrease proportionally in order to leave the value of the right-hand side of Eq. (3.7.14) unaffected.

Another way to understand the same phenomenon is to compare the expressions for the increment in energy $E_f - E_i$ and the increment in momentum $m(v_f - v_i)$ experienced by a particle during acceleration with a force F_a along a distance $z_1 - z_0$ over a time interval $t_f - t_i$

$$E_f - E_i = \int_{z_0}^{z_1} dz\, F_a, \quad m(v_f - v_i) = \int_{t_i}^{t_f} dt\, F_a \qquad (3.7.15)$$

All particles in the beam travel the same distance $z_1 - z_0$ and consequently gain equal amounts of energy. However, the flight time $t_f - t_i$ depends on the initial velocity v_i of a particle. Particles with a relatively large initial velocity will have a relatively short flight time, while particles with a relatively small initial velocity will have a relatively long flight time. Accordingly, fast particles gain less velocity than slow particles, causing a reduction of the velocity spread.

The decrease in velocity spread, due to the acceleration, can also be presented as a consequence of Liouville's theorem. It was found that the application of this theorem to the subspace Γ_2 of axial coordinates implies that the volume $\Delta V_2 = \Delta r_\| \Delta v_\|$ is invariant; see Eq. (3.6.1). It was also found that the length $\Delta r_\|$ is directly proportional to the axial velocity $v_\|$. Thus $\Delta r_\|$ increases during acceleration. As $\Delta r_\| \Delta v_\|$ is invariant, the axial velocity spread $\Delta v_\|$ decreases accordingly, q.e.d. Finally, we recall that Eq. (3.6.1) is equivalent to the adiabatic equation for a one-dimensional ideal gas, which is given by Eq. (3.7.5) (taking $f = 1$).

Thus, in the thermodynamical description of the beam, one may envision the reduction of the axial velocity spread as a "cooling effect" caused by the adiabatic expansion of the "gas" of charged particles during acceleration.

In general, Liouville's theorem, applied to the appropriate phase space, expresses that an expansion of the spatial volume must coincide with a compression of the velocity space. Clearly, this principle does not only affect the longitudinal beam temperature, but it affects the transverse beam temperature as well. Initially, the transverse beam temperature $T_\perp$ is identical to the cathode temperature T_{ca}, as is expressed by the second equation of (3.7.9). As the particles flow through the system, their lateral separation changes with the beam radius. Conservation of emittance, which is a consequence of Liouville's theorem, implies that the transverse beam temperature $T_\perp$, being directly proportional to the transverse velocity spread, increases with decreasing beam radius and vice versa. To illustrate this point, consider the beam geometry of Fig. 3.4 and the corresponding emittance plots of Fig. 3.5. The crossover plane is an image of the source, while the plane P is an image of the entrance pupil. For this particular combination of planes, emittance invariance can be expressed as $\Delta r_p' r_p = \Delta r_c' r_c$. Using Eq. (3.7.3) and $\Delta v_\perp = v_\parallel \Delta r'$, one finds for the relation between the transverse beam temperatures in these planes

$$T_{\perp p} = \frac{\Delta r_p'^2}{\Delta r_c'^2} T_{\perp c} = \frac{r_c^2}{r_p^2} T_{\perp c}. \tag{3.7.16}$$

For other planes than the set of planes considered here, this relation holds only approximately, due to edge effects, as can be seen from Fig. 3.5. Thus, in general

$$T_\perp \approx \frac{r_c^2}{r_0(z)^2} T_{\perp c}. \tag{3.7.17}$$

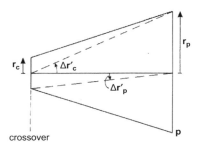

Fig. 3.4 Schematic view of a beam section between a crossover and a plane P, which is an image of the entrance pupil of the system.

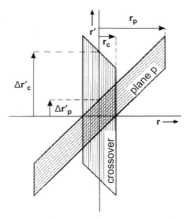

Fig. 3.5 Emittance plots corresponding to the crossover plane and the plane P, shown in Fig. 3.4.

Notice that the transverse beam temperature is identical in all points of a plane perpendicular to the beam axis. Thus, the transverse beam temperature $T_\perp$ depends on the axial position z but not on the radial coordinate r. By combining Eqs. (3.7.4) and (3.7.17) one obtains

$$T_\perp \approx \frac{eV}{2k_B} \frac{(\alpha_0 r_c)^2}{r_0(z)^2}. \tag{3.7.18}$$

This equation relates the transverse beam temperature $T_\perp$, at an arbitrary location within the beam segment, to the value of the geometrical parameters. Alternatively, this relation can be expressed in terms of the normalized brightness β_n, given by Eq. (3.6.4), and the current density $J = I/\pi r_0(z)^2$

$$\beta_n = \frac{I}{V \pi \alpha_0^2 \pi r_c^2} \approx \frac{eJ}{2\pi k_B T_\perp}, \tag{3.7.19}$$

in which it was assumed that the current density is constant in every cross section of the beam. (Accordingly, differential brightness and total brightness are not distinguished.) As β_n is conserved, the ratio of J and $T_\perp$ is conserved too. In other words, conservation of normalized brightness (or emittance) implies that $T_\perp$ is directly proportional to J. Eq. (3.7.19) is equivalent to Langmuir's equation for the maximum current density obtainable in the spot of a cathode-ray tube; see Langmuir (1937). Langmuir's result pertains to the case of a half Maxwell-Boltzmann distribution and therefore differs by a factor 2. In addition, we note that Eq. (3.7.19), unlike Langmuir's equation, requires that $eV/k_B T \gg 1$.

When the total emittance is conserved through the entire system, Eq. (3.7.17) can be generalized to include all segments of the beam.

General beam properties 49

As a result, one can relate the transverse beam temperature $T_\perp$, at an arbitrary location in the beam, to the temperature of the cathode T_{ca}, using

$$T_\perp \approx \frac{r_{ca}^2}{r_0(z)^2} T_{ca}, \qquad (3.7.20)$$

where r_{ca} is the radius of the beam at the cathode. It should be emphasized that this equation presupposes conservation of the *total* emittance. Accordingly, lens aberrations, beam limiting apertures and particle interactions arc assumed to be absent.

Eqs. (3.7.13), (3.7.18), and (3.7.20) yield the main results of this section. From Eq. (3.7.20) it follows that the transverse beam temperature is large in crossovers and small in the dilute parts of the beam. In practical beams, the maximum value of r_0/r_{ca} is of the order 10^2. Thus, the corresponding reduction of the transverse beam temperature, relative to the cathode temperature T_{ca}, is of the order 10^4. According to Eq. (3.7.13), the longitudinal beam temperature $T_\parallel$ is of the order of 10^{-5} times T_{ca}. Thus, disregarding the effect of particle interactions, one may in general assume that $T_\parallel \ll T_\perp$ even in the dilute parts of the beam. The difference between both temperatures will be the largest in the crossover area, where the beam radius is the smallest.

3.8. Thermodynamic limits for particle interaction effects

In the previous section it was demonstrated that nonequilibrium situations in particle beams may occur as a consequence of the expansion and compression of the beam volume under influence of the acceleration in the gun area and the focussing action of lenses. Coulomb interactions between the beam particles were disregarded so far. However, they do affect the velocity distribution of the particles, as well as their spatial distribution, and thus change the macroscopic properties of the beam. In the thermodynamical description of the beam, the particle interactions cause a relaxation toward equilibrium. Within this model one can estimate the effect of particle interactions by assuming that the relaxation is complete, which means that thermodynamic equilibrium is established during the flight. As we are primarily interested in the principle aspects involved, the analysis will be restricted to the generation of energy spread (the Boersch effect).

The speed of the relaxation process depends on the frequency of the collisions and the average impact per collision. Consequently, the relaxation times for the different interaction effects depend on the particle density in the beam and the velocity distribution of the particles. The question whether equilibrium will be established or not can be answered by comparing the appropriate relaxation times with the time of flight T_f, given by Eq. (3.2.3). Anticipating on the results of Sections 4.8 and 7.7, we mention that the equilibrium condition is in general not reached in practical beams. Consequently, equilibrium thermodynamics, as will be employed here, is not suited for an accurate quantitative calculation of the particle interaction effects, and the results merely serve to determine upper limits.

In Section 3.7, it was demonstrated that the longitudinal beam temperature $T_\parallel$ is, in general, expected to be negligible compared to the maximum transverse beam temperature $T_\perp$, which occurs in the crossover areas of the beam. Now, assume that thermodynamic equilibrium is reached in the crossover. The corresponding equilibrium temperature T_e will be given by

$$T_e = \frac{2}{3} T_{c\perp} = \frac{2eV}{3k_B} \langle \alpha^2 \rangle, \tag{3.8.1}$$

utilizing Eq. (3.7.4). The factor 2/3 stems from the fact that the transverse beam temperature corresponds to two degrees of freedom and the longitudinal to one. The corresponding normal root mean square (rms) energy spread $\langle \Delta E_\parallel^2 \rangle^{1/2}$ can be found by employing the non-relativistic relation

$$\frac{\langle \Delta E_\parallel^2 \rangle^{1/2}}{eV} \approx \frac{2 \langle \Delta v_\parallel^2 \rangle^{1/2}}{v_\parallel} = \left(\frac{2k_B T_\parallel}{eV} \right)^{1/2}, \tag{3.8.2}$$

utilizing Eqs. (3.7.14) and (3.7.3). The first step presupposes that $\langle \Delta E_\parallel^2 \rangle^{1/2} \ll eV$. From Eqs. (3.8.1) and (3.8.2), one finds that the equilibrium condition leads to an energy spread

$$\frac{\langle \Delta E_\parallel^2 \rangle^{1/2}}{eV} \approx \frac{2}{3^{1/2}} \langle \alpha^2 \rangle^{1/2} = (2/3)^{1/2} \alpha_0. \tag{3.8.3}$$

In the last step we used $\langle \alpha^2 \rangle = \alpha_0^2/2$, which pertains to the case of uniform angular distribution. For a Gaussian angular distribution one should use $\langle \alpha^2 \rangle = 2\sigma_\alpha^2$. Eq. (3.8.3) imposes an upper limit to the energy spread generated by conversion of kinetic energy from the lateral degrees of freedom to the longitudinal degree of freedom. For the numerical example of Section 3.2 ($V = 10\,\text{kV}$, $\alpha_0 = 10\,\text{mR}$), one obtains $\langle \Delta E_\parallel^2 \rangle^{1/2} = 82$ eV.

Eq. (3.8.3) does not impose a real upper limit to the total generated energy spread under all circumstances as is suggested by some authors; for instance, see Crewe (1987a), Knauer (1979a), and Rose and Spehr (1983). It just refers to the transition of kinetic energy from one degree of freedom to another (*relaxation of kinetic energy*) and does not account for the conversion of potential energy into random velocity components (*relaxation of potential energy*).

In order to clarify this statement, consider the extreme case of a monochromatic homocentric cylindrical beam. In such a beam all particles move initially with identical velocity along parallel trajectories. Observed in the

General beam properties

frame of reference moving with the beam, the particles are at rest. Consequently, there is no internal kinetic energy to be converted. Indeed, Eq. (3.8.3) predicts a zero upper limit for the generated energy spread, since $\alpha_0 = 0$. In other words, according to this analysis the Boersch effect does not occur in such a beam.

However, some energy spread can be generated in such a beam, due to the relaxation of potential energy. The particles entering the beam all have the same axial velocity but are otherwise randomly distributed over the beam volume. The mutual potential energy of a pair of particles depends on their relative position. As these positions are assumed to be randomly distributed, the potential energy will vary for each pair of particles. Near neighbors can have a substantial potential energy. Due to the Coulomb interaction during the flight, part of this potential energy will be redistributed into kinetic energy. As the associated velocity components are fully stochastic, the potential energy relaxation causes an increase of the velocity spread. The increase of the *axial* velocity spread corresponds to the Boersch effect.

It would be convenient to have an expression for the upper limit of the Boersch effect caused by potential energy relaxation, similar to Eq. (3.8.3), specifying the upper limit for the relaxation of kinetic energy. In order to derive such an expression, we have to introduce the concept of Debeye screening first. This will be done in the next section. As the subject of screening is of general importance to the theory of particle interactions, we will investigate its applicability to a system of identical charged particles in some detail. In Section 3.10 we will return to the main line of this chapter and consider the thermodynamic limit for the process of potential energy relaxation.

3.9. Debeye screening

We wish to calculate the effective potential field of a fixed charged particle surrounded by other charged particles that are statistically distributed. This problem was first considered by Debeye and Hückel (1923). They demonstrated that the potential field of an ion in a strong electrolyte is effectively screened by the cloud of particles surrounding it. The effective force range of the ion is therefore confined within a certain characteristic length, which is determined by the charge density and temperature of the medium. This length is called the Debeye screening length λ_D. Here we will study the applicability of this concept to a nonneutralized system of identical charged particles.

Let N be the total number of particles in the system, n the average particle density, e the particle charge, and m the particle mass. Consider a test particle located in the origin ($\mathbf{r} = 0$) of a coordinate system. The Coulomb potential field of the test particle is given by

$$\Phi(\mathbf{r}) = C_C/r, \tag{3.9.1}$$

where $C_C = e/4\pi\varepsilon_0$ (note: $C_0 = eC_C$). The other particles are referred to as field particles. The spatial distribution of the field particles around the origin will be affected by the presence of the electrostatic field of the test particle and consequently will deviate from a uniform distribution. This redistribution of the particles around the origin creates an electrostatic field, which partly compensates the original field of the test particle. The effective screened potential $\Phi_s(\mathbf{r})$ of the test particle is now defined as the sum of the original and the induced potential. The Poisson equation for this potential can be written as

$$\nabla^2 \Phi_s(\mathbf{r}) = \frac{e}{\varepsilon_0} \delta(\mathbf{r}) + \frac{e}{\varepsilon_0} n_d(\mathbf{r}), \tag{3.9.2}$$

where $\delta(\mathbf{r}) = \delta(x)\delta(y)\delta(z)$ is the three-dimensional delta function. The quantity $n_d(\mathbf{r})$ is the local deviation from the average particle density n, which is induced by the test particle

$$n_d(\mathbf{r}) = n_m(\mathbf{r}) - n \quad \text{with} \quad n_m(\mathbf{r}) = \left\langle \sum_{i=1}^{N} e\delta(\mathbf{r} - \mathbf{r}_i) \right\rangle, \tag{3.9.3}$$

where the brackets denote that the local microscopic distribution $n_m(\mathbf{r})$ is obtained by ensemble averaging. Presupposing thermodynamic equilibrium, $n_m(r)$ will obey Maxwell-Boltzmann statistics (as usual we ignore quantum mechanical effects, assuming that the particle density is low enough):

$$n_m(\mathbf{r}) = n e^{-e\Phi_s(\mathbf{r})/k_B T}. \tag{3.9.4}$$

Substitution into Eq. (3.9.3) yields

$$n_d(\mathbf{r}) = n\left(e^{-e\Phi_s(\mathbf{r})/k_B T} - 1\right) \approx -\frac{ne\Phi_s(\mathbf{r})}{k_B T} \tag{3.9.5}$$

The expansion of the exponential function requires $e\Phi_s(\mathbf{r})/k_B T \ll 1$, which means that the potential energy of the particles is assumed to be much smaller than their kinetic energy. Substitution of Eq. (3.9.5) into the Poisson Eq. (3.9.2) yields

$$\nabla^2 \Phi_s(\mathbf{r}) + \Phi_s(\mathbf{r})/\lambda_D^2 = \frac{e}{\varepsilon_0} \delta(\mathbf{r}), \tag{3.9.6}$$

where λ_D is the Debeye screening length, defined as

$$\lambda_D = \left(\frac{\varepsilon_0 k_B T}{e^2 n}\right)^{1/2}. \tag{3.9.7}$$

The solution of Eq. (3.9.6), with the boundary condition that $\Phi_s(\mathbf{r})$ vanishes at infinity, is

$$\Phi_s(\mathbf{r}) = \frac{C_C}{r} e^{-r/\lambda_D}. \tag{3.9.8}$$

Thus, for distances $r \ll \lambda_D$ the effective potential is essentially equal to the bare Coulomb potential, given by Eq. (3.9.1), while it is strongly reduced for distances $r \geq \lambda_D$. Thus, roughly speaking, one may say that according to Eq. (3.9.8) the effective interaction range of the test particle is limited to the Debeye distance λ_D.

We will continue this section by considering the validity of the screening concept for nonneutralized systems. We have assumed that the main effect of the Coulomb interaction is a microscopic redistribution of the particles, while the macroscopic dimensions of the system remain unchanged. Accordingly, we considered the average density n to be constant. Clearly, for a nonneutralized system of charged particles in a field free space, this cannot be entirely true. The collective Coulomb repulsion will cause the system to expand under influence of its own space charge, leading to a lower average density. A way out of this problem is to assume that the time scale of the collective process is much longer then the relaxation time associated with the discrete particle effects.

Let us investigate whether such an assumption is justified or not. In order to estimate the time scale of the collective process, we consider a sphere of radius R within the particle gas. The force F acting on a particle located on the surface of the sphere is given by

$$F = \frac{e^2 n R}{3\varepsilon_0}, \tag{3.9.9}$$

as follows with Gauss's theorem. We now demand that the displacement of the particle due to the force F during some time interval T_f, which is representative for the time scale of the system, is small compared to R:

$$\frac{F T_f^2}{2m} \ll R. \tag{3.9.10}$$

For the case of a particle beam, T_f should be chosen equal to the time of flight, which is given by Eq. (3.2.3). Substitution of Eq. (3.9.9) into Eq. (3.9.10) yields

$$T_f \ll \frac{\sqrt{6}}{\omega_p}, \tag{3.9.11}$$

where ω_p is given by

$$\omega_p = \left(\frac{e^2 n}{m \varepsilon_0}\right)^{1/2}. \tag{3.9.12}$$

This quantity is known as the plasma frequency.

Now, let us estimate the time scale associated with the discrete particle effects. The derivation of the screened potential relies upon the assumption that the density distribution of the field particles, under influence of the electrostatic field of the test particle, obeys Maxwell-Boltzmann statistics; see Eq. (3.9.4). The relaxation time required for the system to establish this equilibrium condition can roughly be estimated as the time in which a particle with a typical thermal velocity $v \approx (k_B T/m)^{1/2}$ travels over the Debeye distance λ_D. Thus, thermodynamic equilibrium requires

$$T_f \gg \frac{\lambda_D}{(k_B T/m)^{1/2}}. \tag{3.9.13}$$

Substitution of the expression for λ_D, given by Eq. (3.9.7), yields

$$T_f \gg \frac{1}{\omega_p}, \tag{3.9.14}$$

where ω_p is again given by Eq. (3.9.12). Debeye screening requires the relaxation time $1/\omega_p$ to be short compared to the timescale of the system T_f.

Clearly, the constraints of Eqs. (3.9.11) and (3.9.14) cannot be fulfilled simultaneously. This analysis shows that, in a system of identical charged particles, collective effects and discrete particle effects take place on the same time scale characterized by the quantity $1/\omega_p$. Thus, one may not assume that the redistribution of nearest neighbors occurs much faster than the spreading under influence of the average space charge force. Nevertheless, Debeye screening may occur for $T_f \gg 1/\omega_p$ but the average density appearing in Eq. (3.9.7) cannot be treated as a constant.

We end up with some additional remarks regarding the screening concept in general. The approximation used in Eq. (3.9.5) requires that $e\Phi_s(\mathbf{r})/k_B T \ll 1$, which says that the average potential energy per particle should be small compared to the average kinetic energy per particle. Let us rewrite this constraint in terms of the characteristic physical quantities of the system. It is to be expected that the final result for the screened potential $\Phi_s(\mathbf{r})$, given by Eq. (3.9.8), depends critically on this condition, only for those values of $\mathbf{r}$ for which $\Phi_s(\mathbf{r})$ differs significantly from the bare Coulomb potential, say for $\mathbf{r} \geq \lambda_D$. Thus, utilizing Eq. (3.9.7), one should demand

$$1 \gg \frac{e\Phi_s(\lambda_D)}{k_B T} \approx \frac{eC_C}{\lambda_D k_B T} = \frac{1}{4\pi \lambda_D^3 n} \approx \frac{1}{N_D}, \tag{3.9.15}$$

where $N_D = (4/3)\pi \lambda_D^3 n$, which is the number of particles contained within a sphere with radius λ_D, called the Debeye sphere. Eq. (3.9.15) states that the constraint $e\Phi_s(\mathbf{r})/k_B T \ll 1$ requires that the number of particles N_D within the Debeye sphere is large. It should be noted that this condition is required anyhow to justify the use of Maxwell-Boltzmann statistics (which pertains to a macroscopic system).

The screening according to Eq. (3.9.8) cannot be taken literally for all particles at all times. In the derivation given above, the test particle is conceived as a fixed force centre around which there exists a spherical symmetric Maxwell-Boltzmann density distribution of all other particles. In reality, the test particle is not fixed but follows some irregular curved trajectory due to the interaction with the other particles. Thus, if an equilibrium condition is established around a test particle at all at some point in time, it will be broken up shortly after that when it changes its course due to some strong collision. Notice also that a density distribution cannot be a spherical symmetric Maxwell-Boltzmann distribution for two different force centres simultaneously. From these arguments, it should be concluded that the screening described by Eq. (3.9.8) will not occur for every particle at all times. Nevertheless, if the system exists long enough, every particle will on the average experience a deficiency of particles in its immediate neighborhood, leading to some kind of screening, which can roughly be characterized by Eq. (3.9.8).

Finally, taking as a numerical example the case of electrons with density $n = 10^{16} \text{m}^{-3}$ and temperature $T = 10^3 \text{K}$, one obtains the following values: Debeye length $\lambda_D = 2 \times 10^{-5}$ m, number of particles in Debeye sphere $N_D = (4/3)\pi\lambda_D^3 n = 4 \times 10^2$, plasma frequency $\omega_p = 6 \times 10^9 \text{s}^{-1}$. Thus, such a system fulfills the requirement that N_D is a large number and some kind of screening will occur after a time interval long compared to $1/\omega_p = 2 \times 10^{-10}$s.

3.10. Potential energy relaxation

We will exploit the analysis of the previous section to determine an upper limit for the effect of potential energy relaxation. Consider the same system of identical particles as in the previous section, but now assume that they are initially all at rest. Let them be uniformly distributed over a volume V. The total potential energy U_i in the initial situation, is given by

$$U_i = \frac{1}{2} \sum_{\substack{i,j=1 \\ i \neq j}}^{N} \frac{C_0}{r_{ij}} \approx \frac{1}{2} n^2 \int_V d\mathbf{r}_i \int_V d\mathbf{r}_j \frac{C_0}{r_{ij}}, \qquad (3.10.1)$$

where $r_{ij} = |\mathbf{r}_i - \mathbf{r}_j|$, the distance between particle i and j. For a large volume V, one may ignore edge effects and approximate

$$U_i \approx \frac{1}{2} n^2 V \int_V d\mathbf{r} \frac{C_0}{r}. \qquad (3.10.2)$$

Notice that the integral in Eq. (3.10.2) diverges when one takes V infinitely large. In other words, the total potential energy per unit of volume (U_i/V) increases with the size of the system. This is a consequence of the long range of the Coulomb force.

Now, assume that the particles start interacting. As a result, part of the initial potential energy U_i will be converted into thermal energy, which is related to the random components of motion. At the same time, the system will expand under influence of its average space charge. This expansion corresponds to a systematic motion and has no direct impact on the thermal energy of the particles. As we are primarily interested in obtaining an *upper limit* for the resulting temperature of the system, we will ignore the average space charge effect entirely and assume that system will reach thermodynamic equilibrium while maintaining its average density n.

In thermodynamic equilibrium, the microscopic density distribution $n_m(\mathbf{r})$ obeys Maxwell-Boltzmann statistics and is given by Eq. (3.9.4). Accordingly, the total potential energy U_e in the equilibrium situation can be expressed as

$$U_e \approx \frac{1}{2} n^2 V \int_V d\mathbf{r} \frac{C_0}{r} e^{-e\Phi_s(r)/k_B T}. \tag{3.10.3}$$

Conservation of energy implies that the difference in potential energy between the initial and the equilibrium situation should be identical to the resulting total thermal energy of the system

$$\frac{3}{2} n V k_B T = U_i - U_e, \tag{3.10.4}$$

using that $N = nV$. Substitution of Eqs. (3.10.2) and (3.10.3) into Eq. (3.10.4) yields

$$\frac{3}{2} n V k_B T = \frac{1}{2} n^2 V \int_V d\mathbf{r} \frac{C_0}{r} \left(1 - e^{-e\Phi_s(r)/k_B T}\right). \tag{3.10.5}$$

As the volume V is large, one may approximate this equation as

$$\frac{3}{2} n V k_B T = \frac{1}{2} n^2 V \int_0^\infty 4\pi r^2 \, dr \frac{C_0}{r} \left(1 - e^{-e\Phi_s(r)/k_B T}\right), \tag{3.10.6}$$

employing spherical coordinates. The integral in Eq. (3.10.6) is finite, despite the fact that we took V infinitely large. This is a result of the fact that the screened potential $\Phi_s(r)$ decreases exponentially for $r \gg \lambda_D$, as can be seen from Eq. (3.9.8).

By combining Eqs. (3.9.7), (3.9.8), and (3.10.6) and substituting $x = r/\lambda_D$, one obtains

$$\int_0^\infty \left[1 - \exp\left(\frac{-1}{4\pi n \lambda_D^3} \frac{\exp(-x)}{x}\right)\right] x \, dx = 3. \tag{3.10.7}$$

This equation was solved numerically. It yields

$$4\pi n \lambda_D^3 = a \quad \text{with} \quad a = 0.08703. \tag{3.10.8}$$

With Eq. (3.9.7), one finds for the final thermal energy per particle

$$\frac{3}{2} k_B T = \frac{3}{2} (4\pi a^2)^{1/3} C_0 n^{1/3}, \tag{3.10.9}$$

which depends on the average particle density n only.

Let us now return to the case of a homocentric cylindrical particle, beam. From Eqs. (3.10.9) and (3.8.2), one obtains as an upper limit for the energy spread generated by potential energy relaxation

$$\frac{\langle \Delta E_\parallel^2 \rangle^{1/2}}{eV} = (32\pi a^2)^{1/6} \frac{C_0^{1/2} n^{1/6}}{(eV)^{1/2}}. \tag{3.10.10}$$

Substitution of the expression for the density n, given by Eq. (3.2.5), yields

$$\frac{\langle \Delta E_\parallel^2 \rangle^{1/2}}{eV} = C_{PE} \frac{J^{1/6}}{V^{7/12}}, \quad C_{PE} = \frac{a^{1/3}}{2^{1/4} \pi^{1/3}} \frac{m^{1/12} e^{1/4}}{\varepsilon_0^{1/2}}, \tag{3.10.11}$$

where $J\ (=I/\pi r_0^2)$ is the current density in the beam. For the case of electrons, one finds $C_{PE} = 5.368 \times 10^{-3}$ in SI-units. Taking as a numerical example $J = 100\ A/m^2$ and $V = 10\,kV$, Eq. (3.10.11) yields $\langle \Delta E_\parallel^2 \rangle^{1/2} = 0.54$ eV.

CHAPTER FOUR

The many body problem of particles interacting through an Inverse Square force law

Contents

4.1.	Introduction	59
4.2.	Vlasov equation	60
4.3.	Fokker–Planck equation	63
4.4.	Aspects of the diffusion approximation	67
4.5.	Calculation of coefficients of dynamical friction and diffusion	69
4.6.	Coulomb logarithm	75
4.7.	Discussion of the Fokker–Planck approach	79
4.8.	Validity of the Fokker–Planck approach for particle beams	81
4.9.	Holtsmark distribution	86
4.10.	Conclusions	89

4.1. Introduction

In the previous chapter, the Boltzmann equation was introduced as the fundamental equation for the time development of the single-particle distribution function $f(\mathbf{r}, \mathbf{v}, t)$ in Γ_6 space. Its implications were studied for the specific case of a system of identical particles, which do not interact. In this chapter, the more complex problem is considered of a system of identical particles, interacting through a force that is proportional to the inverse square of the distance between the particles. This type of N-body problem is extensively studied in plasma physics as well as in stellar dynamics. The fundamental aspects of the approaches taken in these fields are outlined, starting from the Boltzmann equation. Their applicability to the particular problem of Coulomb interactions in particle beams is investigated.

The main part of this chapter is concerned with the so-called Fokker–Planck approach. This approach relies on the assumption that the action of the fluctuating component of the interaction force can be described in terms of a diffusion process in velocity space. This assumption is justified for systems of high particle density operating at a high temperature near thermodynamic equilibrium. Accordingly, the Fokker–Planck approach is suited for plasmas, in which these conditions are usually fulfilled. It is shown that the Fokker–Planck approach is, in general, not suited to describe the impact of the statistical interactions in particle beams. The main reason for this failure is that a particle beam is relatively "cold" compared to a plasma and usually operates at a low density, while the particles have a short flight time relative to the characteristic relaxation times involved. This also indicates that thermodynamic equilibrium is not reached in practical beams. Accordingly, Debeye screening usually does not occur.

The Fokker–Planck approach may be adequate to evaluate the energy spread generated in the vicinity of a narrow crossover in a beam of relatively high density, where the conditions of a plasma are, to some extent, reproduced. The root mean square (rms) energy spread can be determined from the velocity-diffusion coefficient used in the Fokker–Planck equation. The resulting expression is, apart from a numerical factor, indentical to the one presented by Zimmermann (1970) and Knauer (1979a) for a cylindrical beam segment.

The last part of this chapter is concerned with the Holtsmark distribution, which is the probability distribution of the electric field strength acting on some point in a gas composed of randomly distributed point charges. It is shown that the distribution of velocities generated in an extended homocentric cylindrical beam segment is Holtsmarkian, provided that the perturbations are small. This knowledge can be exploited to calculate the energy spread and the angular deflections generated in a nearly homocentric cylindrical beam of relatively low density. The dependency of the energy spread on the experimental parameters is made explicit.

The method centered around the Holtsmark distribution is suited to describe the effect of potential energy relaxation in a homocentric cylindrical beam of low particle density, whereas the Fokker–Planck approach can be used to describe the effect of kinetic energy relaxation in a beam segment with a narrow crossover of high particle density. A more refined model will be presented in Chapter 5, which is suited to calculate the impact of the statistical interactions, in any type of beam geometry, for any practical particle density. It combines some of the features of the two approaches that are discussed here.

4.2. Vlasov equation

Consider a system of N identical particles in which the force $\mathbf{F}_{ij}$ exerted by particle j on particle i can be expressed as

$$\mathbf{F}_{ij} = C_0 \frac{\mathbf{r}_{ij}}{r_{ij}^3}, \qquad (4.2.1)$$

where the relative position $\mathbf{r}_{ij} = \mathbf{r}_j - \mathbf{r}_i$ and C_0 is a physical constant, which depends on the type of interaction. For a system of identical charged particles, $C_0 = e^2/4\pi\varepsilon_0$, where e denotes the particle charge. For a stellar system, $C_0 = -Gm^2$, where G is the gravitational constant and m the "particle" mass. Clearly, in practical stellar systems the mass varies per star and C_0 depends on the pair of stars considered; thus $C_0 = C_{0ij}$. A similar complication arises in multicomponent plasmas. Although the calculations for such systems are more elaborate than those for a system consisting of identical particles, the fundamental issues involved are the same. For simplicity we concentrate on the case of identical charged particles. The calculations presented here are nonrelativistic. Accordingly, magnetic interactions are ignored.

The total force on particle i is equal to the sum of the forces exerted by all other particles:

$$\mathbf{F}_i = C_0 \sum_{j \neq i}^{N} \frac{\mathbf{r}_{ij}}{r_{ij}^3}, \qquad (4.2.2)$$

assuming that external forces are absent. Due to the long range of the interaction force, the number of particles that significantly contribute to the total force $\mathbf{F}_i$ on particle i, will, in general, be large. In other words, every particle interacts with many other particles simultaneously. Most of these interactions will be weak, which means that they cause only small changes in the velocity of the colliding particles. Strong collisions, which correspond to large changes in velocity, will be relatively rare. Consequently, the trajectories in coordinate space follow smoothly curved lines with continuously varying curvature. It should be noted that this situation is quite different from the one encountered in non-ionized gases. At normal densities, the interaction between neutrals is limited to infrequent, strong collisions between pairs. The collisions are well separated in time, which means that collisions involving more than two particles simultaneously are rare. Accordingly, the particles follow a broken line in space. This motion can adequately be characterized by quantities as the "mean free path" (average distance between two collisions) and the "collision frequency" (number of collisions per time unit), which are of central importance to the kinetic theory of non-ionized gases. For instance, see Chapman and Cowling (1970). It might be clear that such concepts are less suited for the description of ionized gases. Thus, the theory for plasmas and stellar dynamics requires a different approach than the one taken in the classical kinetic theory of gases.

As every particle interacts simultaneously with many other particles, both near and remote, it appears to be justifiable to distinguish between the influence of the system as a whole and the influence of the immediate local neighborhood. The former will be a smoothly varying function of position and time, which can be expressed in terms of a potential

$\Phi(\mathbf{r}, t)$, while the latter will be subject to relatively rapid fluctuations. Accordingly, the total interaction force on particle i is divided into two components

$$\mathbf{F}_i = \mathbf{F}_{av,i} + \mathbf{F}_{f,i} = -\frac{\partial}{\partial \mathbf{r}_i} \Phi(\mathbf{r}_i, t) + C_0 \sum_j \frac{\mathbf{r}_{ij}}{r_{ij}^3}, \qquad (4.2.3)$$

in which the summation in the second term on the right-hand side is now limited to neighbors. The potential $\Phi(\mathbf{r}, t)$ corresponds to the smoothed-out distribution of charge. It can be related to the single-particle distribution function $f(\mathbf{r}, \mathbf{v}, t)$ as

$$\Phi(\mathbf{r}, t) = C_0 \int d\mathbf{v}_1 \int d\mathbf{r}_1 \frac{f(\mathbf{r}_1, \mathbf{v}_1, t)}{|\mathbf{r} - \mathbf{r}_1|} \qquad (4.2.4)$$

This potential is a function of the absolute position within the system $\mathbf{r}$ and time t only and is, in that respect, equivalent to an external, time dependent, electrostatic potential. Thus, the first component in Eq. (4.2.3) can be treated as an external force (which depends, however, on the average charge distribution). The second component can not directly be expressed in terms of the macroscopic properties of the system. Formally, the summation over j refers to near neighbors. However, it is difficult to define an adequate criterion to distinguish near neighbors from the other particles in the system. In fact, the second term just represents the remaining force, produced by a specific configuration of particles, after substraction of the average force $\mathbf{F}_{av,i}$ produced by the smoothed-out distribution of charge. One usually identifies this component with the force exerted during collisions. We prefer the more general term "fluctuating force component" and denote it as $\mathbf{F}_f$.

Using Eq. (4.2.3), Boltzmann's Eq. (3.5.2) can now be expressed as

$$\frac{df}{dt} = \frac{\partial f}{\partial t} + \mathbf{v} \cdot \frac{\partial f}{\partial \mathbf{r}} - \frac{1}{m} \frac{\partial \Phi}{\partial \mathbf{r}} \cdot \frac{\partial f}{\partial \mathbf{v}} = \left[\frac{\partial f}{\partial t}\right]_f, \qquad (4.2.5)$$

in which $[\partial f / \partial t]_f$ is the rate of change of $f(\mathbf{r}, \mathbf{v}, t)$ due to the fluctuating component of the interaction force. Disregarding this term ($[\partial f / \partial t]_f = 0$), Eq. (4.2.5) becomes identical to the so-called nonrelativistic Vlasov equation for a single component plasma:

$$\frac{df}{dt} = \frac{\partial f}{\partial t} + \mathbf{v} \cdot \frac{\partial f}{\partial \mathbf{r}} - \frac{1}{m} \frac{\partial \Phi}{\partial \mathbf{r}} \cdot \frac{\partial f}{\partial \mathbf{v}} = 0. \qquad (4.2.6)$$

In combination with Eq. (4.2.4), the Vlasov equation offers a self-consistent description of the system in collisionless continuum approximation, determining the time development of the single-particle distribution function $f(\mathbf{r}, \mathbf{v}, t)$, under influence of the average interaction force. Clearly, the Vlasov equation does not account for the fluctuations due to the discreteness of the particles, referred to as collisional effects. Thus it is unable to describe an approach of the system toward equilibrium. In stellar dynamics, the Vlasov equation (usually referred to as collisionless Boltzmann equation) is used to describe the evolution of stellar systems that can be regarded as collisionless, which means that their characteristic relaxation times are large compared to the total time scale of the universe ($\approx 10^{10}$ years). In particular, it is used to describe the dynamics of Galaxies, see for instance Chandrasekhar (1942) and Contoupolos (1972). In plasma physics, one usually deals with neutralized, multicomponent systems. In such systems, the average interaction force $\mathbf{F}_{av}$ is responsible for the occurrence of collective plasma oscillations.

4.3. Fokker–Planck equation

Let us now consider the fluctuations in the interaction force. Eq. (4.2.3) expresses that the fluctuating component of the total force, acting on a test particle, is related to the distribution of its neighbor field particles, which is a rapidly changing function of time. A rough estimation of the correlation time of the fluctuations τ_f is given by the ratio between the average separation of the particles and their average relative velocity:

$$\tau_f = n^{-1/3}/\langle v^2 \rangle^{1/2} \approx n^{-1/3}\left(\frac{m}{k_B T}\right)^{1/2}, \qquad (4.3.1)$$

where n is the average particle density and m is the particle mass. The relation $\langle v^2 \rangle \approx k_B T/m$ was utilized, assuming thermal equilibrium. In terms of collisions, τ_f can be regarded as a measure for the average collision duration. As a numerical example, consider the case of electrons in a plasma with density $n = 10^{16} m^{-3}$ and temperature $T = 10^5$ K. From Eq. (4.3.1) one finds $\tau_f \approx 4.10^{-12}$ s. This time will be short compared to the time in which any of the macroscopic parameters of the system change appreciable. Thus, in general, the fluctuations in the interaction force will occur with extreme rapidity compared to the rate of change of the average force $\mathbf{F}_{av}$, which is related to the time development of the (macroscopic) single-particle distribution function $f(\mathbf{r}, \mathbf{v}, t)$, as expressed by Eqs. (4.2.3) and (4.2.4).

From these considerations, it appears to be reasonable to assume that there exists a time interval Δt, which is long compared to the correlation time of the fluctuations τ_f but short enough to neglect the variation in $\mathbf{F}_{av}$. The velocity change of the test particle, during this time interval, can then be expressed as

$$\Delta \mathbf{v} = \mathbf{F}_{av}(t)\,\Delta t + \delta \mathbf{v}(t;\,\Delta t), \qquad \delta \mathbf{v}(t;\,\Delta t) = \int_t^{t+\Delta t} \mathbf{F}_f(t')\,dt' \qquad (4.3.2)$$

For the conditions stated ($\Delta t \gg \tau_f$), the displacements $\delta \mathbf{v}(t;\,\Delta t)$ and $\delta \mathbf{v}(t+\Delta t;\,\Delta t)$ experienced by the test particle in successive time intervals are expected to show no correlation. To put it more generally, the displacements experienced in disjoint time intervals can be considered as statistically independent. Thus, on this time scale the motion in velocity space can be visualized as a succession of random microscopic jumps superimposed on a smooth systematic motion, determined by $\mathbf{F}_{av}$. It should be noted that the choice of the time scale is here of crucial importance. On a finer time scale, Δt becomes of the same order as the correlation time of the fluctuations (or collision duration) τ_f. Consequently, the displacements experienced during successive time intervals will be correlated, since they are caused by the interaction within the same complexion of neighbors (or in other words the same collision). Using a more coarse grained time scale implies that the variation in $\mathbf{F}_{av}$ during the time interval Δt can no longer be neglected, while the size of jumps $\delta\mathbf{v}$ due to $\mathbf{F}_f$ can no longer be regarded as microscopically small. Consequently, the separation between $\mathbf{F}_{av}$ and $\mathbf{F}_f$ becomes useless. The constraints on the time interval Δt can be summarized as

$$\Delta t \gg \tau_f, \quad f(\mathbf{r}, \mathbf{v}, t + \Delta t) \approx f(\mathbf{r}, \mathbf{v}, t), \quad \delta v(t; \Delta t) \ll (k_B T/m)^{1/2}. \quad (4.3.3)$$

It is essential to the theory presented in the remainder of this section that a time interval Δt exists that fulfills these requirements.

Although the velocity displacements $\delta\mathbf{v}$ experienced in disjoint time intervals are uncorrelated, there may be a systematic tendency for velocities to jump in a specific direction. Such should be expected when a test particle has an excess velocity relative to the field particles surrounding it. In that case, the field particles will on the average decelerate the test particle. This means that the average of the velocity displacements $\delta\mathbf{v}$ experienced by the test particle during a certain period of time will be non-zero, while the resulting velocity change is directed opposite to the velocity of the test particle. This phenomenon is referred to as dynamical friction.

Under the conditions stated by the Eqs. (4.3.3), the time development of the single-particle distribution function $f(\mathbf{r}, \mathbf{v}, t)$ under influence of the fluctuating component of the interaction force can be described by the integral equation

$$f_f(\mathbf{r}, \mathbf{v}, t + \Delta t) = \int d\,\delta\mathbf{v}\, W_{\Delta t}(\mathbf{r}, \mathbf{v} - \delta\mathbf{v};\, \delta\mathbf{v}) f_f(\mathbf{r}, \mathbf{v} - \delta\mathbf{v}, t), \quad (4.3.4)$$

where $W_{\Delta t}(\mathbf{r}, \mathbf{v}; \delta\mathbf{v})$ represents the transition probability that $\mathbf{v}$ changes by an increment $\delta\mathbf{v}$, during the time interval Δt. This function is normalized such that

$$\int d\,\delta\mathbf{v}\, W_{\Delta t}(\mathbf{r}, \mathbf{v}; \delta\mathbf{v}) = 1. \quad (4.3.5)$$

The assumption underlying Eq. (4.3.4) is that the motion of a particle in phase space, from its location $(\mathbf{r}, \mathbf{v})$ at time t to $(\mathbf{r}, \mathbf{v} + \delta\mathbf{v})$ at time $t + \Delta t$, depends only on its instantaneous position in phase space $(\mathbf{r}, \mathbf{v})$ and is entirely independent of its previous coordinates. In general, a stochastic process with this characteristic, namely that what happens at a given instant of time depends only on the state of the system at that time, is called a Markoff process. The validity of the Markoff assumption depends on the choice of the time interval Δt, as was discussed previously.

The transition probability function $W_{\Delta t}(\mathbf{r}, \mathbf{v}; \delta\mathbf{v})$ represents the essential stochastic nature of the interaction process. Notice that it accounts for jumps in velocity only. Clearly, the fluctuations in the interaction force cause random spatial displacements $\delta\mathbf{r}$ within the time interval Δt as well. However, these displacements are of the order Δt^2 and can thus be neglected if Δt is small enough. Spatial displacements without the time interval Δt are represented through the change in velocity $\delta\mathbf{v}$ and need not be considered separately. Basically, this approach expresses that the trajectories of the particles in coordinate space can adequately be described as a concatenation of straight lines, provided that Δt is chosen short enough. In this context, it should be added that one often studies extended homogeneous systems in which the microscopic spatial displacements can be disregarded anyhow (from a macroscopic point of view).

Eq. (4.3.4) is true irrespective of the size of the jump $\delta\mathbf{v}$. We will now exploit the fact that the jumps in velocity are expected to be small. Expanding Eq. (4.3.4) in Taylor series with respect to t and $\mathbf{v}$, one finds (notation: $f = f_f(\mathbf{r}, \mathbf{v}, t); W_{\Delta t} = W_{\Delta t}(\mathbf{r}, \mathbf{v}; \delta\mathbf{v})$)

$$f + \Delta t \frac{\partial f}{\partial t} + \cdots = \int d\,\delta\mathbf{v} \left[1 - \sum_k \delta v_k \frac{\partial}{\partial v_k} + \frac{1}{2} \sum_{k,l} \delta v_k \delta v_l \frac{\partial^2}{\partial v_k \partial v_l} + \cdots \right] W_{\Delta t} f$$

$$= f - \sum_k \frac{\partial}{\partial v_k} \left[\int d\,\delta\mathbf{v}\, \delta v_k\, W_{\Delta t} \right] f$$

$$+ \frac{1}{2} \sum_{k,l} \frac{\partial^2}{\partial v_k \partial v_l} \left[\int d\,\delta\mathbf{v}\, \delta v_k\, \delta v_l\, W_{\Delta t} \right] f + \cdots,$$

in which the summation indices $k, l = 1, 2, 3$ refer to the components of $\delta\mathbf{v}$. In the second step, the normalization of Eq. (4.3.5) was exploited. Defining the so-called first and second jump moments as

$$\langle \delta v_k \rangle = \int d\,\delta\mathbf{v}\, \delta v_k\, W_{\Delta t}, \quad \langle \delta v_k\, \delta v_l \rangle = \int d\,\delta\mathbf{v}\, \delta v_k\, \delta v_l\, W_{\Delta t}, \qquad (4.3.6)$$

one may write

$$\left[\frac{\partial f}{\partial t}\right]_f \approx \sum_k \frac{\partial}{\partial v_k} \left[\frac{\langle \delta v_k \rangle}{\Delta t} f\right] + \frac{1}{2} \sum_{k,l} \frac{\partial^2}{\partial v_k \partial v_l} \left[\frac{\langle \delta v_k\, \delta v_l \rangle}{\Delta t} f\right]. \qquad (4.3.7)$$

Eq. (4.3.7) relates the short-time stochastic behavior of the test particle, expressed in the jump moments of its velocity, to the long-time behavior of the single-particle distribution function $f(\mathbf{r}, \mathbf{v}, t)$. The principal advantage of this approximation above the original problem expressed by Eq. (4.3.4) is that it directly relates the rate of change of the single-particle distribution function $f(\mathbf{r}, \mathbf{v}, t)$ to the values of the jump moments $\langle \delta v_k \rangle$ and $\langle \delta v_k\, \delta v_l \rangle$ ($k, l = 1, 2, 3$), which can be obtained without detailed knowledge of the entire transition probability function $W_{\Delta t}(\mathbf{r}, \mathbf{v}; \delta \mathbf{v})$. The jump moments can be estimated directly by studying the collisional processes between a test particle and its surrounding field particles over the time interval Δt. This is a problem of particle dynamics, which will be considered in Section 4.5.

From the preceding analysis, it follows that the right-hand side of Eq. (4.2.5) can be expressed as (using the conventions of vector and tensor analysis)

$$\left[\frac{\partial f}{\partial t}\right]_f = -\frac{\partial J}{\partial \mathbf{v}}, \quad J = -\beta \mathbf{v} f - \mathbf{D} \cdot \frac{\partial f}{\partial \mathbf{v}}, \tag{4.3.8}$$

where β is the so-called coefficient of dynamical friction and $\mathbf{D}$ is the diffusion tensor (which is a tensor of the second rank with elements D_{kl}). The vector J is called the collision flux. It represents the flux in velocity space due to the fluctuating component of the interaction force. By comparison with Eq. (4.3.7), it follows that β and $\mathbf{D}$ are related to the jump moments, defined by the Eqs. (4.3.6), as

$$-\beta v_k = \frac{\langle \delta v_k \rangle}{\Delta t}, \quad D_{kl} = \frac{\langle \delta v_k\, \delta v_l \rangle}{2\,\Delta t}. \tag{4.3.9}$$

The dynamical friction force is related to the systematic part of the velocity jumps of the test particle. The notation $(-\beta \mathbf{v})$ expresses that this force is expected to be proportional to minus the velocity of the particle, similar to the friction force given by Stokes law. The elements of the diffusion tensor D_{kl} are related to the purely stochastic part of the velocity jumps of the test particle. In Eq. (4.3.8), it was assumed that the tensor $\mathbf{D}$ is independent of the velocity $\mathbf{v}$. In general, it should be emphasized that the concept of dynamical friction and diffusion relies on the assumption that β and $\mathbf{D}$ are a function of the macroscopic properties of the system only. This is equivalent to the assumption that both $f(\mathbf{r}, \mathbf{v}, t)$ and $W_{\Delta t} = W_{\Delta t}(\mathbf{r}, \mathbf{v}; \delta \mathbf{v})$ are slowly varying functions of $\mathbf{r}$ and $\mathbf{v}$. The properties of β and $\mathbf{D}$ and their dependency on the macroscopic parameters will be discussed in Section 4.5.

Eqs. (4.3.8) and (4.3.9) constitute the main results of this section. They describe the impact of the fluctuations in the interaction force (usually referred to as collisional effects) in so-called diffusion approximation. Combining Eqs. (4.2.5) and (4.3.8), assuming that $\mathbf{D}$ is isotropic ($\mathbf{D}_{kl} = D \delta_{kl}$), one finds

$$\frac{df}{dt} = \frac{\partial f}{\partial t} + \mathbf{v} \cdot \frac{\partial f}{\partial \mathbf{r}} - \frac{1}{m}\frac{\partial \Phi}{\partial \mathbf{r}} \cdot \frac{\partial f}{\partial \mathbf{v}} = \beta \frac{\partial \mathbf{v} f}{\partial \mathbf{v}} + D \frac{\partial^2 f}{\partial v^2}, \qquad (4.3.10)$$

which is an equation of the Fokker–Planck form. In combination with the Eqs. (4.2.4) and (4.3.9), the Fokker–Planck Eq. (4.3.10) offers a self-consistent description of the system in diffusion approximation, determining the time development of the single-particle distribution function $f(\mathbf{r}, \mathbf{v}, t)$, under influence of the combined action of the average interaction force and the fluctuating force component, related to the discreteness of the particles. The remaining task is to calculate the coefficient of dynamical friction β and diffusion coefficient D, as functions of the macroscopic properties of the plasma.

For a general discussion of the Fokker–Planck equation, the reader is referred to van Kampen (1981). A comprehensive review on related topics as random walks, diffusion, Brownian motion and stellar dynamics is given by Chandrasekhar (1943). The Fokker–Planck equation and its application in plasma physics is discussed by Trubnikov (1965), Sivukhin (1966) and Ichimaru (1973).

4.4. Aspects of the diffusion approximation

The Fokker–Planck Eq. (4.3.10) presupposes that the action of the fluctuating component of the interaction force can adequately be described as a diffusion process in velocity space. In this section, we will elucidate some aspects of this so-called diffusion approximation.

The rate of change of the single-particle distribution function $f(\mathbf{r}, \mathbf{v}, t)$ due to the fluctuations in the interaction force is given by Eq. (4.3.8). In order to clarify the structure of this equation, it is instructive to consider an ordinary spatial diffusion process in an atomair gas. Let $\rho(\mathbf{r})$ be the mass density of the gas at position $\mathbf{r}$, $\mathbf{J}_m$ the total mass flow at this location and $\mathbf{D}_r$ the diffusion tensor. The mass flow $\mathbf{J}_m$ is given by

$$\mathbf{J}_m = \mathbf{v}\rho - \mathbf{D}_r \cdot \frac{\partial \rho}{\partial \mathbf{r}}. \qquad (4.4.1)$$

The first term represents the mass flow due to convection. The second term specifies the contribution of diffusion. When $\mathbf{D}_r$ is isotropic, it expresses that the mass flow due to diffusion is proportional to minus the gradient of the mass density. This is known as Fick's law. Continuity requires

$$\frac{\partial \rho}{\partial t} = -\frac{\partial \mathbf{J}_m}{\partial \mathbf{r}}. \qquad (4.4.2)$$

The so-called general diffusion equation follows by substituting Eq. (4.4.1) into Eq. (4.4.2). Here, we merely want to show that Eqs. (4.4.1) and (4.4.2) have the same structure as Eq. (4.3.8). The quantities $\mathbf{v}$ and $\mathbf{D}_r$ appearing in Eq. (4.4.1) can be expressed in terms of the microscopic spatial jump $\delta \mathbf{r}$ of a test particle experienced during the period Δt,

$$v_k = \frac{\langle \delta r_k \rangle}{\Delta t}, \quad D_{kl} = \frac{\langle \delta r_k \, \delta r_l \rangle}{2 \, \Delta t}, \qquad (4.4.3)$$

similar to Eq. (4.3.9). One easily verifies that Eqs. (4.4.1)–(4.4.3) are completely analogous to Eqs. (4.3.8) and (4.3.9). Formally, the last set of equations can be obtained from the first set by replacing the spatial coordinates by velocities: $\mathbf{r} \to \mathbf{v}$ (thus $r_k \to v_k$ and $r_l \to v_l$).

In the derivation of Eq. (4.3.8), the expansion in $\delta \mathbf{v}$ was truncated after the second order term. This implies that the jumps in velocity $\delta \mathbf{v}$ should be small. This condition is expressed by the last constraint in Eq. (4.3.3). Clearly, the smaller one takes the time period Δt, the smaller the corresponding velocity jump $\delta \mathbf{v}$. However, the Markoff assumption requires that Δt is chosen large enough to guarantee that the velocity displacements experienced in successive time intervals are not correlated. This condition is expressed by the first constraint in Eq. (4.3.3). In fact, it says that the velocity displacements experienced in disjoint time-intervals should not be generated by the same collision. The combination of the first and the third constraint in Eq. (4.3.3) implies that the effect of strong collisions cannot be taken into account within the diffusion approximation, since a strong collision results in a large velocity displacement $\delta \mathbf{v}$ within the correlation time τ_f.

Another way to visualize the same problem is the following. Eq. (4.3.8) expresses that the increment of the number of particles contained within a volume element in velocity space is equal to the net flux through the volume boundaries. This is accurate when the path of the test particle in velocity space can be regarded as continuous, despite the fact that one is effectively sampling with a period Δt. The continuity condition is violated during strong collision, in which the velocity of the particle can change appreciably within the period Δt. Let the velocity of the test particle at time t be $\mathbf{v}$. If a strong collision occurs during the next time interval Δt, the particle is in fact "annihilated" at point $\mathbf{v}$ and "created" at some remote point $\mathbf{v} + \delta \mathbf{v}$, without passing through the intermediate points in velocity space. Accordingly, we conclude, once again, that the effect of strong collisions cannot be taken into account by means of Eq. (4.3.8).

The stationary solution of the diffusion equation is a Gaussian distribution. This follows straightforwardly from Eq. (4.3.8). For simplicity, let us consider a homogeneous isotropic system, for which $D_{kl} = D\delta_{kl}$, $f(\mathbf{r}, \mathbf{v}, t) = nf(\mathbf{v}, t)$ and D and β are constant. A stationary solution f_s implies $[\partial f_s / \partial t]_f = 0$. By integration of the individual equations for each of the three components of velocity, one finds for the total velocity distribution

$$f_s(\mathbf{v}) = Ce^{-\beta v_x^2/2D} e^{-\beta v_y^2/2D} e^{-\beta v_z^2/2D} = Ce^{-\beta \mathbf{v}^2/2D}, \qquad (4.4.4)$$

where $C = (\beta/2\pi D)^{3/2}$ as is required by normalization. Thus, the stationary velocity distribution is a three-dimensional Gaussian distribution. This could be expected on the basis of the so-called central limit theorem, which says that the probability distribution of the sum of a great number of independent identical stochastic variables (in this case the sum of all microscopic velocity jumps experienced by the test particle during a certain period of time) will be Gaussian. Notice that no stationary solution exists when the dynamical friction coefficient β is zero. Without dynamical friction, there is no tendency to equilibrium.

From statistical mechanics, one knows that the equilibrium solution must be Maxwellian:

$$f_e(\mathbf{v}) = \left[\frac{m}{2\pi k_B T}\right]^{3/2} e^{-m\mathbf{v}^2/2k_B T}, \qquad (4.4.5)$$

which is identical to the stationary solution given by Eq. (4.4.4), provided that

$$D/\beta = k_B T/m. \qquad (4.4.6)$$

Thus, the assumption that the stationary solution of the diffusion Eq. (4.3.8) is Maxwellian implies that the coefficient of dynamical friction β and the diffusion coefficient D are related according to Eq. (4.4.6). It expresses that the equilibrium condition implies a balance between fluctuations and damping, represented by D and β, respectively. Eq. (4.4.6) is an example of the so-called fluctuation-dissipation theorem.

It should be noted that it is not trivial to identify the stationary solution of the diffusion equation, given by Eq. (4.4.4), with the equilibrium solution of Eq. (4.4.5). In fact, it is a principle task of the theory to demonstrate that an arbitrary initial state of the system will eventually ($t \to \infty$) lead to the stationary equilibrium solution given by Eq. (4.4.5). We will not pursue this problem and will assume that this is indeed the case.

4.5. Calculation of coefficients of dynamical friction and diffusion

In the previous sections, the calculation of the evolution of the single-particle distribution function $f(\mathbf{r}, \mathbf{v}, t)$, under influence of the fluctuating component of the interaction force, was reduced to the evaluation of the coefficient of dynamical friction β and the diffusion tensor $\mathbf{D}$, which are related to the first and second jump moments of the particle velocities, respectively, as expressed by Eq. (4.3.9). In this section, we will consider the dynamical problem of calculating these jump moments, defined by Eq. (4.3.6).

In plasma physics, one usually identifies the effect of the fluctuating component of the interaction force, acting on a particle during the time interval Δt with the impact of all two-particle collisions experienced by the particle in that period. We will follow this approach. Consider a homogeneous system of identical charged particles with density n. Assume that the velocity distribution $f(\mathbf{v})$ is isotropic and that the velocities show no correlation with the position within the system (molecular chaos). Let us focus on a single test particle and calculate its interaction with the other particles in the system, called field particles. First, we will determine the impact of a collision with a single field particle. This leads to an expression for the velocity shift of the test particle as function of the relative initial position and velocity of the two particles. Next, we will determine the mean velocity shift and mean square velocity shift, experienced by the test particle over a period of time Δt, by averaging over all possible relative coordinates of test and field particles, taking their probability distribution into account. It is assumed that the system operates near thermodynamic equilibrium, which implies that the velocity distribution of the particles is (nearly) Maxwellian.

Consider a collision between the test particle and a single field particle. Let us denote the initial position and velocity of the test particle as $\mathbf{r}_t$ and $\mathbf{v}_t$, respectively, and the corresponding coordinates of the field particle as $\mathbf{r}_f$ and $\mathbf{v}_f$. As the interaction force is central, the motion of the particles takes place in a plane, referred to as the orbital plane. The collision can best be described in terms of the relative position $\mathbf{r} = \mathbf{r}_f - \mathbf{r}_t$ and relative velocity $\mathbf{v} = \mathbf{v}_f - \mathbf{v}_t$. The relative position vector follows a hyperbola in the orbital plane; see Fig. 4.1. The deflection angle χ specifies the angle between the directional asymptotes of the hyperbola. It is given by:

$$\tan(\chi/2) = \frac{b_\perp}{b} \quad \text{and} \quad b_\perp = \frac{2C_0}{mv^2}, \qquad (4.5.1)$$

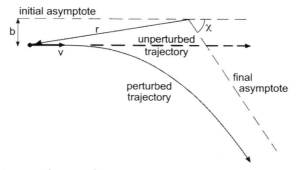

Fig. 4.1 Relative coordinates of two particles with interaction (perturbed trajectory) and without interaction (unperturbed trajectory). The initial and final asymptote indicate the *direction* of the initial and the final relative velocity, respectively. The parameter b is the impact parameter and χ is the deflection angle for a complete collision.

where $v = |\mathbf{v}_f - \mathbf{v}_t|$ is the magnitude of the relative velocity and b is the impact parameter, which is defined as the closest distance of approach for the imaginary case that the particles do not interact. The quantity $b_\perp$ is the value of the impact parameter that leads to a 90 degrees deflection ($\chi = \pi/2$). For a derivation of Eq. (4.5.1), the reader is referred to Section 6.4. It should be emphasized that Eq. (4.5.1) pertains to a complete collision—that is, a collision for which the initial and final state approach the asymptotic conditions. In such a collision, both the initial and final energy will be entirely kinetic. Thus, conservation of energy implies that only the direction of the relative velocity is changed and not its magnitude. Accordingly, a complete collision is sufficiently specified by the deflection angle χ only (as far as the motion in the orbital plane is concerned).

Let us define the direction of the initial relative velocity $\mathbf{v} = \mathbf{v}_f - \mathbf{v}_t$ as the z-direction. Let φ be the angle between the orbital plane and the x-axis; see Fig. 4.2. The three components of the velocity displacement of the test particle can now be expressed as

$$\Delta v_x = \frac{1}{2} v \sin \chi \cos \varphi = v \frac{\tan(\chi/2)}{1 + \tan^2(\chi/2)} \cos \varphi = \frac{v\sqrt{q}}{1+q} \cos \varphi$$

$$\Delta v_y = \frac{1}{2} v \sin \chi \sin \varphi = \frac{v\sqrt{q}}{1+q} \sin \varphi \qquad (4.5.2)$$

$$\Delta v_z = \frac{1}{2} v (\cos \chi - 1) = v \frac{-\tan^2(\chi/2)}{1 + \tan^2(\chi/2)} = \frac{-v}{1+q},$$

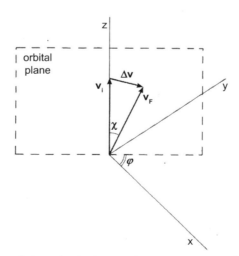

Fig. 4.2 Orientation of the orbital plane with respect to an orthogonal coordinate system, with a z-axis that is directed along the initial relative velocity $\mathbf{v}_i$.

where $q = (b/b_\perp)^2$. The factor $1/2$ is included to account for the fact that the change in the velocity of the test particle is half the change in the relative velocity of the particles, as follows from conservation of momentum (for particles with identical mass).

The number of collisions dN_c experienced by the test particle during the time interval Δt with an impact parameter between b and $b + db$, azimuth between φ and $\varphi + d\varphi$, and relative velocity between $\mathbf{v}$ and $\mathbf{v} + d\mathbf{v}$, is equal to

$$dN_c = nf(\mathbf{v} + \mathbf{v}_t)d\mathbf{v}\, v\, \Delta t\, d\varphi\, b\, db, \qquad (4.5.3)$$

using that $\mathbf{v}_f = \mathbf{v} + \mathbf{v}_t$. Accordingly, the average of an arbitrary collision quantity $A = A(b, v, \varphi)$ (such as $\Delta v_x, \Delta v_y, \Delta v_z, \Delta v_x^2, \ldots$) over all collisions, experienced during time interval Δt, can be expressed as

$$\langle\langle A\rangle\rangle = \int d\mathbf{v} f(\mathbf{v} + \mathbf{v}_t)\langle A(v)\rangle,$$

$$\langle A(v)\rangle = nv\Delta t \int_0^{2\pi} d\varphi \int_0^{b_m} b\, db\, A(b, v, \varphi). \qquad (4.5.4)$$

For convenience of notation, we separated the average over the spatial coordinates b and φ, represented as $\langle A(v)\rangle$, from the total average $\langle\langle A\rangle\rangle$, which also includes an average over the relative velocity $\mathbf{v}$ (the reader should be alert that the notation $\langle\langle..\rangle\rangle$ will also be used to indicate cumulants, e.g., in Section 5.5. Here, it just refers to a double average). The upper integration boundary of the impact parameter b is denoted as b_m. In order to include all field particles, one should take $b_m = \infty$. However, this would lead to a divergence of the b-integral, and one is forced to introduce some finite value for b_m. We will come back to this point later on. From Eqs. (4.5.2) and (4.5.4), one finds by integration

$$\langle \Delta v_x\rangle = \langle \Delta v_y\rangle = 0$$

$$\frac{\langle \Delta v_z\rangle}{\Delta t} = -\pi n b_\perp^2 v^2 \ln(1 + q_m) = -\frac{8\pi n C_0^2}{m^2 v^2}\Lambda_C$$

$$\langle \Delta v_x \Delta v_l\rangle = 0 \quad \text{for } k \neq l \qquad (4.5.5)$$

$$\frac{\langle \Delta v_x^2\rangle}{\Delta t} = \frac{\langle \Delta v_y^2\rangle}{\Delta t} = \frac{1}{2}\pi n b_\perp^2 v^3 \left(\ln(1 + q_m) - \frac{q_m}{1 + q_m}\right) \approx \frac{4\pi n C_0^2}{m^2 v}\Lambda_C$$

$$\frac{\langle \Delta v_z^2\rangle}{\Delta t} = \pi n b_\perp^2 v^3 \frac{q_m}{1 + q_m} \approx \frac{8\pi n C_0^2}{m^2 v},$$

where $q_m = (b_m/b_\perp)^2$ and Λ_C is the so-called Coulomb logarithm, defined as

$$\Lambda_C = \frac{1}{2}\int_0^{q_m} dq \frac{1}{1+q} = \ln\left[(1 + q_m)^{1/2}\right] \approx \ln\left(q_m^{1/2}\right) = \ln(b_m/b_\perp). \qquad (4.5.6)$$

All approximations expressed in Eqs. (4.5.5) and (4.5.6) require $\sqrt{q_m} \gg 1$ (thus $b_m \gg b_\perp$). Notice that the Eq. (4.5.5), as well as Eq. (4.5.2), demonstrate that

$$\langle \Delta v_z \rangle = -\langle \Delta v_x^2 + \Delta v_y^2 + \Delta v_z^2 \rangle / v \approx -\langle \Delta v_x^2 + \Delta v_y^2 \rangle / v, \quad (4.5.7)$$

which follows from the fact that the magnitude of the relative velocity is conserved in every complete collision: $|\mathbf{v} + \Delta \mathbf{v}| = |\mathbf{v}|$.

The Coulomb logarithm Λ_C diverges logarithmically with the maximum value of the impact parameter b_m. The quantities in Eq. (4.5.5) that are proportional to Λ_C diverge accordingly. In plasma physics one usually takes b_m equal to the Debeye screening length, which was introduced in Section 3.9. This choice is somewhat arbitrary, in particular as far as the exact numerical value is concerned. The physical basis of the divergence of the Coulomb logarithm Λ_C is investigated in the next section, leading to the conclusion that different expressions for b_m should be used to treat the cases of neutralized and nonneutralized systems. At this point, it is sufficient to realize that the final results are not very sensitive to the exact value of b_m, since it appears under the logarithm in the expression for Λ_C. Furthermore, it is important to observe that the higher jump moments than the second are finite, even for $b_m \to \infty$. This can directly be verified from the Eqs. (4.5.2) and (4.5.4).

The final step in the calculation of the jump moments is to evaluate the average of the quantities given by Eq. (4.5.5) over the velocity distribution of the field particles, as expressed by the first equation of (4.5.4). Notice that the Coulomb logarithm Λ_C depends on the relative velocity v through $b_\perp$; see Eqs. (4.5.6) and (4.5.1). However, since the exact value of b_m is somewhat arbitrary anyhow, one usually treats Λ_C as a constant by substituting some characteristic value for the relative velocity. We will follow this approach. The orientation of the coordinate system, used so far, depends on the direction of the relative velocity $\mathbf{v} = \mathbf{v}_f - \mathbf{v}_t$. Consequently, it will vary with $\mathbf{v}_f$. Therefore, we now choose a *new coordinate system*, in which the z-axis is in the direction of the test particle $\mathbf{v}_t$. Assuming that the velocity distribution of the field particles is Maxwellian, one finds after a lengthy but straightforward calculation (see, for instance, Chandrasekhar (1942) or Trubnikov (1965))

$$\frac{\langle\langle \Delta v_z \rangle\rangle}{\Delta t} = -\frac{8\pi n C_0^2 \Lambda_C}{m^2 v_T^2} G(v_t/v_T)$$

$$\frac{\langle\langle \Delta v_x^2 \rangle\rangle}{\Delta t} = \frac{\langle\langle \Delta v_y^2 \rangle\rangle}{\Delta t} = \frac{4\pi n C_0^2 \Lambda_C}{m^2 v_t} [\mathrm{erf}(v_t/v_T) - G(v_t/v_T)] \quad (4.5.8)$$

$$\frac{\langle\langle \Delta v_z^2 \rangle\rangle}{\Delta t} = \frac{8\pi n C_0^2 \Lambda_C}{m^2 v_t} G(v_t/v_T),$$

in which $v_T = (2k_B T/m)^{1/2}$ is the thermal velocity and

$$\operatorname{erf}(x) = \frac{2}{\sqrt{\pi}} \int_0^x dt\, e^{-t^2}, \quad G(x) = \frac{\operatorname{erf}(x) - x\{d[\operatorname{erf}(x)]/dx\}}{2x^2}. \tag{4.5.9}$$

The functions $G(x)$ and $\operatorname{erf}(x) - G(x)$ are plotted in Fig. 4.3. For small arguments, one finds by means of power expansion

$$\lim_{x \downarrow 0} G(x) = \frac{2x}{3\sqrt{\pi}}, \quad \lim_{x \downarrow 0} [\operatorname{erf}(x) - G(x)] = \frac{4x}{3\sqrt{\pi}}. \tag{4.5.10}$$

Thus, for $v_t/v_T \ll 1$ Eq. (4.5.8) can be approximated by:

$$\frac{\langle\langle \Delta v_z \rangle\rangle}{\Delta t} = -\frac{16\pi^{1/2} n C_0^2 \Lambda_C}{3m^2 v_T^3} v_t \quad (= -\beta v_t)$$

$$\frac{\langle\langle \Delta v_x^2 \rangle\rangle}{\Delta t} = \frac{\langle\langle \Delta v_y^2 \rangle\rangle}{\Delta t} = \frac{\langle\langle \Delta v_z^2 \rangle\rangle}{\Delta t} = \frac{16\pi^{1/2} n C_0^2 \Lambda_C}{3m^2 v_T} \quad (= 2D). \tag{4.5.11}$$

The diffusion tensor elements can be expressed as $D_{kl} = \delta_{kl} D$, which corresponds to an isotropic diffusion process. This is not surprising, since it was assumed that the velocity distribution of the field particles is spherical symmetric. The diffusion constant D is independent of $\mathbf{v}_t$, while the dynamical friction $(=\langle\langle \Delta v_z \rangle\rangle/\Delta t)$ is proportional to $-\mathbf{v}_t$, as assumed in Eq. (4.3.8). Notice that the Eq. (4.5.11) indeed fulfill relation (4.4.6), justifying, *a posteriori*, the assumption that the equilibrium velocity distribution of the field particles is Maxwellian.

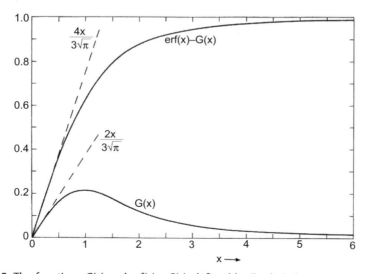

Fig. 4.3 The functions $G(x)$ and $\operatorname{erf}(x) - G(x)$, defined by Eq. (4.5.9).

4.6. Coulomb logarithm

The Coulomb logarithm, defined by Eq. (4.5.6), depends on the ratio $b_m/b_\perp$, where b_m denotes the maximum value of the impact parameter considered in the calculation of the velocity jump moments and $b_\perp$ is the value of the impact parameter corresponding to a 90 degrees deflection, as defined by Eq. (4.5.1). If one takes b_m equal to the Debeye screening length λ_D, given by Eq. (3.9.7), Eq. (4.5.6) transforms to:

$$\Lambda_C \approx \ln(\lambda_D/b_\perp) \approx \ln\left(\frac{2\pi m\langle v^2\rangle (k_B T)^{1/2}\varepsilon_0^{3/2}}{e^3 n^{1/2}}\right) = \ln\left(\frac{6\pi (k_B T\varepsilon_0)^{3/2}}{e^3 n^{1/2}}\right), \tag{4.6.1}$$

using that $\langle v^2\rangle = 3k_B T/m$. Taking as a numerical example the case of electrons in a plasma, with $n = 10^{16}\,m^{-3}$, $T = 10^5\,K$, one finds $\lambda_D/b_\perp = 2\times 10^6$ and $L_C \approx 14$.

The procedure of cutting off the interaction range at the Debeye length is introduced to obtain nondivergent results. This method appears to be rather artificial, and we wish to investigate its physical basis in more detail. The origin of the divergence is not the long range of the Coulomb force, as is often stated, but rather the utilization of Eq. (4.5.1) beyond its range of applicability. This equation specifies the angle of deflection χ in a binary collision, defined as the angle between the asymptotes of the hyperbola, describing the relative motion of the particles in the orbital plane. Accordingly, its applicability requires that the collision is complete within the time interval Δt. This implies that the distance $v\,\Delta t$, over which the collision takes place, should be large compared to the distance of closest approach, which is (for most collisions) of the same magnitude as the impact parameter b. Thus, the applicability of Eq. (4.5.1) requires $v\,\Delta t \gg b$. However, this condition is certainly not fulfilled for very remote collisions—that is, collisions with a large value of the impact parameter b. By using Eq. (4.5.1) to compute the effect of such collisions, one overestimates their contribution to the total deviation of the test particle. This causes the divergence of the results given by the Eq. (4.5.5).

From this analysis the remedy appears to be simple. The usage of Eq. (4.5.1) should be restricted to those impact parameters for which the corresponding collision is nearly complete, say $b \leq c_1 v\,\Delta t$, where $c_1 \ll 1$. For larger values one should employ a more refined particle dynamics, which takes the length of the interaction into account. For large values of b, say $b > c_2 v\,\Delta t$, where $c_2 \gg 1$, the deviations from the unperturbed trajectories will be small, and first order perturbation theory is appropriate to solve the dynamical problem. In this approach, one calculates the interaction force as if the particles were running along their unperturbed trajectories. Nondivergent results for the jump moments can be obtained by using complete collision dynamics for small b-values and first order perturbation dynamics for large b-values. In this approach, one avoids the introduction of an artificial upper limit for b.

Let us consider the first order perturbation approximation for the dynamical problem more closely. As before, we define the direction of the relative velocity $\mathbf{v} = \mathbf{v}_f - \mathbf{v}_t$ as the z-direction. Assume that the test particle is located in the origin of the coordinate system. The unperturbed distance between the particles is now given by $\tilde{r}(t) = [\tilde{z}(t)^2 + b^2]^{1/2} = [(vt)^2 + b^2]^{1/2}$, defining $t = 0$ as the moment of closest approach ($\tilde{r} = b$); see Fig. 4.4. The force acting on the test particle can be expressed in terms of a component parallel to $\mathbf{v}$ (z-direction) and a component perpendicular to $\mathbf{v}$. In first order perturbation approximation, these components are equal to

$$\tilde{F}_\perp(t) = \frac{C_0 b}{\tilde{r}(t)^3}, \quad \tilde{F}_\parallel(t) = \frac{C_0 vt}{\tilde{r}(t)^3}. \tag{4.6.2}$$

For the corresponding components of the velocity shift of the test particle, one finds

$$\Delta \tilde{v}_\perp = \int_{t_0}^{t_0 + \Delta t} dt \frac{\tilde{F}_\perp(t)}{m} = \frac{C_0}{mvb} \left(\frac{z_0 + v \Delta t}{\left[b^2 + (z_0 + v \Delta t)^2\right]^{1/2}} - \frac{z_0}{\left[b^2 + z_0^2\right]^{1/2}} \right)$$

$$\Delta \tilde{v}_\parallel = \int_{t_0}^{t_0 + \Delta t} dt \frac{\tilde{F}_\parallel(t)}{m} = -\frac{C_0}{mv} \left(\frac{1}{\left[b^2 + (z_0 + v \Delta t)^2\right]^{1/2}} - \frac{1}{\left[b^2 + z_0^2\right]^{1/2}} \right),$$

$$\tag{4.6.3}$$

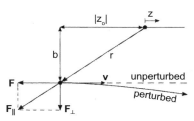

Fig. 4.4 Initial relative coordinates of two particles involved in a weak collision. The force $\mathbf{F}$ acting on the test particle is separated in a component $\mathbf{F}_\parallel$, parallel to the initial relative velocity $\mathbf{v}$, and a component $\mathbf{F}_\perp$, perpendicular to this direction.

where $z_0 = vt_0$. The required x, y, and z components of the velocity shift now follow as $\Delta v_x = \Delta \tilde{v}_\perp \cos\varphi, \Delta v_y = \Delta \tilde{v}_\perp \sin\varphi$, and $\Delta v_z = \Delta \tilde{v}_\parallel$. The resulting expressions constitute an alternative to those of Eq. (4.5.2). They apply to remote collisions—that is, to $b > c_2 v \Delta t$. Notice that both sets of expressions become identical in the limit of weak complete interactions, which corresponds to: $z_0 \to -\infty, (z_0 + v \Delta t) \to \infty, q \to \infty$. The reader might verify that, in this limit, one finds: $\Delta v_\perp \approx 2C_0/mvb$, $\Delta v_\parallel \approx 0$.

One can now perform the first step of the calculation of the first and second order velocity jump moments, which is described by the second equation of (4.5.4) in an alternative way by splitting the b-integration into two parts:

$$\langle A(v) \rangle = n \int_0^{2\pi} d\varphi \left(v \Delta t \int_0^{b_1} b\, db\, A_n(b, v, \varphi) + \int_{b_1}^\infty b\, db \int_{-\infty}^\infty dz_0\, A_r(z_0, b, v, \varphi) \right), \tag{4.6.4}$$

where $b_1 = cv\Delta t$, with $c_1 < c < c_2$. A_n is the expression of the quantity A (which stands for $\Delta v_x, \Delta v_y, \Delta v_z, \Delta v_x^2, \ldots$) obtained for a near (complete) collision, and A_r is the alternative expression for a remote collision obtained by first order perturbation theory. A calculation along these lines was carried out by Sivukhin (1966). We will summarize his findings.

It is sufficient to consider the calculation of $\langle \Delta v_x^2 \rangle = \langle \Delta v_y^2 \rangle = \langle \Delta v_\perp^2 \rangle/2$, since $\langle \Delta v_z^2 \rangle$ can be neglected for $\Lambda_C \gg 1$, while the correct value of $\langle \Delta v_z \rangle$ follows from $\langle \Delta v_\perp^2 \rangle$, using Eq. (4.5.7). Clearly, the first term of Eq. (4.6.4) leads to the expressions for $\langle \Delta v_x^2 \rangle$ and $\langle \Delta v_y^2 \rangle$ given by Eq. (4.5.5), where b_m is now replaced by b_1. The second term in Eq. (4.6.4) can be scaled in such a way that one obtains an expression that is identical to that of the first term but in which Λ_C is replaced by a two dimensional integral, which contains the ratio $b_1/v\Delta t$ as the only free parameter. Accordingly, one should conclude that Eq. (4.5.5) remains true, provided that the Coulomb logarithm is replaced by

$$\Lambda_C = \ln(b_1/b_\perp) + f(v\Delta t/b_1), \tag{4.6.5}$$

where

$$f(u) = \int_0^u \frac{dx}{x^2} \int_{-\infty}^\infty dy \left(\frac{x+y}{[1+(x+y)^2]^{1/2}} - \frac{y}{[1+y^2]^{1/2}} \right)^2$$

By means of an asymptotic analysis, it can be demonstrated that $f(u) \approx \ln(u)$ for $u > 2$. Thus, by using a proper value for the free parameter b_1 (that is $b_1 \ll v\Delta t$), one may approximate Eq. (4.6.5) as

$$\Lambda_C = \ln(v\Delta t/b_\perp). \tag{4.6.6}$$

According to this analysis $b_m = v\,\Delta t$.

We have now removed the divergence in a natural way, but obtained a new problem instead, namely that the Coulomb logarithm Λ_C depends on the time interval Δt. This quantity has been considered as arbitrary so far, the only constraints on it being expressed by the Eq. (4.3.3). At this point, one should realize that Eq. (4.6.3) are valid as long as the particles follow approximately their unperturbed trajectories, which are straight lines. However, the two particles considered are not the only particles in the system. The interaction with the other particles will cause (random) deviations from the unperturbed trajectories. When such a deviation occurs, the interaction between the two particles is effectively broken up and starts again along new trajectories. Let us assume that the average time period between two successive interruptions is τ_0. In general, this will be a small time interval compared to Δt. Thus, the interaction is on the average broken up $\Delta t/\tau_0$ times during the time interval Δt. The corresponding $\Delta t/\tau_0$ deviations, built up during the periods between two successive interruptions, should be regarded as statistically independent. Accordingly, one has to replace the expression for $\langle \Delta v_\perp^2 \rangle_{\Delta t}$, calculated over the time interval Δt, by $(\Delta t/\tau_0) \times \langle \Delta v_\perp^2 \rangle_{\tau_0}$, where $\langle \Delta v_\perp^2 \rangle_{\tau_0}$ is calculated over the time interval τ_0. In this approximation, the Coulomb logarithm Λ_C becomes equal to

$$\Lambda_C \approx \ln\left[\langle v^2 \rangle^{1/2} \tau_0/b_\perp\right] \approx \ln\left[(k_B T/m)^{1/2} \tau_0/b_\perp\right], \qquad (4.6.7)$$

as follows with Eqs. (4.5.5) and (4.6.6). As usual, we replaced the velocity v in the argument of the logarithm by the average value $\langle v^2 \rangle^{1/2}$. Eq. (4.6.7) correspond to $b_m = (k_B T/m)^{1/2} \tau_0$. The remaining problem is now the estimation of τ_0.

In neutralized plasmas, the collective response to the motion of the individual charged particles leads to collective harmonic oscillations of the electrons; see for instance Pines and Bohm (1952) or Ichimaru (1973). The eigen frequency of the system of electrons embedded in the neutralizing background of positive ions is equal to the plasma frequency ω_p, given by Eq. (3.9.12). The organized (collective) motion will interfere with the random (thermal) motion of the individual particles, with a period of approximately $1/\omega_p$. Thus, for neutralized systems, it appears to be reasonable to take $\tau_0 = 1/\omega_p$. Accordingly, one finds $b_m = (k_B T/m)^{1/2}/\omega_p = \lambda_D$, the Debeye screening distance. Thus, the procedure of cutting off the interaction range at the Debeye distance, seems justified for neutralized systems, and the Coulomb logarithm Λ_C should be calculated from Eq. (4.6.1).

On the other hand, for nonneutralized systems, such as the system of identical charged particles considered in this chapter, it appears to be more appropriate to identify τ_0 with the correlation time of the fluctuations in the interaction force τ_f, given by Eq. (4.3.1). Accordingly, one finds $b_m = (k_B T/m)^{1/2}\tau_f = n^{-1/3}$, which is the average distance between the particles in the system. Thus, for nonneutralized systems the Coulomb logarithm Λ_C should be calculated as

$$\Lambda_C \approx \ln\left(\frac{n^{-1/3}}{b_\perp}\right) \approx \ln\left(\frac{m\langle v^2\rangle}{2C_0 n^{1/3}}\right) \approx \ln\left(\frac{3k_B T}{2C_0 n^{1/3}}\right), \qquad (4.6.8)$$

using the same approximations as in Eq. (4.6.1). It should be emphasized that the phenomenon of Debeye screening can occur in nonneutralized systems, as was discussed in Section 3.6. The reason for choosing $b_m = n^{-1/3}$ is not the absence of screening but is rather related to the absence of collective oscillations. Notice that for systems in which the phenomenon of Debeye screening does occur, the distance λ_D will be large compared to the average interparticle distance $n^{-1/3}$ (otherwise the Debeye sphere does not contain many particles, contradicting the premises of the screening concept; see Eq. (3.9.15)). Finally, we note that Eq. (4.6.8) also applies to stellar systems (replacing C_0 by Gm^2), which should be regarded as nonneutralized; see Chandrasekhar (1941, 1942). Similar results (but differing by a numerical factor) are given by Chandrasekhar and von Neumann (1942, 1943). These results are, however, obtained by entirely different reasoning.

4.7. Discussion of the Fokker–Planck approach

In the previous sections, the main issues relating to the Fokker–Planck type of Eq. (4.3.10) were discussed, stressing the aspects concerning the fluctuating component of the interaction force. In this section, we will concentrate on the relationship between the various concepts introduced so far, with the objective to clarify the global philosophy of the Fokker–Planck approach.

The fundamental presupposition of the Fokker–Planck approach is that the action of the fluctuating component of the interaction force can be described in terms of a diffusion process in velocity space, characterized by the coefficient of dynamical friction β and the diffusion constant D. It is assumed that these quantities are determined by the macroscopic properties of the system only. This approach is justified when there exists a time interval Δt, which is long compared to the correlation time of the fluctuations τ_f but small enough to neglect the variation in the macroscopic properties of the system represented by $f(\mathbf{r}, \mathbf{v}, t)$ and small enough to consider the jumps in the velocity $\delta\mathbf{v}(t; \Delta t)$ as microscopic. These constraints on Δt are expressed by Eq. (4.3.3). Spatial displacements within the interval Δt are ignored on the argument that they are of the second order in Δt

and thus negligible compared to the first order shift $\delta \mathbf{v}\, \Delta t$. The stationary solution of the diffusion equation is a Gaussian (velocity) distribution, which is consistent with the assumption that the velocity of the test particle is built up by a large number of uncorrelated microscopic shifts $\delta \mathbf{v}$ (central limit theorem). The stationary solution is Maxwellian, provided that β and D are related by Eq. (4.4.6). Therefore, the model is suited to describe the impact of particle interactions in a system that operates near thermodynamic equilibrium.

The diffusion approximation reduces the problem of determining the effect of the fluctuations in the interaction force on the single particle probability function $f(\mathbf{r}, \mathbf{v}, t)$ to the calculation of the coefficient of dynamical friction β and the diffusion constant D, which are related to the first and second velocity jump moments, respectively. These jump moments, which are specified by Eq. (4.5.11), are proportional to the Coulomb logarithm Λ_C. For neutralized systems (such as plasmas), this quantity is given by Eq. (4.6.1) and for nonneutralized systems (such as systems of identical charged particles and stellar systems) it is given by Eq. (4.6.8). The argument of the logarithm, appearing in these expressions, may be off by a factor, say 2–5. If one denotes this factor as f, the relative error in Λ_C can be expressed as $1 + \ln(f)/\Lambda_C$. Taking as an example $\Lambda_C = 10$, one finds a relative error in Λ_C of 7% for $f = 2$ and 16% for $f = 5$. Clearly, the sensitivity of Λ_C for the error factor f decreases with increasing values of Λ_C.

The Coulomb logarithm Λ_C is an important quantity within the framework of the Fokker–Planck approach. The validity of the diffusion approximation requires that $\Lambda_C \gg 1$. This can be understood from the following arguments:

- One should recall that the first and second jump moments, given by Eq. (4.5.11), diverge proportionally to Λ_C, while higher moments are finite, irrespective of the value of Λ_C (since they are independent of b_m). Accordingly, the higher jump moments can be neglected relative to the first and second one, provided that Λ_C is large enough. This condition guarantees that the time evolution of the single particle distribution function $f(\mathbf{r}, \mathbf{v}, t)$ is indeed dominated by the first and second jump moments, as assumed in the Fokker–Planck approach.
- The Coulomb logarithm is a measure for the relative contribution of (weak) remote interactions compared to (strong) near interactions. Defining all interactions with $b > 2b_\perp$ as remote and all interactions with $b < 2b_\perp$ as near, one finds for the ratio of their contributions to the Coulomb logarithm (see Eq. (4.5.6)).

$$\frac{\Lambda_C, \text{remote}}{\Lambda_C, \text{near}} = \int_4^{q_m} \frac{dq}{1+q} \bigg/ \int_0^4 \frac{dq}{1+q} = \frac{\ln(1 + q_m) - \ln 5}{\ln 5} \qquad (4.7.1)$$

$$\approx 1.2 \Lambda_C - 1 \approx \Lambda_C,$$

assuming that $\Lambda_C \gg 1$. This equation shows that the total contribution of remote interactions is approximately Λ_C times larger than the contribution of near interactions. Thus, when $\Lambda_C \gg 1$, the jumps in the velocity of a test particle will, in general, be small and its motion in velocity space can be considered as quasi-continuous. We recall that this condition is essential for the validity of the diffusion approximation.

The calculation of the coefficient of dynamical friction β and the diffusion constant D is based on a reduction of the N-body problem to two-particle interactions. In this procedure, the interaction between the test particle and a particular field particle is calculated, as if they are the only particles in the system. The total deviation of the test particle due to the interactions with all field particles during the time interval Δt is computed as the sum of all separate two-particle effects occurring in that period. This procedure, which is sometimes referred to as the binary interaction approximation, would be justified when all collisions would be well separated in time. However, in reality many of the two-particle interactions occur simultaneously and not successively, due to the long range of the interaction force. This raises the question whether the binary interaction approximation is valid or not.

In defense of the binary interaction approximation, it should be noted that most interactions are weak, provided that $\Lambda_C \gg 1$. The interactions that occur simultaneously are therefore predominantly remote interactions, causing small deviations in the velocities of the particles involved. These deviations can be estimated by means of first order perturbation dynamics, in which one evaluates the impact of the interaction force experienced along the unperturbed trajectories. This is expressed by Eq. (4.6.3). As forces are additive, the reduction to binary interactions is justified for this type of interactions. In general, one may say that the binary interaction approximation is justified as long as strong interactions are rare ($\Lambda_C \gg 1$). Notice that this condition is required anyhow to justify the diffusion approximation.

4.8. Validity of the Fokker–Planck approach for particle beams

In this section, we will study whether the Fokker–Planck approach is suited to describe the effect of statistical interactions in particle beams or not. To that intent, let us consider a rotational symmetric beam segment in drift space with a uniform density distribution in every cross section of the beam. The macroscopic condition of the beam segment is specified by the values of the experimental parameters I, V, ΔE_0 (or alternatively the cathode temperature T_{ca}), L, r_c, α_0, S_c and S_i, defined in Section 3.2. First, we will express the characteristic theoretical quantities introduced in this chapter in terms of the experimental parameters. Next, we will investigate whether the various assumptions underlying the Fokker–Planck approach are justified for normal operating conditions.

The physical quantities to be examined are the correlation time of the fluctuations in the interaction force τ_f, the Debeye length λ_D, the plasma frequency ω_p and the Coulomb logarithm Λ_C, defined by the Eqs. (4.3.1), (3.9.7), (3.9.12) and (4.6.8), respectively. Given the type of particles, these quantities depend on two parameters only: the particle density n and the beam temperature T. The particle density n is a function of the axial position z, through the beam radius $r_0(z)$, as expressed by Eq. (3.2.5). The temperature in a rotational symmetric beam of non-interacting particles was studied in Section 3.7. It was concluded that one should, in general, distinguish two temperatures, $T_\parallel$ and $T_\perp$, corresponding to the longitudinal and lateral degrees of freedom, respectively. These temperatures are specified in terms of the experimental parameters by Eqs. (3.7.13) and (3.7.18). Assuming, for the moment, that it is useful to distinguish different values for τ_f, λ_D and Λ_C in axial and lateral direction, one finds from Eqs. (4.3.1), (3.9.7), (3.9.12) and (4.6.8)

$$\tau_{f\parallel} = \frac{2\pi^{1/3}(meV)^{1/2}}{\lambda^{1/3} k_B T_{ca}} r_0(z)^{2/3}, \qquad \tau_{f\perp} = \frac{2^{1/2}\pi^{1/3} m^{1/2}}{\lambda^{1/3}(eV)^{1/2}\alpha_0 r_c} r_0(z)^{5/3}$$

$$\lambda_{D\parallel} = \left(\frac{\pi\varepsilon_0}{4e^3 V \lambda}\right)^{1/2} k_B T_{ca} r_0(z) \qquad \lambda_{D\perp} = \left(\frac{\pi\varepsilon_0 V \alpha_0^2 r_c^2}{2e\lambda}\right)^{1/2}$$

$$\omega_p = \left(\frac{e^2 \lambda}{\pi\varepsilon_0 m}\right)^{1/2} \frac{1}{r_0(z)} \qquad (4.8.1)$$

$$\Lambda_{C\parallel} = \ln\left(\frac{\pi^{4/3}\varepsilon_0(k_B T_{ca})^2 r_0(z)^{2/3}}{e^3 V \lambda^{1/3}}\right), \qquad \Lambda_{C\perp} = \ln\left(\frac{2\pi^{4/3}\varepsilon_0 V \alpha_0^2 r_c^2}{e\lambda^{1/3} r_0(z)^{4/3}}\right),$$

where λ is the linear particle density, given by Eq. (3.2.4). Notice that all quantities depend on the axial coordinate z, except $\lambda_{D\perp}$. Defining N_D as the total number of particles in the Debeye box with sides $\lambda_{D\perp}$, $\lambda_{D\perp}$, and $\lambda_{D\parallel}$, one obtains

$$N_D = n\lambda_{D\perp}^2 \lambda_{D\parallel} = \frac{\pi^{1/2}\varepsilon_0^{3/2} V^{1/2}\alpha_0^2 r_c^2}{4e^{5/2}\lambda^{1/2}} \frac{k_B T_{ca}}{r_0(z)}. \qquad (4.8.2)$$

Eqs. (4.8.1) and (4.8.2) express the physical quantities that are relevant to the Fokker–Planck approach in terms of the experimental parameters of the beam.

Let us now consider the same numerical example as in Section 3.2—that is, an electron beam with $I = 1\,\mu A$, $V = 10\,kV$, $L = 0.1\,m$, $S_c = 0.5$, $r_c = 1\,\mu m$, and $\alpha_0 = 10\,mR$, and take $k_B T_{ca} = 0.5\,eV$

(equal to the transverse temperature in the crossover). From Eqs. (4.8.1) and (4.8.2) one obtains:

$$\tau_{f\|c} = 1.5 \times 10^{-9}\,\text{s}, \quad \tau_{f\|1} = 9.2 \times 10^{-8}\,\text{s}$$
$$\tau_{f\perp c} = 4.9 \times 10^{-12}\,\text{s}, \quad \tau_{f\perp 1} = 1.5 \times 10^{-7}\,\text{s}$$
$$\tau_{D\|c} = .032\,\mu\text{m}, \quad \tau_{D\|1} = 16\,\mu\text{m}$$
$$\tau_{D\perp} = 9.1\,\mu\text{m},$$
$$1/\omega_{pc} = 3.1 \times 10^{-11}\,\text{s}, \quad 1/\omega_{p1} = 1.5 \times 10^{-8}\,\text{s}$$
$$\Lambda_{C\|c} = 2.0 \times 10^{-5}, \quad \Lambda_{C\|1} = 7.2 \times 10^{-2}$$
$$\Lambda_{C\perp c} = 6.2, \quad \Lambda_{C\perp 1} = 7.8 \times 10^{-3}$$
$$N_{Dc} = .89, \quad N_{D1} = 1.8 \times 10^{-3},$$

where the subscript c refers to the conditions in the crossover and the subscript 1 to the conditions at the entrance plane of the beam segment. In the calculation of $\Lambda_{C\|c}$, $\Lambda_{C\|1}$ and $\Lambda_{C\perp 1}$, the arguments of the logarithm turned out to be smaller than 1, and we used the full expression $\Lambda_C = \cdot \ln[(1+q_m)^{1/2}]$, instead of the usual approximation $\Lambda_C = \ln[q_m^{1/2}]$; see Eq. (4.5.6).

Let us investigate what can be learned from these numbers. The values found for the correlation time of the fluctuations in the interaction force τ_f, indicate that most collisions take place in the crossover, mainly due to the transverse motion of the particles. In order to get an impression of the average number of collisions experienced by a particle during its flight through the beam segment, these values should be compared to the flight time T_f, given by Eq. (3.2.3). For the quantities referring to the crossover, it is more appropriate to compare them to the flight time through the crossover area T_{fc}. As a measure for this quantity, we use

$$T_{fc} \approx 2r_c/\alpha_0 v_\|, \tag{4.8.3}$$

which is the time that a particle travels over the distance $2r_c/\alpha_0$. For the numerical example, one finds from Eqs. (3.2.3) and (4.8.3)

$$T_f = 1.7\,10^{-9}\,\text{s}, \quad T_{fc} = 3.4\,10^{-12}\,\text{s}.$$

Thus, the correlation time of the fluctuations τ_f cannot be considered as short compared to the appropriate flight time (e.g., $T_f/\tau_{f\|t} = 1.8 \times 10^{-2}$, $T_{fc}/\tau_{f\perp c} = 0.69$). Accordingly, the number of collisions is, in general, not a large number.

The inverse of the plasma frequency ω_p is the relevant measure for the relaxation time of the system to reach thermodynamic equilibrium, as was shown in Section 3.9. From the numbers of the numerical example, one finds that thermodynamic equilibrium will, in general, not be reached

($T_f \omega_{p1} = T_{fc} \omega_{pc} = 0.11$). This implies that Debeye screening, which presupposes thermodynamic equilibrium, will, in general, not occur. In this context, it should be noted that the condition that the number of particles in the Debeye box N_D is large, which is expressed by Eq. (3.9.15), is not fulfilled either.

The validity of the Fokker–Planck approach requires that the Coulomb logarithm is large, as was pointed out in the previous section. Clearly, the only parameter fulfilling this condition is $\Lambda_{C\perp c}$.

In general, one sees that the basic requirements of the Fokker–Planck approach, expressed by the Eq. (4.3.3), are not fulfilled. The correlation time of the fluctuations τ_f is not small compared to the total time scale of the system T_f or T_{fc}. Accordingly, the macroscopic condition of a bunch of particles in the beam, determining the coefficient of dynamical friction β and diffusion constant D, changes appreciably during the time interval τ_f. In other words, β and D cannot be considered as constant during the typical period of a collision. In addition, the contribution of strong interactions, corresponding to large jumps in velocity, can, in general, not be ignored. These results can be summarized by the statement that the motion in velocity space cannot be considered as a succession of a large number of microscopic random displacements, superimposed on a smooth systematic motion.

As the number of collisions is, in general, not large, while the effect of strong collisions cannot be ignored, the resulting velocity distribution is not necessarily Gaussian. This is equivalent to the assertion that the higher jump moments than the second cannot be disregarded, which implies that the diffusion approximation is inaccurate.

Finally, we note that for a particle *beam*, one is not only interested in the velocity distribution of the particles, but also in the distribution of stochastic spatial displacements (trajectory displacement effect). As the Fokker–Planck approach neglects the spatial displacements that are generated during the collision (by just taking the resulting velocity change into account), this method is not suited to describe the trajectory displacement effect.

One may object that this is just a single numerical example and dispute the generality of the conclusions. However, examining the dependency of Eq. (4.8.1) on the experimental parameters, one should conclude that the conditions described above do not change drastically within the practical range of operating conditions encountered in the systems of interest (see Section 1.1). The main insufficiency arises from the fact that beams are relatively cold compared to a plasma, while the flight times of the beam particles are short compared to the relevant relaxation times. Accordingly, the number of collisions is limited, contradicting the basic philosophy of the diffusion approximation.

The requirements of the Fokker–Planck approach are, to a certain extent, fulfilled in the crossover area, which is the "hottest" and most dense section of the beam. Thus, the Fokker–Planck approach may give a reasonable approximation for the Boersch effect generated in this part of the beam. From the second equation of (4.5.11) one finds, using that $n\Delta t = J_c \Delta z / ev_\|^2$ and $v_T = (2k_B T_{\perp c}/m)^{1/2}$,

$$\langle \Delta E^2 \rangle^{1/2} = mv_\| \langle \Delta v_\|^2 \rangle^{1/2} = \frac{2^{7/4} \pi^{1/4}}{3^{1/2}} \frac{C_0 m^{1/4}}{e^{1/2}} \frac{(J_c \Delta z)^{1/2}}{(k_B T_{\perp c})^{1/4}} \Lambda_{C\perp c}^{1/2}, \quad (4.8.4)$$

assuming that the major effect comes from the motion in the lateral directions. The Coulomb logarithm $\Lambda_{C\perp}$ is given by Eq. (4.8.1). Substituting $r_0 = r_c$ and $\lambda = \pi r_c^2 J_c / (2e^3 V/m)^{1/2}$ (see Eq. 3.2.4), it can be expressed as

$$\Lambda_{C\perp c} = \ln\left(\frac{2\pi^{4/3} \varepsilon_0 V^{7/6} \alpha_0^2}{e^{1/2} m^{1/6} J_c^{1/3}}\right) = \ln\left[\frac{k_B T_{\perp c}}{C_0} \left(\frac{eV}{m}\right)^{1/6} \left(\frac{e}{J_c}\right)^{1/3}\right], \quad (4.8.5)$$

in which Eq. (3.7.4) was utilized to express the beam semi-angle α_0 in terms of the temperature $T_{\perp c}$. Apart from a numerical factor, Eqs. (4.8.4) and (4.8.5) are identical to the results given by Zimmermann (1970) and Knauer (1979a) for a cylindrical beam segment of length Δz. (Their results are larger by a factor $(3\pi)^{1/2}/2^{5/4} \approx 1.29$. In addition, Zimmermann's argument of the Coulomb logarithm is smaller by a factor 2.22.) The related models of Zimmermann and Knauer will be further discussed in Chapter 15.

Dynamical friction, specified by the first equation of (4.5.11), can be disregarded since the (test) particles do not have a large excess velocity relative to the surrounding (field) particles. Therefore, it is sufficient to consider the second moment of the distribution of velocity shifts only, provided that the distribution is indeed Gaussian, as implicitly assumed within the diffusion approximation.

The next step would be the calculation of the single particle distribution function $f(\mathbf{r}, \mathbf{v}, t)$, as it develops during the time of flight. This function can be obtained by solving the Fokker–Planck Eq. (4.3.10). However, since one is mainly interested in the generated energy spread, one can omit this step. All relevant information is contained in the second jump moment, given by Eq. (4.8.4). In general, one may say that the primary objective of a theory for statistical Coulomb interactions in particle beams is to calculate the transition probability function $W_{\Delta t}$, specified by the Eqs. (4.3.4) and (4.3.5), rather than the evolution of the single-particle distribution function $f(\mathbf{r}, \mathbf{v}, t)$. In the specific case that the diffusion approximation is justified, the transition probability function $W_{\Delta t}$ is Gaussian, and one can suffice to calculate its second moment.

4.9. Holtsmark distribution

We now like to report on a method for the calculation of collisional effects which can, in some cases, be used where the Fokker–Planck approach fails. Consider the extreme case of a monochromatic homocentric cylindrical beam, in which the particles move with identical velocities along parallel rays. Observed in the frame of reference moving with the beam, the particles are initially at rest. Random velocities will be generated due to the relaxation of potential energy, as was discussed in the Sections 3.8 and 3.10. Since the effective beam temperatures are initially zero, the Fokker–Planck approach is clearly not suited to describe the effect of statistical interactions in such a beam.

Let us assume that the particle density is not so high that the beam will significantly be deformed by the space charge effect. We thus postulate that the displacements caused by the Coulomb interaction between the particles in the beam are small compared to the average separation of the particles. For this condition, the generated displacements in velocity and position can be calculated in first order perturbation approximation, as was outlined in Section 4.6; see Eq. (4.6.3). This means that the real interaction force, acting on a particle, is approximated by the force that would act on that particle when all particles would follow their unperturbed trajectories. Within this approximation, the distribution of the deviations caused by the statistical interactions can be related directly to the distribution of the fluctuation component of the interaction force that would occur in the unperturbed beam.

For the specific case of particles that are at rest relative to each other (as in our example), this distribution was first calculated by Holtsmark (1919). He considered the problem of determining the probability $\rho(\mathbf{F})$ that a given electric field strength $\mathbf{F}$ acts at a point in a gas of density n composed of randomly distributed ions. His result can be expressed as

$$\rho(\mathbf{F}) = \frac{1}{2\pi^2 F} \int_0^\infty k \, dk \, \sin(Fk) e^{-(F_n k)^{3/2}} \tag{4.9.1}$$

where $F = |\mathbf{F}|$ and F_n is the normalized field strength, given by

$$F_n = (4/15)^{2/3} 2\pi C_0 n^{2/3}. \tag{4.9.2}$$

The Holtsmark distribution is plotted in Fig. 4.5. A comprehensive discussion of the Holtsmark distribution is given by Chandrasekhar (1941, 1943). A derivation of the Holtsmark distribution, in the form of Eq. (4.9.1), is included in Section 5.8.

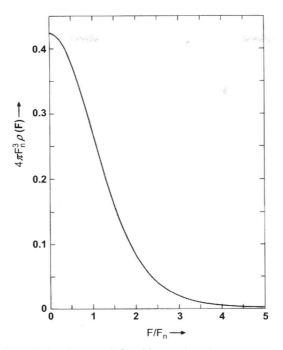

Fig. 4.5 The Holtsmark distribution, defined by Eq. (4.9.1).

The essential observation of this section is the following. For a monochromatic homocentric cylindrical beam of low particle density one may assume that velocity displacement $\Delta \mathbf{v}$, experienced by a particle during the time of flight T_f, is simply given by

$$\Delta \mathbf{v} \approx \mathbf{F} T_f / m, \tag{4.9.3}$$

since $\mathbf{F}$ is constant, due to the fact that the particles are initially at rest relative to each other and do not experience large displacements during the flight.

The distribution of $\mathbf{F}$ is given by Eq. (4.9.1). The distribution of velocity displacements $\Delta \mathbf{v}$, denoted as $\rho_v(\Delta \mathbf{v})$, now follows as

$$\rho_v(\Delta \mathbf{v}) = \left(\frac{m}{T_f}\right)^3 \rho(\mathbf{F} = m\,\Delta \mathbf{v}/T_f). \tag{4.9.4}$$

The factor $(m/T_f)^3$ guarantees normalization of the three-dimensional distribution $\rho_v(\Delta \mathbf{v})$. Eq. (4.9.4) expresses that the distribution of velocity displacements $\rho_v(\Delta \mathbf{v})$ and the distribution of the fluctuating component of the interaction force $\rho(\mathbf{F})$ are entirely correlated. Since $\rho(\mathbf{F})$ is a Holtsmark distribution, $\rho_v(\Delta \mathbf{v})$ is a Holtsmark distribution, too.

The properties of the Holtsmark distribution are quite different from those of a Gaussian distribution. Its asymptotic behavior is specified by

$$\lim_{F \downarrow 0} \rho(\mathbf{F}) = \frac{1}{3\pi^2 F_n^3}, \quad \lim_{F \to \infty} \rho(\mathbf{F}) = \frac{15}{2^{9/2} \pi^{3/2}} \frac{F_n^{3/2}}{F^{9/2}}. \quad (4.9.5)$$

The behavior for $F \to \infty$ shows that the Holtsmark distribution has distinctive non-Gaussian features. Its long tails lead to a divergence of the second and higher moments:

$$\langle F^m \rangle = \int_0^\infty 4\pi F^2 \, dF \, F^m \rho(\mathbf{F}) \to \infty \quad \text{for} \quad m \geq 2. \quad (4.9.6)$$

This divergence is a consequence of the assumption of complete randomness of the particle distribution, in all elements of volume. This assumption is not valid in the very immediate neighborhood of a (test) particle. Accordingly, the Holtsmark distribution predicts to high probabilities for large values of F. Holtsmark (1919) avoided this problem by considering the distribution of the field strength F in a gas, consisting of dipoles or quadrupoles, instead of simple (monopole) ions. Chandrasekhar (1941) took a different approach by postulating that no particles with velocity v occur within a sphere with radius $r = 2C_0 m/v^2$, while the distribution is random for all distances larger than this critical distance.

From Eqs. (4.9.1) and (4.9.2), it follows that the distribution of F scales with $n^{2/3}$. This can directly be understood from the fact that the Coulomb force scales with the inverse square of the average interparticle distance, which is equal to $n^{-1/3}$. For the particular case of a particle beam, one should expect that the interaction effects increase with the 2/3 power of the current density J. More specifically, the (arbitrary) width of the energy distribution ΔE, generated in a homocentric cylindrical beam, scales as

$$\Delta E \sim m v_\| \frac{F_n \Delta t}{m} = F_n \Delta z = \pi (8/15)^{2/3} \frac{C_0 m^{1/3}}{e} \frac{J^{2/3} \Delta z}{V^{1/3}}, \quad (4.9.7)$$

using that $n = J/ev_\|$. We emphasize that this result relies on the assumption that the displacements are small compared to the average interparticle distance $n^{-1/3}$. In addition, it is assumed that the beam is extended (three-dimensional), which guarantees that every test particle is surrounded by a spherical symmetric cloud of field particles.

Clearly, Eq. (4.9.7) shows a different dependency on the experimental parameters than Eq. (4.8.4), obtained for the beam area in the vicinity of a crossover, employing the diffusion approximation (note that $2k_B T_{\perp c} = eV\alpha_0^2$). The former predicts a 2/3 power dependency on the current density and a linear dependency on the length Δz, while the latter is proportional to the square root of both parameters. The difference in the dependency on the length Δz, stems from the fact that the diffusion approach presupposes that the velocity change of a test particle is built up by a large number of independent collisions, while the present analysis assumes that it results from the interaction within a certain complexion of neighbor field particles, which does not change during the time of flight.

4.10. Conclusions

Summarizing the results of this chapter, it appears that the methods employed in plasma physics and stellar dynamics are, in general, not suited to describe all aspects of statistical Coulomb interactions in particle beams for all practical values of the experimental parameters. The diffusion approximation (or Fokker–Planck approach) is suited to calculate the velocity broadening in sections of the beam, characterized by a strong transverse particle motion ("hot" sections) and a high particle density. Such conditions are often encountered in the crossover areas. The analysis of the previous section, which utilizes the Holtsmark distribution, is more appropriate for beam sections with little particle motion and a low particle density. These conditions are encountered in (nearly) homocentric cylindrical beams. Both approaches lead to different types of distributions, while the widths of these distributions follow different dependencies on the experimental parameters. Concerning the fundamental issues involved, it should be realized that the two approaches are based on different physical mechanisms. In the diffusion approximation one determines the transition of kinetic energy by complete collisions from one degree of freedom to another. In this view, the statistical effects are the result of a relaxation of kinetic energy. The approach centered around the Holtsmark distribution is based on the relaxation of potential energy. This implies that the dominant collisions are essentially incomplete.

In the next chapter, the so-called *extended two-particle model* will be introduced, which can, in some respects, be considered as a synthesis of the two approaches that were outlined here. In this method, the solution of the dynamical part of the problem is based on a full calculation of the two-particle collisions without relying on any restricting assumption regarding the completeness or the strength of the interaction. In this approach, both the relaxation of kinetic energy and the relaxation of potential energy are taken into account. The statistical analysis is carried out along the same lines as the derivation of the Holtsmark distribution. This method is sometimes referred to as Markoff's method; see Chandrasekhar (1943). It permits a direct calculation of the entire distribution of displacements in position and velocity without any *a priori* assumption other than that the initial positions of the particles are supposed to be randomly distributed over the beam volume. The extreme examples considered in Section 4.8 (beam segment with a narrow crossover) and Section 4.9 (extended homocentric cylindrical beam segment) are both covered by the extended two-particle model.

CHAPTER FIVE

Concepts of an analytical model for statistical interactions in particle beams

Contents

5.1.	Introduction	91
5.2.	General formulation of the problem	92
5.3.	Reduction of the N-particle problem	96
	5.3.1 First order perturbation approximation	99
	5.3.2 Closest encounter approximation	101
	5.3.3 Extended two-particle approximation	103
	5.3.4 Mean square field fluctuation approximation	104
5.4.	Calculation of the displacement distribution	106
5.5.	Moments and cumulants of the displacement distribution	109
5.6.	On- and off-axis reference trajectories	112
5.7.	Models in one, two, and three dimensions	113
5.8.	Distribution of the interaction force in cylindrical beams	115
5.9.	Representation in the k-domain	120
5.10.	Addition of the effects generated in individual beam segments	127
5.11.	Slice method	131

5.1. Introduction

This chapter discusses the fundamental aspects of an analytical model describing the impact of statistical Coulomb interactions on the properties of a beam of identical charged particles. It starts with a general analysis of this type of N-particle problem. Various methods of reducing the N-particle problem are investigated, leading to the conclusion that the so-called extended two-particle approximation, introduced by van Leeuwen and Jansen (1983), is the least restrictive. The mathematical framework of this model is worked out in detail, demonstrating the feasibility of a direct calculation of the entire distribution of displacements in velocity and position. The model is exploited to determine the distribution of the fluctuating component of the interaction force in an extended beam as well as in a pencil beam. For the former, one finds the Holtsmark distribution. The analysis of this chapter indicates, in general terms, what types of distributions are produced by the statistical interactions. It also predicts the dependency of the width of these distributions on the particle density within the beam. The material of this chapter provides the basic concepts required for the calculation of the Boersch effect, the effect of statistical angular deflections, and the trajectory displacement effect, which are carried out in the next chapters.

5.2. General formulation of the problem

Consider the beam of a probe forming instrument, e.g., a scanning electron microscope or an electron beam pattern generator. Particles are continuously emitted from the cathode and accelerated toward the anode. From there they drift through the remaining part of the column. The direction of the particles is altered by the lenses and the deflectors in the column, depending on their lateral position within the beam and their velocity. As a consequence, each particle experiences a varying density of its surrounding neighbors, the highest density being reached at the location of the crossovers. At the end of the system the particles hit the target, where they are removed from the beam.

We schematize the beam as a succession of beam segments, separated by optical components. We will study the impact of the interaction between the particles for a single beam segment. The object of the theory is to evaluate the statistical effects as functions of the experimental parameters I, V, ΔE_0, L, r_c, α_c, S_c and S_i, defined in Section 3.2. It is assumed that the total effect, generated in the entire beam, can be represented as a sum of the effects generated in the individual beam segments. This is not a trivial assumption, as will be discussed in some detail at the end of this chapter. The beam segments are supposed to be rotational symmetric.

We postulate that no external forces act on the particles during the flight through the individual beam segments, which constitute the beam. Thus, we regard the acceleration region in the gun as well as the lens and deflector areas as infinitely thin. Accordingly, the unperturbed trajectories of the particles can be visualized as straight lines that are broken at the location of the lenses and deflectors. The problem of determining the effect of statistical interactions in the presence of a uniform axial electrostatic field will be discussed in Chapter 10.

As a result of the acceleration in the source, the axial velocity spread is, in general, several orders of magnitude smaller than the lateral velocity spread, as was shown in Section 3.7. For simplicity, we now assume that the initial axial velocity spread can be neglected entirely, and we postulate that the beam is initially monochromatic with respect to the normal energy. This implies that the initial axial velocity is the same for all particles. The validity of this simplifying assumption will be verified in Chapter 10.

The basic approach to the problem is to consider the Coulomb interaction of a single test (or reference) particle with all other particles in the beam, referred to as field (or colliding) particles. Due to this interaction, the test particle will experience a deviation from its unperturbed trajectory. From now on, we will use the term "displacement" to denote any deviation in position or velocity from the unperturbed values. The displacement of the test particle is fully determined by the initial coordinates of the field particles relative to the test particle. Due to the stochastic nature of the beam, another test particle, running along the same trajectory some time later, will be surrounded by a different configuration of field particles and, consequently, experiences a different displacement.

Consider a large set of test particles running successively with identical velocity along a specific trajectory in the beam, called the reference trajectory. Assume that the test particles are well separated and can be considered as independent. Our problem now consists of two parts. First, we wish to determine the total influence of a given set of field particles on the path of a specific test particle. This is called the *dynamical part* of the problem. Next, we want to determine the distribution of displacements in velocity and position of the entire set of test particles successively arriving at the end of the beam segment. This evaluation represents the *statistical part* of the problem. It effectively corresponds to a time average. It is now assumed that this time average can be replaced by an average over all possible configurations of field particles relative to the single test particle, considered in the dynamical part of the problem. This effectively corresponds to an ensemble average. The concepts of time and ensemble average were discussed in Section 3.4.

In general, the outcome of the dynamical and the statistical part of the calculation will, to a certain extent, depend on the choice of the reference trajectory. As we are primarily interested in the effect of statistical interactions, it is advantageous to take the beam axis as reference trajectory. Due to the rotational symmetry of the beam, no systematic forces act on the test particles moving along this central trajectory. Accordingly, the displacements experienced by the particles are purely stochastic, which means that their average displacement is zero. Particles moving along other trajectories in the beam will be subject to the combined action of space charge and statistical effects. The evaluation of the statistical effects experienced along such off-axis trajectories is more involved. Therefore, most calculations are restricted to the on-axis reference trajectory, assuming that the results obtained for this trajectory are representative for other trajectories as well. The validity of this assumption will be verified in

Chapter 10 by evaluating the average effect generated along all possible reference trajectories in the beam.

Fig. 5.1 shows the unperturbed trajectories of a test particle and a field particle at the moment that the field particle passes the x, y-plane of a fixed coordinate system within the beam segment, referred to as the laboratory system. The z-axis of this system coincides with the beam axis. The particles run in the positive z-direction. The x, y-plane is assumed to coincide with the crossover plane, unless specified otherwise. For convenience of notation, the set of relative coordinates of a single field particle with respect to the test particle is abbreviated as

$$\boldsymbol{\xi} = \{x, y, b_z, v_x, v_y\} \quad \text{or} \quad \boldsymbol{\xi} = \{r_\perp, \varphi, b_z, v, \psi\}, \tag{5.2.1}$$

using rectangular or cylindrical coordinates respectively. The different sets of coordinates are related by $x = r_\perp \cos \varphi$, $y = r_\perp \sin \varphi$, $v_x = v \cos \psi$ and $v_y = v \sin \psi$. The quantity $r_\perp$ is the modulus of the projection of the relative position vector $\mathbf{r} = \mathbf{r}_{\text{field}} - \mathbf{r}_{\text{test}}$ on the x, y-plane. The relative velocity $\mathbf{v} = \mathbf{v}_{\text{field}} - \mathbf{v}_{\text{test}}$ is directed perpendicular to the beam axis, since it is assumed that all particles run with identical axial velocity v_z. Accordingly, its modulus v can directly be related to the angle α between the unperturbed velocity of the field particle and the z-direction ($v = v_z \tan \alpha \approx v_z \alpha$). The vector $\boldsymbol{\xi}$ gives a complete specification of the unperturbed trajectory of the field particle relative to that of the test particle. By specifying $\boldsymbol{\xi}_1, \boldsymbol{\xi}_2, \ldots, \boldsymbol{\xi}_{N-1}$ of all $N-1$ field particles, the unperturbed coordinates of all particles can be determined at any moment.

Given the configuration of field particles, specified by the set $\boldsymbol{\xi}_1, \boldsymbol{\xi}_2, \ldots, \boldsymbol{\xi}_{N-1}$, the test particle will experience a certain displacement from the

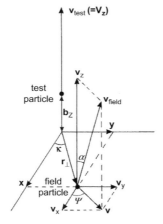

Fig. 5.1 Unperturbed coordinates of an on-axis test particle and a field particle at the moment that the field particle passes the x, y-plane of the laboratory system.

reference trajectory (which is its unperturbed path). The displacement in velocity is expressed in terms of its components Δv_x, Δv_y and Δv_z. The spatial displacement can best be expressed in terms of a virtual displacement in the plane that is optically conjugated to the target plane. The location of this (image) plane is specified by the parameter S_i, see Eq. (3.2.1). The virtual displacement is determined by extrapolating the final position of the test particle toward the image plane by a straight line along its final velocity. The quantities Δx and Δy are the x- and y-component of the virtual radial displacement Δr in the image plane. The trajectory displacement effect is related to the generation of these displacements. The Boersch effect corresponds to the generation of a spread in axial velocities, and the relevant displacement therefore is Δv_z. In general, we will denote the set of relevant displacements by the vector $\Delta \eta$. The reader should be alert that the quantity $\Delta \eta$ may represent a simple scalar (as in the case of the Boersch effect) but is in general a vector.

The dynamical part of the problem now consists of the calculation of $\Delta \eta$ as a function of $\xi_1, \xi_2, \ldots, \xi_{N-1}$:

$$\Delta \eta = \Delta \eta(\xi_1, \xi_2, \ldots, \xi_{N-1}). \tag{5.2.2}$$

The statistical part of the problem is the determination of the probability $P_N(\xi_1, \xi_2, \ldots, \xi_{N-1})$ of the configuration $\xi_1, \xi_2, \ldots, \xi_{N-1}$ and the evaluation of the distribution of displacements $\rho(\Delta \eta)$ from its definition

$$\rho(\Delta \eta) = \int d\xi_1 d\xi_2 \cdots d\xi_{N-1} \, P_N(\xi_1, \xi_2, \ldots, \xi_{N-1}) \delta[\Delta \eta - \Delta \eta(\xi_1, \xi_2, \ldots, \xi_{N-1})],$$

$$\tag{5.2.3}$$

in which $\delta[x]$ is the (multidimensional) delta-Dirac function.

The distribution $\rho(\Delta \eta)$ is normalized when $P_N(\xi_1, \xi_2, \ldots, \xi_{N-1})$ is normalized:

$$\int d\Delta \eta \, \rho(\Delta \eta) = \int d\xi_1 d\xi_2 \cdots d\xi_{N-1} \, P_N(\xi_1, \xi_2, \ldots, \xi_{N-1}). \tag{5.2.4}$$

As we took the beam axis as reference trajectory, the mean value $\langle \Delta \eta \rangle$ of the distribution $\rho(\Delta \eta)$ will be zero. However, the second moment or mean square width $\langle \Delta \eta^2 \rangle$ of the distribution $\rho(\Delta \eta)$ does not vanish:

$$\langle \Delta \eta^2 \rangle = \int d\Delta \eta \, \rho(\Delta \eta) \, \Delta \eta^2$$

$$= \int d\xi_1 d\xi_2 \cdots d\xi_{N-1} \, P_N(\xi_1, \xi_2, \ldots, \xi_{N-1}) \, [\Delta \eta(\xi_1, \xi_2, \ldots, \xi_{N-1})]^2$$

$$\tag{5.2.5}$$

The problem of calculating the effect of statistical interactions now consists of the evaluation of Eqs. (5.2.2) and (5.2.3). The resulting distribution $\rho(\Delta \eta)$ contains all the desired information.

5.3. Reduction of the N-particle problem

A straightforward analytical evaluation of Eqs. (5.2.2) and (5.2.3) does not appear to be feasible, and one is forced to introduce some simplifying assumptions. In this section we wish to study the different alternatives to solve this problem. The discussion is restricted to the fundamental aspects involved. The reader is referred to Chapter 2 for a chronological review of the various approaches discussed in literature and to Chapter 15 for a quantitative comparison of the results of some recent theories.

All approaches discussed here have in common that the initial coordinates of the field particles are assumed to be identical statistically independent quantities

$$P_N(\pmb{\xi}_1, \pmb{\xi}_2, ..., \pmb{\xi}_{N-1}) = \prod_{i=1}^{N-1} P_2(\pmb{\xi}_i), \qquad (5.3.1)$$

where $P_2(\pmb{\xi})$ is the probability that a field particle has coordinates $\pmb{\xi}$ relative to the test particle. For the on-axis reference trajectory in a rotational symmetric beam segment, with a crossover in the x, y-plane, $P_2(\pmb{\xi})$ can be expressed as

$$\int d\pmb{\xi}\, P_2(\pmb{\xi}) = \int_0^\infty dv f_v(v) \int_0^{2\pi} \frac{d\Phi}{2\pi} \int_0^\infty dr_\perp f_r(r_\perp) \int_{-S_c L}^{(1-S_c)L} \frac{db_z}{L}, \qquad (5.3.2)$$

anticipating that the magnitude of the displacement caused by a single field particle $|\Delta\pmb{\eta}|$ depends on the angles ψ and φ only through the relative angle $\Phi = \psi - \varphi$. The crossover-location parameter S_c is defined by Eq. (3.2.1). The distributions $f_r(r_\perp)$ and $f_v(v)$ are related to the current density and angular distribution in the crossover plane respectively. When these distributions are uniform, one should use

$$f_r(r_\perp) = \frac{2r_\perp}{r_c^2}\, \Theta(r_c - r_\perp), \qquad f_v(v) = \frac{2v}{v_0^2}\, \Theta(v_0 - v), \qquad (5.3.3)$$

where $v_0 \approx v_z \alpha_0$ and $\Theta(x)$ is the step function defined by Eq. (3.7.8). In case of Gaussian distributions, one should use

$$f_r(r_\perp) = \frac{2r_\perp}{r_c^2}\, e^{-(r_\perp/r_c)^2}, \qquad f_v(v) = \frac{2v}{v_0^2}\, e^{-(v/v_0)^2}. \qquad (5.3.4)$$

Notice that the σ^2 value of a two-dimensional Gaussian distribution $f(r)$ is equal to half its mean square value: $\sigma^2 = \langle r^2 \rangle / 2$.

Eq. (5.3.2) treats the current density and angular distribution as independent. This assumption is justifiable at the location of a crossover (which is an image of the source or the entrance pupil of the system) provided that the effect of lens aberrations can be neglected. It should be noted, however, that even in the absence of aberrations, this approach still ignores the fact that the illumination of the edge of the crossover may differ from the illumination of the centre of the crossover. This leads to a correlation in the boundaries of the angular and spatial distribution, which may become significant for small beam angles. Furthermore, we emphasize that Eq. (5.3.2) pertains to the on-axis reference trajectory. For an off-axis reference trajectory one may not combine ψ and φ to Φ and one has to account for the fact that the distributions of the relative coordinates r and v are non-rotational symmetric. The symmetry can be restored by considering the average over all off-axis reference trajectories, but the effective distributions of $f_r(r_\perp)$ and $f_v(v)$ differ from those of Eqs. (5.3.3) and (5.3.4).

Eq. (5.3.1) implies that all correlations in the coordinates of the field particles are ignored. In practical systems statistical correlations between the particle coordinates are determined both by the initial probability distribution of the particles leaving the cathode and the dynamics under influence of the Coulomb force during the time of flight. The last "source" of correlation is closely related to the concept of Debeye screening, which was discussed in Section 3.9. Screening of the test particle prerequires that the interaction takes place over a sufficiently long time period to approach thermodynamic equilibrium. It corresponds to a redistribution of the field particles surrounding the test particle under influence of the electrostatic field of the test particle. In terms of the analysis of this chapter, the screening, built up during the flight through the beam sections preceding the one considered, corresponds to correlations within the initial configuration $\xi_1, \xi_2, \ldots, \xi_{N-1}$. Eq. (5.3.1) ignores these correlations as well as those related to the emission process. This simplification may lead to an overestimation of the probability of large displacements $\Delta\eta$ and the tails of the resulting displacement distribution $\rho(\Delta\eta)$ should be viewed with some caution.

Within the approximation of Eq. (5.3.1) one can in principle account for the screening of the test particle in an approximate way by limiting the effective interaction range of the test particle to the Debeye distance. While the coordinates of the field particles are chosen randomly (that is independent of the coordinates of the test particle), it is then assumed that only those field particles interact with the test particle that have an unperturbed trajectory that intersects the Debeye sphere of the test particle. For simplicity, these interactions are calculated from the unscreened (bare) Coulomb potential, while the influence of more distant field particles is neglected entirely. This procedure corresponds to a truncation of the integration boundaries of Eq. (5.3.2). We recall that the analysis of Section 4.9 led to the conclusion that the conditions required for Debeye screening are, in general, not fulfilled in practical beams. Accordingly, we will assume that the Debeye screening does not occur, and we will identify the integration boundaries in Eq. (5.3.2) with the boundaries of the beam.

The dynamical problem for evaluating the displacement of the test particle $\Delta\eta(\xi_1, \xi_2, ..., \xi_{N-1})$ for the configuration of field particles $\xi_1, \xi_2, ..., \xi_{N-1}$ can be expressed as an integral over time

$$\Delta\eta(\xi_1, \xi_2, ..., \xi_{N-1}) = \int_{t_i}^{t_f} dt\, \mathbf{G}_N(\xi_1, \xi_2, ..., \xi_{N-1}, t) \qquad (5.3.5)$$

in which t_i and t_f are the initial and final time of the interaction, corresponding to the moment that the test particle enters and leaves the beam segment respectively. In the case of the Boersch effect $\Delta\eta = \Delta v_z$ and $\mathbf{G}_N$ is directly related to the axial component of the interaction force $\mathbf{F}_z$ acting on the test particle

$$\mathbf{G}_N(\xi_1, \xi_2, ..., \xi_{N-1}, t) = \mathbf{F}_z(\xi_1, \xi_2, ..., \xi_{N-1}, t)/m, \qquad (5.3.6)$$

where m is the particle mass. In the case of the trajectory displacement effect in a rotational symmetric beam, one should consider $\Delta\eta = |\Delta\mathbf{r}| = (\Delta x^2 + \Delta y^2)^{1/2}$. The lateral displacement $\Delta\mathbf{r}$ can be related to the lateral component of the interaction force $\mathbf{F}_\perp$ as

$$\Delta\mathbf{r}(\xi_1, \xi_2, ..., \xi_{N-1}) = \int_{t_i}^{t_f} dt' \int_{t_i}^{t'} dt\, \mathbf{F}_\perp(\xi_1, \xi_2, ..., \xi_{N-1}, t)/m$$

$$= \int_{t_i}^{t_f} dt\, (t_f - t)\, \mathbf{F}_\perp(\xi_1, \xi_2, ..., \xi_{N-1}, t)/m,$$

as follows by changing the order of the integration. Accordingly, the function $\mathbf{G}_N$ is for this case equal to

$$\mathbf{G}_N(\xi_1, \xi_2, ..., \xi_{N-1}, t) = (t_f - t)\mathbf{F}_\perp(\xi_1, \xi_2, ..., \xi_{N-1}, t)/m \qquad (5.3.7)$$

The computation of the interaction force $\mathbf{F}(\xi_1, \xi_2, ..., \xi_{N-1}, t)$ at time t requires the knowledge of the actual positions of all field particles relative to the test particle at that time

$$\mathbf{F}(\xi_1, \xi_2, ..., \xi_{N-1}, t) = -C_0 \sum_{i=1}^{N-1} \frac{\mathbf{r}_i(\xi_1, \xi_2, ..., \xi_{N-1}, t)}{r_i(\xi_1, \xi_2, ..., \xi_{N-1}, t)^3}, \qquad (5.3.8)$$

where $\mathbf{r}_i = \mathbf{r}_{\text{field},i} - \mathbf{r}_{\text{test}}$ and $C_0 = e^2/4\pi\varepsilon_0$, e denoting the particle charge. However, the computation of the positions at time t requires the knowledge of the particle trajectories under influence of the interaction force $\mathbf{F}(\xi_1, \xi_2, ..., \xi_{N-1}, t)$ up to that time. This coupling is the essence of the many particle problem.

So far, we only translated the dynamical problem into the evaluation of $\mathbf{F}(\xi_1, \xi_2, ..., \xi_{N-1}, t)$. At this point, one has to introduce some principle assumptions in order to proceed toward an analytical solution of the N-particle problem. The apparent different alternatives are described next.

5.3.1 First order perturbation approximation

In this approximation it is assumed that the deviations from the unperturbed trajectories are small. The actual interaction force $\mathbf{F}(\boldsymbol{\xi}_1, \boldsymbol{\xi}_2, ..., \boldsymbol{\xi}_{N-1}, t)$ acting on the test particle, which is a function of the actual positions $\mathbf{r}_i(\boldsymbol{\xi}_1, \boldsymbol{\xi}_2, ..., \boldsymbol{\xi}_{N-1}, t)$, is now approximated by the interaction force that would act on the test particle when all particles would follow their unperturbed trajectories. As these trajectories are known beforehand (from the specification of $\boldsymbol{\xi}_1, \boldsymbol{\xi}_2, ..., \boldsymbol{\xi}_{N-1}$) the problem is now decoupled.

The unperturbed position $\tilde{\mathbf{r}}_i$ of a particle i at time t depends only on $\boldsymbol{\xi}_i$ and t. Accordingly, the first order perturbation approximation can be expressed as

$$\mathbf{r}_i(\boldsymbol{\xi}_1, \boldsymbol{\xi}_2, ..., \boldsymbol{\xi}_{N-1}, t) = \tilde{\mathbf{r}}_i(\boldsymbol{\xi}_i, t), \qquad (5.3.9)$$

and Eq. (5.3.8) yields

$$\mathbf{F}(\boldsymbol{\xi}_1, \boldsymbol{\xi}_2, ..., \boldsymbol{\xi}_{N-1}, t) = -C_0 \sum_{i=1}^{N-1} \frac{\tilde{\mathbf{r}}_i(\boldsymbol{\xi}_i, t)}{\tilde{r}_i(\boldsymbol{\xi}_i, t)^3}. \qquad (5.3.10)$$

Consequently, the function $\mathbf{G}_N(\boldsymbol{\xi}_1, \boldsymbol{\xi}_2, ..., \boldsymbol{\xi}_{N-1}, t)$ being proportional to $\mathbf{F}(\boldsymbol{\xi}_1, \boldsymbol{\xi}_2, ..., \boldsymbol{\xi}_{N-1}, t)$, can be written as

$$\mathbf{G}_N(\boldsymbol{\xi}_1, \boldsymbol{\xi}_2, ..., \boldsymbol{\xi}_{N-1}, t) = \sum_{i=1}^{N-1} \mathbf{G}_2(\boldsymbol{\xi}_i, t), \qquad (5.3.11)$$

in which $\mathbf{G}_2$ is the equivalence of $\mathbf{G}_N$ for a beam consisting of only the test particle and a single field particle ($N = 2$). It now follows from Eq. (5.3.5) that the displacement $\Delta \eta$ of the test particle can be expressed as

$$\Delta \eta(\boldsymbol{\xi}_1, \boldsymbol{\xi}_2, ..., \boldsymbol{\xi}_{N-1}) = \sum_{i=1}^{N-1} \Delta \eta_2(\boldsymbol{\xi}_i), \qquad (5.3.12)$$

in which $\Delta \eta_2$ denotes the displacement caused by the interaction with a single field particle. The calculation of $\Delta \eta_2$ is a two-particle problem. In first order perturbation approximation $\Delta \eta_2$ follows from Eq. (5.3.5) and either Eq. (5.3.6) or (5.3.7), taking $N = 2$. This calculation can be carried out analytically, as will be shown in Chapter 6.

It should be emphasized that the reduction of the N-particle dynamical problem to $N-1$ two-particle problems, expressed by Eq. (5.3.12), is entirely justified as long as all particles follow approximately their unperturbed trajectories, which requires that the interaction between the particles be weak. The validity of this approach is a direct consequence of the fact that forces are additive. However, the first order perturbation approximation clearly leads to erroneous results in case some of the interactions cause significant deviations from the unperturbed trajectories of either the test particle or the field particles. Eq. (5.3.9) is no longer true for all particles, and the actual interaction force consequently differs from the interaction force computed from the unperturbed trajectories. Thus, the first order perturbation approximation provides an accurate analysis in case all interactions are weak but breaks down otherwise.

In order to solve the statistical part of the problem it is assumed that the field particles are statistically independent, as expressed by Eq. (5.3.1). Notice that this assumption is consistent with the approximation that the particles run along their unperturbed trajectories (presupposing that no correlations occur as a consequence of the emission process). By substitution of Eqs. (5.3.1) and (5.3.12) into Eq. (5.2.5), one finds for the mean square displacement

$$\langle \Delta \eta^2 \rangle = (N-1) \int d\xi \, P_2(\xi) \, \Delta \eta_2(\xi)^2, \tag{5.3.13}$$

utilizing that $P_2(\xi)$ is normalized and that the mean displacement $\langle \Delta \eta \rangle = 0$. It should be anticipated that a straightforward evaluation of Eq. (5.3.13) in general leads to divergent results, since the expression for $\Delta \eta_2(\xi)$, based on first order perturbation dynamics, overestimates the effect of strong collisions. To prevent this divergence, one should restrict the integration domain to those values of ξ that correspond to weak collisions, assuming that the effect of strong collisions can be neglected.

Loeffler (1969) used an alternative statistical procedure to avoid the divergence of $\langle \Delta \eta^2 \rangle$, while employing the first order perturbation approximation. He argued that this divergence is a consequence of the use of the continuous probability distribution $P_2(\xi)$, which disregards the discrete nature of the beam. In Loeffler's approach, one accounts for the finite number of particles N within the beam by computing the nth largest displacement $\Delta \eta_n$ out of the set of $N-1$ displacements, experienced by the test particle, due to the two-particle interactions with the other $N-1$ particles. This displacement is defined as

$$N(|\Delta \eta_2| > |\Delta \eta_n|) = n - 1/2, \tag{5.3.14}$$

with

$$N(|\Delta \eta_2| > |\Delta \eta_n|) = (N-1) \int d\xi \, P_2(\xi), \tag{5.3.15}$$
$$|\Delta \eta_2(\xi)| > |\Delta \eta_n|$$

in which the integration domain of ξ is limited to that part in which the condition $|\Delta \eta_2(\xi)| > |\Delta \eta_n|$ is fulfilled. The total displacement of the test particle follows by adding up all displacements up to the $(N-1)$th largest one:

$$\langle \Delta \eta^2 \rangle = \sum_{n=1}^{N-1} \Delta \eta_n^2 \tag{5.3.16}$$

Loeffler states that this procedure leads to a median value of the total interaction distribution rather than a mean square displacement. The validity of this statement seems questionable.

5.3.2 Closest encounter approximation

This approach assumes that the displacement of the test particle $\Delta\eta$ is dominated by the single strong interaction with its nearest neighbor field particle, which is defined as the particle with the smallest impact parameter with respect to the test particle. It was proposed by Rose and Spehr (1980) and is also used by Weidenhausen et al. (1985). We recall that the impact parameter b of a two-particle collision is defined as the closest distance of approach in absence of interaction. In terms of the components of $\boldsymbol{\xi}_i$ defined in Eq. (5.2.1), it can be expressed as

$$b(\boldsymbol{\xi}) = \left[b_z^2 + b_r^2\right]^{1/2} = \left[b_z^2 + r^2 \sin^2(\Phi)\right]^{1/2}, \qquad (5.3.17)$$

with $(\Phi = \psi - \varphi)$ as can be seen from Fig. 5.2. Let b_n be the impact parameter of the nearest neighbor and $\boldsymbol{\xi}_n$ the coordinates of this particle relative to the test particle. The closest encounter approximation reduces the N-particle dynamical problem to a two-particle problem using

$$\Delta\eta(\boldsymbol{\xi}_1, \boldsymbol{\xi}_2, \ldots, \boldsymbol{\xi}_{N-1}) = \Delta\eta_2(\boldsymbol{\xi}_n). \qquad (5.3.18)$$

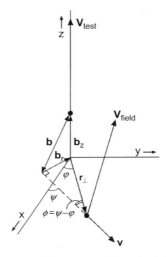

Fig. 5.2 Unperturbed trajectories of a test particle and a field particle in the laboratory system. The impact parameter b is the closest distance of approach in the absence of interaction.

In this model one may not use first order perturbation theory to calculate the two-particle displacement $\Delta\eta_2(\xi_n)$, since this displacement is not necessarily small. Instead one should use the analytical solution of the classical Kepler problem, which is also valid for strong interactions. It should be noted that the usual solution of this problem prerequires the collision to be complete (see Chapter 6), which implies that the initial and final separation of the particles is large compared to the impact parameter b. This assumption is consistent with the closest encounter approach, which focusses on the collision with the smallest value of b. Such a collision will, in general, be complete, provided that the diameter of the crossover is small compared to the diameter of the beam at the start and the end of the beam segment. Thus, the closest encounter approach based on complete collisions seems an adequate approach for beams with a narrow crossover.

In order to solve the statistical part of the problem, one has to determine the probability that an arbitrary field particle, with impact parameter b_n and coordinates ξ_n, is the nearest neighbor of the test particle. Assuming again that the particles are statistically independent, this probability can be expressed as

$$W_2(\xi_n, b_n) = \left[(N-1) \int_{b(\xi_n) = b_n} d\xi_n \, P_2(\xi_n)\right] \times \left[1 - \int_{b(\xi) < b_n} d\xi \, P_2(\xi)\right]^{N-2}. \quad (5.3.19)$$

The first term is identical to the probability that an arbitrary field particle has an impact parameter b_n and coordinates ξ_n, while the second term is identical to the probability that all other field particles have a larger impact parameter. The mean square displacement of the test particle can now be computed as

$$\langle \Delta\eta^2 \rangle = \int db_n \int d\xi_n \, W_2(\xi_n, b_n) \, \Delta\eta_2(\xi_n)^2, \quad (5.3.20)$$

which shows the same structure as Eq. (5.3.13) but is mathematically more complex.

The closest encounter approach presupposes that the main contribution to the total displacement of the test particle stems from a single collision with its nearest neighbor field particle. This assumption is justified when this single collision is strong relative to the interactions with all other field particles. However, it breaks down in case the effect is built up by a large number of weak interactions. Such might be expected for instance for a cylindrical beam at low current density. Clearly, it does not handle the case of multiple strong interactions either.

5.3.3 Extended two-particle approximation

In this approach one follows the same statistical procedure as in the first order perturbation approximation, which is expressed by Eqs. (5.3.1) and (5.3.12). However, the dynamical problem of calculating the two-particle displacement $\Delta\eta_2(\xi)$ is not solved in first order perturbation approximation, but rather by an exact analysis of the Kepler problem, which takes the collision duration into account. Unfortunately, it is not possible to obtain the exact solution of this problem in an explicit form, as will be demonstrated in Chapter 6, and one is forced to evaluate the integral of Eq. (5.3.13) numerically. Since strong interactions are now taken into account properly, this integral will not diverge, as in the first order perturbation approximation. Explicit analytical solutions can be derived for the limiting cases in which it is allowed to use either first order perturbation dynamics or (nearly) complete collision dynamics. This method, here referred to as the extended two-particle approximation, was first presented by van Leeuwen and Jansen (1983).

In this section we will only consider the general philosophy of the extended two-particle model, while the mathematical basis is the subject of the remaining sections of this chapter. The model can, in a way, be regarded as a synthesis of the two methods outlined so far. In case all interactions are weak, it clearly shows the same features as the first order perturbation approximation. It uses the same statistical procedure, while the approach to the dynamical part of the problem leads to identical results for weak interactions. In case the displacement of the test particle is dominated by a single complete collision, it should lead to identical results as obtained with the closest encounter approximation. The dynamics of that single collision is handled similarly. The statistical method is somewhat different, since the extended two-particle model includes the (weak) interaction with the other particles as well. It should be noted that the calculation of these interactions will not be accurate, since the strong interaction causes the test particle to deviate significantly from its unperturbed trajectory. However, since the total displacement of the test particle is dominated by that single strong interaction anyhow (as assumed in the closest encounter approximation), the exact contribution of the weak collision taking place simultaneously is immaterial. Accordingly, one should expect that, for those conditions for which the closest encounter approximation is justified, both methods yield the same results.

From the preceding analysis it follows that the extended two-particle approximation covers all those cases in which either the first order perturbation approximation is justified (multiple weak interactions) or the closest encounter approximation is justified (single strong complete collision).

Furthermore, we note that the exact two-particle analysis provides accurate results even for strong but incomplete collisions, which are not covered by the other methods. The extended two-particle approximation breaks down when the test particle is involved in two or more strong collisions simultaneously or successively in the same beam segment.

5.3.4 Mean square field fluctuation approximation

In the previous approaches, the displacement of the test particle was first calculated for a specific configuration of field particles and averaged over all possible configurations next. We will now consider an approach in which this order is interchanged. The basic assumption of this approach is that the time development of the distribution of field particles surrounding the test particle can be described as a succession of independent states. Accordingly, Eq. (5.3.5) is rewritten as

$$\Delta\eta(\pmb{\xi}_1,\pmb{\xi}_2,\ldots,\pmb{\xi}_{N-1}) = \sum_{s=1}^{N_s} \Delta t_s\, G_N(\pmb{\xi}_{1,s}\cdots\pmb{\xi}_{N-1,s}), \qquad (5.3.21)$$

in which the configuration of field particles in state number s is specified by $\pmb{\xi}_{1,s}, \pmb{\xi}_{2,s}, \ldots, \pmb{\xi}_{N-1,s}$. N_s denotes the total number of independent states encountered by the test particle on its way through the beam segment and Δt_s the duration of state number s (note: $t_i = t_1$, and $t_f = t_{N_s} + \Delta t_{N_s}$). The ensemble average is now carried out over the configuration $\pmb{\xi}_{1,s}, \pmb{\xi}_{2,s}, \ldots, \pmb{\xi}_{N-1,s}$. The mean square displacement follows as

$$\langle \Delta\eta^2 \rangle = \sum_{s=1}^{N_s} \Delta t_n^2 \langle G_N^2 \rangle_s, \qquad (5.3.22)$$

in which

$$\langle G_N^2 \rangle_s = \int d\pmb{\xi}_{1,s}\cdots d\pmb{\xi}_{N-1,s}\, P_{N,s}(\pmb{\xi}_{1,s},\ldots,\pmb{\xi}_{N-1,s})\left[G_N(\pmb{\xi}_{1,s},\ldots,\pmb{\xi}_{N-1,s})\right]^2. \qquad (5.3.23)$$

$P_{N,s}(\pmb{\xi}_{1,s}, \pmb{\xi}_{2,s}, \ldots, \pmb{\xi}_{N-1,s})$ represents the probability of the configuration $\pmb{\xi}_{1,2}, \pmb{\xi}_{2,s}, \ldots, \pmb{\xi}_{N-1,s}$ occurring in state s.

The evolution of the system in the course of time is represented by the set of quantities Δt_s and $P_{N,s}(\pmb{\xi}_{1,s}, \pmb{\xi}_{2,s}, \ldots, \pmb{\xi}_{N-1,s})$, where $s = 1, 2, \ldots, N_s$. The configuration $\pmb{\xi}_{1,s}, \pmb{\xi}_{2,s}, \ldots, \pmb{\xi}_{N-1,s}$ is considered to be fixed during the period Δt_s. Thus, the motion of the field particles relative to the test particle is in fact represented as a series of instantaneous rearrangements, every rearrangement corresponding with the transition to a new independent state. As the configuration of field particles in state s is considered to be fixed during the time interval Δt_s, one may write

$$G_N(\xi_{1,s}, \xi_{2,s}, \ldots, \xi_{N-1,s}) = \sum_{j=1}^{N-1} G_2(\xi_{i,s}) \qquad (5.3.24)$$

since forces are additive. It should be emphasized that Eq. (5.3.24) does not rely on the assumption that the interactions are weak, unlike the similar Eq. (5.3.11).

While the reduction to two-particle interactions given by Eq. (5.3.24) now comes out in a rather natural way, it should be anticipated that difficulties might arise with the determination of the set of quantities Δt_s and $P_{N,s}(\xi_{1,s}, \xi_{2,s}, \ldots, \xi_{N-1,s})$, where $s = 1, 2, \ldots, N_s$. Let us assume that the particles within the state s are statistically independent. Thus, similar to Eq. (5.3.1) it is assumed that

$$P_{N,s}(\xi_{1,s}, \xi_{2,s}, \ldots, \xi_{N-1,s}) = \prod_{j=1}^{N-1} P_{2,s}(\xi_{i,s}), \qquad (5.3.25)$$

where $P_{2,s}(\xi)$ is the probability that a field particle has coordinates ξ relative to the test particle in state s. Combining Eqs. (5.3.23)–(5.3.25) one obtains

$$\langle G_N^2 \rangle_s = (N-1) \int d\xi_s \, P_{2,s}(\xi_s) G_2(\xi_s)^2, \qquad (5.3.26)$$

using that $\langle G_N \rangle_s = 0$. Eqs. (5.3.22) and (5.3.26) now represent the essence of the problem. Eq. (5.3.26) is directly related to the mean square field force occurring in state s, as can be seen from Eqs. (5.3.5)–(5.3.7) (taking $N = 2$). Eq. (5.3.22) expresses that the displacements of the test particle experienced in the different states are assumed to be independent. Accordingly, they are added quadratically. Notice that the individual terms are proportional to Δt_s^2. This quadratic dependency on Δt_s is a consequence of the fact that the displacement experienced in state s does not occur instantaneously but builds up during the entire time interval Δt_s. As a result of this quadratic dependency on Δt_s the outcome of Eq. (5.3.22) depends critically on the choice of the set of intervals Δt_s, where $s = 1, 2, \ldots, N_s$. In other words, the correlation time of the fluctuations is a critical parameter within this model.

Another problem is related to the evaluation of Eq. (5.3.26). As $G_s(\xi_s)$ is proportional to the Coulomb force exerted by the field particle on the test particle, the integral diverges for small interparticle distances if one assumes that distribution of field particles $P_{2,s}(\xi_s)$ is fully random. One is forced to introduce a lower limit for the interparticle distance to prevent this divergence. This complication is a direct consequence of the assumption that the different states are entirely independent. In reality, two particles that are very near to each other in state s will have increased their separation in state $s+1$. Their positions in state $s+1$ are thus correlated. Contrary to the two-particle models, this effect is ignored in this model and the resulting divergence of Eq. (5.3.26) has to be removed artificially.

Models employing the mean square field fluctuation approximation are presented by De Chambost and Hennion (1979), Sasaki (1984, 1986) and Massey et al. (1981). The last model pertains specifically to a cylindrical beam geometry. It deviates from the model described here in the sense that the different states $s = 1, 2, \ldots, N_s$ are not treated as independent but, oppositely, as entirely correlated. Accordingly, the summation of the contributions of the different states is performed by using

$$\langle \Delta \eta^2 \rangle^{1/2} = \sum_{s=1}^{N_s} \Delta t_s \langle \mathbf{G}_N^2 \rangle_s^{1/2} \tag{5.3.27}$$

instead of Eq. (5.3.22). In this model the choice of the time intervals Δt_s (with $s = 1, 2, \ldots, N_s$) is not critical, due to the linear dependency on this parameter. For low particle densities this model yields the same results as the first order perturbation approximation, provided that the divergence related to strong interactions is solved in the same way.

5.4. Calculation of the displacement distribution

We will now concentrate on the so-called extended two particle approximation, which appears to be the least restrictive one of the four alternative approaches outlined in the previous section. In this section we will demonstrate that this model permits a direct calculation of the entire displacement distribution $\rho(\Delta \eta)$. We will follow the analysis given by van Leeuwen and Jansen (1983).

The two-particle approximation relies on the assumption that the total displacement of the test particle can be computed as the sum of the displacements due to pair collisions, as expressed by Eq. (5.3.12). In addition, it is assumed that the field particles are statistically independent, as expressed by Eq. (5.3.1). By substitution of these relations into Eq. (5.2.3), one obtains for the displacement distribution

$$\rho(\Delta \eta) = \int d\boldsymbol{\xi}_1 \, d\boldsymbol{\xi}_2 \cdots d\boldsymbol{\xi}_{N-1} \prod_{i=1}^{N-1} P_2(\boldsymbol{\xi}_i) \, \delta\left(\Delta \eta - \sum_{i=1}^{N-1} \Delta \eta_2(\boldsymbol{\xi}_i) \right). \tag{5.4.1}$$

Let n be the number of components of the displacement $\Delta \eta$. The δ-function is then n-dimensional and can be represented as

$$\delta(x) = \frac{1}{(2\pi)^n} \int d\mathbf{k} \, e^{i\mathbf{k} \cdot \mathbf{x}}, \tag{5.4.2}$$

in which both **x** and **k** are n-component vectors. Utilizing this representation, Eq. (5.4.1) transforms to

$$\rho(\Delta\boldsymbol{\eta}) = \frac{1}{(2\pi)^n} \int d\mathbf{k}\, e^{i\mathbf{k}\cdot\Delta\boldsymbol{\eta}} A_N(\mathbf{k}), \tag{5.4.3}$$

in which

$$A_N(\mathbf{k}) = \left[\int d\boldsymbol{\xi}\, P_2(\boldsymbol{\xi}) e^{-i\mathbf{k}\cdot\Delta\boldsymbol{\eta}_2(\boldsymbol{\xi})}\right]^{N-1}. \tag{5.4.4}$$

A further reduction is possible by making use of the fact that N is, in general, a large number. Utilizing the normalization of $P_2(\boldsymbol{\xi})$, one may rewrite Eq. (5.4.4) as

$$A_N(\mathbf{k}) = \left[1 - \int d\boldsymbol{\xi}\, P_2(\boldsymbol{\xi})\left(1 - e^{-i\mathbf{k}\cdot\Delta\boldsymbol{\eta}_2(\boldsymbol{\xi})}\right)\right]^{N-1}. \tag{5.4.5}$$

The integral in Eq. (5.4.5) will be small for most $\boldsymbol{\xi}$, in particular for those values of $\boldsymbol{\xi}$ that correspond to a large separation of the test and the field particle ($\Delta\boldsymbol{\eta}_2(\boldsymbol{\xi}) \to 0$). Recalling that $P_2(\boldsymbol{\xi})$ is normalized, it becomes clear that the integral will fall off proportional to N, since the total displacement of the test particle is mainly caused by a few neighbor field particles. This can also be observed from Eq. (5.3.2), which is inversely proportional to the beam length L (which is proportional to N). We make the dependency on N explicit by defining

$$P_2(\boldsymbol{\xi}) = \frac{1}{L} p(\boldsymbol{\xi}) = \frac{\lambda}{N} p(\boldsymbol{\xi}), \tag{5.4.6}$$

in which λ is the linear particle density defined by Eq. (3.2.4). Accordingly, Eq. (5.4.5) transforms to

$$A_N(\mathbf{k}) = \left[1 - \frac{\lambda}{N} \int d\boldsymbol{\xi}\, p(\boldsymbol{\xi})\left(1 - e^{-i\mathbf{k}\cdot\Delta\boldsymbol{\eta}_2(\boldsymbol{\xi})}\right)\right]^{N-1} \approx e^{-\lambda p(\mathbf{k})}, \tag{5.4.7}$$

using that N is a large number. The function $p(\mathbf{k})$ is here defined as

$$p(\mathbf{k}) = \int d\boldsymbol{\xi}\, p(\boldsymbol{\xi})\left(1 - e^{-i\mathbf{k}\cdot\Delta\boldsymbol{\eta}_2(\boldsymbol{\xi})}\right). \tag{5.4.8}$$

Combining Eqs. (5.4.3) and (5.4.7), the displacement distribution $\rho(\Delta\boldsymbol{\eta})$ becomes

$$\rho(\Delta\boldsymbol{\eta}) = \frac{1}{(2\pi)^n} \int d\mathbf{k}\, e^{i\mathbf{k}\cdot\Delta\boldsymbol{\eta} - \lambda p(\mathbf{k})}. \tag{5.4.9}$$

Eqs. (5.4.8) and (5.4.9) constitute the major results of this section. One sees that the function $p(\mathbf{k})$ is fully equivalent to $\rho(\Delta\boldsymbol{\eta})$. The evaluation of $p(\mathbf{k})$

involves only two-particle dynamics, represented by $\Delta\eta_2$, and two-particle statistics, represented by $p(\xi)$.

It is often useful to separate the two-particle analysis from the $p(\mathbf{k})$ transform. For that purpose, the two-particle distribution function $\rho_2(\Delta\eta)$ is defined as

$$\rho_2(\Delta\eta) = \int d\xi\, p(\xi)\, \delta[\Delta\eta - \Delta\eta_2(\xi)], \qquad (5.4.10)$$

which is basically the displacement distribution for a system containing only two particles, the test particle and a single field particle. Eq. (5.4.8) can now be expressed as

$$p(\mathbf{k}) = \int d\Delta\eta\, \rho_2(\Delta\eta)\left(1 - e^{-i\mathbf{k}\cdot\Delta\eta}\right). \qquad (5.4.11)$$

The evaluation of the displacement distribution $\rho(\Delta\eta)$ now consists of three steps. The first step, expressed by Eq. (5.4.10), is the calculation of the two-particle distribution function $\rho_2(\Delta\eta)$, employing two-particle dynamics (represented by $\Delta\eta_2(\xi)$) and two-particle statistics (represented by $p(\xi)$). The second step, expressed by Eq. (5.4.11), is the transformation of $\rho_2(\Delta\eta)$ to the k-domain, which results in the function $p(\mathbf{k})$. The third step, expressed by Eq. (5.4.9) can be considered as a back transformation of $p(\mathbf{k})$ to the $\Delta\eta$ domain, which leads to the wanted N-particle displacement distribution $\rho(\Delta\eta)$. As far as the dependency on the experimental parameters is concerned, it should be noted that the two-particle distribution function $\rho_2(\Delta\eta)$ and its transform $p(\mathbf{k})$ are entirely determined by geometrical parameters only, which are represented by $p(\xi)$. The particle density, expressed in terms of λ, enters the model in the evaluation of $\rho(\Delta\eta)$, using Eq. (5.4.9). Accordingly, a particular beam geometry requires only a single computation of the function $p(\mathbf{k})$, which is sufficient to evaluate the distribution $\rho(\Delta\eta)$ for a range of λ-values.

We emphasize that, given a certain value of λ, each of the three functions $\rho_2(\Delta\eta)$, $p(\mathbf{k})$, and $\rho(\Delta\eta)$ contains all relevant information. Accordingly, these functions should be regarded as equivalent representations of the problem, related by simple Fourier-type of integral transformations. The three-step approach of the model is not only convenient from a mathematical point of view, but it clarifies the physical aspects of the problem as well. In fact, we will see that some of the characteristic features of $\rho(\Delta\eta)$ can best be described in terms of the corresponding properties of $p(\mathbf{k})$ and $\rho_2(\Delta\eta)$.

With respect to the relation between the functions $\rho_2(\Delta\eta)$, $p(\mathbf{k})$ and $\rho_2(\Delta\eta)$, one may add that, given the displacement distribution $\rho(\Delta\eta)$, the function $p(\mathbf{k})$ can be retrieved using the inverse Fourier transformation of Eq. (5.4.9):

Concepts of an analytical model for statistical interactions in particle beams

$$e^{-\lambda p(\mathbf{k})} = \int d\,\Delta\eta\, \rho(\Delta\eta)\, e^{-i\mathbf{k}\cdot\Delta\eta}. \tag{5.4.12}$$

Similarly, one can retrieve $\rho_2(\Delta\eta)$ from $p(\mathbf{k})$, using

$$\rho_2(\Delta\eta) = \frac{1}{(2\pi)^n} \int d\mathbf{k}\,[L - p(\mathbf{k})]\, e^{-i\mathbf{k}\cdot\Delta\eta}. \tag{5.4.13}$$

The length L appears in this expression due to the fact that $\rho_2(\Delta\eta)$ is not normalized to 1 but to L. This can be seen from Eqs. (5.4.6) and (5.4.10):

$$\int d\,\Delta\eta\, \rho_2(\Delta\eta) = \int d\xi\, p(\xi) = L \int d\xi\, P_2(\xi) = L. \tag{5.4.14}$$

The practical merit of Eqs. (5.4.12) and (5.4.13) is limited, since one is primarily interested in determining $\rho(\Delta\eta)$ through $p(\mathbf{k})$ and $\rho_2(\Delta\eta)$, instead of the other way around.

In the remainder of this chapter the features of the model will be further investigated, emphasizing the fundamental aspects. The relation between the characteristic properties of the functions $\rho_2(\Delta\eta)$, $p(\mathbf{k})$ and $\rho(\Delta\eta)$ will be discussed in the next section. The aspects of the model related to the choice of the reference trajectory (on- or off-axis) are considered in Section 5.6. In the presentation given so far, the dimension n of the displacement $\Delta\eta$ was left unspecified. Section 5.7 presents the basic equations for $n = 1$, 2 and 3. These results are employed in Section 5.8 to determine the distribution of the fluctuating component of the interaction force in a cylindrical beam. The significance of the representation in the k-domain is the subject of Section 5.9. Finally, it will be shown in Section 5.10 that the model gives a clear indication how to add the results obtained for the individual beam segments.

5.5. Moments and cumulants of the displacement distribution

In this section we will define the moments and cumulants of the displacement distribution $\rho(\Delta\eta)$. We will also investigate their relation with the moments of the two-particle distribution $\rho_2(\Delta\eta)$ as well as the relation with the properties of the function $p(\mathbf{k})$. The material of this section may appear to be rather "technical." However, the resulting relations are of great significance to the model and will be used extensively.

For convenience of notation the analysis of this section is restricted to the case of an one-dimensional displacement $\Delta\eta$ ($n = 1$). However, all results can straightforwardly be extended to cover the general case of a multidimensional displacement $\Delta\eta$ ($n > 1$).

The mth moment of the displacement distribution $\rho(\Delta\eta)$ is defined as

$$\langle \Delta\eta^m \rangle = \int d\Delta\eta\, \rho(\Delta\eta)\, \Delta\eta^m. \tag{5.5.1}$$

Clearly, the zero order moment ($m = 0$) is related to the normalization of $\rho(\Delta\eta)$ and should be equal to 1. Substitution of Eq. (5.4.9) into Eq. (5.5.1), taking $m = 0$, yields

$$\begin{aligned}
1 &= \int d\Delta\eta\, \rho(\Delta\eta) = \int d\Delta\eta \int \frac{dk}{2\pi}\, e^{ik\cdot\Delta\eta - \lambda p(k)} \\
&= \int dk\, \delta(k)\, e^{-\lambda p(k)} = [e^{-\lambda p(k)}]_{k=0},
\end{aligned} \tag{5.5.2}$$

utilizing the representation of the δ-function given by Eq. (5.4.2). Eq. (5.5.2) states that the normalization of $\rho(\Delta\eta)$ requires that $p(k = 0) = 0$. Notice that the property $p(k=0) = 0$ is guaranteed by the definition of $p(k)$, see Eq. (5.4.8).

Similarly, one finds for the first moment ($m = 1$) of $\rho(\Delta\eta)$, using Eqs. (5.5.1) and (5.4.9)

$$\begin{aligned}
\langle \Delta\eta \rangle &= \int d\Delta\eta\, \Delta\eta\, \rho(\Delta\eta) = \int d\Delta\eta\, \Delta\eta \int \frac{dk}{2\pi}\, e^{ik\cdot\Delta\eta - \lambda p(k)} \\
&= -i \int d\Delta\eta \int \frac{dk}{2\pi} \left[\frac{\partial}{\partial k} e^{ik\cdot\Delta\eta}\right] e^{-\lambda p(k)}.
\end{aligned}$$

Integration by parts, using that $p(k) \to \infty$ at the k-integration boundaries, results in

$$\begin{aligned}
\langle \Delta\eta \rangle &= i \int d\Delta\eta \int \frac{dk}{2\pi} \left[\frac{\partial}{\partial k} e^{-\lambda p(k)}\right] e^{ik\cdot\Delta\eta} \\
&= i \int dk\, \delta(k) \frac{\partial}{\partial k} e^{-\lambda p(k)} = i \left[\frac{\partial}{\partial k} e^{-\lambda p(k)}\right]_{k=0},
\end{aligned} \tag{5.5.3}$$

again using Eq. (5.4.2). By repetition of the same mathematical procedure it can in general be proven that the mth moment of $\rho(\Delta\eta)$ is given by

$$\langle \Delta\eta^m \rangle = i^m \left[\frac{\partial^m}{\partial k^m} e^{-\lambda p(k)}\right]_{k=0}. \tag{5.5.4}$$

Thus, all moments of $\rho(\Delta\eta)$ are related to the properties of $p(k)$ at $k = 0$. Accordingly, one may write

$$e^{-\lambda p(k)} = \sum_{m=0}^{\infty} \frac{k^m}{m!} \left[\frac{\partial^m}{\partial k^m} e^{-\lambda p(k)}\right]_{k=0} = \sum_{m=0}^{\infty} \frac{(-ik)^m}{m!} \langle \Delta\eta^m \rangle \tag{5.5.5}$$

Thus, the function $\exp.[-\lambda p(k)]$ can be expressed as a power expansion in k, in which the coefficients are the moments of the distribution $\rho(\Delta\eta)$. A function with this characteristic is called "moment generating function," for instance, see van Kampen (1981). Notice that Eq. (5.5.5) can also be derived directly from Eq. (5.4.12), by power expansion of the exponential function.

Taking the logarithm on both sides of Eq. (5.5.5), one finds by means of power expansion (using $\ln(1+x) = x - x^2/2 + \cdots + (-1)^{j+1} x^j/j$)

$$-\lambda p(k) = \sum_{j=1}^{\infty} \frac{(-1)^{j+1}}{j} \left[\sum_{m=1}^{\infty} \frac{(-ik)^m}{m!} \langle \Delta \eta^m \rangle \right]^j = \sum_{m=1}^{\infty} \frac{(-ik)^m}{m!} \langle\langle \Delta \eta^m \rangle\rangle \quad (5.5.6)$$

in which $\langle\langle \Delta \eta^m \rangle\rangle$ is called the mth cumulant of $p(\Delta \eta)$. All cumulants are implicitly defined by Eq. (5.5.6). The first four cumulants are related to the moments of $p(\Delta \eta)$ as

$$\langle\langle \Delta \eta \rangle\rangle = \langle \Delta \eta \rangle$$
$$\langle\langle \Delta \eta^2 \rangle\rangle = \langle \Delta \eta^2 \rangle - \langle \Delta \eta \rangle^2$$
$$\langle\langle \Delta \eta^3 \rangle\rangle = \langle \Delta \eta^3 \rangle - 3\langle \Delta \eta^2 \rangle \langle \Delta \eta \rangle + 2\langle \Delta \eta \rangle^3$$
$$\langle\langle \Delta \eta^4 \rangle\rangle = \langle \Delta \eta^4 \rangle - 4\langle \Delta \eta^3 \rangle \langle \Delta \eta \rangle - 3\langle \Delta \eta^2 \rangle^2 + 12\langle \Delta \eta^2 \rangle \langle \Delta \eta \rangle^2 - 6\langle \Delta \eta \rangle^4.$$
(5.5.7)

Notice that the first cumulant and first moment are identical, while the second cumulant is identical to the mean square spread (or variance) of the distribution. According to Eq. (5.5.6), the function $-\lambda p(k)$ can be regarded as the "cumulant generating function" of the distribution $p(\Delta \eta)$.

So far, we concentrated on the displacement distribution $p(\Delta \eta)$ and its relation with the function $p(k)$ We now like to investigate the relation with the properties of the two-particle distribution $p_2(\Delta \eta)$. The mth moment of the this distribution is defined as

$$\langle \Delta \eta_2^m \rangle = \int d\Delta \eta \, p_2(\Delta \eta) \, \Delta \eta^m = \int d\xi \, p(\xi) \, \Delta \eta_2^m(\xi), \quad (5.5.8)$$

in which we utilized the definition $p_2(\Delta \eta)$, given by Eq. (5.4.10). The moment generating function of $p_2(\Delta \eta)$ is $p(k)$, as follows by power expansion of the exponential function in Eq. (5.4.11)

$$-p(k) = \sum_{m=1}^{\infty} \frac{(-ik)^m}{m!} \langle \Delta \eta_2^m \rangle. \quad (5.5.9)$$

By comparison of Eqs. (5.5.6) and (5.5.9), one sees that the cumulants of $p(\Delta \eta)$ are directly related to the moments of $p_2(\Delta \eta)$ as

$$\langle\langle \Delta \eta^m \rangle\rangle = \lambda \langle \Delta \eta_2^m \rangle, \quad (5.5.10)$$

which is an important result. For $m=1$ and $m=2$ it yields, employing Eqs. (5.5.7) and (5.5.8)

$$\langle\langle\Delta\eta\rangle\rangle = \langle\Delta\eta\rangle = \lambda \int d\xi\, p(\xi)\, \Delta\eta_2(\xi)$$

$$\langle\langle\Delta\eta^2\rangle\rangle = \langle\Delta\eta^2\rangle - \langle\Delta\eta\rangle^2 = \lambda \int d\xi\, p(\xi)\, \Delta\eta_2(\xi)^2. \qquad (5.5.11)$$

For the on-axis reference trajectory, one finds $\langle\Delta\eta\rangle = 0$, and the second equation becomes identical to Eq. (5.3.13) obtained by direct calculation. (Note: $\lambda p(\xi) = NP_2(\xi) \approx (N-1)\, P_2(\xi)$; see Eq. (5.4.6).)

5.6. On- and off-axis reference trajectories

Eq. (5.4.11), defining the function $p(\mathbf{k})$, can be split up into a real and an imaginary part (using $1 - \exp(-ix) = 2\sin^2(x/2) + i\sin(x)$)

$$p(\mathbf{k}) = p_r(\mathbf{k}) + i p_i(\mathbf{k}), \qquad (5.6.1)$$

in which

$$p_r(k) = 2\int d\Delta\eta\, \rho_2(\Delta\eta) \sin^2(\mathbf{k}\cdot\Delta\eta/2)$$

$$p_i(\mathbf{k}) = \int d\Delta\eta\, \rho_2(\Delta\eta) \sin(\mathbf{k}\cdot\Delta\eta). \qquad (5.6.2)$$

By comparison with Eq. (5.5.6), these functions can be identified as (restricting ourselves again to the one-dimensional case)

$$-\lambda p_r(k) = \sum_{j=1}^{\infty} \frac{(-1)^j k^{2j}}{(2j)!} \langle\langle\Delta\eta^{2j}\rangle\rangle = -\frac{k^2}{2}\langle\langle\Delta\eta^2\rangle\rangle + \frac{k^4}{24}\langle\langle\Delta\eta^4\rangle\rangle + \cdots$$

$$-\lambda p_i(k) = \sum_{j=1}^{\infty} \frac{(-1)^j k^{2j-1}}{(2j-1)!} \langle\langle\Delta\eta^{2j-1}\rangle\rangle = -k\langle\Delta\eta\rangle + \frac{k^3}{6}\langle\langle\Delta\eta^3\rangle\rangle + \cdots, \qquad (5.6.3)$$

using that $\langle\langle\Delta\eta\rangle\rangle = \langle\Delta\eta\rangle$. In general, $p_r(-k) = p_r(k)$, while $p_i(-k) = -p_i(k)$, as can be seen from either Eq. (5.6.2) or (5.6.3). Thus, $p_r(k)$ is a symmetric function and $p_i(k)$ is an antisymmetric function in k.

For the on-axis reference trajectory in a rotational symmetric beam segment, the imaginary part $p_i(k)$ will vanish and $p(\mathbf{k}) = p_r(\mathbf{k})$. In that case, all statistical interaction effects are represented by the real part of $p(\mathbf{k})$. For an off-axis reference trajectory $p_i(\mathbf{k})$ does not vanish and affects the displacement distribution $\rho(\Delta\eta)$, which can now be expressed as

$$\rho(\Delta\eta) = \frac{1}{(2\pi)^n} \int d\mathbf{k}\, e^{i\mathbf{k}\cdot(\Delta\eta - \lambda p_i(\mathbf{k})\mathbf{k}/k^2) - \lambda p_r(\mathbf{k})}, \qquad (5.6.4)$$

as follows from Eqs. (5.4.9) and (5.6.1). Notice that $\rho(\Delta\eta)$ is always real as a consequence of the symmetry properties $p_r(-\mathbf{k}) = p_r(\mathbf{k})$ and $p_i(-\mathbf{k}) = -p_i(\mathbf{k})$. The first term of the expansion of $\lambda p_i(\mathbf{k})$ is equal to $\mathbf{k}\cdot\langle\Delta\eta\rangle$, as can be seen from Eq. (5.6.3). By ignoring higher order terms, Eq. (5.6.4) transforms to

$$\rho(\Delta\eta) \approx \frac{1}{(2\pi)^n} \int d\mathbf{k}\, e^{i\mathbf{k}\cdot(\Delta\eta - \langle\Delta\eta\rangle) - \lambda p_r(\mathbf{k})}. \qquad (5.6.5)$$

The equation shows that the first order effect of an off-axis reference trajectory is the occurrence of a uniform shift in $\rho(\Delta\eta)$ by a value $\langle\Delta\eta\rangle$. This shift corresponds (by definition) to the space charge effect. In the approximation of Eq. (5.6.5), the effect of statistical interactions is entirely represented by the real part of $p(\mathbf{k})$ only. Accordingly, the statistical interactions, experienced along some off-axis reference trajectory, can be described by the same set of equations as used for the on-axis reference trajectory. However, even in this approximation, the resulting shifted distribution $\rho(\Delta\eta - \langle\Delta\eta\rangle)$ may differ from the one obtained for the on-axis reference trajectory, due to the differences in $p(\xi)$, affecting $\rho(\Delta\eta - \langle\Delta\eta\rangle)$ through $p_r(\mathbf{k})$. By including nonvanishing higher order terms in the expansion of $p_i(\mathbf{k})$, the statistical displacement distribution $\rho(\Delta\eta - \langle\Delta\eta\rangle)$ becomes asymmetric (which means $\rho(\Delta\eta - \langle\Delta\eta\rangle) \neq \rho(-\Delta\eta + \langle\Delta\eta\rangle)$), as can be seen from Eq. (5.6.4).

In the remaining of this chapter we will ignore the imaginary part of $p(\mathbf{k})$ by stating $p(\mathbf{k}) = p_r(\mathbf{k})$. As we limit ourselves here to the effect of statistical interactions, this approach is exact for the on axis reference trajectory and correct up to (but not including) third order terms in $\mathbf{k}$ for other reference trajectories.

5.7. Models in one, two, and three dimensions

The main relations of the model are represented by Eqs. (5.4.9)–(5.4.11), defining the functions $\rho(\Delta\eta)$, $\rho_2(\Delta\eta)$ and $p(\mathbf{k})$ respectively. The dimension n of the displacement $\Delta\eta$ was left unspecified so far. We will now consider these equations for the cases $n = 1$, 2 and 3 specifically.

Using scalar notation, Eqs. (5.4.9)–(5.4.11) can be expressed as

$$\rho_2(\Delta\eta_1, \ldots, \Delta\eta_n) = \int d\xi\, p(\xi)\, \delta[\Delta\eta_1 - \Delta\eta_{2,1}(\xi)] \cdots \delta[\Delta\eta_n - \Delta\eta_{2,n}(\xi)] \qquad (5.7.1)$$

$$p(k_1, \ldots, k_n) = \int d\Delta\eta_1 \cdots d\Delta\eta_n\, \rho_2(\Delta\eta_1, \ldots, \Delta\eta_n) \left(1 - e^{-i(k_1\cdot\Delta\eta_1 + \cdots + k_n\cdot\Delta\eta_n)}\right) \qquad (5.7.2)$$

$$\rho(\Delta\eta_1, \ldots, \Delta\eta_n) = \frac{1}{(2\pi)^n} \int dk_1 \cdots dk_n\, e^{i(k_1\cdot\Delta\eta_1 + \cdots + k_n\cdot\Delta\eta_n) - \lambda p(k_1, \ldots, k_n)}, \qquad (5.7.3)$$

where $\Delta\eta_1, \Delta\eta_2, \ldots, \Delta\eta_n$ are the components of $\Delta\eta$; $\Delta\eta_{2,1}, \Delta\eta_{2,2}, \ldots, \Delta\eta_{2,n}$ the components of $\Delta\eta_2$ and $k_1, k_2, \ldots, k_n$ the components of $\mathbf{k}$.

For $n = 1$ these equations transform to

$$p_2(\Delta\eta) = \int d\xi\, p(\xi)\, \delta[\Delta\eta - \Delta\eta_2(\xi)] \tag{5.7.4}$$

$$p(k) = 2 \int_{-\infty}^{\infty} d\Delta\eta\, p_2(\Delta\eta) \sin^2(k\Delta\eta/2) \tag{5.7.5}$$

$$p(\Delta\eta) = \frac{1}{\pi} \int_0^{\infty} dk \cos(k\Delta\eta)\, e^{-\lambda p(k)}. \tag{5.7.6}$$

In Eq. (5.7.5) we ignored the imaginary part of $p(k)$, as justified by the discussion of the previous section. Accordingly, $p(k) = p(-k)$, which was used to obtain Eq. (5.7.6). This set of equations will be used to describe the Boersch effect, taking $\Delta\eta = \Delta v_z$.

For $n = 2$ Eqs. (5.7.1)–(5.7.3) can be reduced to one-dimensional integrals, provided that the distributions $\rho(\Delta\eta_1, \Delta\eta_2)$ and $\rho_0(\Delta\eta_1, \Delta\eta_2)$ are functions of $\Delta\eta = (\Delta\eta_1^2 + \Delta\eta_2^2)^{1/2}$ only. This is to be expected in case of rotational symmetry. The two-particle distribution of $\Delta\eta$ is given by

$$\rho_2(\Delta\eta) = 2\pi\Delta\eta\, p_2(\Delta\eta) = \int d\xi\, p(\xi)\, \delta[\Delta\eta - \Delta\eta_2(\xi)]. \tag{5.7.7}$$

Employing cylindrical coordinates the function $p(k)$ can be expressed as

$$p(k) = p(\mathbf{k}) = \int_0^{\infty} \Delta\eta\, d\Delta\eta \int_0^{2\pi} d\varphi\, \frac{p_2(\Delta\eta)}{2\pi\Delta\eta} \left(1 - e^{-ik\Delta\eta\cos\varphi}\right),$$

which is identical to

$$p(k) = \int_0^{\infty} d\Delta\eta\, p_2(\Delta\eta)[1 - J_0(k\Delta\eta)], \tag{5.7.8}$$

utilizing the integral representation of the Bessel function of the zero order J_0,

$$J_0(x) = \int_0^{2\pi} \frac{d\varphi}{2\pi} e^{ix\cos\varphi}. \tag{5.7.9}$$

By similar reasoning, one finds for the displacement distribution

$$\rho(\Delta\eta) = 2\pi\Delta\eta\, p(\Delta\eta) = \Delta\eta \int_0^{\infty} k\, dk\, J_0(k\Delta\eta)\, e^{-\lambda p(k)}. \tag{5.7.10}$$

Eqs. (5.7.7), (5.7.8) and (5.7.10) will be used to describe the trajectory displacement effect in a rotational symmetric beam, taking $\Delta\eta = \Delta r$.

Finally, for $n = 3$ we consider the situation that the distributions $\rho(\Delta\eta_1, \Delta\eta_2, \Delta\eta_3)$ and $p_2(\Delta\eta_1, \Delta\eta_2, \Delta\eta_3)$ are functions of $\Delta\eta = (\Delta\eta_1^2 + \Delta\eta_2^2 + \Delta\eta_3^2)^{1/2}$ only, which is to be expected in case of spherical symmetry. Employing spherical coordinates, Eqs. (5.7.1)–(5.7.3) transform to

Concepts of an analytical model for statistical interactions in particle beams   115

$$\rho_2(\Delta\eta) = 4\pi\Delta\eta^2\,p_2(\Delta\eta) = \int d\xi\, p(\xi)\,\delta[\Delta\eta - \Delta\eta_2(\xi)] \qquad (5.7.11)$$

$$p(k) = p(\mathbf{k}) = \int_0^\infty d\Delta\eta\, \rho_2(\Delta\eta)\left(1 - \frac{\sin(k\Delta\eta)}{k\Delta\eta}\right) \qquad (5.7.12)$$

$$\rho(\Delta\eta) = 4\pi\Delta\eta^2\,p(\Delta\eta) = \frac{2\Delta\eta}{\pi}\int_0^\infty k\,dk\,\sin(k\Delta\eta)\,e^{-\lambda p(k)}. \qquad (5.7.13)$$

This set of equations will be used to derive the Holtsmark distribution, previously introduced by Eq. (4.9.1).

5.8. Distribution of the interaction force in cylindrical beams

We will now employ the basic equations of the extended two-particle model to compute the probability distribution of the fluctuating electrostatic interaction force, acting on a test particle, located on the axis of a cylindrical beam of charged particles. Doing so we make a little detour from the main line of this chapter, which is primarily concerned with the fundamental concepts of the model. This intermezzo serves as a demonstration of the practical merits of this approach, while it also provides some tools to explain the results obtained for the Boersch effect (discussed in Chapter 7) and the trajectory displacement effect (discussed in Chapters 8 and 9).

We will assume that the beam can be regarded as infinitely long compared to the average separation of the particles. If one assumes, in addition, that the diameter of the beam is infinitely large, the problem simplifies substantially, since the distribution of field particles surrounding the test particle may then be considered as spherical symmetric. Accordingly, the wanted probability distribution of the interaction force is expected to follow the Holtsmark distribution. First, we will investigate whether this result can be reproduced within the model, and next we will study the more complex problem of a beam with a finite, possibly small, diameter.

Consider a test particle surrounded by an infinitely large spherical symmetric cloud of field particles with density n_d. The appropriate model to treat this three dimensional case consists of Eqs. (5.7.11)–(5.7.13). Instead of the probability distribution of displacements $\Delta\eta$ we will determine the distribution of the interaction force $\mathbf{F}$, thus we take $\Delta\eta = \mathbf{F}$. In addition, we have to replace the linear particle density λ by the three-dimensional particle density n_d, which is the relevant measure for the particle density in the present case. Both quantities can be related through the total number of particles N

$$\lambda L = N = (4/3)\pi L^3 n_d, \qquad (5.8.1)$$

in which L now denotes the radius of the sphere of field particles ($L \to \infty$). The magnitude of the force exerted by a field particle, located at a distance r from the test particle, is given by

$$F = \Delta \eta_2 = C_0/r^2, \qquad C_0 = e^2/4\pi\varepsilon_0. \tag{5.8.2}$$

From Eq. (5.7.11), one finds for the two-particle distribution of the interaction force (using that $p(\xi)$ is normalized to L)

$$\rho_2(F) = L \int_0^\infty \frac{4\pi r^2\, dr}{(4/3)\pi L^3}\, \delta(F - C_0/r^2) = \frac{3C_0^{3/2}}{2L^2 F^{5/2}}. \tag{5.8.3}$$

Substitution of this expression into Eq. (5.7.12) yields

$$p(k) = \frac{3C_0^{3/2} k^{3/2}}{2L^2} \int_0^\infty dx\, \frac{x - \sin(x)}{x^{7/2}} = \frac{\pi^{1/2}(2C_0)^{3/2}}{5L^2} k^{3/2}, \tag{5.8.4}$$

in which we used the identity

$$\int_0^\infty dx\, \frac{x - \sin(x)}{x^{7/2}} = \frac{8}{15} \int_0^\infty dx\, \frac{\cos(x)}{x^{1/2}} = \frac{8}{15} \sqrt{\frac{\pi}{2}}, \tag{5.8.5}$$

which follows after several integration by parts. By means of Eq. (5.8.1), one can rewrite Eq. (5.8.4) as

$$\lambda p(k) = (F_n k)^{3/2}, \qquad F_n = (4/15)^{2/3} (2\pi C_0) n_d^{2/3}. \tag{5.8.6}$$

Finally, the desired distribution follows from Eq. (5.7.13)

$$\rho(F) = \frac{1}{2\pi^2 F} \int_0^\infty k\, dk\, \sin(kF)\, e^{-(F_n k)^{3/2}}, \tag{5.8.7}$$

which is indeed identical to the Holtsmark distribution introduced in Section 4.9; see Eqs. (4.9.1) and (4.9.2). The 3/2 power dependency on k is characteristic for the Holtsmark distribution.

Now consider a beam with finite radius r_0. As the distribution $\rho(\mathbf{F})$ is not expected to be spherical symmetric, one has to determine the distribution of the axial force component $F_\parallel$ and the lateral force component $F_\perp$ separately. Let us consider $\rho(F_\perp)$ first. The two-dimensional variant of the model is given by Eqs. (5.7.7), (5.7.8) and (5.7.10). In terms of the coordinates represented by ξ (see Eq. (5.2.1) and Fig. 5.1), one finds

$$\Delta \eta_2 = F_\perp = \frac{C_0 r}{(r^2 + z^2)^{3/2}} = \frac{C_0}{r^2} \frac{1}{\left(1 + (z/r)^2\right)^{3/2}} \tag{5.8.8}$$

analogous to Eq. (5.8.2). The corresponding two-particle force distribution follows from Eq. (5.7.7), assuming a uniform current density distribution within the beam

$$\rho_2(F_\perp) = \int_0^\infty 2\,dz \int_0^{r_0} \frac{2r\,dr}{r_0^2} \delta\left(F_\perp - \frac{C_0}{r^2}\frac{1}{\left(1+(z/r)^2\right)^{3/2}}\right). \tag{5.8.9}$$

By substituting

$$s = \frac{z}{r}, \qquad t = \frac{C_0}{r^2}\frac{1}{(1+s^2)^{3/2}}, \tag{5.8.10}$$

this transforms to

$$\rho_2(F_\perp) = \frac{2C_0^{3/2}}{r_0^2} \int_0^\infty \frac{ds}{(1+s^2)^{9/4}} \int_{t_0}^\infty \frac{dt}{t^{5/2}} \delta[F_\perp - t], \tag{5.8.11}$$

where $t_0 = C_0/r_0^2(1+s^2)^{3/2}$. Using the definition of the δ-function this becomes

$$\rho_2(F_\perp) = \frac{2C_0^{3/2}}{r_0^2 F_\perp^{5/2}} \int_0^\infty \frac{ds}{(1+s^2)^{9/4}} \Theta\left(F_\perp - \frac{C_0}{r_0^2(1+s^2)^{3/2}}\right), \tag{5.8.12}$$

where $\Theta(x)$ is the step function. A further reduction in general terms does not appear to be feasible, and we will restrict the analysis to two extreme cases: $r_0 \to \infty$ and $r_0 \to 0$. In the classification of Section 3.3, the former represents an (extreme) extended cylindrical beam, while the latter corresponds to an (extreme) pencil beam. For $r_0 = 0$, all particles are in a row on the beam axis. For $r_0 \to \infty$, the conditions underlying the Holtsmark distribution are reproduced, and one should expect to obtain its two-dimensional variant.

For $r_0 \to \infty$, the argument of the step function in Eq. (5.8.12) is positive for all positive $F_\perp$, and one directly obtains

$$\rho_2(F_\perp) = \frac{C_0^{3/2}}{r_0^2 F_\perp^{5/2}} I_1, \tag{5.8.13}$$

in which I_1 represents

$$I_1 = \int_0^\infty \frac{2\,ds}{(1+s^2)^{9/4}} = \frac{12\sqrt{\pi}\,\Gamma(3/4)}{5\Gamma(1/4)} = 1.4378,$$

where $\Gamma(x)$ is the Gamma function. The function $p(k)$ now follows from Eq. (5.7.8)

$$p(k) = \frac{C_0^{3/2}}{r_0^2} I_1 I_2 k^{3/2}, \tag{5.8.14}$$

in which I_2 represents

$$I_2 = \int_0^\infty dx\, \frac{1 - J_0(x)}{x^{5/2}} = \frac{4}{9}\int_0^\infty dx\, \frac{J_0(x)}{x^{1/2}} = \frac{2\sqrt{2}\,\Gamma(1/4)}{9\Gamma(3/4)} = 0.92982,$$

employing integration by parts (using $dJ_0(x)/dx = -J_1(x)$ and $d[xJ_1(x)]/dx = xJ_0(x)$). By substituting $n_d = \lambda/\pi r_0^2$ one obtains

$$\lambda p(k) = \pi C_0^{3/2} n_d I_1 I_2 k^{3/2} = (F_n k)^{3/2}, \qquad (5.8.15)$$

equivalent to Eq. (5.8.6). From Eq. (5.7.10), one finds for the distribution of the lateral interaction force in an extended cylindrical beam ($r_0 \to \infty$).

$$\rho(F_\perp) = \frac{1}{2\pi} \int_0^\infty k\, dk\, J_0(kF_\perp) e^{-(F_n k)^{3/2}}, \qquad (5.8.16)$$

which is the two-dimensional variant of the Holtsmark distribution given by Eq. (5.8.7). It specifies the distribution of the interaction force component in a plane perpendicular to the beam axis. (In fact, the same result follows for a plane of arbitrary orientation, since the distribution of field particles is for $r_0 \to \infty$ effectively spherical symmetric.)

Now, consider Eq. (5.8.12) for $r_0 \to 0$—that is, for a cylindrical pencil beam. The argument of the step function is only positive for large values of s, $s > s_0$, where

$$s_0 = \left[(C_0/r_0^2 F_\perp)^{2/3} - 1 \right]^{1/2}. \qquad (5.8.17)$$

The integral in Eq. (5.8.12) can now be approximated as

$$\int_{s_0}^\infty \frac{ds}{(1+s^2)^{9/4}} \approx \int_{s_0}^\infty \frac{ds}{s^{9/2}} = \frac{2}{7 s_0^{7/2}}.$$

Accordingly, one obtains for the two-particle distribution of $F_\perp$ in a pencil beam

$$p_2(F_\perp) = \frac{4(C_0 r_0)^{3/2}}{7 F_\perp^{4/3}}. \qquad (5.8.18)$$

The function $p(k)$ follows from Eq. (5.7.8)

$$p(k) = (4/7)(C_0 r_0)^{1/3} I_3 k^{1/3}, \qquad (5.8.19)$$

in which I_3 represents

$$I_3 = \int_0^\infty dx\, \frac{1 - J_0(x)}{x^{4/3}} = 3 \int_0^\infty dx\, \frac{J_1(x)}{x^{1/3}} = \frac{18\Gamma(5/6)}{2^{1/3}\Gamma(1/6)} = 2.8972,$$

as follows by integration by parts. The 1/3 power dependency on k expressed by Eq. (5.8.19) is characteristic for the distribution of the transverse component of the interaction force in a pencil beam. From Eq. (5.7.10) one finds for this distribution.

$$\rho(F_\perp) = \frac{1}{2\pi} \int_0^\infty k\, dk\, J_0(kF_\perp) e^{-(4/7)(C_0 r_0)^{1/3} I_3 \lambda k^{1/3}}. \qquad (5.8.20)$$

Notice that the width of this distribution tends to zero for $r_0 \to 0$. The physical reason is obvious: for $r_0 = 0$ all particles are on a row and the interaction force does not have a transverse component $\mathbf{F}_\perp$.

Let us now consider the distribution of the axial force component $p(F_\parallel)$. As the calculation follows the same lines of thought as the calculation of $p(\mathbf{F}_\perp)$, we will not present it in detail but only give the main results. From Eq. (5.7.4) one finds for the two-particle force distribution, analogous to Eq. (5.8.9),

$$p_2(F_\parallel) = \int_{-\infty}^{\infty} dz \int_0^{r_0} \frac{2r\,dr}{r_0^2} \delta\left(F_\parallel - \frac{C_0}{r^2} \frac{(z/r)}{(1+z/r)^2)^{3/2}}\right), \qquad (5.8.21)$$

in which $F_\parallel$ carries from $-\infty$ to $+\infty$ (note: $p_2(|F_\parallel|) = 2p_2(F_\parallel)$). Similar to the derivation of Eq. (5.8.11), this can be transformed to

$$p_2(F_\parallel) = \frac{C_0^{3/2}}{r_0^2 |F_\parallel|^{5/2}} \int_0^{\infty} \frac{s^{3/2}\,ds}{(1+s^2)^{9/4}} \Theta\left(|F_\parallel| - \frac{C_0 s}{r_0^2(1+s^2)^{3/2}}\right). \qquad (5.8.22)$$

We will again investigate the extreme cases $r_0 \to \infty$ and $r_0 \to 0$. For $r_0 \to \infty$, the argument of the step function is positive for all positive $F_\parallel$, and Eq. (5.8.22) yields

$$p_2(F_\parallel) = \frac{C_0^{3/2}}{r_0^2 |F_\parallel|^{5/2}} \int_0^{\infty} \frac{s^{3/2}\,ds}{(1+s^2)^{9/4}} = \frac{2C_0^{3/2}}{5r_0^2 |F_\parallel|^{5/2}}. \qquad (5.8.23)$$

The function $p(k)$ follows from Eq. (5.7.5),

$$p(k) = \frac{2^{3/2} C_0^{3/2}}{5r_0^2} k^{3/2} \int_0^{\infty} dx \frac{\sin^2(x)}{x^{5/2}} = \frac{2^{7/2} \pi^{1/2} C_0^{3/2}}{15 r_0^2} k^{3/2}, \qquad (5.8.24)$$

showing the 3/2 power dependency on k characteristic for the Holtsmark distribution. Substituting $n_d = \lambda/\pi r_0^2$ one obtains again Eq. (5.8.15). By means of Eq. (5.7.6), one finds for the distribution of the axial force component in an extended cylindrical beam

$$p(F_\parallel) = \frac{1}{\pi} \int_0^{\infty} dk \cos(kF_\parallel) e^{-(F_n k)^{3/2}}, \qquad (5.8.25)$$

which is the one-dimensional variant of the Holtsmark distribution given by Eq. (5.8.7).

Finally, consider Eq. (5.8.22) for $r_0 \to 0$, which is for a cylindrical pencil beam. Analogous to Eq. (5.8.18), one finds in this limit

$$p_2(F_\parallel) = \frac{C_0^{1/2}}{2|F_\parallel|^{3/2}}. \qquad (5.8.26)$$

The function $p(k)$ follows from Eq. (5.7.5)

$$p(k) = (2C_0)^{1/2} k^{1/2} \int_0^\infty dx \frac{\sin^2(x)}{x^{3/2}} = (2\pi C_0)^{1/2} k^{1/2}. \quad (5.8.27)$$

This k-dependency is characteristic for the distribution of the axial component of the interaction force in a pencil beam. From Eq. (5.7.6), one finds for this distribution

$$\rho(F_\|) = \frac{1}{\pi} \int_0^\infty dk \, \cos(kF_\|) \, e^{-(2\pi C_0)^{1/2} \lambda k^{1/2}}. \quad (5.8.28)$$

The width of the distribution tends to a nonzero finite value for $r_0 \to 0$, in agreement with the fact that the longitudinal component of the interaction force does not vanish in case all particles are on a row.

5.9. Representation in the *k*-domain

We will now proceed with the discussion of the fundamental aspects of the model. From the equations constituting the model (see Section 5.7), it is clear that all information regarding the interaction distribution $\rho(\Delta\eta)$ is also present in the function $p(\mathbf{k})$ as well as in the two-particle distribution $\rho_2(\Delta\eta)$. In this section, we wish to demonstrate that the representation by the function $p(\mathbf{k})$ in the k-domain provides a simple way of describing most of the characteristic properties of $\rho(\Delta\eta)$.

Suppose the quantity $\lambda p(\mathbf{k})$ is, for all $\mathbf{k}$, given by

$$\lambda p(\mathbf{k}) = A_\gamma |\mathbf{k}|^\gamma, \quad (5.9.1)$$

where γ is a numerical constant and A_γ a quantity that depends on the experimental parameters. As $p(\mathbf{k} = 0) = 0$, the corresponding distribution $\rho(\Delta\eta)$ is normalized; see Eq. (5.5.2). In the representation of Eq. (5.9.1), the small k-behavior of $p(\mathbf{k})$ can be expressed as

$$\gamma = 2, \quad A_2 = \langle\langle \Delta\eta^2 \rangle\rangle/2n = \lambda \langle \Delta\eta_2^2 \rangle / 2n \quad (n = 1, 2, 3), \quad (5.9.2)$$

as follows from Eq. (5.6.2) or (5.6.3), assuming that $p(\mathbf{k}) = p_r(\mathbf{k})$ and that $\rho(\Delta\eta)$ is symmetric. In the second step, we used that the second cumulant $\langle\langle \Delta\eta^2 \rangle\rangle$ of $\rho(\Delta\eta)$ is identical to $\lambda \langle \Delta\eta_2^2 \rangle$, where $\langle \Delta\eta_2^2 \rangle$ is the second moment of $\rho_2(\Delta\eta)$; see Eq. (5.5.10) (with $m = 2$). Notice that $\langle\langle \Delta\eta^2 \rangle\rangle = \langle \Delta\eta^2 \rangle$ since $\langle \Delta\eta \rangle = 0$; see Eq. (5.5.7). For the general case of an n-component displacement $\Delta\eta$, one finds from Eqs. (5.7.3), (5.9.1) and (5.9.2)

$$\rho(\Delta\eta) = \rho(\Delta\eta_1, \Delta\eta_2, \ldots, \Delta\eta_n) = \frac{1}{(2\pi)^n} \prod_{j=1}^n \int dk_j e^{ik_j \cdot \Delta\eta_j - A_2 k_j^2},$$

using that $k^2 = k_1^2 + \cdots + k_n^2$. Integration yields

$$\rho(\Delta\eta) = \prod_{j=1}^{n} \frac{1}{(4\pi A_2)^{1/2}} e^{-\Delta\eta_j^2/4A_2} = \frac{1}{(4\pi A_2)^{n/2}} e^{-\Delta\eta^2/4A_2}, \qquad (5.9.3)$$

which is an n-dimensional Gaussian distribution. The mean square of this distribution $\langle\Delta\eta^2\rangle$ is equal to $2nA_2$, which is consistent with Eq. (5.9.2). We conclude that Gaussian behavior prevails when the integral for $\rho(\Delta\eta)$ is dominated by the small k-behavior of $p(\mathbf{k})$. This will be true for large values of the linear particle density λ.

As a practical measure for the width of the displacement distribution we will often use the Full Width at Half Maximum (FWHM). In some applications, one is also interested in the value of the median Full Width (FW_{50}), which is the smallest width containing 50% of the particles. For the distribution of Eq. (5.9.3), one finds

$$FWHM_{\Delta\eta}/A_2^{1/2} = 4(\ln 2)^{1/2} = 3.3302 \qquad (n = 1, 2, 3) \qquad (5.9.4)$$

$$FW_{50_{\Delta\eta}}/A_2^{1/2} = \begin{cases} 4\,\mathrm{erf}^{-1}(1/2) = 1.9078 & (n=1) \\ 4(\ln 2)^{1/2} = 3.3302 & (n=2) \\ 4.3506 & (n=3) \end{cases}$$

in which $\mathrm{erf}^{-1}(x)$ denotes the inverse of the Fresnel error function.

Anticipating on the results of Chapter 7, we note that a linear power dependency in k ($\gamma = 1$) is relevant for the theory of the Boersch effect. We will therefore examine this case. By substituting Eq. (5.9.1), with $\gamma = 1$, into Eqs. (5.7.6), (5.7.10) and (5.7.13), which specify the distribution $\rho(\Delta\eta)$ in one, two, and three dimensions respectively, one finds, after integration

$$\rho(\Delta\eta) = \frac{1}{\pi A_1 \left[1 + (\Delta\eta/A_1)^2\right]} \qquad (n = 1) \qquad (5.9.5)$$

$$\rho(\Delta\eta) = \frac{1}{2\pi A_1^2 \left[1 + (\Delta\eta/A_1)^2\right]^{3/2}} \qquad (n = 2) \qquad (5.9.6)$$

$$\rho(\Delta\eta) = \frac{1}{\pi^2 A_1^3 \left[1 + (\Delta\eta/A_1)^2\right]^2} \qquad (n = 3), \qquad (5.9.7)$$

which are known as Lorentz (or Cauchy) distributions. The FWHM and FW_{50} values of these distributions are given by

$$FWHM_{\Delta\eta}/A_1 = 2\left(2^{2/(n+1)} - 1\right)^{1/2} \quad (n = 1, 2, 3)$$

$$FW_{50_{\Delta\eta}}/A_1 = \begin{cases} 2 & (n=1) \\ 2\sqrt{3} = 3.4641 & (n=2) \\ 4.5389 & (n=3) \end{cases} \qquad (5.9.8)$$

The Lorentzian distributions do not have finite second or higher moments, as can be verified from Eq. (5.5.1). This is generally true for all distributions corresponding to $\gamma < 2$. We will come back to this point at the end of this section. In terms of Eq. (5.9.1), one can represent the Lorentzian distributions (5.9.5), (5.9.6) and (5.9.7) in the k-domain as

$$\gamma = 1, \quad A_1 = FWHM_{\Delta\eta}/2\left(2^{2/(n+1)} - 1\right)^{1/2} \quad (n = 1, 2, 3). \quad (5.9.9)$$

The cases $\gamma = 2$ (Gaussian distribution) and $\gamma = 1$ (Lorentzian distribution) appear to be the only cases for which Eq. (5.7.3) can be evaluated analytically.

The Holtsmark distribution corresponds to the intermediate case $\gamma = 3/2$, as can be seen from Eq. (5.8.6). The Holtsmark distribution pertains to the fluctuating component of the total interaction force **F** in an extended homocentric cylindrical beam segment. In such a beam the displacement of the test particle $\Delta\eta$ experienced during the flight is directly proportional to **F**, or a component of **F**, provided that all particles follow approximately their unperturbed trajectories; see Eq. (4.9.3). The displacement distribution $\rho(\Delta\eta)$ generated in a homocentric cylindrical beam is therefore expected to be Holtsmarkian for small current densities, which are, however, large enough to consider the beam as extended. In Chapters 7, 8 and 9, it will be demonstrated that the Holtsmark distribution also appears for other beam geometries, provided that the beam is extended and that the perturbations experienced by the particles are small.

According to Eq. (5.8.4), the Holtsmark distribution can, in the k-domain, be represented as

$$\gamma = 3/2, \quad A_{3/2} = (4/15)(2\pi C_0)^{3/2} n_d \quad (n = 1, 2, 3). \quad (5.9.10)$$

The corresponding displacement distributions in one, two, and three dimensions are given by Eqs. (5.8.25), (5.8.16) and (5.8.7) respectively. The $FWHM$ and FW_{50} values of these distributions can be retrieved from the data of Table 5.1.

Other k-dependencies are found in beams that have a small transverse dimension compared to the axial separation of the particles, referred to as pencil beams. In the limit of low particle density, the distribution of displacements generated in a pencil beam shows again the same form as the distribution(s) of the interaction force or one of its components. According to Eqs. (5.8.27) and (5.8.19), these distributions can, in the k-domain, be represented as

$$\begin{aligned} \gamma &= 1/2, & A_{1/2} &= (2\pi C_0)^{1/2} \lambda & (n &= 1) \\ \gamma &= 1/3, & A_{1/3} &= (4/7) I_3 (C_0 r_0)^{1/3} \lambda & (n &= 2), \end{aligned} \quad (5.9.11)$$

in which $I_3 = 2.8972$. The corresponding distribution for $n=1$ (axial force component) is given by Eq. (5.8.28) and for $n=2$ (lateral force component) by Eq. (5.8.20). These distributions are relevant for the Boersch effect and the trajectory displacement effect generated in low-density pencil beams. The $FWHM$ and FW_{50} values can be determined from Table 5.1.

The complete list of k-dependencies encountered in the calculation of statistical effects exists of the following γ-values:

$\gamma = 2$ — Gaussian distribution.
$\gamma = 3/2$ — Holstmark distribution,
$\gamma = 1$ — Lorentzian distribution,
$\gamma = 1/2$ — distribution of $F_{\|}$ in a pencil beam, and
$\gamma = 1/3$ — distribution of $F_{\perp}$ in a pencil beam.

The corresponding distributions, in one and two dimensions, are depicted in Figs. 5.3 and 5.4 respectively. These figures clearly show that a small γ-value leads to a distribution with a narrow core and long tails. In Fig. 5.4, we included a plot corresponding to the value $\gamma = 1/2$, which is relevant for the angular deflection distribution in a beam segment with a narrow crossover and low current density, as will be shown in Section 8.3.

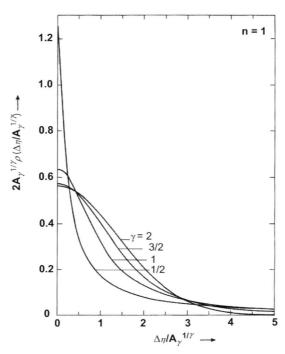

Fig. 5.3 The different types of one-dimensional displacement distributions generated by statistical interactions.

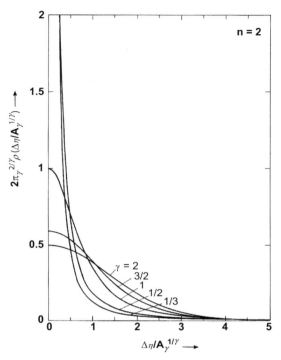

Fig. 5.4 The different types of two-dimensional displacement distributions generated by statistical interactions.

In the remaining part of this section we will investigate how the width and the tails of the displacement distribution $\rho(\Delta\eta)$ depend on the parameter γ and the experimental parameters represented by A_γ. Substitution of Eq. (5.9.1) into Eq. (5.4.9) yields

$$\rho(\Delta\eta) = \frac{1}{(2\pi)^n} \int d\mathbf{k}\, e^{i\mathbf{k}\cdot\Delta\eta - A_\gamma |\mathbf{k}|^\gamma}. \tag{5.9.12}$$

By introducing the scaling

$$\overline{\mathbf{k}} = A_\gamma^{1/\gamma}\mathbf{k}, \qquad \Delta\overline{\eta} = A_\gamma^{-1/\gamma}\Delta\eta, \qquad \overline{\rho}(\Delta\overline{\eta}) = A_\gamma^{n/\gamma}\rho(\Delta\eta), \tag{5.9.13}$$

Eq. (5.9.12) transforms to.

$$\overline{\rho}(\Delta\overline{\eta}) = \frac{1}{(2\pi)^n} \int d\overline{\mathbf{k}}\, e^{i\overline{\mathbf{k}}\cdot\Delta\overline{\eta} - |\overline{\mathbf{k}}|^\gamma}. \tag{5.9.14}$$

Since we have absorbed the parameter A_γ in the scaling, Eq. (5.9.14) does not depend on the experimental parameters. Accordingly, the Full Width at Half Maximum of $\overline{\rho}(\Delta\overline{\eta})$, indicated as $FWHM_\gamma$, is a function of the numerical factor γ only. This function is depicted in Fig. 5.5A, for

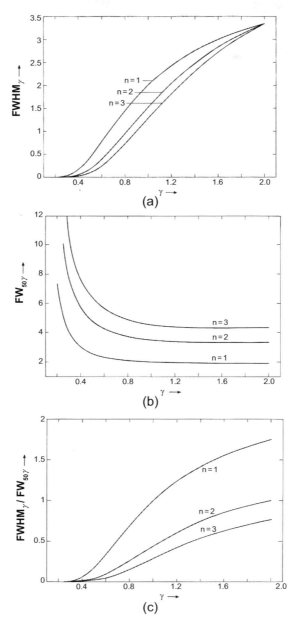

Fig. 5.5 Different width measures of the one-, two- and three-dimensional scaled distribution of displacements as functions of the parameter γ, which defines the shape of the distributions. See Eqs. (5.9.12) and (5.9.15). Panels (a) and (b) give the Full Width at Half Maximum (*FWHM*) and the Full Width median (*FW*$_{50}$) respectively, while the ratio of these quantities is plotted in (c).

Table 5.1 The Full Width at Half Maximum (*FWHM*) and the Full Width median (*FW*$_{50}$) of the different types of distributions generated by statistical interactions, as specified by Eq. (5.9.12), in one, two and three dimensions.

	FWHM$_\gamma$ (= *FWHM*$_{\Delta\eta}$/$A_\gamma^{1/\gamma}$)			*FW*$_{50\gamma}$ (= *FW*$_{50\Delta\eta}$/$A_\gamma^{1/\gamma}$)		
γ	$n=1$	$n=2$	$n=3$	$n=1$	$n=2$	$n=3$
1/3	0.063581	0.013013	0.0049493	3.5751	6.8096	9.1538
1/2	0.44711	0.17997	0.10229	2.5677	4.7478	6.3418
1	2	1.5328	1.2872	2	3.4641	4.5289
3/2	2.8775	2.6554	2.5128	1.9379	3.3206	4.2956
2	3.3302	3.3302	3.3302	1.9078	3.3302	4.3506

$n = 1, 2$ and 3. Table 5.1 gives the corresponding numerical values for the specific cases that are relevant to the theory of statistical interactions. From Eq. (5.9.13), it follows that the unsealed *FWHM* is given by

$$FWHM_{\Delta\eta} = A_\gamma^{1/\gamma} FWHM_\gamma. \quad (5.9.15)$$

This equation expresses the *FWHM* of $\rho(\Delta\eta)$ in terms of the experimental parameters (represented by A_γ) and the parameter γ, which is a constant provided that the shape of $\rho(\Delta\eta)$ remains unchanged. Eq. (5.9.15) is the generalization of Eqs. (5.9.4) and (5.9.8), which apply to the cases $\gamma = 2$ and $\gamma = 1$ respectively. Eq. (5.9.15) can be modified to cover other width measures as well, such as the median width *FW*$_{50}$. The corresponding numerical values of *FW*$_{50\gamma}$ are plotted in Fig. 5.5B. It is important to observe that the ratio *FWHW*$_\gamma$/*FW*$_{50\gamma}$ is not constant but varies with γ, as is shown in Fig. 5.5C.

The parameter A_γ is for all conditions directly proportional to the beam current I, as can be seen from Eqs. (5.9.2), (5.9.10) and (5.9.11), using that λ is directly proportional to I, as expressed by Eq. (3.2.4). Accordingly, the *FWHM* and *FW*$_{50}$ of the energy distribution generated by the Boersch effect and the trajectory displacement distribution are proportional to $I^{1/2}$ in the Gaussian regime ($\gamma = 2$), proportional to $I^{2/3}$ in the Holtsmark regime ($\gamma = 3/2$), and proportional to I in the Lorentzian regime ($\gamma = 1$). For pencil beams the Boersch effect increases with I^2 ($\gamma = 1/2$), and the trajectory displacement effect increases with I^3 ($\gamma = 1/3$).

The fall-off in the tails of the distribution $\rho(\Delta\eta)$ is determined by the small k-behavior, as can be seen from Eq. (5.9.12). For large values of $\Delta\eta = |\Delta\eta|$ one may expand the second part of the exponential function appearing in this equation. By substituting $x = \Delta\eta \mathbf{k}$, one obtains

$$\lim_{\Delta\eta \to \infty} \rho(\Delta\eta) \simeq \frac{1}{(2\pi \Delta\eta)^n} \int d\mathbf{x}\, e^{i\mathbf{x}\Delta\eta/\Delta\eta}\left[1 - A_\gamma(x/\Delta\eta)^\gamma\right]$$

From this equation, it can be understood that the fall-off in the tails is given by

$$\lim_{\Delta\eta\to\infty} \rho(\Delta\eta) \sim \frac{A_\gamma}{\Delta\eta^{n+\gamma}} \qquad (0 < \gamma < 2; \; n = 1, 2, 3). \tag{5.9.16}$$

It should be emphasized that Eq. (5.9.16) does not cover the case $\gamma = 2$, which corresponds to a Gaussian distribution. Notice that Eq. (5.9.16) is in agreement with Eqs. (5.9.5)–(5.9.7), which apply to the case $\gamma = 1$ (Lorentzian distribution) with $n = 1, 2, 3$ respectively. The contribution of the tails to the mth moment of the n-dimensional distribution $\rho(\Delta\eta)$ can now be expressed as

$$\langle \Delta\eta^m \rangle \sim \int_{\Delta\eta_0}^{\infty} \Delta\eta^{n-1} \, d\Delta\eta \, \rho(\Delta\eta) \, \Delta\eta^m \sim \int_{\Delta\eta_0}^{\infty} d\Delta\eta \, \frac{A_\gamma}{\Delta\eta^{1+\gamma-m}}, \tag{5.9.17}$$

in which $\Delta\eta_0$ should be chosen large enough to guarantee the validity of Eq. (5.9.16). Eq. (5.9.17) makes clear that no finite second or higher moments exist for $\gamma < 2$. Only the Gaussian distribution ($\gamma = 2$) has a finite value for the second moment (while the third and higher cumulants of a Gaussian distribution are zero).

Anticipating on the results presented in Chapters 7, 8 and 9, we note that the approximation of the function $p(\mathbf{k})$ by Eq. (5.9.1) will only be accurate within a certain range of $\mathbf{k}$-values. The functions $p(\mathbf{k})$ obtained in the actual calculations for the Boersch effect and the trajectory displacement effect are all proportional to k^2 for small k, proportional to $k^{3/2}$ or k for intermediate k-values, and proportional to $k^{1/2}$ (Boersch effect) or $k^{1/3}$ (trajectory displacement effect) for very large k-values. The particle density, expressed in terms of the linear particle density λ, determines what part of the function $p(\mathbf{k})$ dominates the displacement distribution $\rho(\Delta\eta)$. In general, the small k-behavior is dominant for high values of λ, and the large k-behavior is dominant for small values of λ, as can be understood from Eq. (5.4.9).

5.10. Addition of the effects generated in individual beam segments

The model outlined in the previous sections enables us to calculate the effect of statistical interactions in an individual beam segment, which contains a single crossover or is cylindrical. A practical beam usually consists of several of such beam segments. The question now arises how to determine the effect generated in the total system from the results obtained for the individual segments.

Consider a beam that is partitioned in N_b beam segments. Let the statistical displacement distribution of segment j ($j = 1 \ldots N_b$) be given by

$$\rho_j(\Delta\eta) = \frac{1}{(2\pi)^n} \int d\mathbf{k} \, e^{i\mathbf{k} \cdot \Delta\eta - \lambda_j p_j(\mathbf{k})} \tag{5.10.1}$$

using the representation of Eq. (5.4.9). All information regarding segment j is contained in the quantity $\lambda_j p_j(\mathbf{k})$. We now make the following two assumptions:

(1) The displacements $\Delta\eta_j$ (with $j = 1 \ldots N_b$) experienced by the test particle in the successive beam segments are not correlated, and
(2) All distributions $p_j(\Delta\eta)$ (with $j = 1 \ldots N_b$) can be represented by Eq. (5.9.1), with identical values for the k-dependency parameter γ. This implies that the distributions are assumed to be congruent.

To start with, let us consider a system consisting of two beam segments. Assumption (1) implies that the distribution of statistical effects generated in the total beam can be expressed as a convolution of the distributions corresponding to the individual beam segments

$$\rho(\Delta\eta) = \int d\Delta\eta' \, \rho_1(\Delta\eta') \rho_2(\Delta\eta - \Delta\eta'). \tag{5.10.2}$$

Substitution of the expressions for $\rho_1(\Delta\eta)$ and $\rho_1(\Delta\eta)$, given by Eq. (5.10.1), yields

$$\rho(\Delta\eta) = \int d\Delta\eta' \int \frac{d\mathbf{k}}{(2\pi)^n} \int \frac{d\mathbf{k}'}{(2\pi)^n} e^{i\mathbf{k} \cdot \Delta\eta' + i\mathbf{k}' \cdot (\Delta\eta - \Delta\eta') - \lambda_1 p_1(\mathbf{k}) - \lambda_2 p_2(\mathbf{k}')}. \tag{5.10.3}$$

Similar to the representation of the δ-function given by Eq. (5.4.2), one may write

$$\delta(\mathbf{k} - \mathbf{k}') = \int \frac{d\Delta\eta'}{(2\pi)^n} e^{i\Delta\eta' \cdot (\mathbf{k} - \mathbf{k}')}. \tag{5.10.4}$$

Accordingly, Eq. (5.10.3) transforms to

$$\rho(\Delta\eta) = \frac{1}{(2\pi)^n} \int d\mathbf{k} \, e^{i\mathbf{k} \cdot \Delta\eta - [\lambda_1 p_1(\mathbf{k}) + \lambda_2 p_2(\mathbf{k})]} \tag{5.10.5}$$

By comparison with Eq. (5.10.1), one finds that the total distribution can, in the k-domain, be represented as

$$[\lambda p(\mathbf{k})]_T = \lambda_1 p_1(\mathbf{k}) + \lambda_2 p_2(\mathbf{k}).$$

For a system consisting of N_b beam segments, this expression can be generalized to

$$[\lambda p(\mathbf{k})]_T = \sum_{j=1}^{N_b} \lambda_j p_j(\mathbf{k}) \tag{5.10.6}$$

Thus, assumption (1) implies that the representation of the total displacement distribution in the k-domain can be obtained as the sum of all functions $\lambda p(\mathbf{k})$, corresponding to the individual beam segments.

We like to translate this result in terms of the distributions $\rho_j(\Delta\eta)$, where $j = 1 \ldots N_b$. For that we need assumption (2). Following Eq. (5.9.15), one may express the FWHM of the distribution $\rho_j(\Delta\eta)$ as

$$FWHM_j = A_j^{1/\gamma} FWHM_\gamma, \tag{5.10.7}$$

in which A_j corresponds to the representation in k-space of the distribution of segment j, by means of Eq. (5.9.1). Eq. (5.10.6) implies that the parameter A_T, corresponding to the total effect, is a linear sum of the parameters A_j, corresponding to the individual beam segments. Assumption (2) implies that the parameter $FWHM_\gamma$ has the same numerical value in all segments. Consequently, the FWHM of the total displacement distribution can be expressed as

$$FWHM_T = \left(\sum_{j=1}^{N_b} A_j\right)^{1/\gamma} FWHM_\gamma = \left(\sum_{j=1}^{N_b} \left(A_j^{1/\gamma} FWHM_\gamma\right)^\gamma\right)^{1/\gamma}.$$

By substitution of Eq. (5.10.7), one obtains

$$FWHM_T = \left(\sum_{j=1}^{N_b} FWHM_j^\gamma\right)^{1/\gamma}. \tag{5.10.8}$$

Thus, Gaussian distributions ($\gamma = 2$) should be added quadratically, Lorentzian distributions ($\gamma = 1$) should be added linearly, Holtsmark distributions ($\gamma = 3/2$) should be added with a 3/2 power, etc. Since the power dependency on k does not change by the summation procedure of Eq. (5.10.6), the total distribution will be of the same kind as the distributions generated in the individual beam segments. In other words, the distributions are "stable."

In case assumption (2) is not valid (e.g., not all distributions are congruent), Eq. (5.10.8) can not be applied. However, as long as assumption (1) is valid, Eq. (5.10.6) still holds and the total function $p(\mathbf{k})$ can straightforwardly be determined from the individual contributions. The total displacement distribution follows, as usual, by means of Eq. (5.4.9). Clearly, as the power dependency on k varies, one can not beforehand predict the resulting k-dependency. Accordingly, the shape of the resulting distribution depends on the ratio of the individual contributions, and one can not simply translate Eq. (5.10.6) into a summation rule for the FWHM values. This is only a practical problem and not a principle one, since the resulting FWHM (or any other width) can still be determined from the total $p(\mathbf{k})$ function by means of Eq. (5.4.9).

In general, principle problems do occur when assumption (1) is not valid (which means that the displacements experienced in successive beam segments are correlated). However, there is one exception, namely the extreme case of full correlation. Let us investigate this case in more detail. Full correlation means that the displacement $\Delta\eta_j$ experienced by the test particle in segment j is fully determined by the displacement experienced in the previous segment. Thus, one may write

$$\Delta\eta_j = C_j(\Delta\eta_1), \tag{5.10.9}$$

in which C_j is some (deterministic) function. Accordingly, the total displacement experienced in the entire system is given by

$$\Delta\eta = \sum_{j=1}^{N_b} \Delta\eta_j = \sum_{j=1}^{N_b} C_j(\Delta\eta_1), \tag{5.10.10}$$

and the total displacement distribution can be expressed as

$$\rho(\Delta\eta) = \int d\Delta\eta' \rho_1(\Delta\eta') \delta\left(\Delta\eta - \sum_{j=1}^{N_b} C_j(\Delta\eta')\right). \tag{5.10.11}$$

We now simplify the problem by assuming that

$$\Delta\eta_j = C_j(\Delta\eta_1) = C_j \Delta\eta_1, \tag{5.10.12}$$

which expresses that the displacement in segment j is directly proportional to the displacement in segment 1 (for $j = 2, 3, \ldots, N$). Accordingly, all distributions are assumed to be congruent. Basically, we rely again on the validity of assumption (2). Eq. (5.10.11) now transforms to

$$\rho(\Delta\eta) = \frac{1}{C}\rho_1(\Delta\eta/C), \qquad C = \sum_{j=1}^{N_b} C_j, \tag{5.10.13}$$

which implies

$$FWHM = C\, FWHM_1 = \sum_{j=1}^{N_b} C_j\, FWHM_1 = \sum_{j=1}^{N_b} FWHM_j. \tag{5.10.14}$$

Thus, when the correlation between the displacements experienced by the test particle in the successive beam segments can be expressed by Eqs. (5.10.9) and (5.10.12), one should calculate the total *FWHM* as a linear sum of the *FWHM* corresponding to the individual beam segments.

In all circumstances other than those outlined above, it appears to be impossible to relate the distributions of the separate beam segments uniquely to that of the total system. The basic expectation of our model is that the displacement distributions generated in the different segments are either entirely correlated or not correlated at all. No correlation is expected in case of a succession of segments with narrow crossovers separated by lenses. The lenses cause a redistribution of the trajectories, destroying any correlation between them. (In this respect lens aberrations are expected to have a favorable effect.) Full correlation is expected for a succession of (nearly) cylindrical segments of low current density. The deviations from the unperturbed trajectories are small, and the configuration of field particles surrounding a test particle does not change drastically during the flight. In other words, the "collision" between a test particle and a field particle will be extended throughout all segments, and the contribution of one segment to the total deviation of the test particle is directly proportional to the segment length. Consequently, the separate displacements experienced by the test particle in the successive beam segments are correlated as stated in Eq. (5.10.12), and one may utilize Eq. (5.10.14) to evaluate the *FWHM* of the total displacement distribution.

5.11. Slice method

So far, we assumed that the beam segments were chosen such that the segment boundaries coincide with the location of thin lenses. The reason for this choice is twofold. First, it allows a simple representation of the unperturbed trajectories, which are straight lines due to the absence of optical components within the segment. Second, we expect that the displacements generated in the individual segments, separated by lenses, show no correlation. This allows us to determine the total effect from Eq. (5.10.6) or (5.10.8). Considering the alternative methods of partitioning the beam, it becomes clear that it is not practical to extend the segments over a larger part of the beam and include lenses, primarily due to complications with the representation of the unperturbed trajectories. However, one can easily choose shorter segments than those defined by the distances between the optical components. In this section we will study this alternative in some detail.

Let us subdivide a beam segment between two thin optical components into slices of length Δz; see Fig. 5.6. The displacement distribution $\rho(\Delta\eta)$ is now evaluated per slice. If one assumes that the displacements of the test particle experienced in the individual slices are entirely correlated, one may use Eq. (5.10.14) to determine the *FWHM* of the total displacement distribution, generated in the whole segment. This approach, referred to as the slice method, is justified if the displacements from the unperturbed trajectories are small, which will be true for low current densities.

The slice method has some distinct advantages. By choosing the slices thin enough, one may treat them as cylindrical. Accordingly, one can

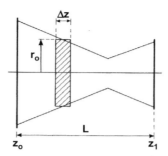

Fig. 5.6 Schematic representation of the subdivision of a beam segment into thin cylindrical slices.

utilize the expression obtained for a cylindrical beam segment to calculate the effect in any beam geometry, using

$$FWHM = \sum_{s=1}^{N_s} FWHM_P(I, V, r_0, \Delta z_s) \approx \int_{z_0}^{z_1} dz \left(\frac{FWHM_P(I, V, r_0, \Delta z)}{\Delta z} \right),$$

(5.11.1)

where $FWHM_P(I, V, r_0, \Delta z)$ is the FWHM of the distribution generated in a cylindrical slice (with Parallel rays) of length Δz, radius r_0, carrying a current I at potential V (supposing a uniform current density distribution). This expression is expected to be linear in Δz. Accordingly, the integrand $FWHM_P(I, V, r_0, \Delta z)/\Delta z$ will only depend on z through the beam radius r_0.

In Chapters 7, 8 and 9 we will use the slice method to relate the results obtained for a beam segment with a crossover to those obtained for a cylindrical beam segment. It should be emphasized that the validity of Eq. (5.11.1) requires the current density to be low enough to guarantee that the displacements from the unperturbed trajectories are small. Returning to the analysis of Section 5.3, one may note that the slice method resembles the procedure of Massey et al. (1981), which is a variant of the "mean square field fluctuation approximation"; see Eq. (5.3.27). However, by considering the *FWHM-v*alue instead of a rms-value we avoid the divergence related to close encounters in a natural way.

CHAPTER SIX

Two-particle dynamics

Contents

6.1.	Introduction	133
6.2.	Basic properties	134
6.3.	Coordinate representation in the orbital plane	138
6.4.	Dynamics of a complete collision	141
6.5.	General analysis of the two-particle dynamical problem	144
6.6.	Numerical approach to the dynamical problem	147
6.7.	Dynamics of a nearly complete collision	148
6.8.	First order perturbation dynamics	153
6.9.	Collisions with zero initial relative velocity	157
6.10.	Expressions for the longitudinal velocity shift Δv_z	160
6.11.	Expressions for the transverse velocity shift $\Delta v_\perp$	162
6.12.	Expressions for the spatial shift Δr	163
6.13.	Coulomb scaling	166
Appendix 6.A. Mathematics of nearly complete collision dynamics		167

6.1. Introduction

This chapter studies the dynamical problem of two particles interacting through the Coulomb force in absence of external fields. The unperturbed motion of the particles in the laboratory system is represented by the so-called geometrical parameters. These parameters are transformed to coordinates in the orbital plane. The calculation of the particle motion in this plane constitutes the well-known Kepler problem, which was solved by Newton (1687). For the repulsive Coulomb force considered here, the solution of this problem predicts that the particles will follow a hyperbolic orbit, which is, in polar coordinates, represented by a function $r(\theta)$. This function gives sufficient information to determine the effect of a collision that is complete in the sense that the particles effectively come from infinity and recede to infinity. However, additional information is required to describe incomplete collisions. Given an arbitrary initial condition, one has to determine each of the polar coordinates r and θ, as well as their derivatives as function of time t. The main part of the chapter is dedicated to the solution of this problem. Integration of the equations of motion results in a function $t(r)$, which implicitly specifies the function $r(t)$ required in the calculation. Unfortunately, the function $t(r)$ cannot be inverted, which prohibits an exact analytical calculation. Three alternative methods are studied to solve this problem. The first method exploits a numerical routine to perform the inversion. The second method is based on a power expansion in r of the exact solution $t(r)$ and applies to nearly complete collisions. The third method is based on first order perturbation theory, which is a valid approximation for weak interactions. The last two methods provide explicit equations that express the velocity and spatial displacements of the particles in terms of the geometrical parameters.

6.2. Basic properties

Consider a test particle moving along the axis of a beam segment of arbitrary geometry. The particle interacts with a single field particle that follows some off-axis trajectory. External forces are assumed to be absent. Fig. 6.1 depicts the unperturbed trajectories of the particles at the moment that the field particle passes the x,y-plane of the laboratory system. The z-axis of the laboratory system coincides with the unperturbed path of the test particle (which is the beam axis). For a beam segment with a crossover, the laboratory system is usually chosen such that the x,y-plane

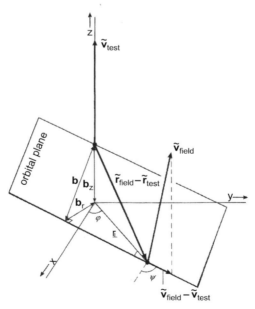

Fig. 6.1 Unperturbed trajectories of a test particle and a field particle in the laboratory system. The orbital plane is spanned by the relative unperturbed velocity vector $\tilde{v}_{field} - \tilde{v}_{test}$ and the relative unperturbed position vector $\tilde{r}_{field} - \tilde{r}_{test}$.

coincides with the crossover plane. The axial location of this plane is specified by the parameter S_c; see Eq. (3.2.1). In this chapter, we will use the parameter S_c in a more general sense to define the axial location $z=0$ of the x,y-plane, irrespective of the beam geometry.

It is assumed that both particles have the same axial velocity. Accordingly, the relative coordinates are completely specified by

$$\xi = (r_\perp, \varphi, b_z, v, \psi), \qquad (6.2.1)$$

as was stated previously by Eq. (5.2.1). This set of parameters defines the unperturbed motion of the particles, which shows only geometrical properties. Therefore, we will here refer to these quantities as the geometrical variables. From the geometrical variables ξ one can obtain the unperturbed coordinates of the particles at any moment. The moment that the field particle passes through the x,y-plane (shown in Fig. 6.1) is defined as $t=0$.

Due to the mutual Coulomb interaction during the flight, the particles will deviate from their unperturbed trajectories. The experienced deviations are determined by the initial relative coordinates of the particles (specified by the geometrical variables and the initial time t_i) and the time of flight T, which is equal to

$$T = t_f - t_i = L/v_z, \qquad (6.2.2)$$

with L the length of the beam segment and v_z the axial velocity of the particles. The initial and final time of the interaction, t_i and t_f, can be expressed as

$$t_i = -S_c T, \quad t_f = (1 - S_c)T, \qquad (6.2.3)$$

in which the start and the end of the interaction is identified with the moments that the field particle passes the entrance and the exit plane of the beam segment respectively.

The variables specifying the deviations from the unperturbed trajectories, here referred to as the dynamical variables, are abbreviated by

$$\Delta\eta = (\Delta x, \Delta y, \Delta v_z), \qquad (6.2.4)$$

using the same notation as in Chapter 5. The quantities Δx and Δy are the virtual spatial displacements in the plane that is optically conjugated to the target plane. This plane is called the image plane or the reference plane. The axial location of the image plane is specified by the quantity S_i, defined by Eq. (3.2.1). The displacements Δx and Δy are the relevant dynamical variables for the calculation of the trajectory displacement effect. For a test particle running along the axis of a rotational symmetric beam it is sufficient to determine the radial displacement $\Delta r = (\Delta x^2 + \Delta y^2)^{1/2}$. The axial velocity displacement Δv_z is the relevant dynamical variable for the calculation of the Boersch effect. In general we will denote all "deviations from the unperturbed trajectory," both in position and velocity, as "displacements" or "shifts."

The basic problem considered in this chapter is to determine the dynamical variables $\Delta\eta$ as functions of the geometrical variables ξ, the initial time t_i, and the interaction time T. The set of geometrical parameters given by Eq. (6.2.1) is clearly not the most general possible, since it assumes that the unperturbed trajectory of the test particle coincides with the beam axis, while it ignores any difference in axial velocity. However, the analysis of this chapter can be generalized to include other cases. The only required modification concerns the transformation from the laboratory system to coordinates in the orbital plane, as will be indicated in the next section.

This section covers the general formulation of the problem. We will first discuss the relation between the coordinate representations in the laboratory system and the center of mass system, and we will next review the physical constants of motion.

The relative position vector $\mathbf{r}$ and the relative velocity vector $\mathbf{v}$ are defined as:

$$\mathbf{r} = \mathbf{r}_c - \mathbf{r}_t, \quad \mathbf{v} = \mathbf{v}_c - \mathbf{v}_t, \qquad (6.2.5)$$

in which the subscript c refers to the field (or colliding) particle and the subscript t to the test particle. The unperturbed coordinates of the particles will be indicated with a tilde (~). Thus, the unperturbed relative position is $\tilde{\mathbf{r}}$ and the unperturbed relative velocity is $\tilde{\mathbf{v}}$. The coordinates referring to the time $t=0$ will be indicated with the subscript 0 and the coordinates referring to the initial time t_i and final time t_f with the subscripts i and f respectively. The unperturbed relative position and velocity at time $t=0$ are related to the geometrical variables as

$$\tilde{\mathbf{r}}_0 = \begin{pmatrix} r_\perp \cos\varphi \\ r_\perp \sin\varphi \\ -b_z \end{pmatrix}, \quad \tilde{\mathbf{v}}_0 = \begin{pmatrix} v\cos\psi \\ v\sin\psi \\ 0 \end{pmatrix}. \qquad (6.2.6)$$

The initial relative coordinates follow from

$$\mathbf{r}_i = \tilde{\mathbf{r}}_0 + t_i \tilde{\mathbf{v}}_0, \quad \mathbf{v}_i = \tilde{\mathbf{v}}_0. \qquad (6.2.7)$$

The unperturbed velocity $\tilde{\mathbf{v}}$ remains constant by definition, as expressed by the second equation. The magnitude $|\tilde{\mathbf{v}}|$ of the unperturbed velocity is identical to the geometrical parameter v (thus $v_i = \tilde{v}_0 = v$).

The center of mass coordinates are defined as

$$\mathbf{R} = \frac{m_c \mathbf{r}_c + m_t \mathbf{r}_t}{m_c + m_t}, \quad \mathbf{V} = \frac{m_c \mathbf{v}_c + m_t \mathbf{v}_t}{m_c + m_t}, \qquad (6.2.8)$$

in which m_c and m_t are the mass of the field and the test particle respectively. We will consider the case of identical particles, thus $m_c = m_t = m$. Accordingly, Eq. (6.2.8) reduce to

$$\mathbf{R} = \frac{1}{2}(\mathbf{r}_c + \mathbf{r}_t), \quad \mathbf{V} = \frac{1}{2}(\mathbf{v}_c + \mathbf{v}_t). \tag{6.2.9}$$

At $t = 0$ the center of mass coordinates are given by

$$\mathbf{R}_0 = \frac{1}{2}\begin{pmatrix} r_\perp \cos\varphi \\ r_\perp \sin\varphi \\ b_z \end{pmatrix}, \quad \mathbf{V} = \frac{1}{2}\begin{pmatrix} v\cos\psi \\ v\sin\psi \\ 2v_z \end{pmatrix}. \tag{6.2.10}$$

As external forces are assumed to be absent, the total momentum of the two-particle system is conserved. Consequently, the velocity of the center of mass remains constant irrespective of the interaction between the particles, and Eq. (6.2.10) applies to both the unperturbed and the perturbed trajectories. Accordingly, the position $\mathbf{R}$ of the center of mass at time t is (independent of the interaction) given by

$$\mathbf{R}(t) = \mathbf{R}_0 + t\mathbf{V}, \tag{6.2.11}$$

and the coordinates of field and test particle at time t can be retrieved from

$$\begin{aligned} \mathbf{r}_c(t) &= \mathbf{R}(t) + \frac{1}{2}\mathbf{r}(t) & \mathbf{v}_c(t) &= \mathbf{V}(t) + \frac{1}{2}\mathbf{v}(t) \\ \mathbf{r}_t(t) &= \mathbf{R}(t) - \frac{1}{2}\mathbf{r}(t), & \mathbf{v}_t(t) &= \mathbf{V}(t) - \frac{1}{2}\mathbf{v}(t). \end{aligned} \tag{6.2.12}$$

If one knows the relative coordinates $r(t)$ and $\mathbf{v}(t)$ as functions of time, one can determine the corresponding coordinates of the individual particles by means of the Eqs. (6.2.10)–(6.2.12).

The absence of external forces also implies that the total energy and the angular momentum of the two-particle system are conserved. The total kinetic energy (in the laboratory system) is defined as:

$$E_{kin,L} = \frac{1}{2}m_c v_c^2 + \frac{1}{2}m_t v_t^2 = \frac{1}{2}(m_c + m_t)V^2 + \frac{1}{2}\frac{m_c m_t}{m_c + m_t}v^2 \tag{6.2.13}$$

The first term on the right-hand side of Eq. (6.2.13) corresponds to the kinetic energy of the center of mass, while the second term corresponds to the kinetic energy observed in a reference system moving with the center of mass (called center of mass system). The first term is constant and can be disregarded in the further analysis. The total energy in the center of mass system is equal to (using $m_c = m_t = m$)

$$E = \frac{1}{4}mv^2 + \frac{C_0}{r}, \tag{6.2.14}$$

in which $C_0 = e^2/4\pi\varepsilon_0$. The value of E is conserved during the encounter and can be determined from the initial coordinates given by Eq. (6.2.7)

$$E = \frac{1}{4}mv_i^2 + \frac{C_0}{r_i}. \tag{6.2.15}$$

The angular momentum J_L relative to the origin of the laboratory system is defined as

$$\mathbf{J}_L = m_c \mathbf{r}_c \times \mathbf{v}_c + m_t \mathbf{r}_t \times \mathbf{v}_t = (m_c + m_t)\mathbf{R} \times \mathbf{V} + \frac{m_c m_t}{m_c + m_t} \mathbf{r} \times \mathbf{v}. \quad (6.2.16)$$

The first term, specifying the angular momentum of the center of mass, is constant as follows by substitution of Eq. (6.2.11). Since $\mathbf{J}_L$ is conserved as a whole, the second term, specifying the angular momentum relative to the center of mass, is constant, too. Using $m_c = m_t = m$ one finds for this term

$$\mathbf{J} = \frac{1}{2} m \mathbf{r} \times \mathbf{v}. \quad (6.2.17)$$

The vector $\mathbf{J}$ can be determined from the initial coordinates. Using Eqs. (6.2.7) and (6.2.6) one obtains

$$\mathbf{J} = \frac{1}{2} m \mathbf{r}_i \times \mathbf{v}_i = \frac{1}{2} m \tilde{\mathbf{r}}_0 \times \tilde{\mathbf{v}}_0 = \frac{1}{2} m v \begin{pmatrix} b_z \sin\varphi \\ -b_z \cos\varphi \\ r_\perp \sin\Phi \end{pmatrix}, \quad (6.2.18)$$

with $\Phi = \psi - \varphi$. The modulus of this vector is

$$J = \frac{1}{2} m b v_i = \frac{1}{2} m b \tilde{v}_0, \quad (6.2.19)$$

where b is the impact parameter, which is given by

$$b = \left(b_z^2 + b_r^2 \right)^{1/2} = \left(b_z^2 + r_\perp^2 \sin^2\Phi \right)^{1/2}, \quad (6.2.20)$$

as can be seen from Fig. 6.1. Conservation of angular momentum in the center of mass system implies that the relative motion of the particles takes place in a plane perpendicular to the vector $\mathbf{r} \times \mathbf{v}$, which is the plane through $\mathbf{r}$ and $\mathbf{v}$. This plane is called the orbital plane and is indicated in Fig. 6.1. From Eqs. (6.2.14) and (6.2.17) one sees that the relative coordinates $\mathbf{r}$ and $\mathbf{v}$ obey the same relations as the coordinates of a single particle with reduced mass $m/2$ moving under influence of a fixed force center. The two-particle problem is thus reduced to this single particle problem.

6.3. Coordinate representation in the orbital plane

In order to describe the motion in the orbital plane it is convenient to construct an orthonormal reference system in this plane

$$\hat{a} = -\frac{\tilde{\mathbf{v}}_0}{|\tilde{\mathbf{v}}_0|}, \quad \hat{b} = \frac{\tilde{\mathbf{r}}_0 - (\tilde{\mathbf{r}}_0 \cdot \hat{a}) \cdot \hat{a}}{|\tilde{\mathbf{r}}_0 - (\tilde{\mathbf{r}}_0 \cdot \hat{a}) \cdot \hat{a}|}. \quad (6.3.1)$$

The vector $\hat{a}$ is directed opposite to the unperturbed relative velocity and the vector $\hat{b}$ points toward the perihelion (point of closest approach)

Two-particle dynamics   139

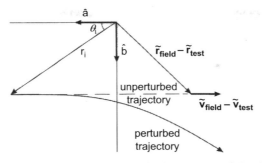

Fig. 6.2 The relative perturbed and unperturbed trajectory of the field and the test particle and the orthonormal coordinate system $(\hat{a}, \hat{b})$ in the orbital plane.

of the unperturbed trajectory; see Fig. 6.2. By substitution of Eq. (6.2.6) one obtains for the vectors $\hat{a}$ and $\hat{b}$ in terms of the coordinates of the laboratory system

$$\hat{a} = -\begin{pmatrix} \sin\psi \\ \cos\psi \\ 0 \end{pmatrix}, \quad \hat{b} = \frac{1}{b}\begin{pmatrix} r_\perp(\cos\varphi - \cos\psi\cos\Phi) \\ r_\perp(\sin\varphi - \sin\psi\cos\Phi) \\ -b_z \end{pmatrix}, \quad (6.3.2)$$

with $\Phi = \psi - \varphi$. These equations constitute the coordinate transformation from the $(\hat{a}, \hat{b})$ coordinate system to the laboratory system. We recall that the analysis is restricted to the specific case of an on-axis test particle, which has the same axial velocity as the field particle. For the general case one can still use Eq. (6.3.1) to perform the transformation from laboratory system to coordinates in the orbital plane, but the representation of $\tilde{\mathbf{r}}_0$, $\tilde{\mathbf{v}}_0$, $\mathbf{R}$ and $\mathbf{V}$ differs from Eqs. (6.2.6) and (6.2.10). Accordingly, Eq. (6.3.2) become different, too. We emphasize that the only modification required to treat the general case concerns the coordinate transformation from the laboratory system to the $(\hat{a}, \hat{b})$ coordinate system in the orbital plane, which is in all cases fully specified by Eqs. (6.3.1) and (6.2.5).

The unperturbed relative trajectory can in the $(\hat{a}, \hat{b})$ coordinate system be represented as:

$$\tilde{\mathbf{r}}(t) = (a_0 - \tilde{v}_0 t)\hat{a} + b\hat{b}, \quad (6.3.3)$$

where

$$a_0 = (\tilde{\mathbf{r}}_0 \cdot \hat{a}) = -r_\perp \cos\Phi$$
$$b = (\tilde{\mathbf{r}}_0 \cdot \hat{b}) = \left(b_z^2 + r_\perp^2 \sin^2\Phi\right)^{1/2}. \quad (6.3.4)$$

The impact parameter b was already introduced by Eq. (6.2.20) and is here repeated for convenience. Notice that the unperturbed relative trajectory is (by definition) directed parallel to the $\hat{a}$-vector, as is depicted in Fig. 6.2.

We will use the polar coordinates r and θ to describe the perturbed relative trajectory in the orbital plane. The initial polar coordinates are given by

$$\begin{aligned} r_i &= \left[(a_0 - \tilde{v}_0 t_i)^2 + b^2\right]^{1/2} \\ \theta_i &= \frac{1}{2}\pi(1 + \sigma_i) - \sigma_i \arcsin(b/r_i) \\ \sigma_i &= \mathrm{sign}(\tilde{v}_0 t_i - a_0) \end{aligned} \qquad (6.3.5)$$

with $\mathrm{sign}(x) = +1$ for $x \geq 0$ and $\mathrm{sign}(x) = -1$ for $x < 0$. The sign-parameter σ_i is negative in case the collision starts before the perihelion and is positive otherwise. The case $\sigma_i = -1$ is depicted in Fig. 6.2.

The dynamical problem now consists of the calculation of the polar coordinates r and θ as function of time under influence of the mutual Coulomb repulsion. This problem will be considered in the next sections. The remaining part of this section is concerned with the back transformation to the laboratory system. From the polar coordinates $r(t)$ and $\theta(t)$ and their derivatives $\dot{r}(t)$ and $\dot{\theta}(t)$ at time t one can retrieve the relative position vector $\mathbf{r}(t)$ and relative velocity vector $\mathbf{v}(t)$ in the laboratory system by using

$$\begin{aligned} \mathbf{r}(t) &= a(t)\hat{a} + b(t)\hat{b} \\ \mathbf{v}(t) &= \dot{a}(t)\hat{a} + \dot{b}(t)\hat{b}, \end{aligned} \qquad (6.3.6)$$

where the vectors $\hat{a}$ and $\hat{b}$ are given by Eq. (6.3.2) and

$$\begin{aligned} a(t) &= r(t)\cos\theta(t) \\ b(t) &= r(t)\sin\theta(t) \\ \dot{a}(t) &= \dot{r}(t)\cos\theta(t) - r(t)\dot{\theta}(t)\sin\theta(t) \\ \dot{b}(t) &= \dot{r}(t)\sin\theta(t) + r(t)\dot{\theta}(t)\cos\theta(t). \end{aligned} \qquad (6.3.7)$$

The coordinates of the individual particles follow through the center of mass motion by means of Eq. (6.2.12).

The displacement from the unperturbed trajectory at time t follows from Eq. (6.3.6) by substracting the unperturbed coordinates given by Eq. (6.3.3)

$$\begin{aligned} \Delta\mathbf{r}(t) &= \Delta a(t)\hat{a} + \Delta b(t)\hat{b} \\ \Delta\mathbf{v}(t) &= \Delta\dot{a}(t)\hat{a} + \Delta\dot{b}(t)\hat{b}, \end{aligned} \qquad (6.3.8)$$

in which

$$\begin{aligned} \Delta a(t) &= a(t) - (a_0 - \tilde{v}_0 t), & \Delta b(t) &= b(t) - b \\ \Delta\dot{a}(t) &= \dot{a}(t) + \tilde{v}_0, & \Delta\dot{b}(t) &= \dot{b}(t). \end{aligned} \qquad (6.3.9)$$

The virtual spatial displacement $\Delta \mathbf{r}_r$ in the reference plane follows using

$$\Delta \mathbf{r}_r(t) = \Delta \mathbf{r}(t) - (t - t_r)\Delta \mathbf{v}(t), \quad t_r = (S_i - S_c)T. \quad (6.3.10)$$

See Eqs. (6.2.3) and (3.2.1). Finally, one can determine the displacements of the individual particles by means of

$$\Delta \mathbf{r}_c(t) = \frac{1}{2}\Delta \mathbf{r}(t), \quad \Delta \mathbf{v}_c(t) = \frac{1}{2}\Delta \mathbf{v}(t)$$

$$\Delta \mathbf{r}_t(t) = -\frac{1}{2}\Delta \mathbf{r}(t), \quad \Delta \mathbf{v}_t(t) = -\frac{1}{2}\Delta \mathbf{v}(t), \quad (6.3.11)$$

as follows from Eq. (6.2.12), using that the center of mass motion is identical for the perturbed and the unperturbed trajectories.

6.4. Dynamics of a complete collision

In terms of the polar coordinates r and θ, Eqs. (6.2.14) and (6.2.17) become

$$E = \frac{1}{4}m\left(\dot{r}^2 + r^2\dot{\theta}^2\right) + \frac{C_0}{r} \quad (6.4.1)$$

$$J = \frac{1}{2}mr^2\dot{\theta}. \quad (6.4.2)$$

The quantities E and J are constants of motion. Their values are specified by the Eqs. (6.2.15) and (6.2.19). The relative motion in the orbital plane has four degrees of freedom. Accordingly, it is fully determined by the values of E, J, and the initial polar coordinates r_i and θ_i. The initial polar coordinates are specified by the Eq. (6.3.5).

The Eqs. (6.4.1) and (6.4.2) can be simplified by introducing the microscaled variables

$$\rho = r/d_s, \quad \tau = t/t_s, \quad (6.4.3)$$

where the microscale distance d_s and the microscale time t_s are defined as:

$$d_s = J/\sqrt{mE}, \quad t_s = J/2E. \quad (6.4.4)$$

Employing this scaling, Eqs. (6.4.1) and (6.4.2) transform to

$$1 = \left(\frac{d\rho}{d\tau}\right)^2 + \rho^2\left(\frac{d\theta}{d\tau}\right)^2 + \frac{2}{\rho\sqrt{q}} \quad (6.4.5)$$

$$1 = \rho^2\left(\frac{d\theta}{d\tau}\right), \quad (6.4.6)$$

where q is defined as

$$q = \frac{4J^2 E}{C_0^2 m}. \quad (6.4.7)$$

Elimination of the differential $d\tau$ from Eqs. (6.4.5) and (6.4.6) yields

$$d\theta = \frac{d\rho}{\rho(\rho^2 - 2\rho/\sqrt{\bar{q}} - 1)^{1/2}}, \qquad (6.4.8)$$

which is a nonlinear first order differential equation in ρ and θ. Integration yields

$$\rho = \frac{\sqrt{\bar{q}}}{\varepsilon \cos(\theta - \theta_p) - 1}, \qquad \varepsilon = (1 + \bar{q})^{1/2}, \qquad (6.4.9)$$

The parameter ε is known as the eccentricity. The integration constant θ_p corresponds to the perihelion angle. By removing the scaling in Eq. (6.4.9) by means of Eqs. (6.4.3) and (6.4.4), one finds

$$r = \frac{r_p(\varepsilon - 1)}{\varepsilon \cos(\theta - \theta_p) - 1}, \qquad (6.4.10)$$

where r_p is the perihelion distance

$$r_p = r(\theta_p) = \frac{J}{\sqrt{mE}}\sqrt{\frac{\varepsilon + 1}{\varepsilon - 1}}. \qquad (6.4.11)$$

Eq. (6.4.10) describes the relative trajectory of two identical charged particles interacting through the Coulomb force. It is the same as the trajectory equation of a single particle with reduced mass $m/2$ moving under influence of a fixed force center. The trajectory Eq. (6.4.10) represents a hyperbolic orbit, as is depicted in Fig. 6.3. The orbit is symmetrical with respect to the line through the origin and the perihelion, called apse line.

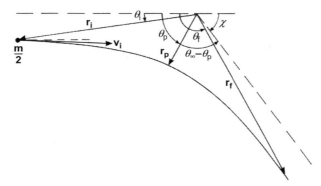

Fig. 6.3 The trajectory of a particle with reduced mass $m/2$, moving under influence of a fixed force center. The polar coordinates with the subscripts i, p and f refer to the initial, the perihelion and the final situation respectively. The initial and final directional asymptotes correspond to the polar angles $\theta_{-\infty}(= 0)$ and θ_{∞} respectively.

The asymptotes of the orbit can be found from Eq. (6.4.10) by taking $r \to \infty$. The polar angles of the initial and final asymptote are denoted as $\theta_{-\infty}$ and θ_{∞} respectively. By definition $\theta_{-\infty} = 0$, while θ_{∞} follows from

$$1/\varepsilon = \cos(\theta_{\infty} - \theta_p) = \cos(\theta_{\infty}/2), \qquad (6.4.12)$$

where the symmetry of the trajectory with respect to the apse line was used. Accordingly, one finds

$$\cos(\theta_{\infty}) = 2\cos^2(\theta_{\infty}/2) - 1 = \frac{1-q}{1+q}$$

$$\sin(\theta_{\infty}) = [1 - \cos^2(\theta_{\infty})]^{1/2} = \frac{2\sqrt{q}}{1+q}, \qquad (6.4.13)$$

while the deflection angle χ can be expressed as ($\chi = \pi - \theta_{\infty}$)

$$\tan\left(\frac{\chi}{2}\right) = \tan\left(\frac{\pi - \theta_{\infty}}{2}\right) = \left(\frac{\cos^2(\theta_{\infty}/2)}{1 - \cos^2(\theta_{\infty}/2)}\right)^{1/2} = \frac{1}{\sqrt{q}}. \qquad (6.4.14)$$

By substitution of Eqs. (6.2.15) and (6.2.19) into Eq. (6.4.7) one obtains for the parameter q

$$q = \left(\frac{mbv_i^2}{2C_0}\right)^2 \left(1 + \frac{4C_0}{mv_i^2 r_i}\right). \qquad (6.4.15)$$

For the asymptotic condition $r_i \to \infty$ this results in

$$q = q_c = \left(\frac{mbv_i^2}{2C_0}\right)^2. \qquad (6.4.16)$$

Substitution into Eq. (6.4.14) yields the well-known expression for the deflection angle of a complete collision

$$\tan\left(\frac{\chi}{2}\right) = \frac{b_\perp}{b}, \quad b_\perp = \frac{2C_0}{mv_i^2}. \qquad (6.4.17)$$

The initial and final state of a complete collision approach the asymptotic conditions. Accordingly, the initial and final energy are both entirely kinetic, and conservative of energy implies that the magnitude of the relative velocity before and after the collision is the same. Thus the dynamical effect of the encounter is a directional change of the relative velocity that is fully specified by the deflection angle χ.

Eq. (6.4.17) expresses the deflection angle χ as function of the impact parameter b and the initial relative velocity v_i. As an alternative one often describes the effect of two-particle collisions in terms of a differential cross section $\sigma(\chi, v_i)$, which is the ratio of the flux of (test) particles scattered into a solid angle $2\pi \sin(\chi)\, d\chi$ and the incoming flux per unit area; see for instance Gryzinski (1964a,b,c). From Eq. (6.4.17) one obtains for the corresponding differential cross section

$$\sigma(\chi, v_i) = \frac{2\pi b\; db}{2\pi\, \sin(\chi)\, d\chi} = \left(\frac{C_0}{mv_i^2}\right)^2 \frac{1}{\sin^4(\chi/2)}, \qquad (6.4.18)$$

which is known as Rutherfords scattering law. We will not use the concept of the differential cross section in the further analysis. Instead we will express the displacements of the test particle directly in terms of the geometrical parameters.

6.5. General analysis of the two-particle dynamical problem

Let us now consider the general problem of a collision starting with an arbitrary (not necessary large) separation r_i and relative velocity v_i, which takes place over some finite time T. Neither the initial nor the final situation necessarily approaches the asymptotic state, and the collision is in general incomplete. Accordingly, it is not sufficient to determine the asymptotic deflection angle χ only. In fact, one needs to know each of the polar coordinates r and θ, as well as their derivatives $\dot{r}$ and $\dot{\theta}$, explicitly as function of time t.

By eliminating $d\theta/d\tau$ from Eqs. (6.4.5) and (6.4.6) one obtains

$$\frac{d\rho}{d\tau} = \frac{H(\rho)}{\rho} \operatorname{sign}(\tau - \tau_p), \qquad (6.5.1)$$

with τ_p the (scaled) time of perihelion passage. The function $H(\rho)$ is defined as:

$$H(\rho) = \left(\rho^2 - \frac{2\rho}{\sqrt{q}} - 1\right)^{1/2}. \qquad (6.5.2)$$

Integration of Eq. (6.5.1) yields

$$|\tau - \tau_p| = T(\rho) = H(\rho) + \frac{1}{\sqrt{q}} \ln\left[\sqrt{\frac{q}{q+1}}\left(H(\rho) + \rho - \frac{1}{\sqrt{q}}\right)\right]. \qquad (6.5.3)$$

This equation specifies the time τ relative to the time of perihelion passage τ_p as function of the scaled separation of the particles ρ. However, actually one requires ρ as function of time. Therefore, one has to evaluate the inverse function of $T(\rho)$, denoted as $\rho(|\tau - \tau_p|)$. Unfortunately, this inversion cannot be performed analytically without introducing approximations. This problem will be discussed in detail in the next sections. For the moment it is assumed that one can indeed evaluate the function

$\rho(|\tau - \tau_p|)$, either by numerical inversion or by using some analytical approximation.

The calculation now proceeds as follows: The initial value ρ_i, obtained from Eqs. (6.3.5) and (6.4.3), specifies the initial (scaled) time τ_i relative to the time of perihelion passage τ_p,

$$\tau_i - \tau_p = \sigma_i T(\rho_i), \tag{6.5.4}$$

in which the sign parameter σ_i is defined by the last equation of (6.3.5). The final time τ_f relative to τ_p follows by adding the total scaled time of flight τ_T

$$\tau_f - \tau_p = \tau_i - \tau_p + \tau_T = \sigma_i T(\rho_i) + \tau_T, \qquad \tau_T = T/t_s \tag{6.5.5}$$

The final value ρ_f is now determined by evaluating $\rho(|\tau_f - \tau_p|)$, for which one requires the analysis of the next sections. In order to take care of the signs we use the auxiliary sign parameters σ_i and σ_f, defined as

$$\begin{aligned}\sigma_i &= \text{sign}(-r_i \cdot \hat{a}) = \text{sign}(\tilde{v}_0 t_i - a_0) \\ \sigma_f &= \text{sign}(\sigma_i T(\rho_i) + \tau_T).\end{aligned} \tag{6.5.6}$$

(For convenience we repeated the definition of σ_i, presented in Eq. (6.3.5)). The following cases are now distinguished:

$\sigma_i = -1$, $\sigma_f = +1$: Collision starts before the perihelion and ends after it ($\tau_i - \tau_p < 0$ and $\tau_f - \tau_p > 0$).

$\sigma_i = -1$, $\sigma_f = -1$: Collision starts before the perihelion and ends before it it ($\tau_i - \tau_p < 0$ and $\tau_f - \tau_p < 0$).

$\sigma_i = +1$, $\sigma_f = +1$: Collision starts after the perihelion and ends after it ($\tau_i - \tau_p > 0$ and $\tau_f - \tau_p > 0$).

The final value ρ_f is in all cases given by

$$\rho_f = \rho\left[\sigma_f(\sigma_i T(\rho_i) + \tau_T)\right] = \rho\left[|\sigma_i T(\rho_i) + \tau_T|\right]. \tag{6.5.7}$$

The corresponding polar angle θ_f follows by means of the ray Eq. (6.4.9) and can be expressed as

$$\theta_f = \theta_i - \sigma_i A(\rho_i) + \sigma_f A\left(\rho_f\right), \tag{6.5.8}$$

where the function $A(x)$ is defined as

$$A(x) = \arccos\left[(1 + \sqrt{\bar{q}}/x)/\varepsilon\right]. \tag{6.5.9}$$

The remaining quantities to be determined are the final values of the derivatives of the polar coordinates ρ and θ with respect to time τ. By means of Eq. (6.5.1) one obtains

$$\left(\frac{d\rho}{d\tau}\right)_f = \frac{\sigma_f H(\rho_f)}{\rho_f}, \qquad (6.5.10)$$

in which the function $H(\rho)$ is given by Eq. (6.5.2). The derivative of the polar angle follows from the conservation of angular momentum, expressed by Eq. (6.4.6),

$$\left(\frac{d\theta}{d\tau}\right)_f = \frac{1}{\rho_f^2}. \qquad (6.5.11)$$

Finally, the unscaled polar coordinates at t_f are retrieved as

$$r_f = d_s \rho_f, \quad \dot{r}_f = \frac{d_s}{t_s}\left(\frac{d\rho}{d\tau}\right)_f, \quad \dot{\theta}_f = \frac{1}{t_s}\left(\frac{d\theta}{d\tau}\right)_f, \qquad (6.5.12)$$

in which the scale parameters d_s and t_s are defined by Eq. (6.4.4). Notice that the (unscaled) angle θ_f is directly given by Eq. (6.5.8). This completes the calculation of the dynamics in the orbital plane.

Next, the final polar coordinates in the orbital plane r_f, $\dot{r}_f$, θ_f and $\dot{\theta}_f$ have to be transformed to coordinates in the laboratory system. This is done by performing the following steps. The polar coordinates are transformed to the rectangular coordinates of the $(\hat{a}, \hat{b})$ system in the orbital plane by means of Eq. (6.3.7). The final relative position $\mathbf{r}_f$ and velocity $\mathbf{v}_f$ in terms of the coordinates of the laboratory system now follow with Eq. (6.3.6). The final coordinates of the individual particles can then be determined through the center of mass motion, using Eqs. (6.2.11) and (6.2.12). The complete calculation scheme is summarized by the flow diagram of Fig. 6.4.

The calculation outlined above can be carried out analytically apart from one step, namely the evaluation of the polar coordinate r as function of time, which is implicitly given by Eq. (6.5.7). We will consider three alternative methods to solve this problem. In the first method the inversion is performed numerically. Section 6.6 is concerned with an algorithm that serves this purpose. In the second method, the inverted function is approximated by expanding Eq. (6.5.3) in powers of ρ. This approach is valid when both ρ_i and ρ_f are large compared to unity, which is the case for complete or nearly complete collisions. This approximation is studied in Section 6.7. The third method is based on first order perturbation theory, which constitutes a valid approximation for weak interactions. It is described in Section 6.8. Special care has to be taken to deal with the case $v_i = 0$, which means that the particles are initially at rest in the center of mass system. This problem will be considered separately in Section 6.9.

Two-particle dynamics 147

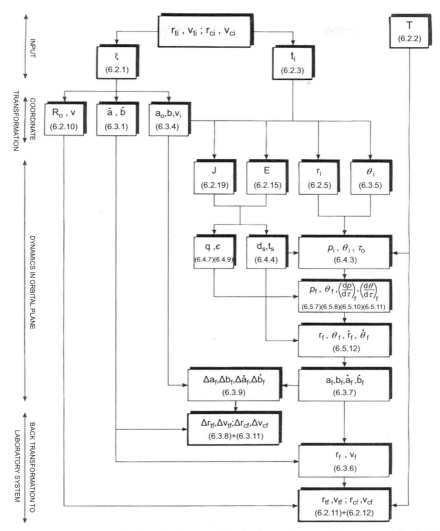

Fig. 6.4 Flow diagram for the calculation of the final coordinates of the test and the field particle from their initial coordinates and the time of flight T. The equation numbers in the boxes refer to the steps in the calculation that yield the quantity (or quantities) indicated in the same box.

6.6. Numerical approach to the dynamical problem

Given the value of time $T = |\tau - \tau_p|$ one has to determine the value of ρ which fulfills (reproducing Eq. (6.5.3))

$$T = T(\rho) = H(\rho) + \frac{1}{\sqrt{q}} \ln\left[\sqrt{\frac{q}{q+1}}\left(H(\rho) + \rho - \frac{1}{\sqrt{q}}\right)\right], \qquad (6.6.1)$$

in which $H(\rho)$ is given by Eq. (6.5.2). For $T=0$ the solution is $\rho=\rho_p$, where

$$\rho_p = \frac{1}{\sqrt{q}}\left[1 + (1+q)^{1/2}\right], \qquad (6.6.2)$$

the scaled perihelion distance. One easily verifies that $T(\rho_p) = 0$ from the fact that $H(\rho_p) = 0$, while the argument of the logarithm in Eq. (6.6.1) becomes equal to 1. For $\rho > \rho_p$ the value of $T(\rho)$ increases monotonically with ρ.

As a first approximation we ignore the logarithm in Eq. (6.6.1) and solve the eq. $T = H(\rho_1)$. This gives

$$\rho_1 = \frac{1}{\sqrt{q}}\left(1 + \left[1 + q(1+T^2)\right]^{1/2}\right). \qquad (6.6.3)$$

It provides the exact solution for $T=0$ but overestimates ρ for $T > 0$. Starting with $\rho = \rho_1$ one can determine the exact solution of Eq. (6.6.1) numerically, using the iteration process

$$\rho_{k+1} = \rho_k + \frac{T - T(\rho_k)}{[dT/d\rho]_{\rho_k}}, \qquad (6.6.4)$$

which is known as the Newton-Raphson method; for instance, see Pipes and Harvill (1983). Using $dT/d\rho = \rho/H(\rho)$ one obtains

$$\rho_{k+1} = \rho_k + \frac{H(\rho_k)}{\rho_k}[T - T(\rho_k)]. \qquad (6.6.5)$$

This process is convergent for any start value $\rho > \rho_p$. A good start value is provided by Eq. (6.6.3).

6.7. Dynamics of a nearly complete collision

In this section, we will determine the inverse function of $T(\rho)$, denoted as $\rho(T)$, by expanding Eq. (6.5.3) into powers of ρ. Using the series representations

$$\begin{aligned}(1+\delta)^{1/2} &= 1 + \frac{1}{2}\delta - \frac{1}{8}\delta^2 + \frac{1}{16}\delta^3 + O(\delta^4) \\ \ln(1+\delta) &= \delta - \frac{1}{2}\delta^2 + O(\delta^3)\end{aligned} \qquad (6.7.1)$$

(for $\delta < 1$) one obtains with Eq. (6.5.2)

$$H(\rho) = \rho - \frac{1}{\sqrt{q}} - \frac{q+1}{2q}\rho^{-1} - \frac{q+1}{2q^{3/2}}\rho^{-2} + O(\rho^{-3})$$

$$\ln\left[\sqrt{\frac{q}{q+1}}\left(H(\rho) + \rho - \frac{1}{\sqrt{q}}\right)\right] = \ln\left[\sqrt{\frac{q}{q+1}}2\rho\right] - \frac{1}{\sqrt{q}}\rho^{-1} \qquad (6.7.2)$$

$$- \frac{q+3}{4q}\rho^{-2} + O(\rho^{-3}).$$

Accordingly, Eq. (6.5.3) can be written as

$$T(\rho) = \rho + \frac{1}{\sqrt{q}}\left[\ln\left(\sqrt{\frac{q}{q+1}}2\rho\right) - 1\right] - \frac{q+3}{2q}\rho^{-1} - \frac{3q+5}{4q^{3/2}}\rho^{-2} + O(\rho^{-3}). \tag{6.7.3}$$

The successive terms will decrease in magnitude provided that $\rho \gg 1$ and that q is positive (nonzero). The inverse function of $T(\rho)$ can now be approximated as

$$\rho(T) = T - \frac{1}{\sqrt{q}}\left[1 - \frac{1}{\sqrt{q}}T^{-1}\right]\left[\ln\left(\sqrt{\frac{q}{q+1}}2T\right) - 1\right]$$
$$+ \frac{q+3}{2q}T^{-1} + O\left(\frac{\ln T}{T^2}\right). \tag{6.7.4}$$

Fig. 6.5 compares the exact relation between ρ and T, given by Eq. (6.5.3) with the approximation of Eq. (6.7.4). It also shows the zero-order approximation obtained by ignoring the terms in Eq. (6.7.4) proportional to T^{-1}. The figure shows that Eq. (6.7.4) yields accurate results for $\rho/\rho_p - 1 \gtrsim 0.5$, thus $\rho \gtrsim 1.5\rho_p$.

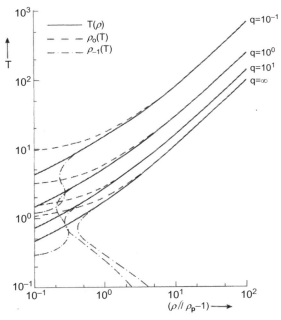

Fig. 6.5 The function $T(\rho)$, defined by Eq. (6.5.3) (and Eq. (6.6.1)), and the approximations for its inverse function $\rho(T)$ that are used for nearly complete collisions ($T \gg 1$). The approximating functions $\rho_0(T)$ and $\rho_{-1}(T)$ are accurate up to and including terms of the order T^0 and T^{-1} respectively and are obtained from Eq. (6.7.4).

The calculation that was outlined in Section 6.5 can now be performed analytically. We have done so for the case $\sigma_i = -1$ and $\sigma_f = +1$ (interaction starts before the perihelion and ends after it), as is described in some detail in Appendix 6.A. In general terms the calculation proceeds as follows: The final scaled polar coordinates ρ_f and θ_f and their derivatives $[d\rho/d\tau]_f$ and $[d\theta/\tau]_f$ are determined as functions of the initial scaled polar coordinates ρ_i and θ_i and the scaled time of flight τ_T. By substitution of these expressions into Eq. (6.3.7) and removing the scaling defined by Eqs. (6.4.3) and (6.4.4), one obtains the final values for the coordinates a and b and their derivatives $\dot{a}$ and $\dot{b}$. The deviations from the unperturbed trajectories, represented by Δa, Δb, $\Delta \dot{a}$ and $\Delta \dot{b}$, then follow with Eq. (6.3.9). This way one eventually finds (see Appendix 6.A)

$$\Delta a_f \simeq \frac{2(Tv_i - r_i)}{1+q_c} - \frac{1-q_c}{1+q_c}\left[2h\Lambda_c - h\frac{Tv_i}{r_i}\right] + \frac{2\sqrt{q_c}}{1+q_c}\left[b - \frac{2b}{1+q_c}\left(\frac{Tv_i}{r_i} - 1\right)\right]$$

$$\Delta b_f \simeq \frac{2\sqrt{q_c}}{1+q_c}\left[Tv_i - r_i - 2h\Lambda_c + \frac{2h}{1+q_c}\left(\frac{Tv_i}{r_i} - 1\right)\right]$$

$$\Delta \dot{a}_f \simeq \frac{2v_i}{1+q_c} + \frac{1-q_c}{1+q_c}\left[\frac{hv_i}{r_i} - \frac{hv_i}{Tv_i - r_i}\right] - \frac{4q_c}{(1+q_c)^2}\frac{hv_i}{r_i}$$

$$\Delta \dot{b}_f \simeq \frac{2\sqrt{q_c}}{1+q_c}\left[v_i + \frac{hv_i}{r_i} - \frac{2}{1+q_c}\frac{hv_i}{Tv_i - r_i}\right],$$

(6.7.5)

in which Λ_c and h are defined as

$$\Lambda_c = \ln\left(\sqrt{\frac{q_c}{q_c+1}}\frac{2}{b}\sqrt{r_i(Tv_i - r_i)}\right) - 1, \quad h = \frac{2C_0}{mv_i^2} \qquad (6.7.6)$$

and q_c is given by Eq. (6.4.16) (note: $\sqrt{q_c} = b/h$). The predictions of Eq. (6.7.5) were compared with the exact results obtained by numerical calculation, using the method outlined in Section 6.6. It was found that Eq. (6.7.5) indeed yield correct results provided that

$$r_i \gg b \quad \text{and} \quad Tv_i - r_i \gg b \quad \text{and} \quad h/r_i \ll 1 \qquad (6.7.7)$$

The collisions for which these conditions are fulfilled are called nearly complete collisions.

The Eqs. (6.7.5) are valid for arbitrary values of q_c ($q_c \geq 0$) and thus apply to both strong and weak collisions (as long as they are nearly complete). The case $q_c = 0$ corresponds to a central collision ($b = 0$), while weak interactions are found for $\sqrt{q_c} \gg 1$, as can be seen from Eq. (6.4.14). Let us investigate both cases. For $\sqrt{q_c} \gg 1$ Eq. (6.7.5) transform to

$$\Delta a_f \cong h\left[\ln\left(\frac{4r_i(Tv_i-r_i)}{b^2}\right)-\frac{Tv_i}{r_i}\right]$$

$$\Delta b_f \cong \frac{2h(Tv_i-r_i)}{b}\left\{1-\frac{2h}{Tv_i-r_i}\left[\ln\left(\frac{4r_i(Tv_i-r_i)}{b^2}\right)-1\right]\right\} \cong \frac{2h(Tv_i-r_i)}{b}$$

$$\Delta \dot{a}_f \cong hv_i\left(\frac{1}{Tv_i-r_i}-\frac{1}{r_i}\right)$$

$$\Delta \dot{b}_f \cong \frac{2hv_i}{b}\left(1-\frac{h}{Tv_i-r_i}\right)\cong \frac{2hv_i}{b}.$$

(6.7.8)

These equations apply to weak nearly complete collisions.

For a central collision ($q_c=0$) one obtains from Eq. (6.7.5)

$$\Delta a_f \cong 2(Tv_i-r_i)-h\left[\ln\left(\frac{4r_i(Tv_i-r_i)}{h^2}\right)-2-\frac{Tv_i}{r_i}\right]$$

$$\Delta \dot{a}_f \cong 2v_i + hv_i\left(\frac{1}{r_i}-\frac{1}{Tv_i-r_i}\right)$$ (6.7.9)

$$\Delta b_f = \Delta \dot{b}_f = 0.$$

The motion is one-dimensional and takes place in the $\hat{a}$-direction. Notice that the expression for $\Delta \dot{a}_f$ also follows directly from conservation of energy.

The coordinates Δa_f, Δb_f, $\Delta \dot{a}_f$ and $\Delta \dot{b}_f$ refer to the condition at time t_f, which is the time that the field particle leaves the beam segment. In order to determine the virtual spatial displacement in the reference plane one has to extrapolate the final coordinates over the time interval $T_1 = t_f - t_r$, as is expressed by Eq. (6.3.10). One can perform this extrapolation for the coordinates in the orbital plane, using

$$\Delta a_r = \Delta a_f - T_1 \Delta \dot{a}_f, \quad \Delta b_r = \Delta b_f - T_1 \Delta \dot{b}_f \quad (6.7.10)$$

Substitution of Eq. (6.7.5) yields

$$\Delta a_r \cong 2\frac{(T-T_1)v_i-r_i}{1+q_c}-h\frac{1-q_c}{1+q_c}\left[2\Lambda_c-\frac{(T-T_1)v_i}{r_i}-\frac{v_iT_1}{Tv_i-r_i}\right]$$

$$+\frac{2q_ch}{1+q_c}\left[1-\frac{2}{1+q_c}\left(\frac{(T-T_1)v_i}{r_i}-1\right)\right]$$

$$\Delta b_r \cong \frac{2\sqrt{q_c}}{1+q_c}\left[(T-T_1)v_i-r_i-2h\Lambda_c+\frac{2h}{1+q_c}\left(\frac{(T-T_1)v_i}{r_i}-1\right)+h\frac{T_1v_i}{Tv_i-r_i}\right]$$

(6.7.11)

These equations specify the virtual spatial displacement in an arbitrary reference plane (defined by T_1), which is required for the calculation of the trajectory displacement effect.

The Eqs. (6.7.5) and (6.7.11) constitute the main results of this section. It should be emphasized that they pertain to a nearly complete collision, which implies that the conditions expressed by Eq. (6.7.7) should be fulfilled. We will conclude this section by bringing the results in an alternative form, which is sometimes more convenient. Eqs. (6.7.5) and (6.7.11) contain the initial distance r_i, which is related to the geometrical parameters through the first equation of (6.3.5). In order to make the dependency on the geometrical parameters explicit we now approximate r_i and its reciprocal value as

$$r_i = a_0 - v_i t_i + \frac{b^2}{2(a_0 - v_i t_i)} + \cdots \approx a_0 - v_i t_i$$

$$\frac{1}{r_i} = \frac{-1}{v_i t_i} + \frac{a_0}{(v_i t_i)^2} + \cdots \approx \frac{-1}{v_i t_i}.$$

(6.7.12)

These approximations rely on the conditions expressed by Eq. (6.7.7). The approximation used in the second equation of (6.7.12) requires in addition that $a_0 \ll v_i t_i$, which implies that the x, y-plane of the laboratory system should be located somewhere near the middle of the beam segment (thus $S_c \approx 1/2$). The quantities t_i and $T_1 = t_f - t_r$ can be expressed in terms of the parameters S_c and S_i and the time of flight T; see Eqs. (6.2.3) and (6.3.10). Furthermore, v_i is equal to the geometrical parameter v. Accordingly, Δa_r and Δb_r, given by Eq. (6.7.11), and the derivatives $\Delta \dot{a}_f$ and $\Delta \dot{b}_f$, given by Eq. (6.7.5), can be expressed as

$$\Delta a_r \cong 2\frac{(S_i - S_c)Tv - a_0}{1 + q_c} - h\frac{1 - q_c}{1 + q_c}\left[2\Lambda_c - \frac{S_i}{S_c} - \frac{1 - S_i}{1 - S_c}\right]$$

$$+ h\frac{2q_c}{1 + q_c}\left[1 - \frac{2}{1 + q_c}\left(\frac{S_i}{S_c} - 1\right)\right]$$

$$\Delta b_r \cong \frac{2\sqrt{q_c}}{1 + q_c}\left[(S_i - S_c)Tv - a_0 - 2h\Lambda_c\right.$$

$$\left. + \frac{2h}{1 + q_c}\left(\frac{S_i}{S_c} - 1\right) + h\frac{1 - S_i}{1 - S_c}\right]$$

(6.7.13)

$$\Delta \dot{a}_f \cong \frac{2v}{1 + q_c} + \frac{1 - q_c}{1 + q_c}\frac{h}{T}\left(\frac{1}{S_c} - \frac{1}{1 - S_c}\right) - \frac{4q_c}{(1 + q_c)^2}\frac{h}{S_c T}$$

$$\Delta \dot{b}_f \cong \frac{2\sqrt{q_c}}{1 + q_c}\left[v + \frac{h}{T}\left(\frac{2}{S_c(1 + q_c)} - \frac{1}{1 - S_c}\right)\right],$$

where Λ_c, which is defined by Eq. (6.7.6), is now expressed as

$$\Lambda_c = \ln\left(\sqrt{\frac{q_c}{q_c+1}\frac{2vT}{b}}\sqrt{S_c(1-S_c)}\right) - 1. \tag{6.7.14}$$

The Eqs. (6.7.13) specify the deviation from the unperturbed trajectories as functions of the geometrical parameters $\boldsymbol{\xi}=(r_\perp, \varphi, b_z, v, \psi)$, the time of flight T, and the parameters S_c and S_i. The quantities a_0, q_c and h are determined by $\boldsymbol{\xi}$ through Eqs. (6.3.4), (6.4.16) and (6.7.6) respectively. Note that the expressions for Δa_r and Δb_r diverge logarithmically (through Λ_c) for $T\to\infty$, while the derivatives $\Delta\dot{a}_f$ and $\Delta\dot{b}_f$, are finite for $T\to\infty$. The values of $\Delta\dot{a}_f$ and $\Delta\dot{b}_f$ for $T\to\infty$ are directly related to the value of the asymptotic polar angle θ_∞, as can be seen from Eq. (6.4.13).

6.8. First order perturbation dynamics

Another method to solve the two-particle dynamical problem by analytical means is given by the so-called first order perturbation approximation, which is valid for weak collisions. This approach was discussed previously in Section 5.3, where we considered the N-particle dynamical problem from a general perspective. A two-particle collision is called weak if either the eccentricity ε is large or the perihelion is already passed at the start of the collision or the perihelion is not reached during the flight. In terms of the parameters defined in the previous sections these conditions can be expressed as

$$\varepsilon = \sqrt{q+1} \gg 1$$

or
$$\sigma_i = +1 \quad \text{and} \quad r_i \gg b \tag{6.8.1}$$

or
$$\sigma_f = -1 \quad \text{and} \quad r_i - Tv_i \gg b.$$

The deflection from the unperturbed trajectory will be small in each of these cases. Accordingly, one may approximate the real force acting on the particles by the force that would act when the particles would follow their unperturbed trajectories. The force that acts on the test particle is given by

$$\tilde{\mathbf{F}}(t) = \frac{C_0(v_i t - a_0)}{\tilde{r}(t)^3}\hat{a} - \frac{C_0 b}{\tilde{r}(t)^3}\hat{b}, \tag{6.8.2}$$

in which the unperturbed distance between the particles $\tilde{r}(t)$ is given by Eq. (6.3.3). In first order perturbation approximation one computes the displacements in position and velocity experienced by the test particle as

$$\Delta \mathbf{v}_{tf} = \int_{t_i}^{t_f} dt \frac{\tilde{\mathbf{F}}(t)}{m}$$

$$\Delta \mathbf{r}_{tf} = \int_{t_i}^{t_f} ds \int_{t_i}^{s} dt \frac{\tilde{\mathbf{F}}(t)}{m} = \int_{t_i}^{t_f} dt (t_f - t) \frac{\tilde{\mathbf{F}}(t)}{m}. \qquad (6.8.3)$$

The right-hand side of the equation for $\Delta \mathbf{r}$ was obtained by changing the order of the integration. Integration of Eq. (6.8.3) yields

$$\Delta \mathbf{v}_{tf} = \frac{C_0}{mv_i} \left\{ \left[\frac{1}{r_i} - \frac{1}{r_f} \right] \hat{a} - \left[\frac{a_i}{br_i} + \frac{Tv_i - a_i}{br_f} \right] \hat{b} \right\}$$

$$\Delta \mathbf{r}_{tf} = \frac{C_0}{mv_i^2} \left\{ \left[\sigma_i \ln\left(\frac{|a| + r_i}{b} \right) - \sigma_f \ln\left(\frac{|Tv_i - a_i| + r_f}{b} \right) + \frac{Tv_i}{r_i} \right] \hat{a} \right. \qquad (6.8.4)$$

$$\left. + \left[\frac{b^2 - a_i(Tv_i - a_i)}{br_i} - \frac{r_f}{b} \right] \hat{b} \right\},$$

in which

$$a_i = a_0 - v_i t_i$$

$$\sigma_i = \mathrm{sign}(-a_i), \qquad \sigma_f = \mathrm{sign}(Tv_i - a_i) \qquad (6.8.5)$$

$$r_i = \left[a_i^2 + b^2 \right]^{1/2}, \quad r_f = \left[(Tv_i - a_i)^2 + b^2 \right]^{1/2},$$

similar to Eqs. (6.3.4), (6.3.5) and (6.5.6).

Utilizing the Eqs. (6.3.8) and (6.3.11) one can express the results of Eqs. (6.8.4) in terms of the coordinates of the $(\hat{a}, \hat{b})$ reference system.

$$\Delta a_f \cong -h \left[\sigma_i \ln\left(\frac{|a_i| + r_i}{b} \right) - \sigma_f \ln\left(\frac{|Tv_i - a_i| + r_f}{b} \right) + \frac{Tv_i}{r_i} \right]$$

$$\Delta b_f \cong -\frac{h}{b} \left(\frac{b^2 - a_i(Tv_i - a_i)}{r_i} - r_f \right)$$

$$\Delta \dot{a}_f \cong -hv_i \left(\frac{1}{r_i} - \frac{1}{r_f} \right) \qquad (6.8.6)$$

$$\Delta \dot{b}_f \cong \frac{hv_i}{b} \left(\frac{a_i}{r_i} + \frac{Tv_i - a_i}{r_f} \right),$$

in which the length h is given by Eq. (6.7.6). Eqs. (6.8.6), which apply to weak interactions, should be compared to Eqs. (6.7.5), which are valid for nearly complete collisions. Both sets of equations should yield the same results for collisions that are both weak and nearly complete. According to Eqs. (6.7.7) and (6.8.5) a nearly complete collision implies that

$$\sigma_i = -1, \quad \sigma_f = +1, \quad a_i \approx r_i, \quad Tv_i - a_i \approx r_f. \qquad (6.8.7)$$

Substitution in Eq. (6.8.6) yields

$$\Delta a_f \cong h\left[\ln\left(\frac{4r_i r_f}{b^2}\right) - \frac{Tv_i}{r_i}\right]$$

$$\Delta b_f \cong \frac{2hr_f}{b}\left(1 - \frac{b^2}{2r_i r_f}\right) \cong \frac{2hr_f}{b}$$

$$\Delta \dot{a}_f \cong hv_i\left(\frac{1}{r_f} - \frac{1}{r_i}\right)$$

$$\Delta \dot{b} \cong \frac{2hv_i}{b},$$

(6.8.8)

which is indeed identical to Eq. (6.7.8) provided that one may identify r_f with $Tv_i - r_i$. For a weak nearly complete collision both quantities are in fact related by

$$r_f = Tv_i - r_i + \frac{1}{2}b^2\left(\frac{1}{r_f} - \frac{1}{r_i}\right) + \cdots \approx Tv_i - r_i, \qquad (6.8.9)$$

as follows by eliminating a_i from the expanded expressions for r_i and r_f given by Eq. (6.8.5). We have just demonstrated that there are two routes to obtain Eq. (6.8.8). In Section 6.7 we started by considering a nearly complete collision, which fulfills Eq. (6.7.7), and took the limit $\sqrt{q_c} \to \infty$ next, which refers to weak interactions. In this section, we interchanged this order, thus starting by assuming that the interaction is weak and considering the limit of complete collisions next.

A weak central collision corresponds to $b = 0$ and $\sigma_i = \sigma_f$ (the perihelion is not passed during the collision). Accordingly, one finds from Eqs. (6.8.6) and (6.8.5) for this case

$$\Delta a_f \cong -h\left[\sigma_i \ln\left(\frac{r_i}{r_f}\right) + \frac{Tv_i}{r_i}\right], \quad \Delta \dot{a}_f \cong -hv_i\left(\frac{1}{r_i} - \frac{1}{r_f}\right)$$

$$\Delta b_f = \Delta \dot{b}_f = 0, \quad r_i = |a_i|, \quad r_f = |Tv_i - a_i|.$$

(6.8.10)

The reader should notice that Eqs. (6.7.9) and (6.8.10) both apply to a central collision. However, the former refers to a nearly complete collision in which the perihelion is passed ($r_i \gg h$, $r_f \gg h$, $\sigma_i = -1$, $\sigma_f = +1$), while the latter refers to a weak incomplete collision in which the perihelion is not reached ($r_i \approx r_f$, $\sigma_i = \sigma_f$).

The virtual displacements in the reference plane follow again by means of Eq. (6.7.10). Substitution of Eq. (6.8.6) gives

$$\Delta a_r \cong -h\left[\sigma_i \ln\left(\frac{|a_i|+r_i}{b}\right) - \sigma_f \ln\left(\frac{|Tv_i - a_i|+r_f}{b}\right) + \frac{(T-T_1)v_i}{r_i} + \frac{T_1 v_i}{r_f}\right]$$

$$\Delta b_r \cong -\frac{h}{b}\left[r_i - r_f - \frac{(T-T_1)v_i a_i}{r_i} + \frac{(v_i T - a_i)T_1 v_i}{r_f}\right],$$

(6.8.11)

which should be compared with Eq. (6.7.11).

The Eqs. (6.8.6) and (6.8.11) constitute the main results of this section. They are analogous to Eqs. (6.7.5) and (6.7.11), derived for a nearly complete collision. One would like to bring Eqs. (6.8.6) and (6.8.11) in the same form as Eq. (6.7.13). However, for a weak collision one may not approximate r_i by means of Eq. (6.7.12), since b is not necessarily small compared to $Tv_i - a_0$. Thus, one has to use the full expression for both r_i and r_f given by Eq. (6.8.5). One may replace the quantities t_i and $T_1 = t_f - t_r$ in Eqs. (6.8.5), (6.8.6) and (6.8.11) by the parameters S_i and S_c and the time of flight T, using Eqs. (6.2.3) and (6.3.10). The Eq. (6.8.5) transform to (denoting v_i as v)

$$a_i = S_c v T + a_0$$
$$\sigma_i = \text{sign}[-S_c Tv - a_0], \qquad \sigma_f = = \text{sign}[(1 - S_c)Tv - a_0] \qquad (6.8.12)$$
$$r_i = \left[(S_c Tv + a_0)^2 + b^2\right]^{1/2}, \quad r_f = = \left\{((1 - S_c)Tv - a_0)^2 + b^2\right\}^{1/2},$$

where a_0 and the impact parameter b are given by Eq. (6.3.4). Eqs. (6.8.6) and (6.8.11) transform accordingly. As the modifications are minor, we leave this to the reader.

We finish this section with a note regarding the fundamental aspects of the first order perturbation approximation that was utilized here. Formally speaking, the first order perturbation method corresponds to an expansion of Eq. (6.5.3) in the parameter $1/\sqrt{q}$. Utilizing Eq. (6.7.1) one can express Eq. (6.5.3) for $(1/\sqrt{q}) \to 0$ as

$$T(\rho) = (\rho^2 - 1)^{1/2} + \frac{1}{\sqrt{q}} \left[\ln(\rho^2 - 1)^{1/2} - \frac{1}{2} - \rho + \frac{\rho}{(\rho^2 - 1)^{1/2}} - \frac{\rho^2}{2(\rho^2 - 1)} \right]$$
$$+ \frac{1}{q} \left[\frac{2\rho}{\rho - 1} \left(1 - \frac{(\rho^2 - 1)^{1/2}}{\rho}\right) - \frac{\rho^2}{2(\rho^2 - 1)^{1/2}} \right] + O\left(q^{-3/2}\right).$$
$$(6.8.13)$$

The first term corresponds to the unperturbed trajectory, as can be verified by removing the scaling (notice that for $1/\sqrt{q} = 0$ one may approximate $d_s = b$ and $v_s = v_i$). In Fig. 6.5 the solid line for $q = \infty$ depicts the relation between ρ and T given by this term. In first order perturbation approximation one includes the term in Eq. (6.8.13), which is linear in $1/\sqrt{q}$, and ignores higher order terms. This approach constitutes an alternative for the derivation of Eq. (6.8.6) outlined above, which was, however, preferred for its simplicity.

Two-particle dynamics   157

6.9. Collisions with zero initial relative velocity

In this section, we will consider the case $v_i = 0$. This means that the particles are initially at rest in the center of mass system. During the time of flight T part of their potential energy will be converted into kinetic energy. The generated relative velocity will be directed along the line joining the two particles. The particle motion is thus limited to a line rather than to a plane. Consequently, the representation in the $(\hat{a}, \hat{b})$ reference system, defined by Eq. (6.3.1), is inapplicable and the analysis of the previous sections should be modified.

From Eq. (6.2.1) it follows that the set of relevant geometrical variables reduces to (taking $v = 0$)

$$\xi = (r_\perp, \varphi, b_z) \tag{6.9.1}$$

See also Fig. 6.1. As the relative motion of the particles takes place along the line joining their centers, it is sufficient to calculate the relative position r and its derivative $\dot{r}$. Taking $\dot{\theta} = 0$ for all time t Eq. (6.4.1) becomes

$$E = \frac{1}{4}m\dot{r}^2 + \frac{C_0}{r}. \tag{6.9.2}$$

The energy E in the center of mass is conserved during the interaction and is thus identical to its initial value

$$E = C_0/r_i, \quad r_i = \left(b_z^2 + r_\perp^2\right)^{1/2}, \tag{6.9.3}$$

analogous to Eqs. (6.2.15) and (6.3.5). To simplify the equations we introduce the microscaling quantities

$$d_s = \frac{C_0}{E}(= r_i), \quad v_s = 2\sqrt{\frac{E}{m'}}, \quad t_s = \frac{m^{1/2}C_0}{2E^{3/2}}, \tag{6.9.4}$$

which will be used instead of the quantities defined by Eq. (6.4.4). One sees that d_s is equal to the initial distance r_i while v_s is the relative velocity for $r \to \infty$. Employing this scaling one can rewrite Eq. (6.9.2) as

$$1 = \left(\frac{d\rho}{d\tau}\right)^2 + \frac{1}{\rho}. \tag{6.9.5}$$

Integration yields

$$|\tau - \tau_i| = T(\rho) = \sqrt{\rho^2 - \rho} + \frac{1}{2}\ln\left[2\sqrt{\rho^2 - \rho} + 2\rho - 1\right], \tag{6.9.6}$$

which should be compared with Eq. (6.5.3).

Clearly, we face the same problem as before. The function $T(\rho)$ defined by Eq. (6.9.6) cannot be inverted analytically without introducing approximations. We will investigate the same approaches to this problem as described in Sections 6.6–6.8, namely numerical inversion, power expansion for large ρ (which is justified for nearly complete collisions), and first order perturbation theory (which is justified for weak collisions).

The numerical approach described in Section 6.6 is also applicable to the present problem. The iteration process of Eq. (6.6.5) can again be used, where the function $H(\rho)$ now represents

$$H(\rho) = \sqrt{\rho^2 - \rho}. \tag{6.9.7}$$

A suitable start value of the iteration process is given by $\rho_1 = 1$.

A collision is called half-complete if the final scaled distance $\rho_f \gg 1$, which means $r_f \gg r_i$. For this condition one may expand Eq. (6.9.6) in a power series in ρ. Using Eq. (6.7.1) one finds

$$T(\rho) = \rho + \frac{1}{2}[\ln(4\rho) - 1] - \frac{3}{8}\frac{1}{\rho} - \frac{5}{32}\frac{1}{\rho^2} + O(\rho^{-3}). \tag{6.9.8}$$

Accordingly, the inverted function can be approximated as

$$\rho(T) = T - \frac{1}{2}\left(1 - \frac{1}{2T}\right)[\ln(4T) - 1] + \frac{3}{8}\frac{1}{T} + O\left(\frac{\ln T}{T^2}\right) \tag{6.9.9}$$

Eqs. (6.9.8) and (6.9.9) are analogous to Eqs. (6.7.3) and (6.7.4) respectively. Eq. (6.9.9) gives the final scaled distance ρ as function of the scaled interaction time T. In addition, one requires the derivative of ρ with respect to time. Substitution of Eq. (6.9.9) into Eq. (6.9.5) yields

$$\left(\frac{d\rho}{d\tau}\right) = 1 - \frac{1}{2T}\left[1 + \frac{1}{2T}\left(\ln(4T) - \frac{1}{2}\right)\right] + O(T^{-3}). \tag{6.9.10}$$

Eqs. (6.9.9) and (6.9.10) provide a complete specification of the final situation as function of the scaled time T. The next step is to remove the scaling using Eqs. (6.4.3) and (6.9.4),

$$r_f \simeq \frac{2AT}{r_i^{1/2}} - \frac{r_i}{2}\left(1 - \frac{r_i^{3/2}}{4AT}\right)\left[\ln\left(\frac{8AT}{r_i^{3/2}}\right) - 1\right] + \frac{3}{16}\frac{r_i^{3/2}}{AT}$$

$$v_f \simeq \frac{2A}{r_i^{1/2}} - \frac{r_i}{2T}\left\{1 + \frac{r_i^{3/2}}{4AT}\left[\ln\left(\frac{8AT}{r_i^{3/2}}\right) - \frac{1}{2}\right]\right\}, \tag{6.9.11}$$

in which $A = (C_0/m)^{1/2}$ and T the time of flight given by Eq. (6.2.2). As the initial relative velocity v is zero the change in relative velocity is simply given by

$$\Delta v_f = v_f. \tag{6.9.12}$$

The virtual displacement in the reference plane follows from

$$\Delta r_r = r_f - r_i - T_1 v_f, \tag{6.9.13}$$

with $T_1 = t_f - t_r = (1 - S_i)T$; see Eqs. (6.3.10) and (6.2.3). Substitution of Eq. (6.9.11) into Eq. (6.9.13) yields

$$\Delta r_r = S_i \frac{2AT}{r_1^{1/2}} - \frac{r_i}{2}\left[S_i + \ln\left(\frac{8AT}{r_i^{3/2}}\right)\right] + \frac{r_i^{5/2}}{8AT}\left[(2-S_i)\ln\left(\frac{8AT}{r_i^{3/2}}\right) + \frac{S_i}{2}\right] \quad (6.9.14)$$

The velocity and spatial displacement of the test particle follow from

$$\Delta v_{\perp t} \cong v_f \frac{r_\perp}{2r_i}, \quad \Delta v_{zt} \cong v_f \frac{b_z}{2r_i}, \quad \Delta r_{\perp t} \cong \Delta r_r \frac{r_\perp}{2r_i}, \quad (6.9.15)$$

We emphasize that Eqs. (6.9.11) and (6.9.14) apply to nearly half-complete collisions, which implies $T \gg 1$. Using unscaled quantities this condition can be expressed as

$$T \gg r_i^{3/2}/2A, \quad (6.9.16)$$

as follows with the last equation of (6.9.4).

Finally, we will employ perturbation theory to determine the velocity shift of the test particle for the case $v=0$. In this approach one assumes that the relative distance r changes very little during the interaction, thus $r_f \gg r_i$. In terms of the scaled coordinates used above this means $\rho \approx 1$. Power expansion of Eq. (6.9.6) about $\rho=1$ gives, using Eq. (6.7.1),

$$T(\rho) = 2(\rho - 1)^{1/2} + \frac{1}{3}(\rho - 1)^{3/2} + O\left[(\rho - 1)^2\right], \quad (6.9.17)$$

analogous to Eq. (6.8.13), derived for the general case $v \neq 0$ (note, however, that in the general case ρ can be large compared to 1). By inversion one obtains

$$\rho(T) = 1 + \frac{1}{4}T^2 - \frac{1}{48}T^4 + O(T^5). \quad (6.9.18)$$

The derivative of ρ with respect to T follows with Eq. (6.9.5),

$$\left(\frac{d\rho}{dT}\right) = \frac{1}{2}T - \frac{1}{12}T^3 + O(T^4). \quad (6.9.19)$$

By removing the scaling in Eqs. (6.9.18) and (6.9.19), using Eqs. (6.4.3) and (6.9.4), one finds

$$r_f \cong r_i + \frac{A^2 T^2}{r_i^2} - \frac{1}{3}\frac{A^4 T^4}{r_i^5}$$
$$v_f \cong 2\frac{A^2 T}{r_i^2} - \frac{4}{3}\frac{A^4 T^3}{r_i^5}, \quad (6.9.20)$$

where $A=(C_0/m)^{1/2}$. The virtual displacement in the reference plane follows with Eq. (6.9.13)

$$\Delta r_r = (2S_i - 1)\frac{A^2 T^2}{r_i^2} - \left(\frac{4S_i}{3} - 1\right)\frac{A^4 T^4}{r_i^5}. \quad (6.9.21)$$

First order perturbation dynamics is achieved by truncating the expansions in Eqs. (6.9.20) and (6.9.21) after the terms proportional to (A^2/r_i^2). This quantity corresponds to the acceleration of the individual particles at $t = t_i$. The spatial displacement $\Delta r_{\perp t}$ and the velocity displacements $\Delta v_{\perp t}$ and Δv_{zt} experienced by the test particle can be determined by means of Eq. (6.9.15).

6.10. Expressions for the longitudinal velocity shift Δv_z

In this section, we will present some explicit expressions for the axial velocity shift of the test particle Δv_z in terms of the geometrical parameters. We will employ the results obtained in Sections 6.7–6.9. Expressions for the transverse velocity shift $\Delta v_\perp$ and the spatial shift Δr will be presented in Sections 6.11 and 6.12 respectively.

By utilizing Eqs. (6.3.8) and (6.3.11) one can express the transformation of the displacements of the test particle from coordinates in the $(\hat{a}, \hat{b})$ reference system to coordinates in the laboratory system as

$$\begin{aligned}\Delta \mathbf{r}_{tr} &= -\left(\Delta a_r \hat{a} + \Delta b_r \hat{b}\right)/2 \\ \Delta \mathbf{v}_{tf} &= -\left(\Delta \dot{a}_f \hat{a} + \Delta \dot{b}_f \hat{b}\right)/2,\end{aligned} \quad (6.10.1)$$

in which $\hat{a}$ and $\hat{b}$ are given by Eq. (6.3.2). $\Delta \mathbf{r}_{tr}$ is the virtual lateral displacement in the reference plane, and $\Delta \mathbf{v}_{tf}$ is the final velocity displacement. The latter is the relevant dynamical parameter for the Boersch effect, while the former is relevant for the trajectory displacement effect. For convenience of notation we will from now on omit the indices t (for test particle), r (for reference plane), and f (for final). The reader should be alert that all results apply to the coordinates of the test particle at $t=t_f$, while the spatial displacements are extrapolated to the reference plane.

Eqs. (6.3.2) express the vectors $\hat{a}$ and $\hat{b}$ in terms of the coordinates of the laboratory system. The vector $\hat{a}$ has no z-component due to the fact that the particles have the same axial velocity. Consequently, the velocity shift Δv_z is entirely determined by the quantity $\Delta \dot{b}_f$ only

$$\Delta v_z = -\Delta \dot{b}_f \, b_z/2b, \quad (6.10.2)$$

as follows from Eqs. (6.3.2) and (6.10.1). For a nearly complete collision one finds with Eqs. (6.7.13), (6.4.16), and (6.7.6)

$$\Delta v_z \simeq \frac{mv^3 b_z}{2C_0(1+q_c)}\left[1 + \frac{2C_0}{mv^3 T}\left(\frac{2}{S_c(1+q_c)} - \frac{1}{1-S_c}\right)\right], \quad (6.10.3)$$

where q_c is given by Eq. (6.4.16). A complete collision corresponds to $T \to \infty$. In this limit Eq. (6.10.3) becomes

$$\Delta v_z \simeq \frac{mv^3 b_z}{2C_0(1+q_c)} \qquad (6.10.4)$$

independent of T and S_c.

The effect of a weak collision can be determined from Eqs. (6.10.2), (6.8.6), (6.8.12) and (6.7.6)

$$\Delta v_z \simeq \frac{C_0 b_z}{mvb^2}\left[\frac{S_c Tv + a_0}{\left[(S_c Tv + a_0)^2 + b^2\right]^{1/2}} + \frac{(1-S_c)Tv - a_0}{\left\{((1-S_c)Tv - a_0)^2 + b^2\right\}^{1/2}}\right], \qquad (6.10.5)$$

where a_0 and b are given by Eq. (6.3.4). For a collision that is both weak and complete one obtains

$$\Delta v_z \simeq \frac{2C_0 b_z}{mvb^2}, \qquad (6.10.6)$$

as follows from Eq. (6.10.4), taking $q_c \gg 1$, as well as from Eq. (6.10.5), using $T \to \infty$.

The case $v = 0$ requires a separate treatment, as was discussed in Section 6.9. For a half-complete collision one obtains from Eqs. (6.9.11), (6.9.15) and (6.9.3)

$$\Delta v_z \simeq \left(\frac{C_0}{m}\right)^{1/2} \frac{b_z}{\left(b_z^2 + r_\perp^2\right)^{3/4}} - \frac{b_z}{4T} \quad (v=0), \qquad (6.10.7)$$

in which all terms of Eq. (6.9.11) were included up to and including those of the order T^{-1}. In the limit $T \to \infty$ one may ignore the second term in Eq. (6.10.7). The remaining term expresses that the initial potential energy is completely converted into kinetic energy.

For a weak collision only a small fraction of the initial potential energy will be converted into kinetic energy. From Eqs. (6.9.20), (6.9.15) and (6.9.3) one obtains for this case

$$\Delta v_z \simeq \frac{C_0}{m} \frac{b_z T}{\left(b_z^2 + r_\perp^2\right)^{3/2}} \quad (v=0), \qquad (6.10.8)$$

in which only the most significant term of Eq. (6.9.20) was included. Eq. (6.10.8) presupposes that the force acting on the test particle is approximately constant during the time of flight T (first order perturbation theory). Eqs. (6.10.7) and (6.10.8) are relevant for a monochromatic homocentric cylindrical beam sections in which the particles are initially at rest in the frame of reference moving with the beam.

Most theories on the Boersch effect that are based on two-particle dynamics utilize Eq. (6.10.4) or Eq. (6.10.6). Loeffler (1969) started from Eq. (6.10.5) but used Eq. (6.10.6) in the actual calculation. (We note that Loeffler's eq. (4) follows with $\Delta E = mv_z \Delta v_z$, $\varphi = \Phi - \pi$, $\alpha L_1 = S_c T v$, and $\alpha L_2 = (1 - S_c) T v$). All expressions yield $\Delta v_z = 0$ for $b_z = 0$. This can be understood from the fact that the orbital plane is then oriented perpendicular to the beam axis; see Fig. 6.1. Consequently, the displacements have no component in the axial direction.

6.11. Expressions for the transverse velocity shift $\Delta v_\perp$

Using Eqs. (6.10.1) and (6.3.2) one can express the transverse velocity shift $\Delta v_\perp$ as

$$\Delta v_\perp = \left(A_v^2 + B_v^2 \frac{[r_\perp \sin(\Phi)]^2}{b^2} \right)^{1/2}, \quad A_v = |\Delta \dot{a}_f/2|, \quad B_v = |\Delta \dot{b}_f/2|. \tag{6.11.1}$$

In case $r_\perp = 0$ one simply finds $\Delta v_\perp = A_v$ and $\Delta v_\perp$ is entirely determined by $\Delta \dot{a}_f$ only. This can be understood from the fact that the orbital plane is, for $r_\perp = 0$, directed parallel to the beam axis; see Fig. 6.1.

For a nearly complete collision one obtains by substitution of Eq. (6.7.13) into Eq. (6.11.1), using Eq. (6.7.6)

$$\Delta v_\perp \cong \frac{v}{1 + q_c} \left\{ \left[1 + \frac{2C_0}{mv^3 T} \left[\frac{1 - q_c}{2} \left(\frac{1}{S_c} - \frac{1}{1 - S_c} \right) - \frac{2q_c}{1 + q_c} \frac{1}{S_c} \right] \right]^2 \right.$$

$$\left. + q_c \left(\frac{r_\perp \sin(\Phi)}{b} \right)^2 \left[1 + \frac{2C_0}{mv^3 T} \left(\frac{2}{S_c(1 + q_c)} - \frac{1}{1 - S_c} \right) \right]^2 \right\}^{1/2}, \tag{6.11.2}$$

with q_c given by Eq. (6.4.16). A complete collision corresponds to $T \to \infty$. In this limit Eq. (6.11.2) becomes

$$\Delta v_\perp \cong \frac{v}{1 + q_c} \left[1 + q_c \left(\frac{r_\perp \sin(\Phi)}{b} \right)^2 \right]^{1/2} \tag{6.11.3}$$

independent of T and S_c.

For a weak collision one finds from Eqs. (6.11.1), (6.8.6), (6.8.12), and (6.7.6)

$$\Delta v_\perp \cong \frac{C_0}{mvb} \left[\left(\frac{b}{r_i} - \frac{b}{r_f}\right)^2 + \left(\frac{r_\perp \sin(\Phi)}{b}\right)^2 \left(\frac{S_c T v + a_0}{r_i} + \frac{(1-S_c)Tv - a_0}{r_f}\right)^2 \right]^{1/2},$$

(6.11.4)

with $r_i = [(S_c T v + a_0)^2 + b^2]^{1/2}$, $r_f = \{[(1-S_c)Tv - a_0]^2 + b^2\}^{1/2}$, and a_0 and b given by Eq. (6.3.4). Note that $\Delta v_\perp = 0$ in case $S_c = 1/2$ and $r_\perp = 0$. Thus weak collisions in a point crossover ($r_c = 0$) that is located in the middle of a beam segment ($S_c = 1/2$) do not result in angular deflections of the test particle. This effect is due to the symmetry of the trajectories with respect to the crossover. The angular displacement experienced by the test particle in the first half of the beam segment is exactly canceled during its flight through the second half of the beam segment. For a collision that is both weak and complete one obtains

$$\Delta v_\perp \cong \frac{2C_0 r_\perp \sin(\Phi)}{mvb^2}, \quad (6.11.5)$$

as follows from Eq. (6.11.3) taking $q_c \gg 1$, as well as from Eq. (6.11.4), taking $T \to \infty$.

The case $v = 0$ requires a separate treatment. For a half-complete collision it follows from Eqs. (6.9.11), (6.9.15), and (6.9.3)

$$\Delta v_\perp \cong \left(\frac{C_0}{m}\right)^{1/2} \frac{r_\perp}{\left(b_z^2 + r_\perp^2\right)^{3/4}} - \frac{r_\perp}{4T} \quad (v = 0) \quad (6.11.6)$$

in which all terms of Eq. (6.9.11) were included up to and including those of the order T^{-1}. For a weak collision one obtains from Eqs. (6.9.20), (6.9.15) and (6.9.3)

$$\Delta v_\perp \cong \frac{C_0}{m} \frac{r_\perp T}{\left(b_z^2 + r_\perp^2\right)^{3/2}} \quad (v = 0) \quad (6.11.7)$$

in which the most significant term of Eq. (6.9.20) was included only.

6.12. Expressions for the spatial shift Δr

Using Eqs. (6.10.1) and (6.3.2) one can express the virtual radial displacement of the test particle in the reference plane as

$$\Delta r = \left(A_r^2 + B_r^2 \frac{[r_\perp \sin(\Phi)]^2}{b^2}\right)^{1/2}, \quad A_r = |\Delta a_r/2|, \; B_r = |\Delta b_r/2| \quad (6.12.1)$$

similar to Eq. (6.11.1) referring to the transverse velocity shift $\Delta v_\perp$. In ease $r_\perp = 0$ one finds $\Delta r = A_r$, and Δr is entirely determined by Δa_r only.

For a nearly complete collision the quantities A_r and B_r follow directly from Eqs. (6.7.13), (6.4.16), and (6.7.6)

$$A_r \cong \left| \frac{(S_i - S_c)Tv - a_0}{1 + q_c} - \frac{b}{2\sqrt{q_c}} \frac{1 - q_c}{1 + q_c} \left[2\Lambda_c - \frac{S_i}{S_c} - \frac{1 - S_i}{1 - S_c} \right] \right.$$
$$\left. + \frac{b\sqrt{q_c}}{1 + q_c} \left[1 - \frac{2}{1 + q_c} \left(\frac{S_i}{S_c} - 1 \right) \right] \right|$$

$$B_r \cong \left| \frac{\sqrt{q_c}}{1 + q_c} [(S_i - S_c)Tv - a_0] \right.$$
$$\left. - \frac{2b}{1 + q_c} \left[\Lambda_c - \frac{1}{1 + q_c} \left(\frac{S_i}{S_c} - 1 \right) - \frac{1}{2} \frac{1 - S_i}{1 - S_c} \right] \right|,$$

(6.12.2)

where q_c is given by Eq. (6.4.16), Λ_c by Eq. (6.7.14), and a_0 and b by Eq. (6.3.4). It should be noted that the assumption that the collision is nearly complete implies that $S_c \approx 1/2$; see Eq. (6.7.7). Eq. (6.12.2) simplify considerably for the case $S_c = S_i$, which implies that the reference plane is identical to the x,y-plane of the laboratory system (which coincides with the crossover plane)

$$A_r \cong \left| \frac{a_0}{1 + q_c} + \frac{b}{\sqrt{q_c}} \frac{1 - q_c}{1 + q_c} [\Lambda_c - 1] - \frac{b\sqrt{q_c}}{1 + q_c} \right|$$
$$B_r \cong \left| \frac{a_0 \sqrt{q_c}}{1 + q_c} + \frac{2b}{1 + q_c} \left[\Lambda_c - \frac{1}{2} \right] \right|$$

$(S_c = S_i)$ (6.12.3)

A further reduction is achieved for the case of a point crossover ($r_c = 0$ thus $r_\perp = 0$). From Eqs. (6.12.1), (6.12.3) and (6.7.14) one obtains for $r_\perp = 0$.

$$\Delta r \cong \left| \frac{b_z}{\sqrt{q_c}} \left\{ \frac{1 - q_c}{1 + q_c} \left[\ln \left(\sqrt{\frac{q_c}{q_c + 1}} \frac{2vT}{b_z} \sqrt{S_c(1 - S_c)} \right) - 2 \right] - \frac{q_c}{1 + q_c} \right\} \right|$$

$(r_\perp = 0)$, (6.12.4)

previously published by van Leeuwen and Jansen (1983) (their Eq. (4.9) follows by substituting $S_c = 1/2$).

For a weak collision the quantities A_r and B_r follow from Eqs. (6.8.11), (6.8.12) and (6.7.6)

$$A_r \cong \frac{C_0}{mv^2} \left| \sigma_i \ln \left(\frac{|S_c vT + a_0| + r_i}{b} \right) - \sigma_f \ln \left(\frac{|(1 - S_c)vT - a_0| + r_f}{b} \right) \right.$$
$$\left. + vT \left(\frac{S_i}{r_i} + \frac{1 - S_i}{r_f} \right) \right|$$

(6.12.5)

$$B_r \cong \frac{C_0}{mv^2 b} \left| r_i - r_f - vT \left[\frac{S_i(S_c vT + a_0)}{r_i} - \frac{(1 - S_i)[(1 - S_c)vT - a_0]}{r_f} \right] \right|,$$

where r_i, r_f, σ_i and σ_f are given by Eq. (6.8.12) and a_0 and b by Eq. (6.3.4). For the case of a point crossover ($r_c = 0$ thus $r_\perp = 0$) one finds from Eqs. (6.12.1), (6.12.5) and (6.8.12)

$$\Delta r \cong \left| \frac{C_0}{mv^2} \ln\left(\frac{[S_c vT + r_i][(1-S_c)vT + r_f]}{b_z^2}\right) - vT\left(\frac{S_i}{r_i} + \frac{1-S_i}{r_f}\right) \right| (r_\perp = 0), \tag{6.12.6}$$

with $r_i = [(S_c vT)^2 + b_z^2]^{1/2}$ and $r_f = \{[(1-S_c)vT]^2 + b_z^2\}^{1/2}$. Eq. (6.12.6) becomes independent of S_i for $S_c = 1/2$, which implies $r_i = r_f$. This is a consequence of the fact that the transverse velocity of the test particle is not changed ($\Delta v_\perp = 0$), as can be verified from Eq. (6.11.4). Its final transverse velocity is identical to its initial transverse velocity, and the resulting effect is a radial shift only. Accordingly, the virtual displacement of the test particle Δr is independent of the choice of the reference plane. It should be emphasized that this phenomenon occurs only for a point crossover located in the middle of the beam segment; thus $S_c = 1/2$ and $r_c = 0$ (implying $r_\perp = 0$).

For a collision that is both weak and nearly complete it can be proven that

$$\Delta r \cong \frac{2C_0}{mv^2} \left\{ \left[\ln\left(\frac{2vT\sqrt{S_c(1-S_c)}}{b}\right) - \frac{1}{2}\left(\frac{S_i}{S_c} + \frac{1-S_i}{1-S_c}\right) \right]^2 + \left[\frac{[vT(S_i - S_c) + r_\perp \cos(\Phi)r_\perp \sin(\Phi)]^2}{b^2}\right] \right\}^{1/2}. \tag{6.12.7}$$

The result follows from Eqs. (6.12.1) and (6.12.2), taking $\sqrt{q_c} \to \infty$, as well as from Eqs. (6.12.1) and (6.12.5), taking $r_i = S_c vT + a_0$ and $r_f = (1-S_c)vT - a_0$ with $a_0 = -r_\perp \cos(\Phi)$.

The case $v = 0$ requires a separate treatment as was discussed in Section 6.9. For a half-complete collision one obtains from Eqs. (6.9.14), (6.9.15) and (6.9.3)

$$\Delta r \cong \left| \frac{(C_0/m)^{1/2} r_\perp TS_i}{\left(b_z^2 + r_\perp^2\right)^{3/4}} - \frac{r_\perp}{4}\left[S_i + \ln\left(\frac{8(C_0/m)^{1/2}T}{\left(b_z^2 + r_\perp^2\right)^{3/4}}\right)\right] \right| \quad (v = 0) \tag{6.12.8}$$

in which the two most significant terms of Eq. (6.9.14) were included only. For a weak collision one finds from Eqs. (6.9.21), (6.9.15) and (6.9.3)

$$\Delta r \cong \left| \frac{C_0}{m} \frac{r_\perp T^2(S_i - 1/2)}{\left(b_z^2 - r_\perp^2\right)^{3/2}} \right| \quad (v = 0) \tag{6.12.9}$$

in which the most significant term of Eq. (6.9.20) was included only. Eq. (6.12.9) shows that a weak collision has (in first order perturbation approximation) no resulting effect in case $S_i = 1/2$. Thus the virtual displacement in a reference plane located in the middle of the beam segment is zero.

6.13. Coulomb scaling

In the analytical as well as in the numerical calculations on Coulomb interactions it is often convenient to scale the equations. Scaling usually simplifies the notation and may lead to a reduction of the number of independent parameters. In the succeeding chapters we will frequently use the so-called Coulomb or δ,ν-scaling, in which length and velocity are scaled with δ and ν respectively, where

$$\delta = \left(\frac{4C_0 T^2}{m}\right)^{1/3}, \quad \nu = \left(\frac{4C_0}{mT}\right)^{1/3}. \tag{6.13.1}$$

The aim of this scaling is to establish equations that do not explicitly depend on T, C_0, or m. For an electron beam with $V = 10\,\text{kV}$ and $L = 0.1\,\text{m}$ one finds $\delta = 14.2\,\mu\text{m}$ and $\nu = 8.44 \times 10^3\,\text{m/s}$, which can be considered as typical values. Quantities scaled with δ and/or ν are indicated with an asterisk.

From Eq. (6.13.1) it can directly be seen that the quantities δ and ν fulfill the relations

$$\delta/\nu = T, \quad \delta\nu^2 = 4C_0/m \tag{6.13.2}$$

Thus, the scale-measure for time δ/ν is identical to the time of flight T. The total energy E and angular momentum J in the center of mass system, given by Eqs. (6.2.15) and (6.2.19) respectively, can now be expressed as

$$E^* = \frac{4E}{m\nu^2} = v^{*2} + 1/r_i^*$$
$$J^* = \frac{2J}{m\,\delta\nu} = b^* v^*, \tag{6.13.3}$$

where $v^* = v/\nu$, $b^* = b/\delta$ and the scaled initial separation r_i^* is given by

$$r_i^* = \frac{r_i}{\delta} = \left[(S_c v^* + a_0^*)^2 + b^{*2}\right]^{1/2}, \tag{6.13.4}$$

as follows from Eq. (6.8.12). The parameter q, defined by Eq. (6.4.7), and its equivalent q_c for a complete collision ($r_i \to \infty$) can be expressed as

$$q = 4J^{*2}E^* = q_c(1 + 1/v^{*2}r_i^*), \quad q_c = (2b^* v^{*2})^2, \tag{6.13.5}$$

which are the scaled versions of Eqs. (6.4.15) and (6.4.16).

It can straightforwardly be demonstrated that by employing the δ,ν-scaling the basic equations presented in Sections 6.2–6.4 reduce to a form which does not contain T, C_0 and m explicitly. As a consequence one may formally express the displacements experienced by the test particle as

$$\Delta\mathbf{r} = \delta \Delta\mathbf{r}^*(r_\perp^* \varphi, b_z^*, v^*, \psi, S_c, S_i)$$
$$\Delta\mathbf{v} = \nu \Delta\mathbf{v}^*(r_\perp^* \varphi, b_z^*, v^*, \psi, S_c, S_i), \qquad (6.13.6)$$

with $r_\perp^* = r_\perp/\delta$, $b_z^* = b_z/\delta$, $v^* = v/\nu$. The reader may verify that the functions $\Delta\mathbf{r}^*$ and $\Delta\mathbf{v}^*$ are indeed independent of T, C_0 and m by introducing the δ,ν-scaling in the expressions for Δv_z, $\Delta v_\perp$ and Δr presented in Sections 6.10, 6.11 and 6.12.

Appendix 6.A. Mathematics of nearly complete collision dynamics

This appendix is concerned with the analytical calculation of the dynamics of a nearly complete collision. Given the initial polar coordinates ρ_i, and θ_i and the scaled time of flight τ_T we shall compute the final scaled polar coordinates ρ_f, θ_f and their derivatives $[d\rho/d\tau]_f$ and $[d\theta/\tau]_f$. These quantities will then be used to determine the final values of the coordinates a and b and their derivatives $\dot{a}$ and $\dot{b}$. Finally, the displacements of the particles follow in terms of the quantities Δa, Δb, $\Delta \dot{a}$ and $\Delta \dot{b}$. Since the calculation pertains to a nearly complete collision it is assumed that the conditions stated by Eqs. (6.7.7) are fulfilled.

Substitution of Eqs. (6.7.3) and (6.7.4) into Eq. (6.5.7) yields

$$\rho_f = \tau_T - \rho_i - \frac{2\Lambda}{\sqrt{q}}\left(1 - \frac{1}{\sqrt{q}}\frac{1}{\tau_T - \rho_i}\right) + \frac{q+3}{2q}\left(\frac{1}{\rho_i} + \frac{1}{\tau_T - \rho_i}\right)$$
$$+ O\left(\frac{\ln \rho_i}{\rho_i^2}\right) + O\left(\frac{\ln(\tau_T - \rho_i)}{(\tau_T - \rho_i)^2}\right), \qquad (6.A.1)$$

where Λ is defined as

$$\Lambda = \ln\left(\sqrt{\frac{q}{q+1}} 2\sqrt{\rho_i(\tau_T - \rho_i)}\right) - 1. \qquad (6.A.2)$$

Eq. (6.A.1) gives the final polar coordinate ρ_f as function of its initial value ρ_i and the scaled time of flight τ_T.

The objective of the calculation is to evaluate Eq. (6.3.7) at $t = t_f$, which leads to the final coordinates a_f and b_f and their derivatives $\dot{a}_f$ and $\dot{b}_f$. In addition to the final polar distance ρ_f, one requires the final polar angle θ_f and the final values of the derivatives $[d\rho/d\tau]_f$ and $[d\theta/d\tau]_f$. Following the calculation scheme outlined in Section 6.5 (see also Fig. 6.4) we shall express these quantities in terms of power series in ρ_i, θ_i and $\tau_T - \rho_i$, similar to Eq. (6.A.1).

In order to obtain a consistent calculation one should beforehand specify to which order in ρ_i, θ_i and $\tau_T - \rho_i$ the various equations are to be expanded. We introduce the following notation to classify the order of a term containing some combination of the quantities ρ_i, θ_i and $\tau_T - \rho_i$

$$O_{-n} = O\left(\frac{\theta_i^l}{\rho_i^k (\tau_T - \rho_i)^m}\right), \quad n = k + l + m \qquad (6.A.3)$$

(notice that $\theta_i \sim 1/\rho_i$). This classification does not account for logarithmic dependencies, and a term O_{-n} may contain a factor $[\ln(\rho_i)]^r [\ln(\tau_T - \rho_i)]^s$, with r and s some finite positive number.

Our aim is to determine the coordinates a_f and b_f up to and including terms of O_{-1}. Accordingly, the expression for the polar distance ρ_f in Eq. (6.A.1) was developed to this order. Since the highest order term occurring in the expansion of ρ_f is of O_1, one should compute $\sin(\theta_f)$ and $\cos(\theta_f)$ up to and including terms of O_{-2}. This guarantees that all terms of O_{-1} will be present in a_f and b_f, as can be seen from Eq. (6.3.7). The derivatives $\dot{a}_f$ and $\dot{b}_f$ have to be expanded up to and including terms of O_{-2}, because they are multiplied with terms of O_1 in the extrapolation of the final coordinates to virtual coordinates in the reference plane. As a consequence one should determine $[d\rho/d\tau]_f$ up to and including terms of O_{-2} and $[d\theta/d\tau]_f$ up to and including terms O_{-3}; see again Eq. (6.3.7).

The final polar angle θ_f is specified by Eqs. (6.5.8) and (6.5.9). By using the series expansion

$$\arccos(x + \delta) = \arccos(x) - \frac{\delta}{(1-x^2)^{1/2}} - \frac{\delta^2 x}{2(1-x^2)^{3/2}} + O(\delta^3)$$

(for $\delta < 1$) one obtains

$$\theta_f = \theta_i + \arccos\left(\frac{1-q}{1+q}\right) - \left(\frac{1}{\rho_i} + \frac{1}{\rho_f}\right) - \frac{1}{2\sqrt{q}}\left(\frac{1}{\rho_i^2} + \frac{1}{\rho_f^2}\right) + O_{-3}, \qquad (6.A.4)$$

using that $2\arccos(1/\varepsilon) = \arccos[(1+q)/(1-q)]$, as follows from the definition of the eccentricity ε, given by the second equation of (6.4.9), and the dynamical result of Eq (6.4.13). Instead of the polar angle θ_f itself, one rather needs the related functions $\sin(\theta_f)$ and $\cos(\theta_f)$. By using the relations

$$\cos\left[\arccos\left(\frac{1-q}{1+q}\right) + \delta\right] = \frac{1-q}{1+q}\cos\delta - \frac{2\sqrt{q}}{1+q}\sin\delta$$

$$\sin\left[\arccos\left(\frac{1-q}{1+q}\right) + \delta\right] = \frac{2\sqrt{q}}{1+q}\cos\delta + \frac{1-q}{1+q}\sin\delta$$

Two-particle dynamics

(as follows with $\sin\{\arccos[(1-q)/(1+q)]\} = 2\sqrt{q}/(1+q)$) and the expansions

$$\cos(\delta) = 1 - \delta^2/2 + O(\delta^4), \quad \sin(\delta) = \delta + O(\delta^3),$$

one finds from Eq. (6.A.4)

$$\cos(\theta_f) = \frac{1-q}{1+q}P - \frac{2\sqrt{q}}{1+q}Q, \quad \sin(\theta_f) = \frac{2\sqrt{q}}{1+q}P + \frac{1-q}{1+q}Q, \quad (6.A.5)$$

where P and Q stand for

$$P = 1 - \frac{1}{2}\theta_i^2 - \frac{1}{2}\left(\frac{1}{\rho_i} + \frac{1}{\rho_f}\right)^2 + \theta_i\left(\frac{1}{\rho_i} + \frac{1}{\rho_f}\right) + O_{-3}$$

$$Q = \theta_i - \left(\frac{1}{\rho_i} + \frac{1}{\rho_f}\right) - \frac{1}{2\sqrt{q}}\left(\frac{1}{\rho_i^2} + \frac{1}{\rho_f^2}\right) + O_{-3}. \quad (6.A.6)$$

Notice that Eq. (6.A.5) transform to Eq. (6.4.13) for $P=1$ and $Q=0$. These values for P and Q are indeed found from Eq. (6.A.6) in the limit $\rho_i \to \infty$, $\rho_f \to \infty$ and $\theta_i \to 0$. In order to evaluate the Eq. (6.A.6) one needs an expression for $1/\rho_f$ in terms of ρ_i and τ. From Eq. (6.A.1) it follows (using $(1+\delta)^{-1} = 1 - \delta + O(\delta^2)$)

$$\frac{1}{\rho_f} = \frac{1}{\tau_T - \rho_i} + \frac{1}{(\tau_T - \rho_i)^2}\frac{2\Lambda}{\sqrt{q}} + O_{-3}. \quad (6.A.7)$$

Substitution into the Eq. (6.A.6) yields

$$P = 1 - \frac{1}{2}\left(\frac{1}{\rho_i} + \frac{1}{\tau_T - \rho_i}\right)^2 - \frac{1}{2}\theta_i^2 + \theta_i\left(\frac{1}{\rho_i} + \frac{1}{\tau_T - \rho_i}\right) + O_{-3}$$

$$Q = \theta_i - \left(\frac{1}{\rho_i} + \frac{1}{\tau_T - \rho_i}\right) - \frac{1}{2\sqrt{q}}\left(\frac{1}{\rho_i^2} + \frac{1+4\Lambda}{(\tau_T - \rho_i)^2}\right) + O_{-3}. \quad (6.A.8)$$

According to the second equation of (6.3.5) the initial polar angle θ_i is related to ρ_i and the scaled impact parameter β as

$$\theta_i = \arcsin(\beta/\rho_i) = \beta/\rho_i + O_{-3}, \quad \beta = b/d_s. \quad (6.A.9)$$

Substitution into Eq. (6.A.8) yields

$$P = 1 - \frac{1}{2}\left(\frac{\beta-1}{\rho_i} - \frac{1}{\tau_T - \rho_i}\right)^2 + O_{-3}$$

$$Q = \frac{\beta-1}{\rho_i} - \frac{1}{\tau_T - \rho_i} - \frac{1}{2\sqrt{q}}\left(\frac{1}{\rho_i^2} + \frac{1+4\Lambda}{(\tau_T - \rho_i)^2}\right) + O_{-3} \quad (6.A.10)$$

which gives the final values $\sin\theta_f$ and $\cos\theta_f$ in terms of ρ_i, τ_T, and β, using Eq. (6.A.5).

The next step is to determine the derivatives of $[d\rho/d\tau]_f$ and $[d\theta/d\tau]_f$. Combining Eq. (6.5.10) and the first equation of (6.7.2) one obtains

$$\left(\frac{d\rho}{d\tau}\right)_f = \frac{H(\rho_f)}{\rho_f} = 1 - \frac{1}{\sqrt{q}}\rho_f^{-1} - \frac{q+1}{2q}\rho_f^{-2} + O_{-3}. \tag{6.A.11}$$

With Eq. (6.A.7) it now follows that

$$\left(\frac{d\rho}{d\tau}\right)_f = 1 - \frac{1}{(\tau_T - \rho_i)}\frac{1}{\sqrt{q}} - \frac{1}{(\tau_T - \rho_i)^2}\frac{4\Lambda + q + 1}{2q} + O_{-3}. \tag{6.A.12}$$

Substitution of Eq. (6.A.7) into Eq. (6.5.11) yields

$$\left(\frac{d\theta}{d\tau}\right)_f = \frac{1}{(\tau_T - \rho_i)^2} + \frac{1}{(\tau_T - \rho_i)^3}\frac{4\Lambda}{\sqrt{q}} + O_{-4} \tag{6.A.13}$$

The Eqs. (6.A.1), (6.A.5), (6.A.10), (6.A.12) and (6.A.13) constitute a complete set of expressions from which the final values of the scaled polar coordinates and their derivatives with respect to the scaled time τ can be determined from the initial value ρ_i, the scaled impact parameter β (or alternatively the polar angle θ_i), and the time of flight τ_T. The only additional physical parameter appearing in the expressions is the eccentricity related constant q, which specifies the strength of the interaction, as can be understood from Eq. (6.4.14). This equation indicates that large values of q correspond to weak interactions.

The second part of the calculation consists of the evaluation of the coordinates a and b, and their derivatives $\dot{a}$ and $\dot{b}$. We will denote their microscaled equivalents as α, β, $\dot{\alpha}$ and $\dot{\beta}$. By scaling Eq. (6.3.7) transform to (for $t = t_f$)

$$\alpha_f = a_f/d_s = \rho_f \cos(\theta_f)$$
$$\beta_f = b_f/d_s = \rho_f \sin(\theta_f)$$
$$\dot{\alpha}_f = \dot{a}_f/v_s = \left(\frac{d\rho}{d\tau}\right)_f \cos(\theta_f) = \rho_f \left(\frac{d\theta}{d\tau}\right)_f \sin(\theta_f) \tag{6.A.14}$$
$$\dot{\beta}_f = \dot{b}_f/v_s = \left(\frac{d\rho}{d\tau}\right)_f \sin(\theta_f) + \rho_f \left(\frac{d\theta}{d\tau}\right)_f \cos(\theta_f)$$

with $v_s = d_s/t_s$; see Eq. (6.4.4). By substitution of Eqs. (6.A.5) into Eq. (6.A.14) one obtains

$$\alpha_f = \frac{1-q}{1+q}A - \frac{2\sqrt{q}}{1+q}B, \quad \beta_f = \frac{2\sqrt{q}}{1+q}A + \frac{1-q}{1+q}B$$
$$\dot{\alpha}_f = \frac{1-q}{1+q}C - \frac{2\sqrt{q}}{1+q}D, \quad \dot{\beta}_f = \frac{2\sqrt{q}}{1+q}C + \frac{1-q}{1+q}D, \tag{6.A.15}$$

in which A, B, C and D are defined as

$$A = \rho_f P, \qquad\qquad B = \rho_f Q$$
$$C = \left(\frac{d\rho}{d\tau}\right)_f P - \rho_f \left(\frac{d\theta}{d\tau}\right)_f Q, \quad D = \left(\frac{d\rho}{d\tau}\right)_f Q + \rho_f \left(\frac{d\theta}{d\tau}\right)_f P. \qquad (6.A.16)$$

Substitution of Eqs. (6.A.1), (6.A.10), (6.A.12) and (6.A.13) yields

$$A = \tau_T - \rho_i - \frac{2\Lambda}{\sqrt{q}} + \frac{1}{\rho_i}\left(\frac{3(q+1)}{2q} - \beta\right)$$
$$+ \frac{1}{\tau_T - \rho_i}\frac{4\Lambda + 3}{2q} - \frac{\tau_T - \rho_i}{2\rho_i^2}(\beta - 1)^2 + O_{-2}$$

$$B = -1 + \frac{\beta - 1}{\rho_i}\left(\tau_T - \rho_i - \frac{2\Lambda}{\sqrt{q}}\right) - \frac{1}{2\sqrt{q}}\left(\frac{1}{\tau_T - \rho_i} + \frac{\tau_T - \rho_i}{\rho_i^2}\right) + O_{-2}.$$

$$C = 1 - \frac{1}{\tau_T - \rho_i}\frac{1}{\sqrt{q}} - \frac{(\beta - 1)^2}{2\rho_i^2} - \frac{1}{(\tau_T - \rho_i)^2}\frac{4\Lambda + 1}{q} - \frac{2(\beta - 1)}{\rho_i(\tau_T - \rho_i)} + O_{-3}$$

$$D = \frac{\beta - 1}{\rho_i} - \frac{1}{2\sqrt{q}}\left(\frac{1}{\rho_i^2} + \frac{2(\beta - 1)}{\rho_i(\tau_T - \rho_i)} - \frac{1}{(\tau_T - \rho_i)^2}\right) + O_{-3}.$$

$$(6.A.17)$$

The Eqs. (6.A.15) and (6.A.17) specify the final coordinates in the $(\hat{a}, \hat{b})$ reference system as function of the initial coordinates ρ_i and β, the time of flight τ_T and the parameter q. All coordinates are still scaled, and the next step is to remove this scaling. The scale parameters d_s and t_s are given by Eq. (6.4.4). At this point it should be noted that d_s and t_s depend on the energy E and therefore on the initial distance r_i; see Eq. (6.2.15). This is also the case for the parameter q, as can be seen from Eq. (6.4.15). These dependencies can be made explicit by using

$$d_s = \frac{J}{\sqrt{mE}} = b\left(1 + \frac{2h}{r_i}\right)^{-1/2} = b\left(1 - \frac{h}{r_i} + \frac{3}{2}\frac{h^2}{r_i^2} + O_{-3}\right)$$

$$v_s = \frac{d_s}{t_s} = \frac{2\sqrt{E}}{\sqrt{m}} = v_i\left(1 + \frac{2h}{r_i}\right)^{1/2} = v_i\left(1 + \frac{h}{r_i} - \frac{1}{2}\frac{h^2}{r_i^2} + O_{-3}\right) \qquad (6.A.18)$$

$$q = q_c\left(1 + \frac{2h}{r_i}\right),$$

in which h and q_c are defined as

$$h = \frac{2C_0}{mv_i^2}, \quad q_c = \left(\frac{b}{h}\right)^2 = \left(\frac{mbv_i^2}{2C_0}\right)^2. \qquad (6.A.19)$$

Substitution of Eqs. (6.A.18) into Eqs. (6.A.15) and (6.A.17) finally gives (after some reorganization of the resulting terms)

$$a_f = \frac{1-q_c}{1+q_c}\left[Tv_i - r_i - 2h\Lambda_c + h\frac{Tv_i}{r_i}\right]$$
$$+ \frac{2\sqrt{q_c}}{1+q_c}\left[b - \frac{2b}{1+q_c}\left(\frac{Tv_i}{r_i} - 1\right)\right] + O_{-1}$$
$$b_f = \frac{2\sqrt{q_c}}{1+q_c}\left[Tv_i - r_i - 2h\Lambda_c + \frac{2h}{1+q_c}\frac{Tv_i}{r_i}\right]$$
$$- \frac{1-q_c}{1+q_c}\left[b + \frac{2b}{1+q_c}\right] + O_{-1}$$
(6.A.20)
$$\dot{a}_f = \frac{1-q_c}{1+q_c}\left[v_i + \frac{hv_i}{r_i} - \frac{hv_i}{Tv_i - r_i}\right] - \frac{4q_c}{(1+q_c)^2}\frac{hv_i}{r_i} + O_{-2}$$
$$\dot{b}_f = \frac{2\sqrt{q_c}}{1+q_c}\left[v_i + \frac{hv_i}{r_i}\frac{2}{1+q_c} - \frac{hv_i}{Tv_i - r_i}\right] + O_{-2},$$

in which Λ_c is defined as

$$\Lambda_c = \ln\left(\sqrt{\frac{q_c}{q_c+1}}\frac{2}{b}\sqrt{r_i(Tv_i - r_i)}\right) - 1, \qquad (6.A.21)$$

similar to Eq. (6.A.2). The expansions of Eq. (6.A.20) include one order less than those of Eq. (6.A.17). We did carry out the expansion one order further, but the resulting expressions became impractically large and the additional terms are therefore omitted.

The deviations from the unperturbed coordinates follow by Eq. (6.3.9). The Equation for Δa contains the quantity $a_0 - v_i t$ (note: $v_i = \tilde{v}_0$). For $t = t_f$ one may write (using that $\sigma_i = -1$ and $\sigma_f = +1$)

$$a_0 - v_i t_f = a_0 - v_i t_i - v_i T = r_i - v_i T - \frac{b^2}{2r_i} + O_{-2}, \qquad (6.A.22)$$

as follows by expanding and inverting the expression for r_i given by Eq. (6.3.5) and using that $T = t_f - t_i$. By substitution of Eqs. (6.A.20) into Eq. (6.3.9) one obtains

$$\Delta a_f = \frac{2(Tv_i - r_i)}{1+q_c} - \frac{1-q_c}{1+q_c}\left[2h\Lambda_c - h\frac{Tv_i}{r_i}\right]$$
$$+ \frac{2\sqrt{q_c}}{1+q_c}\left[b - \frac{2b}{1+q_c}\left(\frac{Tv_i}{r_i} - 1\right)\right] + O_{-1}$$
$$\Delta b_f = \frac{2\sqrt{q_c}}{1+q_c}\left[Tv_i - r_i - 2h\Lambda_c + \frac{2h}{1+q_c}\left(\frac{Tv_i}{r_i} - 1\right)\right] + O_{-1} \qquad (6.A.23)$$
$$\Delta \dot{a}_f = \frac{2v_i}{1+q_c} + \frac{1-q_c}{1+q_c}\left[\frac{hv_i}{r_i} - \frac{hv_i}{Tv_i - r_i}\right] - \frac{4q_c}{(1+q_c)^2}\frac{hv_i}{r_i} + O_{-2}$$
$$\Delta \dot{b}_f = \frac{2\sqrt{q_c}}{1+q_c}\left[v_i + \frac{hv_i}{r_i}\frac{2}{1+q_c} - \frac{hv_i}{Tv_i - r_i}\right] + O_{-2},$$

which completes the calculation.

CHAPTER SEVEN

Boersch effect

Contents

7.1.	Introduction	173
7.2.	General aspects	174
7.3.	Beam segment with a narrow crossover	177
7.4.	Homocentric cylindrical beam segment	189
7.5.	Beam segment with a crossover of arbitrary dimensions	198
7.6.	Results for Gaussian angular and spatial distributions	213
7.7.	Thermodynamic limits	218

7.1. Introduction

In this chapter, the Boersch effect will be calculated by means of the extended two-particle approach. This approach was outlined in Chapter 5. It utilizes the two-particle dynamics of Chapter 6. The analysis presented here pertains to a rotational symmetric beam segment in drift space.

The chapter starts with a summary of the relevant results obtained so far. This material is then used to calculate the Boersch effect in two specific beam geometries, which are most suited for an analytical treatment of the problem. One is a beam segment with a narrow crossover, and the other is a homocentric cylindrical beam segment. In the former the Boersch effect is generated by relaxation of kinetic energy; in the latter it is generated by relaxation of potential energy. The general case of a beam segment with a crossover of arbitrary dimensions is considered next. The angular and spatial distribution in the crossover are uniform in most calculations. However, Gaussian distribution(s) are considered too, leading to the definition of some effective width measures for which the results found for uniform distributions do apply. For all geometries, the full energy distribution produced by the Boersch effect is computed. Explicit expressions are presented for the Full Width at Half Maximum ($FWHM$) of this distribution, as well as the Full Width median value (FW_{50}) and the root mean square value (rms). The number of independent parameters in the model is reduced by utilizing scaled (dimensionless) quantities. This also simplifies the notation of the intermediate results. In the final results, the scaling is removed in order to make the dependency on the experimental parameters explicit.

A computer program has been developed that can perform every step of the calculation numerically, following the scheme of the extended two-particle approach. First, it calculates the two-particle distribution, utilizing the analytical results for the two-particle dynamical problem. Next, it performs two steps, each corresponding to a Fourier-type of integral transform. This finally yields the many particle velocity displacement distribution, which is equivalent to the energy distribution generated by the Boersch effect. The most surprising result of these investigations is probably that one has to distinguish a variety of regimes, depending on the particle density in the beam and the beam geometry. Each regime corresponds to a different type of energy distribution. In the general case of a beam segment of arbitrary dimensions, one ends up with four regimes: the Gaussian regime, the Lorentzian regime, the Holtsmark regime and the pencil beam regime. The widths of the distinguished energy distributions show different dependencies on the experimental parameters.

Where possible, the numerical calculations are verified by analytical means. Analytical results for the width of the energy distribution can be obtained for all limiting cases where a single regime is dominant. A direct analytical calculation of the behavior of the energy distribution in the transition areas between the regimes seems not possible. The single exception is the transition between the Gaussian and the Lorentzian regime in a beam segment with a narrow crossover. In order to increase the applicability of the model, the numerical data for the $FWHM$ and the FW_{50} of the energy distribution is approximated with analytical expressions, covering the entire range of operating conditions.

7.2. General aspects

The Boersch effect corresponds to a broadening of the distribution of axial velocities $\rho(\Delta v_z)$ due to statistical Coulomb interactions between the beam particles. We will utilize the extended two-particle approach to compute this phenomenon. This model was outlined in Chapter 5. The dynamical part of the problem consists of the calculation of the axial velocity shift Δv_z experienced by the test particle due to the interaction with a single field particle. This problem was studied in Chapter 6. The shift Δv_z can be expressed as a function of the geometrical variables $\xi = (r_\perp, b_z, v, \Phi)$, the time of flight T, and the initial time $t_i = -S_c T$; see Eqs. (6.2.1), (6.2.2) and (6.2.3) respectively and Fig. 5.1 (note: $\Phi = \psi - \varphi$).

Explicit analytical equations can be determined for complete collisions and for weak collisions. For a complete collision we found

$$\Delta v_z \cong \frac{mv^3 b_z}{2C_0 \left[1 + \left(b_z^2 + r_\perp^2 \sin^2\Phi\right)(mv^2/2C_0)^2\right]}. \quad (7.2.1)$$

See Eqs. (6.10.4), (6.4.16), and (6.3.4). The effect of a weak collision can be described by (reproducing Eq. (6.10.5))

$$\Delta v_z \cong \frac{C_0 b_z}{mvb^2} \left[\frac{S_c Tv + a_0}{\left[(S_c Tv + a_0)^2 + b^2\right]^{1/2}} + \frac{(1 - S_c)Tv - a_0}{\left\{((1 - S_c)Tv - a_0)^2 + b^2\right\}^{1/2}}\right] \quad (7.2.2)$$

in which

$$a_0 = -r_\perp \cos\Phi, \qquad b = \left(b_z^2 + r_\perp^2 \sin^2\Phi\right)^{1/2}. \quad (7.2.3)$$

Eqs. (7.2.1) and (7.2.2) apply to the case that the initial relative velocity of the particles is nonzero ($v \neq 0$). In case $v = 0$, we found for a half-complete collision

$$\Delta v_z \cong \left(\frac{C_0}{m}\right)^{1/2} \frac{b_z}{\left(b_z^2 + r_\perp^2\right)^{3/4}} \quad (v = 0), \quad (7.2.4)$$

as follows from Eq. (6.10.7), taking $T \to \infty$. For a weak collision we found (reproducing Eq. (6.10.8))

$$\Delta v_z \cong \frac{C_0}{m} \frac{b_z T}{\left(b_z^2 + r_\perp^2\right)^{3/2}} \quad (v = 0). \quad (7.2.5)$$

Eqs. (7.2.1) and (7.2.2) will be used in the calculation of the Boersch effect generated in a beam segment with a crossover, while Eqs. (7.2.4) and (7.2.5) will be used for a homocentric cylindrical beam segment. Collisions that are neither complete nor weak will be evaluated numerically, following the procedure outlined in Section 6.6.

The statistical part of the problem consists of the evaluation of Eqs. (5.7.4), (5.7.5) and (5.7.6). Taking $\Delta \eta = \Delta v_z$, these equations become

$$p_2(\Delta v_z) = \int_0^{v_0} \frac{2v \, dv}{v_0^2} \int_0^{2\pi} \frac{d\Phi}{2\pi} \int_0^{r_c} \frac{2r_\perp \, dr_\perp}{r_c^2} \int_{-S_c L}^{(1-S_c)L} db_z \, \delta[\Delta v_z - \Delta v_z(v, \Phi, r_\perp, b_z)]$$

$$(7.2.6)$$

$$p(k) = 2 \int_{-\infty}^{\infty} d\Delta v_z \, p_2(\Delta v_z) \sin^2(k \Delta v_z / 2) \quad (7.2.7)$$

$$\rho(\Delta v_z) = \frac{1}{\pi} \int_0^\infty dk \cos(k\Delta v_z) e^{-\lambda p(k)}, \tag{7.2.8}$$

in which we expressed $p(\xi)\, d\xi$, directly in terms v, Φ, $r_\perp$, and b_z, using Eqs. (5.4.6) and (5.3.2). The distribution in v and $r_\perp$ are here taken uniform with a cutoff at $v_0 = \alpha_0 v_z$ and r_c respectively, as prescribed by Eq. (5.3.3). In Section 7.6 we will consider the case that one of these distribution or both of them are Gaussian. Eqs. (7.2.6)–(7.2.8) constitute the basis of our three-step approach, which was discussed in Section 5.4. The function $\rho_2(\Delta v_z)$ is called the two-particle distribution. The function $p(k)$ represents its transform to the k-domain. Both functions are entirely determined by the geometrical properties of the beam only, represented by v_0, r_c, and L. The linear particle density λ enters the model in the last step, which yields the (N-particle) displacement distribution $\rho(\Delta v_z)$.

We recall that our model for statistical interactions relies on the following assumptions:
- The field particles can be considered as statistically independent, as expressed by Eq. (5.3.1).
- The total displacement of the test particle is equal to the sum of all displacements experienced in the two-particle interactions with the individual field particles, as expressed by Eq. (5.3.12).
- Magnetic interactions and relativistic effects can be ignored. Accordingly, the effect is assumed to be entirely the result of Coulomb interactions.

In this chapter we make some further simplifications that facilitate the calculations considerably:
- The reference trajectory, which is the unperturbed trajectory of the test particles, coincides with the beam axis. It is assumed that the effect experienced by the test particles running along this (on-axis) trajectory is representative for other (off-axis) trajectories as well.
- The beam is initially monochromatic with respect to the normal energy of the particles, which implies that all particles run initially with identical axial velocity. It is assumed that the impact of statistical interactions on the properties of the beam does not depend on the initial energy spread, which will be present in practical beams.

In Chapter 10 we will present the results of some further calculations, which do not rely on these simplifications. The results confirm the validity of the approach taken here.

Finally, we emphasize that all calculations are restricted to beams that fulfill the following conditions:
- The beam is rotational symmetric.
- The beam is paraxial, which means that $\tan \alpha_0 \approx \alpha_0$ and r_c are small.
- External forces are absent, and thus the particles fly through drift space.

7.3. Beam segment with a narrow crossover

To indicate whether a crossover should be considered as narrow or not we introduce the characteristic beam geometry quantities K, K_1 and K_2,

$$K = \frac{\alpha_0 L}{2r_c}, \quad K_1 = 2KS_c, \quad K_2 = 2K(1 - S_c), \quad (7.3.1)$$

employing the experimental parameters defined in Section 3.2. K_1 is the ratio between the beam radius at the start of the beam segment and the crossover radius. K_2 is the corresponding ratio referring to the end of the beam segment. A crossover is called narrow if it has the characteristic $K \gg 1$. This condition implies $K_1 \gg 1$ and $K_2 \gg 1$, provided that the crossover is located somewhere near the middle of the beam segment ($S_c \approx 0.5$).

In this section we will calculate the Boersch effect in a beam segment with the characteristic $K_1 \gg 1$ and $K_2 \gg 1$. Physically this implies that the main effect comes from collisions that are complete or nearly complete. Accordingly, one may use Eq. (7.2.1) to determine the shift Δv_z of the test particle caused by the interaction with a single field particle. It should be noted that the initial and final potential energy of two particles involved in a complete collision is effectively zero. A complete collision does not affect the total kinetic energy of the particles but merely leads to directional changes and a redistribution of the total kinetic energy. Accordingly, a calculation based on this type of collisions implicitly assumes that the Boersch effect stems from a conversion of kinetic energy from one degree of freedom into another (more specifically from the lateral degrees of freedom into the longitudinal degree of freedom). This approach is justified for the present case of a narrow crossover as will be shown in Section 7.5, where we treat the general case using full collision dynamics.

The program of this section is to evaluate the two-particle distribution $\rho_2(\Delta v_z)$ from Eqs. (7.2.1) and (7.2.6). This distribution will be used to determine the N-particle displacement distribution $\rho(\Delta v_z)$ by means of Eqs. (7.2.7) and (7.2.8). It is convenient to introduce the following scaling quantities for length and velocity respectively

$$d_0 = \frac{C_0}{m v_0^2}, \quad v_0 = \alpha_0 v_z. \quad (7.3.2)$$

v_0 is the maximum transverse velocity of a particle, and d_0 is the closest distance of approach for two particles, starting out with an infinite separation and relative velocity $2v_0$. The objective of this scaling is to simplify the equations by isolating key combinations of experimental parameters.

For an electron beam with $V = 10\,\text{kV}$ and $\alpha_0 = 10\,\text{mR}$, one finds $d_0 = 0.72\,\text{nm}$ and $v_0 = 5.9 \times 10^5\,\text{m/s}$, which can be considered as typical values. All quantities scaled with d_0 and/or v_0 are headed with a bar.

Employing the d_0, v_0-scaling, Eq. (7.2.1) transforms to

$$\Delta \bar{v}_z = \frac{\Delta v_z}{v_0} = \frac{\bar{v}^3 \bar{b}_z}{2\left[1 + \left(\bar{b}_z^2 + \bar{r}_\perp^2 \sin^2 \Phi\right) \bar{v}^4/4\right]}. \tag{7.3.3}$$

The scaled two-particle distribution $\bar{\rho}_2(\Delta \bar{v}_z)$ follows with Eq. (7.2.6)

$$\bar{\rho}_2(\Delta \bar{v}_z) = \frac{v_0}{d_0} \rho_2(\Delta v_z)$$

$$= \int_0^1 2\bar{v}\,d\bar{v} \int_0^{2\pi} \frac{d\Phi}{2\pi} \int_0^{\bar{r}_c} \frac{2\bar{r}_\perp\,d\bar{r}_\perp}{\bar{r}_c^2} \int_{-\infty}^\infty d\bar{b}_z\,\delta\left[\Delta\bar{v}_z - \Delta\bar{v}_z\left(\bar{v}, \Phi, \bar{r}_\perp, \bar{b}_z\right)\right], \tag{7.3.4}$$

in which $\Delta\bar{v}_z(\bar{v}, \Phi, \bar{r}, \bar{b}_z)$ is given by Eq. (7.3.3) and $\bar{r}_c$ is the scaled crossover radius. In terms of the experimental parameters this parameter can be expressed as

$$\bar{r}_c = \frac{r_c}{d_0} = \frac{8\pi\varepsilon_0}{e}\alpha_0^2 V r_c, \tag{7.3.5}$$

as follows with Eq. (7.3.2). We evaluated the integrals of Eq. (7.3.4) numerically for different values of the scaled crossover radius $\bar{r}_c$. The results are shown in Fig. 7.1.

It would be convenient to evaluate Eq. (7.3.4) analytically. However, this is only possible in some special cases. For a homocentric beam ($\bar{r}_c = 0$), Eq. (7.3.4) reduces to a two-dimensional integral over $\bar{v}$ and $\bar{b}_z$, which can be carried out analytically. For $\bar{r}_c \neq 0$ one can obtain a good analytical approximation when $\bar{r}_c \Delta \bar{v}_z \gg 1$ (large crossover and/or large velocity shift) or, oppositely, when $\bar{r}_c \Delta \bar{v}_z \ll 1$. In order to investigate these individual cases, we will first bring Eq. (7.3.4) in a different form.

One can rewrite Eq. (7.3.4) by using the following identity

$$\int_0^{2\pi} \frac{d\Phi}{2\pi} \int_0^{\bar{r}_c} \frac{2\bar{r}_\perp\,d\bar{r}_\perp}{\bar{r}_c^2} = \frac{4}{\pi} \int_0^1 dy\,\sqrt{1 - y^2},$$

with $y = (\bar{r}_\perp/\bar{r}_c) \sin \Phi$. This is allowed since the expression for $\Delta\bar{v}_z$, given by Eq. (7.3.3), depends on $\bar{r}_\perp$ and Φ only through the combination $\bar{r}_\perp \sin \Phi$. In addition, we substitute

$$z = \bar{b}_z \bar{v}^2/2.$$

Eq. (7.3.4) now transforms to

$$\bar{\rho}_2(\Delta\bar{v}_z) = \frac{16}{\pi} \int_0^1 \frac{d\bar{v}}{\bar{v}} \int_0^1 dy\,\sqrt{1-y^2} \int_0^\infty dz\,\delta\left(\Delta\bar{v}_z - \frac{z\bar{v}}{z^2 + R^2}\right), \quad \Delta\bar{v}_z \geq 0, \tag{7.3.6}$$

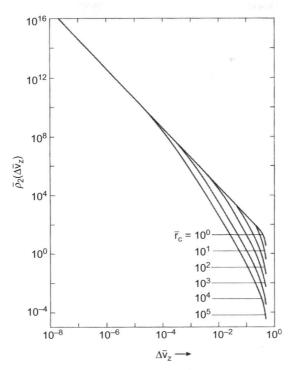

Fig. 7.1 The scaled two-particle distribution $\bar{p}_2(\Delta\bar{v}_z)$ based on complete collisions, for different values of the scaled crossover radius $\bar{r}_c$.

where $R = R(\bar{v}, y)$ is defined as

$$R(\bar{v}, y) = \left(1 + \bar{v}^4 y^2 \bar{r}_c^2 / 4\right)^{1/2}. \tag{7.3.7}$$

As the distribution $\bar{p}_2(\Delta\bar{v}_z)$ is symmetric in $\Delta\bar{v}_z$ it is sufficient to evaluate it for positive values of $\Delta\bar{v}_z$. Accordingly, we limited to z-integration in Eq. (7.3.6) to positive values only. One sees that the second term in the argument of the δ-function has a maximum for $z = R$. We now split the z-integration into the intervals $[0, R]$ and $[R, \infty]$ and substitute $z = R^2/z$ in the second interval. It is here essential to note that the integrand is not affected by this substitution. The z-integral of Eq. (7.3.6) now transforms to (notation: $R = R(\bar{v}, y)$)

$$\int_0^R dz \left(1 + \frac{R^2}{z^2}\right) \delta\left(\Delta\bar{v}_z - \frac{z\bar{v}}{z^2 + R^2}\right) = \int_{u=0}^{u=1} -2Rd\left(\frac{\sqrt{1-u^2}}{u}\right) \delta\left(\Delta\bar{v}_z - \frac{u\bar{v}}{2R}\right). \tag{7.3.8}$$

In the last step, we substituted $u = 2Rz/(z^2 + R^2)$. Eq. (7.3.6) becomes

$$\bar{p}_2(\Delta\bar{v}_z) = \frac{32}{\pi} \int_0^1 \frac{d\bar{v}}{\bar{v}} \int_0^1 dy \sqrt{1-y^2} \int_0^1 \frac{R(\bar{v},y)du}{u^2(1-u^2)^{1/2}} \delta\left(\Delta\bar{v}_z - \frac{u\bar{v}}{2R(\bar{v},y)}\right), \tag{7.3.9}$$

which can be rewritten as (for $0 < \Delta\bar{v}_z < 1/2$)

$$\bar{p}_2(\Delta\bar{v}_z) = \frac{16}{\pi \Delta\bar{v}_z{}^2} \int_{2\Delta\bar{v}_z}^1 d\bar{v} \int_0^{\min\left[1,\left(1-4\Delta\bar{v}_z{}^2/\bar{v}^2\right)^{1/2}/\Delta\bar{v}_z\bar{v}\bar{r}_c\right]}$$

$$\times dy \frac{(1-y^2)^{1/2}}{\left[1 - \Delta\bar{v}_z{}^2(4/\bar{v}^2 - \bar{v}^2 y^2 \bar{r}_c{}^2)\right]^{1/2}}. \tag{7.3.10}$$

Eq. (7.3.9) will be used, later on, to evaluate the function $p(k)$, while Eq. (7.3.10) is most suited to study the special cases mentioned previously. For a homocentric beam ($\bar{r}_c = 0$), Eq. (7.3.10) becomes

$$\bar{p}_2(\Delta\bar{v}_z) = \frac{4(1 - 4\,\Delta\bar{v}_z{}^2)^{1/2}}{\Delta\bar{v}_z{}^2}, \quad (\bar{r}_c = 0, \quad 0 < \Delta\bar{v}_z < 1/2). \tag{7.3.11}$$

For small values of $\bar{r}_c$ and small values of $\Delta\bar{v}_z$ ($\bar{r}_c\Delta\bar{v}_z \ll 1$), one finds from Eq. (7.3.10)

$$\bar{p}_2(\Delta\bar{v}_z) \simeq \frac{4}{\Delta\bar{v}_z{}^2}, \quad (\bar{r}_c\Delta\bar{v}_z \ll 1, \quad 0 < \Delta\bar{v}_z < 1/2). \tag{7.3.12}$$

For the opposite case ($\bar{r}_c\Delta\bar{v}_z \gg 1$), Eq. (7.3.10) can be approximated as

$$\bar{p}_2(\Delta\bar{v}_z) \simeq \frac{8}{\Delta\bar{v}_z{}^3 \bar{r}_c} \ln\left(\frac{1}{2\,\Delta\bar{v}_z}\right), \quad (\bar{r}_c\Delta\bar{v}_z \gg 1, \quad 0 < \Delta\bar{v}_z < 1/2). \tag{7.3.13}$$

The behavior of the two-particle function $\bar{p}_2(\Delta\bar{v}_z)$ described by Eqs. (7.3.11)–(7.3.13) is confirmed by the numerical data plotted in Fig. 7.1.

We now proceed with the calculation of the function $p(k)$ and the displacement distribution $\rho(\Delta v_z)$. According to Eq. (7.2.7), one can express the scaled function $\bar{p}(\bar{k})$ as

$$\bar{p}(\bar{k}) = \frac{p(k)}{d_0} = 2 \int_{-\infty}^{\infty} d\Delta\bar{v}_z \bar{p}_2(\Delta\bar{v}_z) \sin^2(\bar{k}\Delta\bar{v}_z/2), \tag{7.3.14}$$

with $\bar{k} = kv_0$. Substitution of Eq. (7.3.9) and partial integration with respect to u, yields

$$\bar{p}(\bar{k}) = \bar{k}\frac{32}{\pi} \int_0^1 d\bar{v} \int_0^1 dy \sqrt{1-y^2} \int_0^1 du \frac{(1-u^2)^{1/2}}{u} \sin\left(\frac{\bar{k}\bar{v}u}{(4 + \bar{v}^4 y^2 \bar{r}_c{}^2)^{1/2}}\right). \tag{7.3.15}$$

This expression was evaluated numerically for different values of the scaled crossover radius $\bar{r}_c$. The results are shown in Fig. 7.2. One sees that the dependency on $\bar{k}$ is quadratic for small values of $\bar{k}$, particularly in combination with large $\bar{r}_c$. The dependency becomes linear for large $\bar{k}$-values.

The integrals in Eq. (7.3.15) can be simplified for the extreme cases $\bar{k} \to 0$ and $\bar{k} \to \infty$. For $\bar{k} \to 0$, one may replace the sin-function by its argument and carry out the u-integration, which yields a factor $\pi/4$. The integration over $\bar{v}$ can be performed next and one obtains

$$\bar{p}(\bar{k}) = \frac{1}{2}\bar{p}_2(\bar{r}_c)\bar{k}^2, \quad \bar{p}_2(\bar{r}_c) = \frac{8}{\bar{r}_c}\int_0^1 dy \frac{\sqrt{1-y^2}}{y} \sinh^{-1}\left(\frac{\bar{r}_c y}{2}\right), \quad (\bar{k} \to 0)$$

(7.3.16)

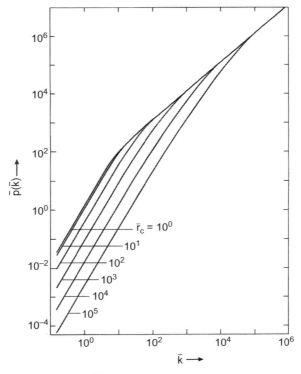

Fig. 7.2 The scaled function $\bar{p}(\bar{k})$ corresponding to the distribution of axial velocity displacements generated by complete collisions, for different values of the scaled crossover radius $\bar{r}_c$.

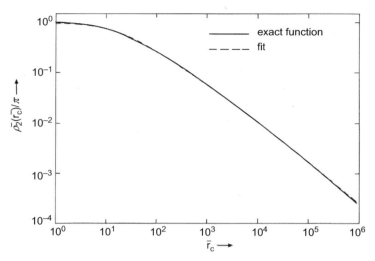

Fig. 7.3 The function $\bar{p}_2(\bar{r}_c)$ defined by Eq. (7.3.16) and its approximation given by Eq. (7.3.17). This function gives the $\bar{r}_c$-dependency of the function $\bar{p}(\bar{k})$, plotted in Fig. 7.2, for $\bar{k} \to 0$.

(note that $\sinh^{-1}(x) = \ln[x + (1+x^2)^{1/2}]$). For a homocentric beam $\bar{r}_c = 0$, and one finds $\bar{p}_2 = \pi$. The function $\bar{p}_2(\bar{r}_c)/\pi$ is drawn in Fig. 7.3. For later use, we fitted this function with

$$\bar{p}_{2a}(\bar{r}_c) = \frac{\pi}{1 + \pi\bar{r}_c/\{2\ln[.8673(114.6 + \bar{r}_c)]\}^2}, \quad (7.3.17)$$

which has the same asymptotic behavior for $\bar{r}_c \to 0$ and $\bar{r} \to \infty$ as $\bar{p}_2(\bar{r}_c)$. For comparison, we included the fit $\bar{p}_{2a}(\bar{r}_c)/\pi$ in Fig. 7.3. The accuracy of the fit is better than 2% for all $\bar{r}_c$ values.

In the limit $\bar{k} \to \infty$ one can simplify Eq. (7.3.15) by exploiting the following representation of the δ-function:

$$\delta(x) = \lim_{k \to \infty} \frac{\sin(kx)}{\pi x}. \quad (7.3.18)$$

See, for comparison Eq. (5.4.2). The u and $\bar{v}$ integration can now successively be performed (note that the u-integration carries only over half the δ-function), while the y-integration yields again a factor $\pi/4$. Accordingly, one finds

$$\bar{p}(\bar{k}) = \bar{p}_\infty \bar{k}, \quad \bar{p}_\infty = 4\pi, \quad (\bar{k} \to \infty). \quad (7.3.19)$$

The derivation of Eq. (7.3.19) makes clear that the large k-behavior of $\bar{p}(\bar{k})$ comes from the small $\Delta\bar{v}_z$ behavior of the two-particle distribution $\bar{p}_2(\Delta\bar{v}_z)$ described by Eq. (7.3.12).

Boersch effect

In order to achieve a simple expression for the function $\bar{p}(\bar{k})$, we approximate Eq. (7.3.15) by

$$\bar{p}_a(\bar{k}) = \bar{p}_\infty \sqrt{\bar{k}^2 + [\bar{p}_\infty/\bar{p}_2(\bar{r}_c)]^2} - \bar{p}_\infty^2/\bar{p}_2(\bar{r}_c), \qquad (7.3.20)$$

in which $\bar{p}_2(\bar{r}_c)$ and $\bar{p}_\infty$ are given by Eqs. (7.3.16) and (7.3.19) respectively. The function $p_a(\bar{k})$ has the same asymptotic behavior as the original function $\bar{p}(\bar{k})$. Fig. 7.4 shows both functions for different values of $\bar{r}_c$. One sees that the fit is accurate for small and large $\bar{k}$-values but shows a significant deviation from the real function for intermediate values of $\bar{k}$, especially in combination with large values of r_c.

The final step consists of the calculation of the displacement distribution $\rho(\Delta v_z)$. By scaling Eq. (7.2.8), one obtains

$$\bar{\rho}(\Delta \bar{v}_z) = v_0 \rho(\Delta v_z) = \frac{1}{\pi} \int_0^\infty d\bar{k} \cos(\bar{k}\Delta\bar{v}_z) e^{-\bar{\lambda}\bar{p}(\bar{k})}, \qquad (7.3.21)$$

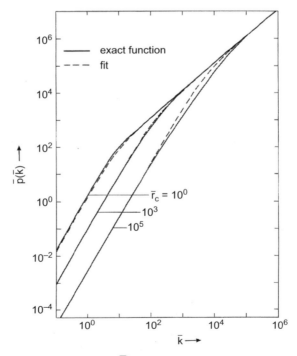

Fig. 7.4 The exact scaled function $\bar{p}(\bar{k})$ for complete collisions and its approximation given by Eq. (7.3.20).

where $\bar{\lambda}$ is the scaled linear particle density, which is given by

$$\bar{\lambda} = \lambda\, d_0 = \frac{C_0}{em}\frac{I}{\alpha_0^2 v_z^3} = \frac{m^{1/2}}{2^{7/2}\pi\varepsilon_0 e^{1/2}}\frac{I}{\alpha_0^2 V^{3/2}}, \qquad (7.3.22)$$

as follows with Eqs. (3.2.4) and (7.3.2).

The advantage of the approximation given by Eq. (7.3.20) is that it allows the $\bar{k}$-integration to be done analytically. Substitution into Eq. (7.3.21) and integration yields

$$\bar{p}_a(\Delta\bar{v}_z) = \frac{\alpha\beta\exp(\alpha\beta)}{\pi(\Delta\bar{v}_z^2 + \alpha^2)^{1/2}}\, K_1\!\left[\beta(\Delta\bar{v}_z^2 + \alpha^2)^{1/2}\right], \qquad (7.3.23)$$

where K_1 is the modified Bessel function and α and β are defined as

$$\alpha = \bar{\lambda}\bar{p}_\infty, \qquad \beta = \bar{p}_\infty/\bar{p}_2(\bar{r}_c) \qquad (7.3.24)$$

The shape of the distribution $\bar{p}_a(\Delta\bar{v}_z)$, given by Eq. (7.3.23), depends on the values of $\bar{\lambda}$ and $\bar{r}_c$ through α and β. One should distinguish two extreme cases.

(I) Gaussian regime. For $\bar{\lambda} \gg \bar{p}(\bar{r}_c)/\bar{p}_\infty^2$ (or $\alpha\beta \gg 1$) and $\Delta\bar{v}_z \ll \bar{\lambda}\bar{p}_\infty$ (or $\Delta\bar{v}_z \ll \alpha$) one may utilize that

$$\lim_{x\to\infty} K_1(x) = (\pi/2x)^{1/2}\exp(-x)[1 + O(1/x)],$$

in which $O(1/x)$ stands for terms of the order $1/x$. Eq. (7.3.23) now transforms to (approximating $\Delta\bar{v}_z^2 + \alpha^2 \cong \Delta\bar{v}_z^2$)

$$\bar{p}_a(\Delta\bar{v}_z) = \frac{1}{(2\pi)^{1/2}\sigma}\exp\!\left[-\Delta\bar{v}_z^2/2\sigma^2\right], \quad \sigma^2 = \frac{\alpha}{\beta} = \bar{\lambda}\bar{p}_2(\bar{r}_c), \qquad (7.3.25)$$

which is a one-dimensional Gaussian distribution.

(II) Lorentzian regime. For $\bar{\lambda} \ll \bar{p}(\bar{r}_c)/\bar{p}_\infty^2$ (or $\alpha\beta \ll 1$) and $\Delta\bar{v}_z \ll \bar{p}^2(\bar{r}_c)/\bar{p}_\infty$ (or $\Delta\bar{v}_z \ll 1/\beta$) one may use that

$$\lim_{x\to 0} K_1(x) = 1/x + O(x), \qquad \lim_{x\to 0}\exp(x) = 1 + O(x).$$

Accordingly, Eq. (7.3.23) transforms to

$$\bar{p}_a(\Delta\bar{v}_z) = \frac{1}{\pi\alpha\!\left[1 + (\Delta\bar{v}_z/\alpha)^2\right]}, \qquad (7.3.26)$$

which is a one-dimensional Lorentzian distribution.

Summarizing the results so far, we found that the distribution of the scaled axial velocity displacements $\bar{p}(\Delta\bar{v}_z)$ generated in a beam segment with a narrow crossover is Gaussian for large values of the scaled linear particle density $\bar{\lambda}$ and Lorentzian for small $\bar{\lambda}$-values. This behavior is related to the quadratic and linear behavior of the function $\bar{p}(\bar{k})$ for small $\bar{k}$ and large $\bar{k}$ respectively.

The mean square value $\langle \Delta \bar{v}_z^2 \rangle$ of the distribution $\bar{\rho}(\Delta \bar{v}_z)$ can be determined in several ways. One way is to use its definition

$$\langle \Delta \bar{v}_z^2 \rangle = \int d\Delta \bar{v}_z \, \bar{\rho}(\Delta \bar{v}_z) \, \Delta \bar{v}_z^2 \tag{7.3.27}$$

and to substitute Eq. (7.3.23) for $\bar{\rho}(\Delta \bar{v}_z)$. Another way is to determine $\langle \Delta \bar{v}_z^2 \rangle$ from the two-particle distribution $\bar{\rho}_2(\Delta \bar{v}_z)$ given by Eq. (7.3.9), using

$$\langle \Delta \bar{v}_z^2 \rangle = \bar{\lambda} \int d\Delta \bar{v}_z \, \bar{\rho}_2(\Delta \bar{v}_z) \, \Delta \bar{v}_z^2, \tag{7.3.28}$$

which is based on Eq. (5.5.10) (with $m=2$). The usual way to calculate $\langle \Delta \bar{v}_z^2 \rangle$ is to evaluate

$$\langle \Delta \bar{v}_z^2 \rangle = \frac{\bar{\lambda}}{d_0} \int d\boldsymbol{\xi} \, p(\boldsymbol{\xi}) \, [\Delta \bar{v}_z(\boldsymbol{\xi})]^2, \tag{7.3.29}$$

as follows from the second equation of (5.5.11) with $\Delta \eta = \Delta v_z$. The probability distribution $p(\boldsymbol{\xi})$ is defined by Eqs. (5.4.6) and (5.3.2), while $\Delta \bar{v}_z(\boldsymbol{\xi})$ is (for complete collisions) given by Eq. (7.3.3). Finally, one may determine $\langle \Delta \bar{v}_z^2 \rangle$ from the small $\bar{k}$-behaviour of the function $\bar{p}(\bar{k})$. We will utilize this method. Substitution of Eq. (7.3.16) into Eq. (5.5.4) (with $m=2$) directly yields

$$\langle \Delta \bar{v}_z^2 \rangle = \bar{\lambda} \bar{p}_2(\bar{r}_c). \tag{7.3.30}$$

It should be emphasized that this relation is generally true, irrespective of the shape of the distribution $\bar{\rho}(\Delta \bar{v}_z)$.

Other measures for the width of the distribution, in general, do depend on the shape of the distribution. As a practical width measure, we consider the Full Width at Half Maximum (FWHM). The FWHM of the Gaussian distribution (7.3.25) is equal to

$$\overline{FWHM}_G = 2(2\ln 2)^{1/2} \sigma = 2(2\pi \ln 2)^{1/2} P_{CE}(\bar{r}_c) \bar{\lambda}^{1/2}, \tag{7.3.31}$$

where the function $P_{CE}(\bar{r}_c)$ defined as

$$P_{CE}(\bar{r}_c) = [\bar{p}_2(\bar{r}_c)/\pi]^{1/2}. \tag{7.3.32}$$

The subscript CE indicates that the function refers to the Energy distribution generated in a beam segment with a Crossover. The FWHM of the Lorentzian distribution, given by Eq. (7.3.26), is equal to

$$\overline{FWHM}_L = 2\alpha = 2\bar{\lambda} \bar{p}_\infty = 8\pi \bar{\lambda}. \tag{7.3.33}$$

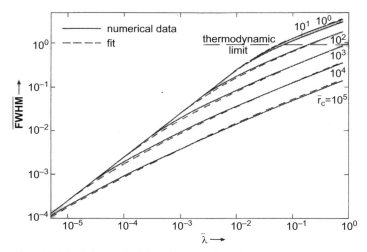

Fig. 7.5 The *FWHM* of the scaled distribution of axial velocity displacements $\bar{p}(\Delta\bar{v}_z)$ based on complete collisions, for different values of the scaled crossover radius $\bar{r}_c$. The depicted fit is defined by Eqs. (7.3.34) and (7.3.35). The thermodynamic limit is given by Eq. (7.7.2) and refers to the relaxation of kinetic energy.

Clearly, Eqs. (7.3.31) and (7.3.33) show different dependencies on the parameters $\bar{\lambda}$ and $\bar{r}_c$ and, therefore, lead to different dependencies on the experimental parameters. The dependency of the *FWHM* of $\bar{p}(\Delta\bar{v}_z)$ on $\bar{\lambda}$ and $\bar{r}_c$ is depicted in Fig. 7.5. The input for this figure was determined numerically using the exact function $\bar{p}(\bar{k})$. Thus, the results do not rely on the approximation contained in Eq. (7.3.20). The thermodynamic limit indicated in Fig. 7.5 will be discussed in Section 7.7.

Anticipating the analysis of Section 7.5, we mention that Eqs. (7.3.31) and (7.3.33) can be interpolated, using

$$\overline{FWHM} = 2(2\pi \ln 2)^{1/2} G_{CE}(\bar{\lambda}, \bar{r}_c) \bar{\lambda}^{1/2}, \qquad (7.3.34)$$

where the function $G_{CE}(\bar{\lambda}, \bar{r}_c)$ is defined as

$$G_{CE}(\bar{\lambda}, \bar{r}_c) = \left[\frac{1}{P_{CE}(\bar{r}_c)^4} + 1.575 \times 10^{-3} \frac{\bar{r}_c^{4/3}}{\bar{\lambda}^{2/3}} + \frac{7.606 \times 10^{-4}}{\bar{\lambda}^2} \right]^{1/4}. \qquad (7.3.35)$$

The function $P_{CE}(\bar{r}_c)$ is defined by Eq. (7.3.32). The fit represented by Eqs. (7.3.34) and (7.3.35) is also plotted in Fig. 7.5. The accuracy of the fit is better than 5% over the entire range of operating conditions.

In the Gaussian regime, $\bar{\lambda} \gg P_{CE}(\bar{r}_c)^2/16\pi$, and one may approximate $G_{CE}(\bar{\lambda},\bar{r}_c) \approx P_{CE}(\bar{r}_c)$. This implies that $\overline{FWHM} \approx \overline{FWHM}_G$, where $\overline{FWHM}_G$ is given by Eq. (7.3.31). In the Lorentzian regime $\bar{\lambda} \ll P_{CE}(\bar{r}_c)^2/16\pi$, and one may approximate $G_{CE}(\bar{\lambda},\bar{r}_c) \approx 6.022\bar{\lambda}^{1/2}$. This implies that $\overline{FWHM} \approx \overline{FWHM}_L$, where $\overline{FWHM}_L$ is given by Eq. (7.3.33). The second term between brackets in the right-hand side of Eq. (7.3.35) is included to improve the quality of the fit. Its origin will be clarified in Section 7.5.

Eqs. (7.3.31) through (7.3.35) provide a complete specification of the $\overline{FWHM}$ of the generated axial velocity distribution $\rho(\Delta v_z)$ in terms of the scaled coordinates defined by Eqs. (7.3.2). In order to make the dependency on the experimental parameters explicit, we will now remove this scaling. The properties of the scaled velocity distribution $\bar{\rho}(\Delta \bar{v}_z)$ can be translated in terms of the corresponding energy distribution, using

$$\Delta E \cong m v_z \Delta v_z = m v_z^2 \alpha_0 \Delta \bar{v}_z = E 2\alpha_0 \Delta \bar{v}_z, \quad (7.3.36)$$

assuming that $v_z \gg \Delta v_z$. Accordingly, Eq. (7.3.34) can be expressed as

$$\frac{FWHM_E}{E} = C_{CGE} G_{CE}(\bar{\lambda},\bar{r}_c) \sqrt{\frac{I}{V^{3/2}}}, \quad C_{CGE} = \left(\frac{8(\ln 2)^2 m}{\varepsilon_0^2 e}\right)^{1/4}. \quad (7.3.37)$$

The ratio $I/V^{3/2}$ is known as the perveance. In the case of electrons, one finds $C_{CGE} = 726.62$ in SI-units (note: the subscript CGE indicates that the constant applies to a beam geometry with a Crossover and refers to the Gaussian Energy distribution). Eq. (7.3.37) yields the main result of this section. It gives the FWHM of the energy distribution generated in a beam segment with a narrow crossover for the entire range of operating conditions. The operating conditions are defined by the parameters $\bar{\lambda}$ and $\bar{r}_c$. Fig. 7.6 indicates which values of $\bar{\lambda}$ and $\bar{r}_c$ apply to practical systems. From the figure one sees that systems employing thermionic emission predominantly operate in the Gaussian regime and in the transition area between the Gaussian and the Lorentzian regime, whereas systems with a field emission gun predominantly operate in the Lorentzian regime. We emphasize that the analysis of this section pertains to a beam segment with a narrow crossover ($K \gg 1$), for which one may assume that the Boersch effect is generated by complete collisions. In Section 7.5, where we treat the general case of a beam with a crossover in which K is not necessarily large, it will be shown that one should in general distinguish four regimes. In addition to the Gaussian regime and the Lorentzian regime, one finds, for small K-values, a Holtsmark regime and a pencil beam regime, corresponding to medium and small values of $\bar{\lambda}$ respectively.

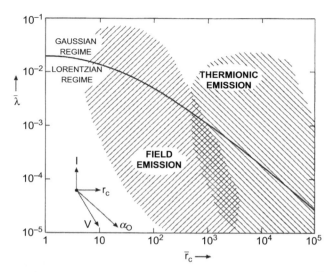

FIG. 7.6 Practical operating conditions in terms of the quantities $\bar{r}_c$ and $\bar{\lambda}$ for a beam segment with a narrow crossover ($K \gg 1$). The solid line separates the regimes leading to a Gaussian and a Lorentzian type of energy distribution. The effect of varying each of the experimental parameters is indicated by the arrows.

We conclude this section by investigating the dependency on the experimental parameters in the different regimes, covered by Eq. (7.3.37). The Gaussian regime corresponds to $\bar{\lambda} \gg P_{CE}(\bar{r}_c)^2/16\pi$, for which one may approximate that $G_{CE}(\bar{\lambda}, \bar{r}_c) \approx P_{CE}(\bar{r}_c)$, as was mentioned before. The function $P_{CE}(\bar{r}_c)$, defined by Eq. (7.3.32), is for all practical purposes best described by the fit given by Eq. (7.3.17). For small values of $\bar{r}_c$, one obtains $P_{CE}(\bar{r}_c) \approx 1$. For these conditions Eq. (7.3.37) transforms to

$$\frac{FWHM_E}{E} = C_{CGE}\sqrt{\frac{I}{V^{3/2}}}, \quad \left(\bar{\lambda} \gg P_{CE}(\bar{r}_c)^2/16\pi, \ \bar{r}_c \leq 10\right). \tag{7.3.38}$$

For large values of $\bar{r}_c$, one finds from Eqs. (7.3.32) and (7.3.17)

$$P_{CE}(\bar{r}_c) \approx \frac{9.20}{(\pi\bar{r}_c)^{1/2}}[1 + .217\ln(1 + \bar{r}_c/114.6)].$$

Accordingly, Eq. (7.3.37) yields

$$\frac{FWHM_E}{E} = C^*_{CGE}\frac{I^{1/2}}{V^{5/4}r_c^{1/2}\alpha_0}[1 + .217\ \ln(1 + \bar{r}_c/114.6)],$$

$$\left(\bar{\lambda} \gg P_{CE}(\bar{r}_c)^2/16\pi, \ \bar{r}_c \geq 100\right), \tag{7.3.39}$$

where the constant C^*_{CGE} is given by

$$C^*_{CGE} = 9.20 \frac{(\ln 2)^{1/2}}{2^{3/4}\pi} \frac{(me)^{1/4}}{\varepsilon_0}$$

In the case of electrons, one finds $C^*_{CGE} = 0.1012$ in *SI*-units.

The Lorentzian regime corresponds to $\bar{\lambda} \ll P_{CE}(\bar{r}_c)^2/16\pi$. Eq. (7.3.37) now transforms to

$$\frac{FWHM_E}{E} = C_{CLE} \frac{I}{\alpha_0 V^{3/2}}, \quad \left(\bar{\lambda} \ll P_{CE}(\bar{r}_c)^2/16\pi\right) \tag{7.3.40}$$

where the constant C_{CLE} is given by

$$C_{CLE} = \left(\frac{2m}{\varepsilon_0^2 e}\right)^{1/2}.$$

In the case of electrons, one finds $C_{CLE} = 3.8085 \times 10^5$ in *SI*-units.

7.4. Homocentric cylindrical beam segment

In this section we will calculate the Boersch effect in a homocentric cylindrical beam. As before we assume that the beam is monochromatic. Consequently, all particles run initially with identical velocities along parallel trajectories. Thus, in the frame of reference moving with the beam the particles are initially at rest. Due to the Coulomb interaction, part of their mutual potential energy will be converted into kinetic energy. This process causes a generation of random velocity components. The generated spread of axial velocities corresponds to the Boersch effect.

From this analysis it becomes clear that "relaxation of potential energy" is the mechanism that generates the Boersch effect in this type of beam geometry. This mechanism should be distinguished from "relaxation of kinetic energy," which is dominant in a beam segment with a narrow crossover, as was discussed in the previous section. In this respect, both geometries should be considered as opposite. Accordingly, different types of collisions are involved. Complete collisions generate most of the effect in a beam segment with a narrow crossover, while weak incomplete collisions are dominant in a homocentric cylindrical beam.

For the case of a cylindrical beam, one cannot apply the scaling defined by Eq. (7.3.2), because $v_0 = 0$. Instead, we will use the δ,ν-scaling, given by Eq. (6.13.1). The scaled two-particle distribution $\rho_2^*(\Delta v_z^*)$ follows from Eq. (7.2.6) (notation: $r^* = r_\perp^*$):

$$\rho_2^*(\Delta v_z^*) = \frac{\nu r_0^{*2}}{\delta}\rho_2(\Delta v_z)$$
$$= \int_0^{r_0^*} 2r^* dr^* \int_0^\infty db_z^* \delta\left[\Delta v_z^* - \Delta v_z^*\left(r^*, b_z^*\right)\right] \quad (\Delta v_z^* > 0). \tag{7.4.1}$$

It is sufficient to consider the case $\Delta v_z^* = (\Delta v_z/\nu) > 0$, thus limiting the b_z^*-integration to positive b_z^*-values. The function $\Delta v_z^*(r^*, b_z^*)$ can be determined analytically for weak collisions and half-complete collisions; see Eqs. (7.2.5) and (7.2.4) respectively. In all other cases, one has to compute $\Delta v_z^*(r^*, b_z^*)$ numerically, following the method that was outlined in Section 6.6. We evaluated Eq. (7.4.1) numerically on the basis of the exact analysis of the function $\Delta v_z^*(r^*, b_z^*)$. The outcome is presented in Fig. 7.7.

It should be anticipated that the function $\rho_2^*(\Delta v_z^*)$ is determined by weak interactions for most of the Δv_z^*-range plotted in Fig. 7.7, while its behavior for very large Δv_z^*-values stems from half-complete collisions.

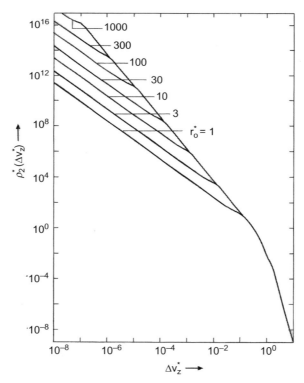

Fig. 7.7 The scaled two-particle distribution $\rho_2^*(\Delta v_z^*)$ for a homocentric cylindrical beam segment, plotted for different values of the scaled beam radius r_0^*.

We will verify this statement by evaluating Eq. (7.4.1) analytically, employing the approximations given by Eqs. (7.2.5) and (7.2.4) respectively. Let us first consider the limit of weak interactions. We recall that Eq. (7.2.5) is based on the first order perturbation approximation

$$\Delta v_z \cong F_\| T/m. \tag{7.4.2}$$

Thus, instead of evaluating $\rho_2^*(\Delta v_z^*)$ directly from Eqs. (7.4.1) and (7.2.5), one may as well determine $\rho_2^*(\Delta v_z^*)$ from the distribution of axial forces $\rho_2(F_\|)$ occurring in a cylindrical beam by using

$$\rho_2^*(\Delta v_z^*) = \frac{\nu r_0^{*2}}{\delta}\rho_2(\Delta v_z) = \frac{\nu r_0^{*2}}{\delta}\frac{m}{T}\rho_2(F_\| = \Delta v_z m/T). \tag{7.4.3}$$

The calculation of $\rho_2(F_\|)$ was carried out in Section 5.8. Eq. (5.8.23) specifies $\rho_2(F_\|)$ for an extended beam ($r_0 \to \infty$). Substitution into Eq. (7.4.3) yields

$$\rho_2^*(\Delta v_z^*) = \frac{1}{20\, \Delta v_z^{*5/2}} \quad \left(\Delta v_z^* \ll 1, r_0^{*2}\Delta v_z^* \gg 1\right) \tag{7.4.4}$$

in which we utilized the definitions of δ and ν given by Eq. (6.13.1).

For a pencil beam ($r_0 \to 0$), one should use Eq. (5.8.26). Substitution into Eq. (7.4.3) yields

$$\rho_2^*(\Delta v_z^*) = \frac{r_0^{*2}}{4\, \Delta v_z^{*3/2}} \quad \left(\Delta v_z^* \ll 1, r_0^{*2}\Delta v_z^* \ll 1\right). \tag{7.4.5}$$

One may verify from Fig. 7.7 that the small and the intermediate Δv_z^*-behavior of $\rho_2^*(\Delta v_z^*)$ indeed follows Eqs. (7.4.4) and (7.4.5).

In order to investigate the large Δv_z^*-behavior of $\rho_2^*(\Delta v_z^*)$, we scale the expression for Δv_z given by Eq. (7.2.4) and substitute the result into Eq. (7.4.1). This way one obtains

$$\rho_2^*\left(\Delta v_z^*\right) = \int_0^{r_0^*} 2r^* dr^* \int_0^\infty db_z^*\, \delta\left(\Delta v_z^* - \frac{b_z^*}{2\left(r^{*2}+b_z^{*2}\right)^{3/4}}\right). \tag{7.4.6}$$

By substituting

$$s = \frac{b_z^*}{r^*}, \quad t = \frac{s}{2r^{*1/2}(1+s^2)^{3/4}} \tag{7.4.7}$$

and carrying out the t-integration, Eq. (7.4.6) transforms to

$$\rho_2^*\left(\Delta v_z^*\right) = \frac{1}{16\, \Delta v_z^{*7}}\int_0^\infty ds\, \frac{s^6}{(1+s^2)^{9/2}}\, \Theta\!\left(\Delta v_z^* - \frac{s}{2r_0^{*1/2}[1+s^2]^{3/4}}\right), \tag{7.4.8}$$

in which $\Theta(x)$ is the step function defined by Eq. (3.7.8).

For an extended beam ($r_0^* \to \infty$), the argument of the step function is positive for all positive values of Δv_z^*, and one obtains

$$\rho_2^*(\Delta v_z^*) = \frac{1}{16 \Delta v_z^{*7}} \int_0^\infty ds \frac{s^6}{(1+s^2)^{9/2}} = \frac{1}{112} \frac{1}{\Delta v_z^{*7}}. \tag{7.4.9}$$

This asymptote gives a good approximation of $\rho_2^*(\Delta v_z^*)$ for $\Delta v_z^* \gtrsim 1$, as can be verified from Fig. 7.7.

For a pencil beam ($r_0^* \to 0$), the argument of the step function is positive for $s > s_0 \approx 1/(4\Delta v_z^{*2} r_0^*)$. Thus, the lower integration boundary of the integral in Eq. (7.4.8) should be replaced by s_0. As $s_0 \gg 1$ one may approximate

$$\rho_2^*(\Delta v_z^*) = \frac{1}{16 \Delta v_z^{*7}} \int_{s_0 = 1/(4\Delta v_z^{*2} r_0^*)}^\infty ds \; 1/s^3 = \frac{r_0^{*2}}{2 \Delta v_z^{*3}}. \tag{7.4.10}$$

However, this asymptote is not reached for normal operating conditions ($r_0^* \gtrsim 1$) and will be disregarded in the remaining analysis. Eqs. (7.4.4), (7.4.5), and (7.4.9) provide a complete analysis of the behavior of the two-particle distribution $\rho_2^*(\Delta v_z^*)$.

The next step is to determine the scaled function $p^*(k^*)$, which follows from Eq. (7.2.7),

$$p^*(k^*) = \frac{r_0^{*2}}{\delta} p(k) = 4 \int_0^\infty d \Delta v_z^* \rho_2^*(\Delta v_z^*) \sin^2(k^* \Delta v_z^*/2), \tag{7.4.11}$$

with $k^* = k\nu$. We evaluated the integral of Eq. (7.4.11) numerically from the numerical data of $\rho_2^*(\Delta v_z^*)$. The results are shown in Fig. 7.8.

The behavior of the function $p^*(k^*)$ can be understood from the analysis of the two-particle distribution $\rho_2^*(\Delta v_z^*)$. Substitution of Eq. (7.4.4) into Eq. (7.4.11) yields

$$p^*(k^*) = \frac{1}{10\sqrt{2}} k^{*3/2} \int_0^\infty dx \frac{\sin^2(x)}{x^{5/2}} = \frac{(2\pi)^{1/2}}{15} k^{*3/2}, \tag{7.4.12}$$

which provides an accurate approximation for $k^* \gtrsim 10$ and $k^* \lesssim 10 r_0^{*2}$. This result can also be obtained from Eq. (5.8.24), which refers to the distribution of axial forces in an extended beam.

Substitution of Eq. (7.4.5) into Eq. (7.4.11) yields

$$p^*(k^*) = \frac{r_0^{*2}}{\sqrt{2}} k^{*1/2} \int_0^\infty dx \frac{\sin^2(x)}{x^{3/2}} = \sqrt{\frac{\pi}{2}} r_0^{*2} k^{*1/2}, \tag{7.4.13}$$

which provides an accurate approximation for $k^* \gtrsim 10 r_0^{*2}$. This result is equivalent to Eq. (5.8.27), referring to the distribution of axial forces in a pencil beam.

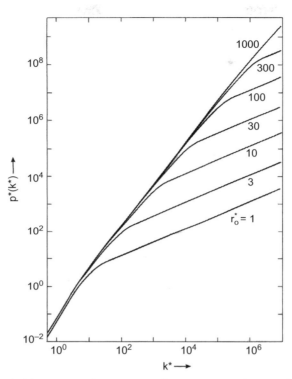

Fig. 7.8 The scaled function $p^*(k^*)$ corresponding to the distribution of axial velocity displacements generated in a homocentric cylindrical beam segment, plotted for different values of the scaled beam radius r_0^*.

For small k^*-values, the function $p^*(k^*)$ shows a quadratic k^*-dependency, which can be described as

$$\lim_{k^* \to 0} p^*(k^*) = \frac{1}{2} p_2^*(r_0^*) k^{*2}, \quad p_2^*(\infty) = 0.1513. \quad (7.4.14)$$

The small k^*-behavior of the function $p^*(k^*)$ is directly related to the second moment $\langle \Delta v_{z2}^{*2} \rangle$ of the two-particle distribution $p_2^*(\Delta v_z^*)$, as follows by expanding the sin-function in Eq. (7.4.11) for small k^*. By comparing this expansion with Eq. (7.4.14) one obtains

$$\langle \Delta v_{z2}^{*2} \rangle = p_2^*(r_0^*). \quad (7.4.15)$$

The quantity $\langle \Delta v_{z2}^{*2} \rangle$ does not stem from either weak or half-complete collisions but is built up over the entire Δv_z^*-range. This prohibits an analytical estimation of its value. We evaluated the function $p_2^*(r_0^*)/p_2^*(\infty)$ numerically. The result is depicted in Fig. 7.9. For later use, we fitted the numerical data with the function

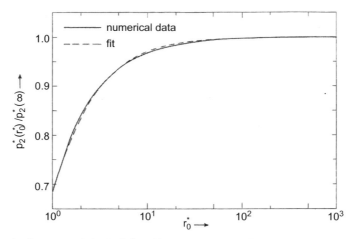

Fig. 7.9 The function $p^*_2(r_0^*)$, defined by Eq. (7.4.14) and its approximation given by Eq. (7.4.16). This function gives the r_0^*-dependency of the function $p^*(k^*)$, plotted in Fig. 7.8, for $k^* \to 0$.

$$p_{2a}^*(r_0^*) = \frac{p_2^*(\infty)}{\left(1 + .240/r_0^{*8/7}\right)^{7/4}} \quad (r_0^* \gtrsim 1), \quad (7.4.16)$$

which is also plotted in Fig. 7.9. The set of Eqs. (7.4.12)–(7.4.14) provides a complete analysis of the behavior of the function $p^*(k^*)$.

The final step consist of the calculation of the displacement distribution $\rho^*(\Delta v_z^*)$. By scaling Eq. (7.2.8), one obtains

$$\rho^*(\Delta v_z^*) = \nu \rho(\Delta v_z) = \frac{1}{\pi} \int_0^\infty dk^* \cos(k^* \Delta v_z^*) e^{-4\lambda^* p^*(k^*)}, \quad (7.4.17)$$

where λ^* is the scaled linear particle density for a cylindrical beam, which is given by

$$\lambda^* = \lambda \frac{\delta}{4r_0^{*2}} = \frac{C_0}{em} \frac{IL^2}{r_0^2 v_z^3} = \frac{m^{1/2}}{2^{7/2} \pi \varepsilon_0 e^{1/2}} \frac{IL}{r_0^2 V^{3/2}}, \quad (7.4.18)$$

analogous to $\bar{\lambda}$, used for a beam segment with a crossover; see Eq. (7.3.22). The second moment of this distribution follows from the function $p^*(k^*)$, using Eq. (5.5.4) (with $m = 2$)

$$\langle \Delta v_z^{*2} \rangle = 4 p_2^*(r_0^*) \lambda^*. \quad (7.4.19)$$

This relation can also be obtained from Eq. (7.4.15), using Eq. (5.5.10) (with $m = 2$).

Boersch effect

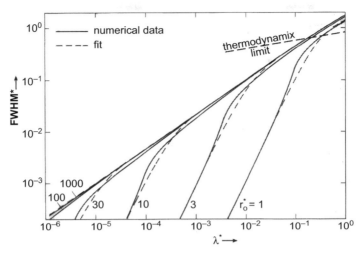

Fig. 7.10 The *FWHM* of the scaled distribution of axial velocity displacements $\rho^*(\Delta v_z^*)$ generated in a homocentric cylindrical beam segment, for different values of the scaled beam radius r_0^*. The depicted fit is defined by Eqs. (7.4.27) and (7.4.28). The thermodynamic limit is given by Eq. (7.7.4) and refers to the relaxation of potential energy.

We evaluated the *FWHM* of $\rho^*(\Delta v_z^*)$ numerically from the numerical data of $p^*(k^*)$. The results are plotted in Fig. 7.10. A key parameter for the interpretation of this data is $\chi_p = 4\lambda^* r_0^{*3}$, which is called the pencil beam factor for a cylindrical beam. It determines whether a beam should be considered as a pencil beam ($\chi_p \ll 1$) or an extended beam ($\chi_p \gg 1$). The physical significance of χ_p becomes evident by removing the scaling. Using Eq. (7.4.18) and Eq. (6.13.1), one obtains

$$\chi_p = 4\lambda^* r_0^{*3} = \lambda r_0. \tag{7.4.20}$$

Thus, χ_p is the ratio between the beam radius r_0 and the average axial separation of the particles $1/\lambda$. The following regimes should be distinguished in the data of Fig. 7.10:

(I) Gaussian regime. For $\lambda^* \gtrsim 1$ and $\chi_p \gg 1$ the distribution $\rho^*(\Delta v_z^*)$ is Gaussian, with a *FWHM* given by

$$\begin{aligned} FWHM_G^* &= 4\left[2p_2^*(\infty)\ln 2\right]^{1/2} P_{PE}(r_0^*)\lambda^{*1/2} \\ &= 1.832\ P_{PE}(r_0^*)\lambda^{*1/2}, \end{aligned} \tag{7.4.21}$$

where the function $P_{PE}(r_0^*)$ is defined as

$$P_{PE}(r_0^*) = \left[p_2^*(r_0^*)/p_2^*(\infty)\right]^{1/2}, \tag{7.4.22}$$

with $p_2^*(r_0^*)$ and $p_2^*(\infty)$ defined by Eq. (7.4.14). The subscript PE indicates that the result refers to the Energy distribution generated in a homocentric cylindrical beam segment in which the particles follow Parallel trajectories. The Gaussian behavior of $\rho^*(\Delta v_z^*)$ stems from the quadratic behavior of $p^*(k^*)$ for small k^* values.

(II) Holtsmark regime. For $\lambda^* \ll 1$ and $\chi_p \gg 1$ the distribution $\rho^*(\Delta v_z^*)$ becomes identical to the one-dimensional variant of the Holtsmark distribution. Substitution of Eq. (7.4.12) into Eq. (7.4.17) yields

$$\rho^*\left(\Delta v_z^*\right) = \frac{1}{\pi} \int_0^\infty dk^* \cos\left(k^* \Delta v_z^*\right) e^{-(4\sqrt{2\pi}/15)\lambda^* k^{*3/2}}, \qquad (7.4.23)$$

analogous to Eq. (5.8.25), specifying the distribution of the axial force component $F_\parallel$ in an extended cylindrical beam. The shape of this type of distribution is depicted in Fig. 5.3 (curve corresponding to $\gamma = 3/2$). The $FWHM$ of the distribution of Eq. (7.4.23) is equal to

$$FWHM_H^* = 2.8775 \left(4\sqrt{2\pi}/15\right)^{2/3} \lambda^{*2/3} = 2.1998\, \lambda^{*2/3}, \qquad (7.4.24)$$

as follows with Eq. (5.9.15) and Table 5.1.

(III) Pencil beam regime. For $\chi_p \ll 1$ the distribution $\rho^*(\Delta v_z^*)$ is determined by the large k^*-behavior specified by Eq. (7.4.13). Substitution into Eq. (7.4.17) yields

$$\rho^*\left(\Delta v_z^*\right) = \frac{1}{\pi} \int_0^\infty dk^* \cos\left(k^* \Delta v_z^*\right) e^{-2\sqrt{2\pi} r_0^{*2} \lambda^* k^{*1/2}}, \qquad (7.4.25)$$

analogous to Eq. (5.8.28), specifying the distribution of the axial force component $F_\parallel$ in a pencil beam. A plot of this type of distribution is given by Fig. 5.3 (curve for $\gamma = 1/2$). The $FWHM$ of the distribution of Eq. (7.4.25) follows again with Eq. (5.9.15) and Table 5.1,

$$FWHM_P^* = .447118\pi\, r_0^{*4} \lambda^{*2} = 11.237\, r_0^{*4} \lambda^{*2}. \qquad (7.4.26)$$

It can directly be verified that Eqs. (7.4.21), (7.4.24), and (7.4.26) provide a complete description of the different regimes shown in Fig. 7.10.

In order to interpolate Eqs. (7.4.21), (7.4.24), and (7.4.26) we use

$$FWHM^* = FWHM_H^* H_{PE}(\lambda^*, r_0^*), \qquad (7.4.27)$$

where $FWHM_H^*$ is given by Eq. (7.4.24) and the function $H_{PF}(\lambda^*, r_0^*)$ is defined as

$$H_{PE}(\lambda^*, r_0^*) = \left\{ \left[1 + \left(\frac{FWHM_H^*}{FWHM_G^*}\right)^6\right]^{1/4} + \left(\frac{FWHM_H^*}{FWHM_P^*}\right)^{3/2} \right\}^{-2/3},$$

which is identical to

$$H_{PE}(\lambda^*, r_0^*) = \left[\left(1 + 2.997 \frac{\lambda^*}{P_{PE}(r_0^*)^6}\right)^{1/4} + 1.386 \frac{1}{\chi_p^2}\right]^{-2/3}, \quad (7.4.28)$$

with $\chi_p = 4\lambda^* r_0^{*3}$, as specified by Eq. (7.4.20). The function $P_{PE}(r_0^*)$ is specified by Eq. (7.4.22) and the fit function given by Eq. (7.4.16). The function $H_{PE}(\lambda^*, r_0^*)$ is similar to the function $G_{CE}(\bar{\lambda}, \bar{r}_c)$, given by Eq. (7.3.35). However, in the present case we took $FWHM_H^*$ as reference, instead of $FWHM_G^*$, because the Holtsmark regime is dominant for practical operating conditions. For comparison, the fit of Eq. (7.4.27) is included in Fig. 7.10.

Eqs. (7.4.21) through (7.4.28) give a complete specification of the FWHM of the generated axial velocity distribution $\rho^*(\Delta v_z^*)$ in terms of the scaled coordinates, defined by Eqs. (6.13.1). The remaining task is to remove the scaling in order to make the dependency on the experimental parameters explicit. The properties of the scaled velocity distribution $\rho^*(\Delta v_z^*)$ can be related to the corresponding energy distribution, using

$$\Delta E \cong m v_z \Delta v_z = E \frac{2\nu}{v_z} \Delta v_z^* = E \left(\frac{4e}{\pi \varepsilon_0} \frac{1}{LV}\right)^{1/3} \Delta v_z^*. \quad (7.4.29)$$

The general expression of Eq. (7.4.27) transforms to

$$\frac{FWHM_E}{E} = C_{PHE} H_{PE}(\lambda^*, r_0^*) \frac{I^{2/3} L}{V^{4/3} r_0^{4/3}},$$

$$C_{PHE} = 2.8775 \frac{m^{1/3}}{(15\pi)^{2/3} \varepsilon_0}. \quad (7.4.30)$$

In the case of electrons, one finds $C_{PHE} = 2.4147$ in SI-units (note: the subscript PHE indicates that the constant applies to a Parallel beam and refers to the Holtsmark type of Energy distribution).

For the Holtsmark regime ($\lambda^* \ll 1$ and $\chi_p \gg 1$), one may approximate $H_{PE}(\lambda^*, r_0^*) = 1$. Accordingly, Eq. (7.4.30) yields directly

$$\frac{FWHM_E}{E} = C_{PHE} \frac{I^{2/3} L}{V^{4/3} r_0^{4/3}}, \quad (\lambda^* \ll 1, \ \chi_p \gg 1). \quad (7.4.31)$$

For the Gaussian regime ($\lambda^* \gtrsim 1$ and $\chi_p \gg 1$), Eq. (7.4.30) transforms to

$$\frac{FWHM_E}{E} = C_{PGE} P_{PE}(r_0^*) \frac{I^{1/2} L^{2/3}}{V^{13/12} r_0}, \quad (\lambda^* \gtrsim 1, \ \chi_p \gg 1). \quad (7.4.32)$$

where the constant C_{PGE} is given by

$$C_{PGE} = \frac{2^{17/12} \left[p_2^*(\infty) \ln 2 \right]^{1/2}}{\pi^{5/6}} \frac{e^{1/12} m^{1/4}}{\varepsilon_0^{5/6}},$$

in which $p_2^*(\infty)$ is defined by Eq. (7.4.14). Eq. (7.4.32) can also be derived directly from Eqs. (7.4.21) and (7.4.29). In the case of electrons, one finds $C_{PGE} = 0.4537$ in *SI*-units.

For the Pencil beam regime ($\chi_p \ll 1$) Eq. (7.4.30) yields

$$\frac{FWHM_E}{E} = C_{PPE} \frac{I^2 L}{V^2}, \qquad (\chi_p \ll 1), \qquad (7.4.33)$$

where the constant C_{PPE} is given by

$$C_{PPE} = 0.44711 \frac{m}{4\varepsilon_0 e^2},$$

which can also be derived directly from Eqs. (7.4.26) and (7.4.29). In the case of electrons, one finds $C_{PPE} = 4.4800 \times 10^{17}$ in *SI*-units.

7.5. Beam segment with a crossover of arbitrary dimensions

In the previous two sections the Boersch effect was calculated for the extreme cases of a beam segment with a narrow crossover and a homocentric cylindrical beam segment respectively. In this section we like to investigate the general case of a beam segment with a crossover in which the parameter K, defined by Eq. (7.3.1), has some arbitrary value between, say, 10^0 and 10^4. Unfortunately, this case is less suited for an analytical treatment, and one has to rely strongly on numerical calculations. In order to establish results that are of practical merit, the final numerical data will be fitted with analytical expressions. Our aim is to obtain a set of expressions that for $K \to \infty$ (narrow crossover) yield the results of Section 7.3 and for $K \to 0$ (homocentric cylindrical beam) yield the results of Section 7.4.

We will again use the scaling of Eq. (7.3.2). The scaled two-particle distribution $\bar{p}_2(\Delta \bar{v}_z)$ is defined by Eq. (7.3.4). As K is not necessarily large, only a fraction of the collisions will be complete. Accordingly, one may not use Eq. (7.3.3) for the entire integration domain of Eq. (7.3.4). The conditions for which Eq. (7.3.3) is valid can be expressed as

$$\bar{r}_c K_1 \bar{v} + \bar{a}_0 \gg \bar{b} \quad \text{and} \quad \bar{r}_c K_2 \bar{v} - \bar{a}_0 \gg \bar{b} \quad \text{and} \quad \bar{r}_c K \bar{v}^3 \gg 2, \qquad (7.5.1)$$

as follows from Eq. (6.7.7), using Eqs. (6.3.5), (6.2.3), (6.7.6), (7.3.1), and (7.3.2). We defined $\bar{a}_0 = a_0/d_0$ and $\bar{b} = b/d_0$, where a_0 and b are given by Eq. (7.2.3). Eq. (7.3.3) predicts too large values of $\Delta \bar{v}_z$ in case the constraints of Eq. (7.5.1) are violated.

For collisions that are weak, one may use Eq. (7.2.2). By scaling, it transforms to

$$\Delta \bar{v}_z \cong \frac{\bar{b}_z}{\bar{v}\bar{b}^2} \left[\frac{\bar{r}_c K_1 \bar{v} + \bar{a}_0}{\left[(\bar{r}_c K_1 \bar{v} + \bar{a}_0)^2 + \bar{b}^2\right]^{1/2}} + \frac{\bar{r}_c K_2 \bar{v} - \bar{a}_0}{\left\{(\bar{r}_c K_2 \bar{v} - \bar{a}_0)^2 + \bar{b}^2\right\}^{1/2}} \right]. \quad (7.5.2)$$

Notice that this expression depends on K_1 and K_2, unlike Eq. (7.3.3). We recall that a collision is weak when the eccentricity ε is large or the perihelion is not reached during the flight or the perihelion is already passed at the start of the collision, as expressed by Eq. (6.8.1). In terms of the sealed parameters used here, these conditions transform to

$$\bar{b}\bar{v}^2 \gg 2$$

or

$$\left[(\bar{r}_c K_1 \bar{v} + \bar{a}_0)^2 + \bar{b}^2\right]^{1/2} \gg \bar{r}_c K \bar{v} \quad (7.5.3)$$

or

$$\left[(\bar{r}_c K_2 \bar{v} - \bar{a}_0)^2 + \bar{b}^2\right]^{1/2} \gg \bar{r}_c K \bar{v}.$$

In the calculation of the two-particle distribution $\bar{p}_2(\Delta \bar{v}_z)$ from Eq. (7.3.4), our computer program determines for every two-particle collision whether all constraints of Eq. (7.5.1) are satisfied or not. If so, the collision is complete and it determines the displacement $\Delta \bar{v}_z$ from Eq. (7.3.3). If not, it tests next whether one or more of the constraints of Eq. (7.5.3) is satisfied or not. If so, the collision is weak and it evaluates the corresponding displacement $\Delta \bar{v}_z$ from Eq. (7.5.2). For collision that are neither complete nor weak, the displacement $\Delta \bar{v}_z$ is computed by means of the numerical approach that was outlined in Section 6.6 and is referred to as full collision dynamics. Clearly, one may use full collision dynamics for all types of collisions, but, if applicable, the analytical expressions are preferred, since their evaluation requires typically five times less computation time.

To start with, we will consider the case that the crossover is located in the middle of the beam segment; thus $S_c = 0.5$ and $K = K_1 = K_2$. At the end of this section, we will indicate how the results can be generalized to include other beam geometries. The two-particle distribution $\bar{p}_2(\Delta \bar{v}_z)$ was evaluated numerically from Eq. (7.3.4) for $K = 0.1, 0.2, 0.5, 1, 2, 5, 10, 20, 50, 100, 200, 500, 1000, 2000, 5000$, and $10,000$. For every K we considered the cases $\bar{r}_c = 10^0, 10^1, 10^2, 10^3, 10^4$, and 10^5. Fig. 7.11a–c depict the results obtained for $K = 1$, $K = 100$ and $K = 10,000$ respectively. These figures should be compared to Fig. 7.1, which is computed on the basis of complete collisions. The plots are identical for large values of $\Delta \bar{v}_z$. However, significant differences occur for small and intermediate

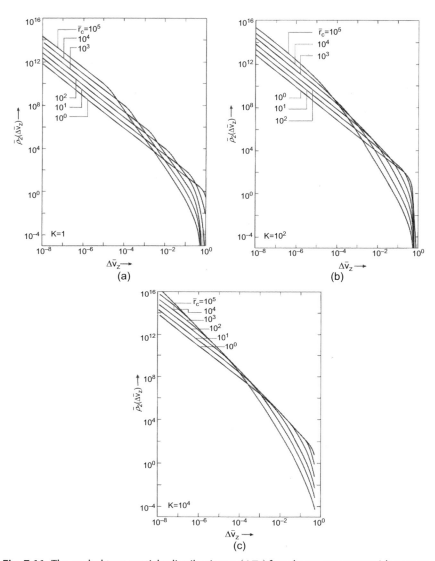

Fig. 7.11 The scaled two-particle distribution $\bar{p}_2(\Delta \bar{v}_z)$ for a beam segment with a crossover in the middle, plotted for different values of the scaled crossover radius $\bar{r}_c$. Panels (a–c) pertain to different values of the beam geometry parameter $K = \alpha_0 L/2 r_c$, as is indicated. The plots should be compared to those of Fig. 7.1, derived on the basis of complete collisions.

$\Delta \bar{v}_z$-values, especially in combination with small $\bar{r}_c$-values. Fig. 7.11 shows that the values obtained for $\bar{p}_2(\Delta \bar{v}_z)$ increase with K and ultimately approach the values of Fig. 7.1 for $K \to \infty$.

Let us investigate whether the results depicted in Fig. 7.11A–C can be understood by analytical means. The reader might verify that

Eq. (7.3.13), derived on the basis of complete collisions, provides an accurate approximation of the numerical data of $\bar{p}_2(\Delta\bar{v}_z)$ for $\Delta\bar{v}_z(K\bar{r}_c/2)^{1/3} \gg 1$. This makes it clear that the large $\Delta\bar{v}_z$-behaviour of $\bar{p}_2(\Delta\bar{v}_z)$ [thus $\Delta\bar{v}_z \gg (2/K\bar{r}_c)^{1/3}$ is determined by strong complete collisions. Small displacements $\Delta\bar{v}_z$ correspond to weak, predominantly incomplete collisions with more distant field particles. The plots of Fig. 7.11 show that the contribution of these collisions to $\bar{p}_2(\Delta\bar{v}_z)$ is very sensitive to the value of the beam geometry parameter K.

In order to analyze the small $\Delta\bar{v}_z$-behaviour in a quantitative way we will evaluate Eq. (7.3.4) on the basis of weak interactions, for which one may use Eq. (7.5.2). An exact solution of this problem does not appear to be feasible, and we will restrict the discussion to some special cases. We will first investigate the limit $K \to 0$ and next the limit $K \gg 1$. For $K \to 0$, one may approximate Eq. (7.5.2) as (taking $K_1 = K_2 = K$)

$$\Delta\bar{v}_z \cong \frac{\bar{b}_z \bar{r}_c K}{\left(\bar{b}_z^2 + \bar{r}_\perp^2\right)^{3/2}}, \quad (K \to 0), \tag{7.5.4}$$

using Eqs. (7.2.3). This results is identical to the scaled version of Eq. (7.2.5), derived for $v = 0$. This expression was used in the previous section to evaluate the contribution of weak interactions in a homocentric cylindrical beam segment. The corresponding results for the two-particle distribution $p_2(\Delta v_z)$ are expressed by Eqs. (7.4.4) and (7.4.5). Apparently, one may use these results to describe the limit $K \to 0$. However, one has to transform the results from the δ,ν-scaling used in Section 7.4 to the d_0,v_0-scaling used here. From Eqs. (6.13.1) and (7.3.2), it follows that the different sets of scaling quantities are related as

$$d_0 = \frac{1}{4v_0^{*2}}\delta, \qquad v_0 = v_0^* \nu$$

$$\delta = (4\bar{r}_c K)^{2/3} d_0, \qquad \nu = \left(\frac{2}{\bar{r}_c K}\right)^{1/3} v_0. \tag{7.5.5}$$

Accordingly, the different sets of parameters can be related using

$$\bar{r}_c = 4v_0^{*2} r_c^*, \qquad K = \frac{v_0^*}{2r_c^*}, \qquad \bar{\lambda} = \frac{r_0^{*2}}{v_0^{*2}}\lambda^*,$$

$$r_c^* = \left(\frac{\bar{r}_c}{16K^2}\right)^{1/3}, \qquad v_0^* = (\bar{r}_c K/2)^{1/3}, \qquad \lambda^* = 4\bar{\lambda}K^2(\bar{r}_c/\bar{r}_0)^2 \tag{7.5.6}$$

in which r_c denotes the crossover radius and r_0 the radius of the cylindrical beam. Clearly, both quantities become identical for $K \to 0$.

With Eqs. (7.5.5) and (7.5.6), one can rewrite Eqs. (7.4.4) and (7.4.5) as

$$\rho_2(\Delta\bar{v}_z) = \frac{2^{5/2}K^{3/2}}{5\bar{r}_c^{1/2}\Delta\bar{v}_z^{5/2}}, \quad \left[\Delta\bar{v}_z(\bar{r}_cK/2)^{1/3} \ll 1, \quad \Delta\bar{v}_z\bar{r}_c^{2/3}/(32K)^{1/3} \gg 1\right]$$

(7.5.7)

$$\bar{\rho}_2(\Delta\bar{v}_z) = \frac{(\bar{r}_cK/2)^{1/2}}{\Delta\bar{v}_z^{3/2}}, \quad \left[\Delta\bar{v}_z(\bar{r}_cK/2)^{1/3} \ll 1, \quad \Delta\bar{v}_z\bar{r}_c^{2/3}/(32K)^{1/3} \ll 1\right],$$

(7.5.8)

using the definition of the scaled two-particle distribution $\bar{\rho}_2(\Delta\bar{v}_z)$ given by Eq. (7.3.4). We note that these Equations can also be obtained directly from Eqs. (7.3.4) and (7.5.4). The asymptotic behavior described by Eq. (7.5.8) agrees with the data for small $\Delta\bar{v}_z$ shown in Fig. 7.11a–c. The behavior predicted by Eq. (7.5.7) becomes only manifest for extreme small K-values in combination with large $\bar{r}_c$-values; see the data for intermediate $\Delta\bar{v}_z$ shown Fig. 7.11a.

For $K \gg 1$ one may approximate Eq. (7.5.2) as (taking again $K_1 = K_2 = K$)

$$\Delta\bar{v}_z \simeq \frac{2\bar{r}_cK}{\bar{b}_z\left[(\bar{r}_cK\bar{v})^2 + \bar{b}_z^2\right]^{1/2}}, \quad (K \gg 1, \bar{b}_z \gg \bar{r}_\perp, \bar{v} \neq 0), \quad (7.5.9)$$

in which it was assumed that $\bar{v} \neq 0$ and $\bar{b}_z \gg \bar{r}_\perp$, thus $b \approx b_z$. This is justified since we are considering weak and therefore remote interactions. From Eqs. (7.3.4) and (7.5.9), one obtains (for $\Delta\bar{v}_z > 0$)

$$\bar{\rho}_2(\Delta\bar{v}_z) = \int_0^1 2\bar{v}\,d\bar{v} \int_0^\infty d\bar{b}_z\,\delta\left(\Delta\bar{v}_z - \frac{2\bar{r}_cK}{\bar{b}_z\left[(\bar{r}_cK\bar{v})^2 + \bar{b}_z^2\right]^{1/2}}\right). \quad (7.5.10)$$

By substituting

$$s = \frac{\bar{b}_z}{\bar{r}_cK\bar{v}}, \quad t = \frac{2}{\bar{r}_cKs(1+s^2)^{1/2}}\frac{1}{\bar{v}^2} \quad (7.5.11)$$

and carrying out the t-integration, Eq. (7.5.10) transforms to

$$\bar{\rho}_2(\Delta\bar{v}_z) = \frac{2^{3/2}}{(\bar{r}_cK)^{1/2}\Delta\bar{v}_z^{5/2}} \int_0^\infty \frac{ds}{s^{3/2}(1+s^2)^{3/4}} \Theta\left(\Delta\bar{v}_z - \frac{2}{\bar{r}_cKs(1+s^2)^{1/2}}\right)$$

(7.5.12)

in which $\Theta(x)$ is the step function. Let s_0 be a value of the integration variable s for which the argument of the step function becomes zero

$$s_0^4 - s_0^2 - (2/\bar{r}_cK\,\Delta\bar{v}_z)^2 = 0.$$

This equation has only a single positive root, which defines the lower integration boundary of Eq. (7.5.12). For $\bar{r}_cK\,\Delta\bar{v}_z \ll 1$, one finds

$s_0 \approx (2/\bar{r}_c K \, \Delta\bar{v}_z)^{1/2}$. Thus $s_0 \gg 1$. Accordingly, the integral in Eq. (7.5.12) yields approximately $1/2s_0^2 = \bar{r}_c K \, \Delta\bar{v}_z/4$ and one retrieves Eq. (7.5.8). For $\bar{r}_c K \, \Delta\bar{v}_z \gg 1$ one finds $s_0 \approx (2/\bar{r}_c K \, \Delta\bar{v}_z)^{1/4}$. Thus $s_0 \ll 1$. Accordingly, the integral in Eq. (7.5.12) is approximately equal to $2/\sqrt{s_0} = (2\bar{r}_c K \, \Delta\bar{v}_z)^{1/2}$ and one obtains

$$\bar{p}_2(\Delta\bar{v}_z) = \frac{4}{\Delta\bar{v}_z^2}, \quad (K \gg 1, \bar{r}_c K \, \Delta\bar{v}_z \gg 1, \Delta\bar{v}_z < 1/2), \tag{7.5.13}$$

which is identical to Eq. (7.3.12). Apparently, this dependency stems from collisions that are both weak and complete. Accordingly, the result does not depend on K. The different regimes of the two-particle distribution $\bar{p}_2(\Delta\bar{v}_z)$ are fully specified by Eqs. (7.3.13), (7.5.7), (7.5.8), and (7.5.13).

We now proceed with the calculation of the function $\bar{p}(\bar{k})$, defined by Eq. (7.3.14). This equation was evaluated numerically from the data of $\bar{p}_2(\Delta\bar{v}_z)$. The results for $K = 1$, 100, and 10,000 are shown in Fig. 7.12a–c respectively. These figures should be compared to Fig. 7.2, derived on the basis of complete collisions. The most significant differences occur for large $\bar{k}$-values. For complete collisions the large $\bar{k}$-behavior is given by Eq. (7.3.19). All curves corresponding to different $\bar{r}_c$-values ultimately become linear in $\bar{k}$ for $\bar{k} \to \infty$. This dependency corresponds to the inverse-square behavior of $\bar{p}(\Delta\bar{v}_z)$ found for small $\Delta\bar{v}_z$; see Eq. (7.3.12). For the general case studied in this section we found that $\bar{p}(\Delta\bar{v}_z)$ becomes ultimately proportional to $1/\Delta\bar{v}_z^{3/2}$ for $\Delta\bar{v}_z \to 0$; see Eq. (7.5.8). Substitution of this expression into Eq. (7.3.14) yields

$$\bar{p}(\bar{k}) = 2(\bar{r}_c K \bar{k})^{1/2} \int_0^\infty dx \, \frac{\sin^2 x}{x^{3/2}} = (4\pi \bar{r}_c K)^{1/2} \bar{k}^{1/2}, \quad (\bar{k} \to \infty), \tag{7.5.14}$$

which is in agreement with the data of Fig. 7.12a–c.

By similar reasoning, one finds that the behavior described by Eq. (7.5.7) leads to a 3/2-power dependency on $\bar{k}$,

$$\bar{p}(\bar{k}) = \frac{8(K\bar{k})^{3/2}}{5\bar{r}_c^{1/2}} \int_0^\infty dx \, \frac{\sin^2 x}{x^{5/2}} = \frac{32\sqrt{\pi} K^{3/2}}{15 \bar{r}_c^{1/2}} \bar{k}^{3/2}. \tag{7.5.15}$$

This dependency becomes manifest for intermediate $\bar{k}$-values in the curves for which $K/\bar{r}_c^{1/2} \ll 1$, see Fig. 7.12a. For increasing K-values the intermediate $\bar{k}$-range ultimately becomes dominated by a linear $\bar{k}$-dependency; see Fig. 7.12c. This dependency corresponds to the $\Delta\bar{v}_z$-behavior described by Eq. (7.5.13). Substitution of this expression into Eq. (7.3.14) yields

$$\bar{p}(\bar{k}) = 8\bar{k} \int_0^\infty dx \, \frac{\sin^2 x}{x^2} = 4\pi \bar{k}, \tag{7.5.16}$$

which is identical to Eq. (7.3.19).

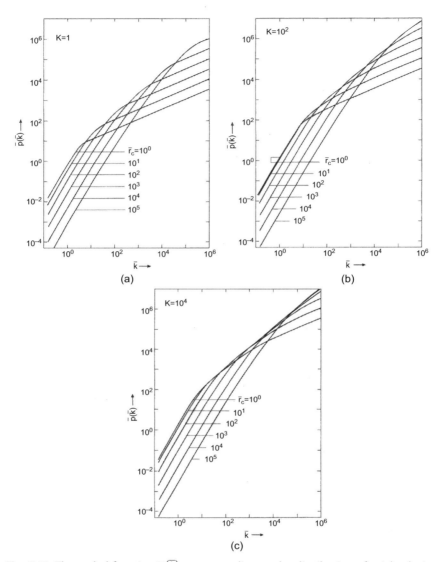

Fig. 7.12 The scaled function $\bar{p}(\bar{k})$ corresponding to the distribution of axial velocity displacements generated in a beam segment with a crossover in the middle, plotted for different values of the scaled crossover radius $\bar{r}_c$. Panels (a–c) pertain to different values of the beam geometry parameter $K = \alpha_0 L/2r_c$, as is indicated. The plots should be compared to those of Fig. 7.2, derived on the basis of complete collisions.

Boersch effect

For small $\bar{k}$-values, all curves shown in Fig. 7.12a–c become quadratic in $\bar{k}$. In order to describe this behavior we define the function $\bar{p}_2(\bar{r}_c, K)$ as

$$\bar{p}(\bar{k}) = \frac{1}{2}\bar{p}_2(\bar{r}_c, K)\bar{k}^2, \quad (\bar{k} \to 0). \tag{7.5.17}$$

The function $\bar{p}_2(\bar{r}_c, K)$ is the generalization of the function $\bar{p}_2(\bar{r}_c)$ defined by Eq. (7.3.16). The function $\bar{p}_2(\bar{r}_c, K)$ is directly related to the second moment of the displacement distribution $\bar{p}(\Delta \bar{v}_z)$, as can be seen from Eq. (7.3.30). We evaluated $\bar{p}_2(\bar{r}_c, K)$ numerically. The results are plotted in Fig. 7.13.

In order to obtain an analytical expression for $\bar{p}_2(\bar{r}_c, K)$, we will consider its behavior for $K \to \infty$ and $K \to 0$. In the limit $K \to \infty$, it should yield the result that was derived on the basis of complete collisions,

$$\lim_{K \to \infty} \bar{p}_2(\bar{r}_c, K) = \bar{p}_2(\bar{r}_c), \tag{7.5.18}$$

in which $\bar{p}_2(\bar{r}_c)$ defined by Eq. (7.3.16). For $K \to 0$, one should retrieve the result for a homocentric cylindrical beam segment, expressed by Eq. (7.4.14). By means of the scale relations of Eqs. (7.5.5) and (7.5.6), one obtains

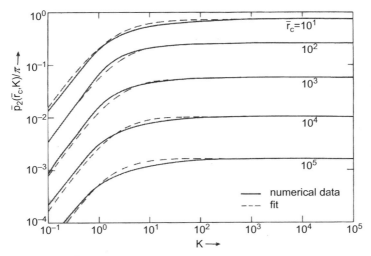

Fig. 7.13 The function $\bar{p}_2(\bar{r}_c, K)$ defined by Eq. (7.5.17) and its approximation given by Eq. (7.5.20). This function gives the dependency on $\bar{r}_c$ and K of the function $\bar{p}(\bar{k})$, plotted in Fig. 7.12a–c, for $\bar{k} \to 0$.

$$\lim_{K \to 0} \bar{p}_2(\bar{r}_c, K) = 16(2K^2/\bar{r}_c)^{2/3} p_2^* \left[(\bar{r}_c/16K^2)^{1/3} \right]. \quad (7.5.19)$$

For $\bar{r}_c \gg 16K$, one may approximate $p_2^* \left[(\bar{r}_c/16K)^{1/3} \right] \approx p_2^*(x) \approx 0.513$; see Eq. (7.4.14). We recall that the functions $\bar{p}_2$ and $\bar{p}_2^*$ can be approximated by the functions $\bar{p}_{2a}$ and p_{2a}^*, defined by Eqs. (7.3.17) and (7.4.16) respectively. As an approximation for the general function $\bar{p}_2(\bar{r}_c, K)$, we use

$$\bar{p}_{2a}(\bar{r}_c, K) = \frac{\pi}{.788 + .6\left(\frac{\bar{r}_c}{K_2}\right)^{2/3} \left[1 + .5\left(\frac{K^2}{\bar{r}_c}\right)^{4/9}\right]^{3/2} + \frac{\pi \bar{r}_c}{\{2 \ln [.8673(114.6 + \bar{r}_c)]\}^2}},$$

(7.5.20)

which shows the asymptotic behavior described by Eqs. (7.5.18) and (7.5.19), provided that $\bar{r}_c \gg 16K$. For comparison, we included a plot of the function given by Eq. (7.5.20) in Fig. 7.13.

Finally, we determined the FWHM of the displacement distribution $\bar{\rho}(\Delta \bar{v}_z)$ from the numerical data of the function $\bar{p}(\bar{k})$ by evaluating the integral of Eq. (7.3.21). The results for $K = 1$, 10, 100, 1000, and 10,000 are plotted in Fig. 7.14a–e. The following regimes should be distinguished.

(I) Gaussian regime. For large values of $\bar{\lambda}$ the distribution $\bar{\rho}(\Delta \bar{v}_z)$ is determined by the small $\bar{k}$- behavior of $\bar{p}(\bar{k})$ expressed by Eq. (7.5.17). The quadratic $\bar{k}$-dependency leads to a Gaussian distribution with a FWHM that is given by Eq. (7.3.31). P_{CE} now denotes a function that depends both on $\bar{r}_c$ and K,

$$P_{CE}(\bar{r}_c, K) = \left[\bar{p}_2(\bar{r}_c, K)/\pi\right]^{1/2}, \quad (7.5.21)$$

where $\bar{p}_2(\bar{r}_c, K)$ can be approximated by Eq. (7.5.20).

(II) Holtsmark regime. For intermediate value of $\bar{\lambda}$ and small K-values, the distribution $\bar{\rho}(\Delta \bar{v}_z)$ is dominated by the 3/2 power dependency of $\bar{p}(\bar{k})$, expressed by Eq. (7.5.15). The corresponding distribution is of the Holtsmark-type, with a FWHM that is equal to

$$FWHM_H = 2.8775\, 8\left(\frac{6\pi}{25}\right)^{1/3} H(K)\, \bar{\lambda}^{2/3} / \bar{r}_c^{1/3}. \quad (7.5.22)$$

The function $H(K)$ can be estimated from the numerical data such as presented in Fig. 7.11a–e. As an approximation of the numerical data, we use the function

$$H(K) = \frac{1}{\left[1 + 9K^{-2} + 2K^{-1/3}\right]^{1/2}}, \quad (7.5.23)$$

Boersch effect 207

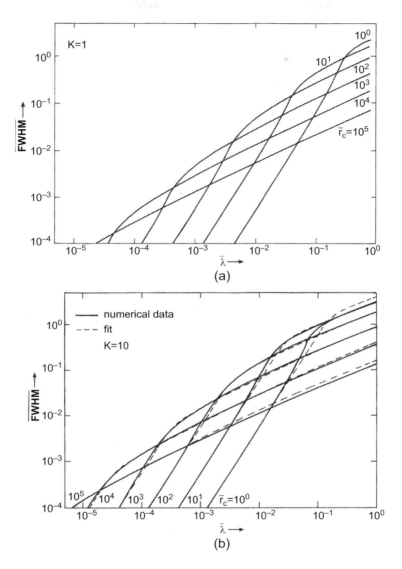

Fig. 7.14 The *FWHM* of the scaled distribution of axial velocity displacements $\bar{p}(\Delta\bar{v}_z)$ generated in a beam segment with a crossover in the middle, plotted for different values of the scaled crossover radius $\bar{r}_c$. Panels (a–e) pertain to different values of the beam geometry parameter $K = \alpha_0 L/2r_c$, as is indicated. The fit depicted in panel b is defined by Eqs. (7.5.30)–(7.5.32). As the quality of the fit is about the same for all K-values, it is omitted in the other figures. Panels (a–e) should be compared to Fig. 7.5, derived on the basis of complete collisions.

(Continued)

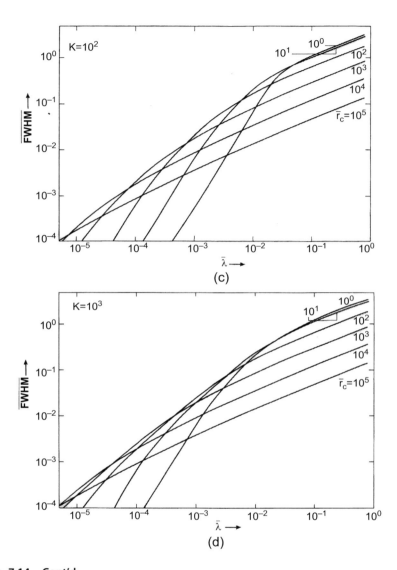

Fig. 7.14—Cont'd

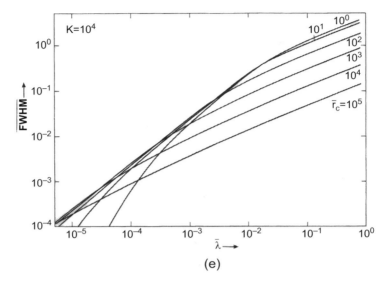

(e)

Fig. 7.14—Cont'd

which has the following asymptotes

$$\lim_{K \to 0} H(K) = K/3, \quad \lim_{K \to \infty} H(K) = 1 - K^{-1/3}. \quad (7.5.24)$$

We will demonstrate that these asymptotes can be derived from the analytical results. The limit $K \to 0$ corresponds to the conditions of a homocentric cylindrical beam, which was studied in the previous section. In the Holtsmark regime, one may use Eq. (7.4.24). By utilizing the scale relations of Eqs. (7.5.5) and (7.5.6), one can rewrite Eq. (7.4.24) in terms of the quantities $\bar{\lambda}$, $\bar{r}_c$, and K. By comparison with Eq. (7.5.22), one finds $H(K) = K/3$, which is in agreement with the first relation of Eq. (7.5.24).

The behavior for $K \to \infty$ can be verified from Eq. (7.4.24) by employing the slice method, which was described in Section 5.11. This method requires that the deviations of the particles from their unperturbed trajectories are small. For the Holtsmark regime, this condition is always fulfilled (when the deviations would be large the resulting distribution is not Holtsmarkian). In order to perform the slice method, one should express the Equations in terms of the experimental parameters. Using Eq. (7.3.36), one can rewrite Eq. (7.5.22) as

$$\frac{FWHM_E}{E} = C_{CHE} H(K) \frac{I^{2/3}}{V^{4/3} r_c^{1/3} \alpha_0},$$

$$C_{CHE} = 2.8775 \frac{2\,3^{1/3} m^{1/3}}{(5\pi)^{2/3} \varepsilon_0} \quad (= 6 C_{PHE}) \quad (7.5.25)$$

In the case of electrons, one finds $C_{CHE} = 14.488$ in SI-units. This equation refers to the energy spread generated in the entire beam segment. According to Eq. (7.4.31), the contribution of a single slice at axial position z of length Δz to the total generated energy distribution can be expressed as

$$\Delta(FWHM_E/E) = C_{PHE} \frac{I^{2/3}}{V^{4/3} r_c^{4/3}} \frac{\Delta}{[1 + \alpha_0 |z - z_c|/r_c]^{4/3}}, \qquad (7.5.26)$$

where z_c denotes the z-coordinate of the crossover. Substitution into Eq. (5.11.1) and integration yields Eq. (7.5.25), where the function $H(K)$ is now given by

$$H(K) = \left(1 - \frac{1}{2(1 + K_1)^{1/3}} - \frac{1}{2(1 + K_2)^{1/3}}\right), \qquad (7.5.27)$$

in which K_1 and K_2 are given by Eq. (7.3.1). Here we study the case $K = K_1 = K_2$. One directly sees that Eq. (7.5.27) satisfies the relations of Eq. (7.5.24). Thus Eq. (7.5.27) shows the same asymptotic behavior for $K \to 0$ and $K \to \infty$ as Eq. (7.5.23), based on the numerical results. However, for intermediate K-values Eq. (7.5.27) leads to somewhat smaller values than predicted by Eq. (7.5.23). This difference is due to the fact that the calculation based on the slice method relies on the assumption that the particle density is uniform in every cross section of the beam. In the numerical evaluation, on the other hand, we assumed a uniform spatial and angular distribution in the crossover; see Eq. (7.2.6). Consequently, the density distribution is not entirely uniform in the planes that do not coincide with the crossover, the lowest density being reached at the edge of the beam.

(III) Lorentzian regime. For intermediate $\bar{\lambda}$ and large K-values, the distribution $\bar{\rho}(\Delta \bar{v}_z)$ is dominated by the linear part of $\bar{p}(\bar{k})$, described by Eq. (7.5.16). Accordingly, it takes on the Lorentzian form, given by Eq. (7.3.26). The corresponding FWHM is given by Eq. (7.3.33). In terms of experimental parameters, it leads to Eq. (7.3.40).

(IV) Pencil beam regime. For small $\bar{\lambda}$ values, the distribution $\bar{\rho}(\Delta \bar{v}_z)$ is determined by the large $\bar{k}$-behavior of the function $\bar{p}(\bar{k})$, which shows a square-root dependency, as expressed by Eq. (7.5.14). The corresponding distribution is characteristic for pencil beams; see Eqs. (5.8.28) and (7.4.25). A beam with a crossover is referred to as a pencil beam if it fulfills the condition $\chi_c \ll 1$. The quantity χ_c is called the pencil beam factor for a beam segment with a crossover

$$\chi_c = 2K \bar{r}_c \bar{\lambda} = \alpha_0 L \lambda, \qquad (7.5.28)$$

which is the ratio between the distance a_0L and the average axial separation of the particles $1/\bar{\lambda}$. The *FWHM* of the distribution $\bar{p}(\Delta \bar{v}_z)$ generated in a pencil beam can be expressed as

$$\overline{FWHM}_P = .44711 \, 4\pi \, K\bar{r}_c\bar{\lambda}^2. \tag{7.5.29}$$

This relation can be obtained from Eq. (7.4.26) using the scaling relations of Eqs. (7.5.5) and (7.5.6). Eq. (7.4.33) gives the *FWHM* in terms of the experimental parameters. Notice that this Equation is independent of the beam diameter r_0 and thus independent of the beam geometry.

In order to interpolate the results obtained for the different regimes, we use

$$\overline{FWHM} = \overline{FWHM}_0 \, G_{CE}(\bar{\lambda}, \bar{r}_c, K)$$
$$\overline{FWHM}_0 = 2(2\pi \ln 2)^{1/2}\bar{\lambda}^{1/2}, \tag{7.5.30}$$

similar to Eq. (7.3.34). The function G_{CE} now also depends on K and is defined as

$$G_{CE}(\bar{\lambda}, \bar{r}_c, K) = \left[\left(\frac{\overline{FWHM}_0}{\overline{FWHM}_G}\right)^4 + \left(\frac{\overline{FWHM}_0}{\overline{FWHM}_H}\right)^4 \right.$$
$$\left. + \left(\frac{\overline{FWHM}_0}{\overline{FWHM}_L}\right)^4 + \left(\frac{\overline{FWHM}_0}{\overline{FWHM}_P}\right)^4 \right]^{-1/4}.$$

By substitution of Eqs. (7.3.31), (7.5.22), (7.3.33), and (7.5.29), one obtains

$$G_{CE}(\bar{\lambda}, \bar{r}_c, K) = \left[\frac{1}{P_{CE}(\bar{r}_c, K)^4} + \frac{A\bar{r}_c^{4/3}}{\bar{\lambda}^{2/3}H(K)^4} + \frac{B}{\bar{\lambda}^2} + \frac{C}{\bar{\lambda}^{-6}\bar{r}_c^{4}K^4} \right]^{-1/4}, \tag{7.5.31}$$

in which the constants A, B, and C are equal to

$$A = 1.575 \times 10^{-3}, \quad B = 7.606 \times 10^{-4}, \quad C = 0.3045. \tag{7.5.32}$$

The function $P_{CE}(\bar{r}_c, K)$ is defined by Eqs. (7.5.21) and (7.5.20), and the function $H(K)$ by Eq. (7.5.23). For $K \to \infty$, one retrieves Eq. (7.3.35), using that $H(\infty) = 1$ and $P_{CE}(\bar{r}_c, \infty) = P_{CE}(\bar{r}_c)$. This explains, *a posteriori*, the form of Eq. (7.3.35). For comparison, the fit of Eqs. (7.5.30) and (7.5.31) is included in Fig. 7.14b ($K = 10$).

Eq. (7.3.37) expresses the final result in terms of the experimental parameters. Eqs. (7.3.37) and (7.5.31) constitute the main results of this chapter. These expressions should yield accurate results for the *FWHM* energy spread generated in a beam segment of arbitrary geometry provided that the theoretical parameters $\bar{\lambda}$, $\bar{r}_c$, and K, defined by Eqs. (7.3.22), (7.3.5), and (7.3.1) respectively, fulfill the conditions

$$\bar{\lambda}K^2 < 1, \quad \bar{r}_c < 10^6, \tag{7.5.33}$$

which is usually the case for practical operating conditions.

Similar results can be derived for the Full Width median *(FW₅₀)* energy spread. Employing Eq. (5.9.15) and Table 5.1, one finds from Eq. (7.5.30)

$$\overline{FW}_{50} = .57288 \, \overline{FWHM}_0 \, G_{CE}(\bar{\lambda}, \bar{r}_c, K). \tag{7.5.34}$$

The function $G_{CE}(\bar{\lambda}, \bar{r}_c, K)$ is again given by Eq. (7.5.31), now taking the constants A, B, and C equal to

$$A = 8.246 \times 10^{-4}, \quad B = 8.192 \times 10^{-5}, \quad C = 3.015 \times 10^{-5} \tag{7.5.35}$$

instead of the values given by Eq. (7.5.32). The results for the individual regimes can be obtained from the expressions for the *FWHM*-values, using

$$\begin{aligned} FW_{50G} &= .57288 \, FWHM_G \\ FW_{50H} &= .67347 \, FWHM_H \\ FW_{50L} &= 1.0000 \, FWHM_L \\ FW_{50P} &= 5.7429 \, FWHM_P, \end{aligned} \tag{7.5.36}$$

as follows from the data of Table 5.1.

Finally, let us consider the case that the crossover is not located in the middle of the beam segment, and thus $S_c \neq 0.5$. Accordingly, and $K_1 \neq K_2$; See Eq. (7.3.1). In good approximation, one may determine the effective value of the function $G_{CE}(\bar{\lambda}, \bar{r}_c, K)$ as

$$G_{CE}(\bar{\lambda}, \bar{r}_c, K)_{eff} = \frac{1}{2} \left[G_{CE}(\bar{\lambda}, \bar{r}_c, K_1) + G_{CE}(\bar{\lambda}, \bar{r}_c, K_2) \right]. \tag{7.5.37}$$

The function $G_{CE}(\bar{\lambda}, \bar{r}_c, K)$ becomes independent of K for large K-values, especially in combination with large $\bar{\lambda}$-values. In that case, Eq. (7.5.37) is trivial. For small and intermediate K and $\bar{\lambda}$-values one may assume that the energy spread generated in the second part of the beam segment (from the crossover to the exit plane) is not affected by the interaction in the first part of the beam segment (from the entrance plane to the crossover), which justifies Eq. (7.5.37).

The validity of Eq. (7.5.37) can also be understood from the Equations applying to the individual regimes. For the pencil beam regime, we found Eq. (7.5.29), which complies with Eq. (7.5.37), since $K = (K_1 + K_2)/2$. The same conclusion follows for the Holtsmark regime; see Eqs. (7.5.22) and (7.5.27). In the Lorentzian regime, the result becomes independent of K; see Eq. (7.3.33), and Eq. (7.5.37) is evident. In the Gaussian regime, the approximation of Eq. (7.5.37) may become inaccurate due to the contribution of strong interactions. However, these errors will be small, since the dependency on K is weak, as can be seen from Eq. (7.5.20).

7.6. Results for Gaussian angular and spatial distributions

The results of the previous sections apply to the case of a uniform spatial and a uniform angular distribution in the crossover; see Eq. (7.2.6). We will now investigate the required modifications in case either the angular or the spatial distribution or both distributions are Gaussian, utilizing Eq. (5.3.4). For a Gaussian angular distribution, one has to replace the v-integration in Eq. (7.2.6) by

$$\int_0^\infty \frac{2v \, dv}{v_0^2} e^{-(v/v_0)^2}, \quad v_0 = \sqrt{2} v_z \sigma_a, \qquad (7.6.1)$$

where σ_a is the σ-value of the angular distribution in the crossover. For this two-dimensional Gaussian distribution $\sigma_a^2 = \langle \alpha^2 \rangle / 2$. For a Guassian spatial distribution, one has to replace the $r_\perp$-integration in Eq. (7.2.6) by

$$\int_0^\infty \frac{2r_\perp \, dr_\perp}{r_c^2} e^{-(r_\perp/r_c)^2}, \quad r_c = \sqrt{2} \sigma_r, \qquad (7.6.2)$$

where σ_r is the σ-value of the spatial distribution in the crossover (with $\sigma_r^2 = \langle r_\perp^2 \rangle / 2$).

Starting from the modified distribution of field particles prescribed by Eqs. (7.6.1) and (7.6.2) one can repeat the program outlined in Sections 7.3, 7.4, and 7.5. As the principle aspects involved remain the same, we will restrict the discussion to the presentation of the main results. We evaluated Eqs. (7.2.6) and (7.2.7) on the basis of complete collisions, using Eq. (7.2.1). Analogous to Eq. (7.3.15), we obtained the following expressions for the scaled function $\bar{p}(\bar{k})$:

- For a Gaussian angular distribution and a Gaussian spatial distribution

$$\bar{p}_{gg}(\bar{k}) = \bar{k} \frac{16}{\sqrt{\pi}} \int_0^\infty d\bar{v} \, e^{-\bar{v}^2} \int_0^\infty dy \, e^{-y^2} \int_0^1 du \frac{(1-u^2)^{1/2}}{u} \sin\left(\frac{\bar{k}\bar{v}u}{(4+\bar{v}^4 y^2 \bar{r}_c^2)^{1/2}}\right). \qquad (7.6.3)$$

- For a Gaussian angular distribution and a uniform spatial distribution

$$\bar{p}_{gu}(\bar{k}) = \bar{k} \frac{32}{\pi} \int_0^\infty d\bar{v} \, e^{-\bar{v}^2} \int_0^1 dy \sqrt{1-y^2}$$
$$\times \int_0^1 du \frac{(1-u^2)^{1/2}}{u} \sin\left(\frac{\bar{k}\bar{v}u}{(4+\bar{v}^4 y^2 \bar{r}_c^2)^{1/2}}\right). \qquad (7.6.4)$$

- For a uniform angular distribution and a Gaussian spatial distribution

$$\bar{p}_{ug}(\bar{k}) = \bar{k} \frac{16}{\sqrt{\pi}} \int_0^1 d\bar{v} \int_0^\infty dy \, e^{-y^2} \int_0^1 du \frac{(1-u^2)^{1/2}}{u} \sin\left(\frac{\bar{k}\bar{v}u}{(4 + \bar{v}^4 y^2 \bar{r}_c^2)^{1/2}}\right).$$

(7.6.5)

As before, we will investigate the behavior for $\bar{k} \to 0$ and for $\bar{k} \to \infty$.

For $\bar{k} \to 0$, Eqs. (7.6.3), (7.6.4), and (7.6.5) show a quadratic dependency on $\bar{k}$, as described by the first equation of (7.3.16). For the function $\bar{p}_2(\bar{r}_c)$ one finds from Eqs. (7.6.3), (7.6.4), and (7.6.5) respectively:

$$\bar{p}_{2gg}(\bar{r}_c) = \frac{4\sqrt{\pi}}{\bar{r}_c} \int_0^\infty dy \frac{\exp(-y^2)}{y} N\left(\frac{\bar{r}_c y}{2}\right) \qquad (7.6.6)$$

$$\bar{p}_{2gu}(\bar{r}_c) = \frac{8}{\bar{r}_c} \int_0^1 dy \frac{\sqrt{1-y^2}}{y} N\left(\frac{\bar{r}_c y}{2}\right) \qquad (7.6.7)$$

$$\bar{p}_{2ug}(\bar{r}_c) = \frac{4\sqrt{\pi}}{\bar{r}_c} \int_0^\infty dy \frac{\exp(-y^2)}{y} \sinh^{-1}\left(\frac{\bar{r}_c y}{2}\right), \qquad (7.6.8)$$

where the function $N(x)$ is defined as

$$N(x) = \int_0^\infty dz \frac{\exp(-z/x)}{(1+z^2)^{1/2}} = \frac{1}{2} \int_0^\infty dz \, \exp[-\sinh(z)/x]. \qquad (7.6.9)$$

For all practical purposes, one may approximate this function by

$$N_a(x) = \frac{1}{1/x + 1/\ln(17.58 + 1.180 \, x)}, \qquad (7.6.10)$$

which has the same asymptotic behavior for $x \to 0$ and $x \to \infty$ as the original function. The functions $\bar{p}_{2uu}(\bar{r}_c)$, $\bar{p}_{2gg}(\bar{r}_c)$ $\bar{p}_{2gu}(\bar{r}_c)$ and $\bar{p}_{2ug}(\bar{r}_c)$, defined by Eqs. (7.3.16), (7.6.6), (7.6.7), and (7.6.8) respectively, are plotted in Fig. 7.15. Similar to Eq. (7.3.17), we approximate these functions by

$$\bar{p}_{2a}(\bar{r}_c) = \frac{\pi}{1 + \pi F \bar{r}_c / \{2 \ln [.8673(E + F\bar{r}_c)]\}^2}, \qquad (7.6.11)$$

in which E and F are numerical constants, which are different for each plot

$$\begin{aligned} E_{gg} &= 40.74, & F_{gg} &= 1.315 \\ E_{gu} &= 49.19, & F_{gu} &= 1.138 \\ E_{ug} &= 100.0, & F_{ug} &= 1.160 \\ (E_{uu} &= 114.6, & F_{uu} &= 1) \end{aligned} \qquad (7.6.12)$$

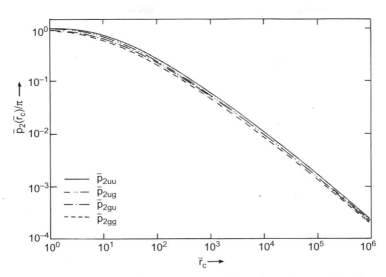

Fig. 7.15 The function $\bar{p}_2(\bar{r}_c)$ for different types of angular and spatial distributions in the crossover. The first subscript refers to the type of angular distribution and the second subscript to the type of spatial distribution (u = uniform, g = Gaussian). The curve $\bar{p}_{2uu}$ is defined by Eq. (7.3.16) and is the same as plotted in Fig. 7.3. The other curves are defined by Eqs. (7.6.6)–(7.6.8).

The constants E_{uu} and F_{uu} correspond to the values used in Eq. (7.3.17). The small $\bar{k}$-behavior of the function $\bar{p}(\bar{k})$, which is specified by the function $\bar{p}_2(\bar{r}_c)$, determines the mean square $\langle \Delta v_z^2 \rangle$ of the velocity distribution $\rho(\Delta v_z)$; see Eq. (7.3.30). The quadratic $\bar{k}$-dependency corresponds to a Gaussian distribution $\rho(\Delta v_z)$. The FWHM value of this distribution is proportional to $[\bar{p}_2(\bar{r}_c)]^{1/2}$; see Eq. (7.3.31).

For $\bar{k} \to \infty$, Eqs. (7.6.3)–(7.6.5) show a linear dependency on $\bar{k}$, as described by Eq. (7.3.19). The constant $\bar{p}_\infty$ varies for different angular and spatial distributions:

$$\bar{p}_{\infty gg} = \bar{p}_{\infty gu} = 2\pi^{3/2}, \quad \bar{p}_{\infty uu} = \bar{p}_{\infty ug} = 4\pi. \tag{7.6.13}$$

The linear $\bar{k}$-behavior corresponds to a Lorentzian distribution $\rho(\Delta v_z)$; see Eqs. (7.3.26) and (7.3.24). Eq. (7.6.13) indicates that the Lorentzian distribution is not affected by the type of spatial distribution but depends only on the angular distribution. This is in agreement with the fact that the FWHM of the corresponding energy distribution does not depend on r_c; see Eq. (7.3.40). The physical argument is that the Lorentzian behavior stems from weak complete collisions, which occur for a large axial separation b_z and a large flight distance Tv. The displacement Δv_z caused by such a collision is insensitive for $r_\perp$, as can be seen from Eq. (7.2.2).

It is often convenient to express the results obtained for non-uniform distributions in terms of the effective width measures α_{eff} and r_{eff}, which are defined as the widths for which the expressions obtained for uniform distributions yield the same results as those obtained within the full calculation, taking the proper distribution(s) into account. From Eqs. (7.6.13) and (7.3.40), it follows that the effective width of a Gaussian angular distribution is, for the Lorentzian regime, given by

$$\alpha_{eff} = \frac{2}{\sqrt{\pi}} \alpha_c = \frac{2\sqrt{2}}{\sqrt{\pi}} \sigma_\alpha = 1.596 \, \sigma_\alpha \quad \text{(Lorentzian regime)}. \quad (7.6.14)$$

For the Gaussian regime, the effective width measures depend on $\bar{r}_c$, as can be seen from Fig. 7.15. We ignore this dependency and approximate

$$\alpha_{eff} \cong 1.6 \, \sigma_\alpha, \quad r_{eff} \cong 1.6 \, \sigma_r \quad \text{(Gaussian regime)}, \quad (7.6.15)$$

as can be understood from Eqs. (7.6.11) and (7.6.12) with the definition of $\bar{r}_c$; see Eq. (7.3.5). The effective width measures of Eq. (7.6.15) yield very accurate results for $\bar{r}_c \gtrsim 100$ but are less suited to describe the results for smaller values of $\bar{r}_c$.

The remaining cases to be investigated pertain to the Holtsmark regime and the pencil beam regime. For those cases, the distribution $\rho(\Delta v_z)$ can be determined from the distribution of the axial interaction force $\rho(F_\parallel)$, occurring in the unperturbed beam. Accordingly, it is sufficient to determine in what way $\rho(F_\parallel)$ is affected by the type of density distribution. The calculation of $\rho(F_\parallel)$ for a cylindrical beam with a uniform density distribution was carried out in Section 5.8. The results are expressed by Eqs. (5.8.25) and (5.8.28). We will reconsider this calculation, now starting from a Gaussian radial density distribution. Using Eq. (7.6.2), one may modify Eq. (5.8.21) as

$$\rho_2(F_\parallel) = \int_{-\infty}^{\infty} dz \int_0^{\infty} \frac{2r \, dr}{r_0^2} e^{(r/r_0)^2} \delta\left(F_\parallel - \frac{C_0}{r^2} \frac{(z/r)}{\left(1 + (z/r)^2\right)^{3/2}} \right). \quad (7.6.16)$$

Analogous to Eq. (5.8.22), this transforms to

$$\rho_2(F_\parallel) = \frac{C_0^{3/2}}{r_0^2 F_\parallel^{5/2}} \int_0^{\infty} ds \frac{s^{3/2}}{(1+s^2)^{9/4}} \exp\left(-\frac{C_0}{F_\parallel r_0^2} \frac{s}{(1+s^2)^{3/2}} \right). \quad (7.6.17)$$

As before, we will investigate the extreme cases $r_0 \to \infty$ and $r_0 \to 0$, which correspond to an extreme extended beam and a pencil beam respectively.

For $r_0 \to \infty$, one may replace the exponential function in Eq. (7.6.17) by unity, and one retrieves Eq. (5.8.23). This expression leads to the one-dimensional variant of the Holtsmark distribution given by Eq. (5.8.25). Thus, one obtains the same equation as for a uniform density distribution. However r_0 now stands for $\sqrt{2}\sigma_r$ instead of the outer beam radius; see Eq. (7.6.2). The *FWHM* of the corresponding energy distribution is, for a homocentric cylindrical beam, given by Eq. (7.4.31) and for a beam segment with a crossover by Eq. (7.5.25). In general, the equations for a uniform distribution are retrieved by taking

$$\alpha_{eff} = \alpha_0 = \sqrt{2}\sigma_a, \quad r_{eff} = r_c = \sqrt{2}\sigma_r \quad \text{(Holtsmark regime)}. \quad (7.6.18)$$

For $r_0 \to 0$, one can approximate the integral of Eq. (7.6.16) by taking $1+s^2 \cong s^2$, which is justified since the contribution to the integral comes from large s-values. This gives

$$p_2(F_\parallel) = \frac{C_0^{3/2}}{r_0^2 F_\parallel^{5/2}} \int_0^\infty ds \frac{1}{s^3} \exp\left(-\frac{C_0}{F_\parallel r_0^2}\frac{1}{s^2}\right) = \frac{C_0^{1/2}}{2F_\parallel^{3/2}}, \quad (7.6.19)$$

which is identical to Eq. (5.8.26). This result does not depend on r_0 and is therefore not affected by the type of current density distribution. This can also be seen from the expression for the *FWHM* of the corresponding energy distribution, which is given by Eq. (7.4.33). The physical explanation is that all particles are effectively on a row. Accordingly, the result depends on the axial distribution of the particles only and not on their radial distribution.

Eqs. (7.6.14), (7.6.15), and (7.6.18) specify, for the different regimes, the values of the effective beam semi-angle α_{eff} of a Gaussian angular distribution and the effective crossover radius r_{eff} of a Gaussian spatial distribution. For beams with a narrow crossover, the generated energy distribution is either Gaussian or Lorentzian. In these regimes one should use $1.60 \times \sigma$ as an effective width for a Gaussian (angular or spatial) density distribution. For a (nearly) cylindrical extended beam, the generated energy distribution is Holtsmarkian. As an effective width for a Gaussian density distribution, one should, in this case, use $\sqrt{2} \times \sigma$. The type of density distribution does not affect the results obtained for a pencil beam.

The *FWHM* of the energy distribution generated in a beam segment with a crossover is, in the case of a uniform angular and spatial distribution, given by Eqs. (7.3.37) and (7.5.31). The different regimes are implicitly covered by these equations, and the question arises how to implement the results of this section. One way is to assume that Eq. (7.6.15) provides a suitable approximation for all regimes. This leads to reasonably accurate results for the Gaussian, the Lorentzian, and the pencil beam regime but introduces a significant error in the Holtsmark regime. A more accurate approach is to use the full equations obtained in this section for Gaussian density distributions instead of the various effective width measures α_{eff} and r_{eff}. Starting from Eqs. (7.6.11), (7.6.12), (7.6.13), and (7.6.18), one finds for each of the four combinations of a uniform or a Gaussian angular distribution and a uniform or a Gaussian spatial distribution (*uu*, *gg*, *gu*, and *ug*) a different version of the function $G_{CE}(\bar{\lambda}, \bar{r}_c, K)$, which is for the case of a uniform angular distribution and a uniform spatial distribution (*uu*) given by Eq. (7.5.31). For the resulting set of equations, the reader is referred to Section 16.4, which summarizes the analytical prescriptions for the calculation of the Boersch effect.

7.7. Thermodynamic limits

We wish to investigate whether the thermodynamic limits to the Boersch effect considered in Chapter 3 are reached for practical operating conditions. We distinguished two kinds of upper limits. One is associated with the "relaxation of kinetic energy" and given by Eq. (3.8.3). The other is associated with the "relaxation of potential energy" and given by Eq. (3.10.11). The former is relevant for a beam segment with a narrow crossover in which one may neglect the contribution of the relaxation of potential energy (as long as complete collisions are dominant). The latter can best be examined in a homocentric beam segment in which relaxation of kinetic energy does not occur.

It is convenient to express the thermodynamic limits in terms of the scaled parameters used in this chapter. Utilizing Eq. (7.3.36) one may rewrite Eq. (3.8.3) as (notation used in this chapter: $E = eV$, $\Delta E = \Delta E_{\parallel}$)

$$\langle \Delta \bar{v}_z^2 \rangle^{1/2} = 1/\sqrt{6}, \tag{7.7.1}$$

which specifies the thermodynamic limit of the relaxation of kinetic energy in a beam with a narrow crossover in terms of the rms value of the distribution of scaled axial velocities $\bar{p}(\Delta \bar{v}_z)$. Assuming that this distribution is Gaussian, the corresponding *FWHM* value is given by

$$\overline{FWHM} = \left(\frac{4\ln 2}{3}\right)^{1/2} = .96135. \tag{7.7.2}$$

This upper limit is indicated in Fig. 7.5. One sees that in order to reach this limit, one requires $\bar{\lambda} \gtrsim 0.05$ and small values of $\bar{r}_c$. From Fig. 7.6, one should conclude that these conditions are beyond the range of normal operation. Thus, for a single crossover, the thermodynamic limit of kinetic energy relaxation will, in general, not be reached. However, Figs. 7.5 and 7.6 also imply that the thermodynamic limit may possibly be reached for a succession of crossovers, in particular when the system employs a field emission gun of high brightness.

Let us now consider the thermodynamic limit of potential energy relaxation. Utilizing Eqs. (7.4.29) and (7.4.18), one can rewrite Eq. (3.10.11) as

$$\langle \Delta v_z^{*2} \rangle^{1/2} = (a/2)^{1/3} \lambda^{*1/6} = 0.3517\ \lambda^{*1/6}, \qquad (7.7.3)$$

which gives the thermodynamic limit of the relaxation of potential energy in a homocentric cylindrical beam segment, in terms of the rms value of the distribution of scaled axial velocities $\rho^*(\Delta v_z^*)$. Assuming that this distribution is Gaussian, the corresponding FWHM value is given by

$$FWHM^* = a^{1/3} 2^{7/6} (\ln 2)^{1/2} \lambda^{*1/6} = 0.8283\ \lambda^{*1/6}. \qquad (7.7.4)$$

This upper limit is indicated in Fig. 7.10. One sees that in order to reach this limit, one requires $\lambda^* \gtrsim 0.4$ and small values of r_c^*. One should conclude that this limit is not reached for normal operating conditions.

As the thermodynamic limit is in general not reached in either of the two extreme beam geometries considered, one may safely assume that the thermodynamic limit (of both kinetic and potential energy relaxation) is, in general, not reached in a single beam segment of any geometry. However, as was mentioned, the thermodynamic limit may possibly be reached in a beam consisting of a succession of beam segments all operating at extreme high particle density.

CHAPTER EIGHT

Statistical angular deflections

Contents

8.1. Introduction	221
8.2. General aspects	222
8.3. Beam segment with a narrow crossover	225
8.4. Homocentric cylindrical beam segment	231
8.5. Beam segment with a crossover of arbitrary dimensions	240
8.6. Application of the slice method	261
8.7. Results for Gaussian angular and spatial distributions	262

8.1. Introduction

In this chapter, the extended two-particle approach will be used to calculate the distribution of transverse velocity displacements $\rho(\Delta v_\perp)$ generated by statistical Coulomb interactions in a rotational symmetric beam segment in drift space. The distribution $\rho(\Delta v_\perp)$ is equivalent to the distribution of angular deflections $\rho(\Delta\alpha)$, since $\Delta\alpha \approx \Delta v_\perp / v_z$. The primary interest of this calculation is to obtain a better physical insight in the trajectory displacement effect, which is the subject of Chapter 9.

The calculation of the distribution of transverse velocity displacements $\rho(\Delta v_\perp)$ is, in many respects, similar to the calculation of the distribution of axial velocity displacements $\rho(\Delta v_z)$, which was presented in Chapter 7. Consequently, the organization of this chapter is quite similar, too. The chapter starts with a summary of the relevant results of Chapter 5 and 6, which cover the basics of the statistical part of the model and the calculation of the two-particle velocity displacement $\Delta v_\perp$ respectively. This material is then used to perform the calculation of the statistical angular deflections in two specific beam geometries, which are most suited for an analytical treatment of the problem. One is a beam segment with a narrow crossover and the other is a homocentric cylindrical beam segment. The general case of a beam segment of arbitrary geometry is considered next. The angular and spatial distribution in the crossover are taken uniform in most calculations, but Gaussian distribution(s) are considered too. For all geometries the full angular deflection distribution is computed.

Explicit expressions are presented for the Full Width at Half Maximum (*FWHM*) of this distribution, as well as the Full Width median value (*FW*$_{50}$) and the root mean square value (rms). Scaling is applied to reduce the number of independent parameters and to simplify the notation of the intermediate results. In the final results, the scaling is removed in order to make the dependency on the experimental parameters explicit.

A computer program has been developed to perform the various steps of the calculation numerically, similar to the program used for the Boersch effect. The results show that one should, in general, distinguish four regimes, each corresponding to a different type of angular deflection distribution: the Gaussian regime, the weak complete collision regime, the Holtsmark regime, and the pencil beam regime. The widths of the various distributions show different dependencies on the experimental parameters. It was found that the location of the crossover strongly influences the results, especially for low particle densities. This is due to the fact that the angular deflection experienced by a particle in the first part of the beam segment (from the entrance plane to the crossover) can partly or entirely be cancelled out in the second part of the beam segment (from the crossover to the exit plane), depending on the type of collision involved and the location of the crossover.

Where possible, the numerical calculations are verified by analytical means. Analytical results for the width of the angular deflection distribution can be obtained for all limiting cases where a single regime is dominant. A direct analytical calculation of the behavior in the transition areas between the different regimes seems not possible. However, the numerical data for the *FWHM* and *FW*$_{50}$ values are approximated by analytical expressions, covering the entire range of operating conditions.

8.2. General aspects

Consider a test particle running in axial direction along the central reference trajectory in a rotational symmetric beam segment. Due to the Coulomb interaction with its neighbor field particles, it experiences a change in transverse velocity $\Delta v_\perp$ and ends up with a spatial displacement Δr_f. The displacements $\Delta v_\perp$ and Δr_f experienced by a large number of test particles will be randomly distributed due to the stochastic nature of the distribution of field particles within the beam volume. In this chapter, we will calculate the distribution $\rho(\Delta \mathbf{v}_\perp)$. The distribution $\rho(\Delta \mathbf{v}_\perp)$ is equivalent to the distribution of angular deflections $\rho(\Delta \boldsymbol{\alpha})$ since, in paraxial approximation, $\Delta \boldsymbol{\alpha} \approx \Delta \mathbf{v}_\perp / \mathbf{v}_z$.

The trajectory displacement effect, which will be studied in the next chapter, corresponds to the occurrence of random lateral displacements Δr. The displacement Δr of a test particle is determined by extrapolating its final perturbed position along its final perturbed velocity toward some reference plane. Clearly, this procedure combines the lateral shift at the end of the beam segment Δr_f and the change in lateral velocity $\Delta v_\perp$ into the single virtual shift Δr. The reference plane is assumed to be optically conjugated to the target plane of the system. The distribution $p(\Delta r)$ can therefore directly be related to the blurring observed at the target. It provides all information that is needed to evaluate the performance of a practical system. The calculation of the distribution $p(\Delta v_\perp)$ is mainly of theoretical importance. Our primary objective is to provide the theoretical means to handle the more complex calculation of the trajectory displacement distribution $p(\Delta r)$.

We will determine the distribution $p(\Delta v_\perp)$ by employing the extended two-particle approach, which was outlined in Chapter 5. The dynamical part of the problem consists of the calculation of the transverse velocity shift $\Delta v_\perp$ experienced by the test particle due to the interaction with a single field particle. This problem was studied in Chapter 6. The shift $\Delta v_\perp$ can be expressed as a function of the geometrical variables $\xi = (r_\perp, b_z, v, \Phi)$, the time of flight T, and the initial time $t_i = -S_c T$; see Eqs. (6.2.1)–(6.2.3) respectively and Fig. 5.1 (note: $\Phi = \psi - \varphi$). Explicit analytical equations can be determined for (nearly) complete collisions and for weak collisions. For a complete collision ($T \to \infty$), we found

$$\Delta v_\perp \cong \frac{v\left[1 + \left(mv^2 r_\perp \sin(\Phi)/2C_0\right)^2\right]^{1/2}}{1 + \left(b_z^2 + r_\perp^2 \sin^2\Phi\right)(mv^2/2C_0)^2}. \tag{8.2.1}$$

See Eqs. (6.11.3), (6.4.16), and (6.3.4). The deviation $\Delta v_\perp$ caused by a weak collision is given by (reproducing Eq. 6.11.4)

$$\Delta v_\perp \cong \frac{C_0}{mvb}\left[\left(\frac{b}{\left[(S_c Tv + a_0)^2 + b^2\right]^{1/2}} - \frac{b}{\left\{[(1-S_c)Tv - a_0]^2 + b^2\right\}^{1/2}}\right)^2 \right.$$
$$\left. + \left(\frac{r_\perp \sin(\Phi)}{b}\right)^2 \left(\frac{S_c Tv + a_0}{\left[(S_c Tv + a_0)^2 + b^2\right]^{1/2}}\right.\right.$$
$$\left.\left. + \frac{(1-S_c)Tv - a_0}{\left\{[(1-S_c)Tv - a_0]^2 + b^2\right\}^{1/2}}\right)^2\right]^{1/2},$$
$$\tag{8.2.2}$$

in which

$$a_0 = -r_\perp \cos\Phi, \quad b = \left(b_z^2 + r_\perp^2 \sin^2\Phi\right)^{1/2}. \tag{8.2.3}$$

Notice that $\Delta v_\perp = 0$ when both $r_\perp = 0$ and $S_c = 1/2$. This implies that weak collisions taking place in a beam segment with a point crossover in the middle do not generate angular deflections. On the other hand, complete collisions do cause angular deflections in such a geometry, as can be seen from Eq. (8.2.1).

Eqs. (8.2.1) and (8.2.2) apply to the case that the initial relative velocity of the particles is nonzero ($v \neq 0$). In case $v = 0$, we found for a half-complete collision (reproducing Eq. 6.11.6)

$$\Delta v_\perp \cong \left(\frac{C_0}{m}\right)^{1/2} \frac{r_\perp}{\left(b_z^2 + r_\perp^2\right)^{3/4}} - \frac{r_\perp}{4T}, \quad (v = 0), \tag{8.2.4}$$

and for a weak collision (reproducing Eq. 6.11.7)

$$\Delta v_\perp \cong \frac{C_0}{m} \frac{r_\perp T}{\left(b_z^2 + r_\perp^2\right)^{3/2}}, \quad (v = 0). \tag{8.2.5}$$

Eqs. (8.2.1) and (8.2.2) will be used in the calculation of statistical angular deflections generated in a beam segment with a crossover, while Eqs. (8.2.4) and (8.2.5) are relevant for a homocentric cylindrical beam segment.

The statistical part of the problem consists of the evaluation of Eqs. (5.7.7), (5.7.8) and (5.7.10), in which $\Delta\eta$ now represents $\Delta v_\perp$

$$p_2(\Delta v_\perp) = \int_0^{v_0} \frac{2v\,dv}{v_0^2} \int_0^{2\pi} \frac{d\Phi}{2\pi} \int_0^{r_c} \frac{2r_\perp\,dr_\perp}{r_c^2} \\ \times \int_{-S_c L}^{(1-S_c)L} db_z \,\delta[\Delta v_\perp - \Delta v_\perp(v, \Phi, r_\perp, b_z)] \tag{8.2.6}$$

$$p(k) = \int_0^\infty d\Delta v_\perp\, p_2(\Delta v_\perp)[1 - J_0(k\Delta v_\perp)] \tag{8.2.7}$$

$$\rho(\Delta v_\perp) = \frac{1}{2\pi} \int_0^\infty k\,dk\, J_0(k\,\Delta v_\perp) e^{-\lambda p(k)}, \tag{8.2.8}$$

in which we expressed $p(\xi)d\xi$ directly in terms v, Φ, $r_\perp$ and b_z, using Eqs. (5.4.6) and (5.3.2). The distribution in v and $r_\perp$ are taken uniform with a cutoff at $v_0 = \alpha_0 v_z$ and r_c respectively, as prescribed by Eq. (5.3.3). Due to the rotational symmetry of the beam and the choice of the central reference trajectory, the displacement distribution $\rho(\Delta \mathbf{v}_\perp)$ will be rotational symmetric. This implies that $\rho(\Delta \mathbf{v}_\perp)$ depends only on the magnitude of the displacement $\Delta v_\perp$ and not on its direction. The probability of a displacement of size $\Delta v_\perp$ is equal to $\rho(\Delta v_\perp) = 2\pi \Delta v_\perp{}^2 \rho(\Delta \mathbf{v}_\perp)$.

The model represented by Eqs. (8.2.6)–(8.2.8) can be considered as the two-dimensional equivalent of the model utilized in the analysis of the Boersch effect; see Eqs. (7.2.6)–(7.2.8). The fundamental aspects involved are the same, and the reader is referred to Section 7.2 for a summary of the underlying assumptions.

8.3. Beam segment with a narrow crossover

In this section, we will calculate the distribution of transverse velocity displacements $\rho(\Delta v_\perp)$ generated in a beam segment with the characteristics $K_1 \gg 1$ and $K_2 \gg 1$. The quantities K_1 and K_2 represent the ratio between the beam radius, at the start and end of the segment respectively, and the crossover radius; see Eq. (7.3.1). The expectation is that in a beam geometry for which $K_1 \gg 1$ and $K_2 \gg 1$, the main contribution to $\rho(\Delta v_\perp)$ stems from collisions that are complete or nearly complete. Accordingly, we will start from Eq. (8.2.1). We emphasize that this approach is only justified for a narrow crossover. The general case of a beam geometry with arbitrary values for K_1 and K_2 will be treated in Section 8.5.

For convenience of notation, we will utilize the scaling quantities d_0 and v_0, defined by Eq. (7.3.2). Employing this scaling, Eq. (8.2.1) transforms to

$$\Delta \bar{v}_\perp = \frac{\Delta v_\perp}{v_0} = \frac{\bar{v}\left[1 + \bar{r}_\perp{}^2 \sin^2(\Phi)\bar{v}^4/4\right]^{1/2}}{1 + \left(\bar{b}_z{}^2 + \bar{r}_\perp{}^2 \sin^2\Phi\right)\bar{v}^4/4}. \quad (8.3.1)$$

The scaled two-particle distribution $\bar{\rho}_2(\Delta \bar{v}_\perp)$ follows from Eq. (8.2.6)

$$\bar{\rho}_2(\Delta \bar{v}_\perp) = \frac{v_0}{d_0} \rho_2(\Delta v_\perp)$$
$$= \int_0^1 2\bar{v}\, d\bar{v} \int_0^{2\pi} \frac{d\Phi}{2\pi} \int_0^{\bar{r}_c} \frac{2\bar{r}_\perp\, d\bar{r}_\perp}{\bar{r}_c{}^2} \int_{-\infty}^\infty d\bar{b}_z\, \delta\big[\Delta \bar{v}_\perp - \Delta \bar{v}_\perp(\bar{v}, \Phi, \bar{r}_\perp, \bar{b}_z)\big],$$
(8.3.2)

where $\Delta \bar{v}_\perp(\bar{v}, \Phi, \bar{r}, \bar{b}_z)$ is given by Eq. (8.3.1). The quantity $\bar{r}_c = r_c/d_0$ is the scaled crossover radius, specified by Eq. (7.3.5). We evaluated Eq. (8.3.2) numerically for different values of $\bar{r}_c$. The results are plotted in Fig. 8.1.

For a number of extreme cases, one can evaluate the two-particle distribution $\bar{\rho}_2(\Delta \bar{v}_\perp)$ by analytical means. Analogous to the derivation of Eq. (7.3.9), one finds from Eqs. (8.3.1) and (8.3.2).

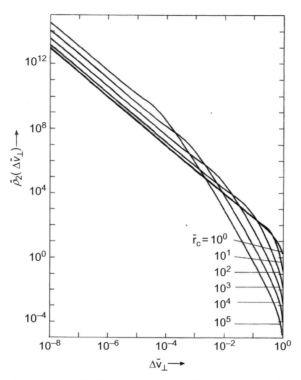

Fig. 8.1 The scaled two-particle distribution $\bar{p}_2(\Delta \bar{v}_\perp)$ based on complete collisions, for different values of the scaled crossover radius $\bar{r}_c$.

$$\bar{p}_2(\Delta\bar{v}_\perp) = \frac{16}{\pi} \int_0^1 \frac{d\bar{v}}{\bar{v}} \int_0^1 dy \sqrt{1-y^2} \int_0^1 \frac{R(\bar{v},y)du}{u^{3/2}(1-u)^{1/2}} \delta\left(\Delta\bar{v}_\perp - \frac{u\bar{v}}{2R(\bar{v},y)}\right), \tag{8.3.3}$$

where the function $R(\bar{v}, y)$ is defined as (reproducing Eq. 7.3.7).

$$R(\bar{v},y) = \left(1 + \bar{v}^4 y^2 \bar{r}_c^2/4\right)^{1/2}. \tag{8.3.4}$$

Eq. (8.3.3) can be rewritten as (for $0 < \Delta\bar{v}_\perp < 1$)

$$\bar{p}_2(\Delta\bar{v}_\perp) = \frac{16}{\pi \Delta\bar{v}_\perp^{3/2}} \int_{\Delta\bar{v}_\perp}^1 d\bar{v} \int_0^{\min\left[1, 2\left(1-\Delta\bar{v}_\perp^2/\bar{v}^2\right)^{1/2}/\Delta\bar{v}_\perp\bar{v}\bar{r}_c\right]} dy \frac{(1-y^2)^{1/2} R(\bar{v},y)^{1/2}}{[\bar{v} - \Delta\bar{v}_\perp R(\bar{v},y)]^{1/2}}. \tag{8.3.5}$$

Eq. (8.3.3) will be used to evaluate the function $p(k)$, while Eq. (8.3.5) is most suited to investigate the asymptotic behavior of $\bar{p}_2(\Delta\bar{v}_\perp)$.

For a point crossover ($\bar{r}_c = 0$), Eq. (8.3.5) becomes

$$\bar{p}_2(\Delta\bar{v}_\perp) = \frac{8(1-\Delta\bar{v}_\perp)^{1/2}}{\Delta\bar{v}_\perp^{3/2}}, \quad (\bar{r}_c = 0, 0 < \Delta\bar{v}_\perp < 1). \tag{8.3.6}$$

This seems the only case that leads to an exact analytical solution. However, a good approximation of $\bar{p}_2(\Delta\bar{v}_\perp)$ can be obtained for $\bar{r}_c \neq 0$ when $\bar{r}_c \Delta\bar{v}_\perp \ll 1$ or oppositely when $\bar{r}_c \Delta\bar{v}_\perp \gg 1$.

For small values of $\Delta\bar{v}_\perp$ and small values of $\bar{r}_c$ ($\bar{r}_c \Delta\bar{v}_\perp \ll 1$), one may approximate Eq. (8.3.5) as

$$\bar{p}_2(\Delta\bar{v}_\perp) = \frac{8}{\Delta\bar{v}_\perp^{3/2}} f_\infty(\bar{r}_c), \quad (\bar{r}_c \Delta\bar{v}_\perp \ll 1,\ 0 < \Delta\bar{v}_\perp < 1), \tag{8.3.7}$$

where the function $f_\infty(\bar{r}_c)$ is defined as

$$f_\infty(\bar{r}_c) = \frac{2}{\pi} \int_0^1 \frac{d\bar{v}}{\bar{v}^{1/2}} \int_0^1 dy (1-y^2)^{1/2} R(\bar{v}, y)^{1/2}. \tag{8.3.8}$$

This function has the following properties

$$\lim_{\bar{r}_c \to \infty} f_\infty(\bar{r}_c) = \frac{8}{15} \sqrt{\frac{2\,\Gamma(3/4)}{\pi\,\Gamma(1/4)}} \bar{r}_c^{1/2} = .14383\ \bar{r}_c^{1/2}$$

$$\lim_{\bar{r}_c \to 0} f_\infty(\bar{r}_c) = 1 \tag{8.3.9}$$

For later use we approximate the function $f_\infty(\bar{r}_c)$ by

$$f_{\infty^a}(\bar{r}_c) = (1 + .02069\,\bar{r}_c)^{1/2}, \tag{8.3.10}$$

which has the same asymptotic behavior for $\bar{r}_c \to 0$ and $\bar{r}_c \to \infty$ as the original function $f_\infty(\bar{r}_c)$.

For large values of $\Delta\bar{v}_\perp$ and large values of $\bar{r}_c$ ($\bar{r}_c \Delta\bar{v}_\perp \gg 1$) Eq. (8.3.5) yields in good approximation.

$$\bar{p}_2(\Delta\bar{v}_\perp) = \frac{128}{\pi} \frac{1}{\bar{r}_c \Delta\bar{v}_\perp^{5/2}} \left(\frac{1}{\Delta\bar{v}_\perp^{1/2}} - 1 \right), \quad (\bar{r}_c \Delta\bar{v}_\perp \gg 1, 0 < \Delta\bar{v}_\perp < 1). \tag{8.3.11}$$

The behavior of the two-particle function $\bar{p}_2(\Delta\bar{v}_\perp)$ described by Eqs. (8.3.6), (8.3.7) and (8.3.11) is in agreement with the numerical data plotted in Fig. 8.1.

We now proceed with the calculation of the function $p(k)$ and the displacement distribution $\rho(\Delta v_\perp)$. By scaling Eq. (8.2.7), one obtains

$$\bar{p}(\bar{k}) = \frac{p(k)}{d_0} = \int_0^\infty d\Delta\bar{v}_\perp \bar{p}_2(\Delta\bar{v}_\perp) \left[1 - J_0(\bar{k}\Delta\bar{v}_\perp)\right], \tag{8.3.12}$$

with $\bar{k} = kv_0$. Substitution of Eq. (8.3.3) and partial integration with respect to u, using that $dJ_0(z)/dz = -J_1(z)$, yields

$$\bar{p}(\bar{k}) = \bar{k}\frac{32}{\pi}\int_0^1 d\bar{v}\int_0^1 dy\sqrt{1-y^2}\int_0^1 du\left(\frac{1}{u}-1\right)^{1/2} J_1\left(\frac{u\bar{k}\bar{v}}{R(\bar{v},y)}\right), \quad (8.3.13)$$

in which J_1 is the first order Bessel function. We evaluated Eq. (8.3.13) numerically for different values of the scaled crossover radius $\bar{r}_c$. The results are plotted in Fig. 8.2. One sees that the function $\bar{p}(\bar{k})$ shows a quadratic $\bar{k}$-dependency for small $\bar{k}$-values and a square-root dependency for large $\bar{k}$-values.

We will investigate the extreme cases $\bar{k} \to 0$ and $\bar{k} \to \infty$ in more detail. For $\bar{k} \to 0$, one may approximate the Bessel function as $J_1(z) \approx z/2$ and carry out the u-integration, which yields a factor $\pi/8$. The $\bar{v}$-integration can be performed next and one obtains

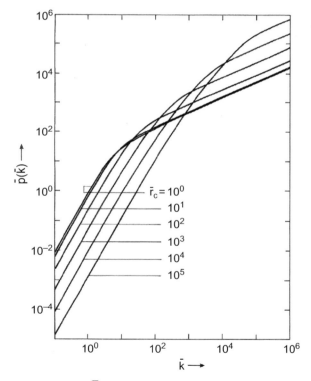

Fig. 8.2 The scaled function $\bar{p}(\bar{k})$ corresponding to the distribution of transverse velocity displacements generated by complete collisions, for different values of the scaled crossover radius $\bar{r}_c$.

$$\bar{p}(\bar{k}) = \frac{1}{4}\bar{p}_2(\bar{r}_c)\bar{k}^2,$$

$$\bar{p}_2(\bar{r}_c) = \frac{8}{\bar{r}_c}\int_0^1 dy\, \frac{\sqrt{1-y^2}}{y}\sinh^{-1}\left(\frac{\bar{r}_c y}{2}\right), \quad (\bar{k} \to 0). \tag{8.3.14}$$

The function $\bar{p}_2(\bar{r}_c)$ is the same function as found for the axial velocity distribution $\rho(\Delta v_z)$; see Eq. 7.3.16. Notice, however, that its definition differs by a factor 2, which is related to the fact that the distribution $\rho(\Delta \mathbf{v}_\perp)$ is two-dimensional, while the distribution $\rho(\Delta v_z)$ is one-dimensional. The function $\bar{p}_2(\bar{r}_c)$ can be approximated by the function $\bar{p}_{2a}(\bar{r}_c)$, which is given by Eq. (7.3.17). See also Fig. 7.3.

For $\bar{k} \to \infty$, one may approximate the u-integral in Eq. (8.3.13), using

$$\lim_{x\to\infty}\int_0^1 du\left(\frac{1}{u}-1\right)^{1/2}J_1(xu) \approx \int_0^1 du\, \frac{J_1(xu)}{u^{1/2}} = \frac{\Gamma(3/4)}{\Gamma(1/4)}\frac{2\sqrt{2}}{\sqrt{x}}, \tag{8.3.15}$$

based on the argument that the contribution to the integral comes from small u-values. Accordingly, one finds for $\bar{k} \to \infty$

$$\bar{p}(\bar{k}) = \bar{p}_\infty(\bar{r}_c)\bar{k}^{1/2}, \quad \bar{p}_\infty(\bar{r}_c) = \frac{32\sqrt{2}\,\Gamma(3/4)}{\Gamma(1/4)}f_\infty(\bar{r}_c), \quad (\bar{k}\to\infty), \tag{8.3.16}$$

where the function $f_\infty(\bar{r}_c)$ is given by Eq. (8.3.8). A good approximation of $f_\infty(\bar{r}_c)$ is provided by Eq. (8.3.10).

The final step consists of the calculation of the distribution $\bar{p}(\Delta \mathbf{v}_\perp)$. By scaling Eq. (8.2.8), one obtains

$$\bar{p}(\Delta \bar{\mathbf{v}}_\perp) = v_0^2 \rho(\Delta \mathbf{v}_\perp) = \frac{1}{2\pi}\int_0^\infty \bar{k}\,d\bar{k}\,J_0(\bar{k}\Delta\bar{v}_\perp)e^{-\bar{\lambda}\bar{p}(\bar{k})}, \tag{8.3.17}$$

where $\bar{\lambda}$ is the scaled linear particle density for a beam segment with a crossover, specified by Eq. (7.3.22). The second moment of this distribution follows directly from the function $\bar{p}(\bar{k})$, using Eqs. (5.5.4) (with $m=2$) and (8.3.14).

$$\langle \Delta\bar{v}_\perp^2\rangle = \bar{p}_2(\bar{r}_c)\bar{\lambda}. \tag{8.3.18}$$

By comparison with Eq. (7.3.30), one sees that $\langle \Delta\bar{v}_\perp^2\rangle = \langle \Delta\bar{v}_z^2\rangle$. It should be emphasized that this result relies on the assumption that complete collisions are dominant.

The FWHM of the distribution $\bar{p}(\Delta \mathbf{v}_\perp)$ given by Eq. (8.3.17), is plotted in Fig. 8.3 as function of $\bar{\lambda}$, for different values of $\bar{r}_c$. Two regimes should be distinguished.

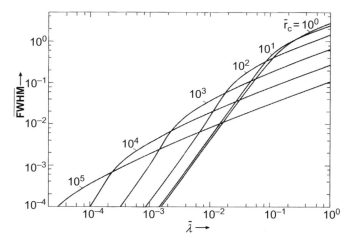

Fig. 8.3 The *FWHM* of the scaled distribution of transverse velocity displacements $\bar{p}(\Delta \bar{\mathbf{v}}_\perp)$ based on complete collisions, for different values of the scaled crossover radius $\bar{r}_c$.

(I) Gaussian regime. For large $\bar{\lambda}$-values, $\bar{p}(\Delta \bar{\mathbf{v}}_\perp)$ becomes equal to a two-dimensional Gaussian distribution. This can be understood from the fact that the integral of Eq. (8.3.17) is for large $\bar{\lambda}$-values dominated by the small $\bar{k}$-behavior of the function $\bar{p}(\bar{k})$, which shows a quadratic dependency; see Eq. (8.3.14). The *FWHM* of $\bar{p}(\Delta \bar{\mathbf{v}}_\perp)$, is in the Gaussian regime, given by

$$\overline{FWHM}_G = 2(\pi \ln 2)^{1/2} P_{CA}(\bar{r}_c) \bar{\lambda}^{-1/2}, \qquad (8.3.19)$$

$\left(\bar{\lambda} \gg 0.17 P_{CA}(\bar{r}_c)^{2/3} / f_\infty(\bar{r}_c)^{4/3}\right)$ where the function $P_{CA}(\bar{r}_c)$ is defined as

$$P_{CA}(\bar{r}_c) = P_{CE}(\bar{r}_c) = \left[\bar{p}_2(\bar{r}_c)/\pi\right]^{1/2} \qquad (8.3.20)$$

identical to Eq. (7.3.32). By removing the scaling by means of Eq. (7.3.2), one finds for the *FWHM* of the corresponding angular deflection distribution (using $FWHM_\alpha = \alpha_0 \, FWHM$)

$$FWHM_\alpha = C_{CGA} P_{CA}(\bar{r}_c) \sqrt{\frac{I}{V^{3/2}}}, \quad C_{CGA} = \left(\frac{(\ln 2)^2 m}{8\varepsilon_0^2 e}\right)^{1/4}. \qquad (8.3.21)$$

In the case of electrons, one finds $C_{CGA} = 256.90$ in *SI*-units. The subscript *CGA* indicates that the constant refers to the Gaussian Angular deflection distribution, generated in a beam segment with a Crossover.

(II) Weak complete collision regime. The distribution $\bar{p}(\Delta \bar{\mathbf{v}}_\perp)$ is for small $\bar{\lambda}$-values determined by the large $\bar{k}$-behavior of the function $\bar{p}(\bar{k})$, which shows a square-root dependency, as can be seen from Eq. (8.3.16). Substitution of Eq. (8.3.16) into Eq. (8.3.17) yields

$$\bar{\rho}(\Delta \bar{\mathbf{v}}_\perp) = \frac{1}{2\pi} \int_0^\infty \bar{k}\,d\bar{k}\, J_0(\bar{k}\,\Delta\bar{v}_\perp) e^{-\bar{\lambda}\bar{p}_\infty(\bar{r}_c)\bar{k}^{1/2}}. \qquad (8.3.22)$$

This type of distribution is depicted in Fig. 5.4 (curve corresponding to $\gamma = 1/2$). Its *FWHM* is given by

$$\overline{FWHM}_w = .17997 \bar{p}_\infty(\bar{r}_c)^2 \bar{\lambda}^2 = 42.105 f_\infty(\bar{r}_c)^2 \bar{\lambda}^2, \qquad (8.3.23)$$

$\left(\bar{\lambda} \ll 0.17 P_{CA}(\bar{r}_c)^{2/3}/f_\infty(\bar{r}_c)^{4/3}\right)$ as follows with Eq. (5.9.15) and Table 5.1. It can be verified that the $\bar{k}$-dependency given by Eq. (8.3.16) stems from weak complete collisions. Hence, this regime is called the "weak complete collision regime." Note that the analysis of the axial velocity distribution $\bar{\rho}(\Delta\bar{v}_z)$ resulted in a function $\bar{p}(\bar{k})$, which is linear for large $\bar{k}$, also due to the presence of weak complete collisions. The corresponding distribution $\bar{\rho}(\Delta\bar{v}_z)$ was found to be Lorentzian; see Eq. (7.3.26). Thus the similarity between the distributions $\bar{\rho}(\Delta\bar{\mathbf{v}}_\perp)$ and $\bar{\rho}(\Delta\bar{v}_z)$ disappears for small $\bar{\lambda}$-values.

The *FWHM* of the corresponding angular deflection distribution follows from Eq. (8.3.23) (using again $FWHM_\alpha = \alpha_0 \overline{FWHM}$)

$$FWHM_\alpha = C_{CWA} f_\infty(\bar{r}_c)^2 \frac{I^2}{\alpha_0^3 V^3}, \qquad (8.3.24)$$

where the constant C_{CWA} is equal to

$$C_{CWA} = 0.17997 \left(\frac{4\Gamma(3/4)}{\pi\Gamma(1/4)}\right)^2 \frac{m}{\varepsilon_0^2 e} = 0.033329 \frac{m}{\varepsilon_0^2 e}.$$

In the case of electrons, one finds $C_{CWA} = 2.4172 \times 10^9$ in *SI*-units.

8.4. Homocentric cylindrical beam segment

In this section, we will determine the distribution of transverse velocity displacements $\rho(\Delta v_\perp)$ generated in a homocentric cylindrical beam segment. The calculation is quite similar to the calculation of the distribution of axial velocity displacements $\rho(\Delta v_z)$ in this type of beam; see Section 7.4. The reader is referred to this section for the qualitative aspects involved.

We will utilize the δ, ν-scaling given by Eq. (6.13.1). The scaled two-particle distribution $\rho_2^*(\Delta v_\perp^*)$ follows from Eq. (8.2.6) (notation: $r^* = r_\perp^*$).

$$\rho_2{}^*(\Delta v_\perp{}^*) = \frac{\nu r_0^{*2}}{\delta}\rho_2(\Delta v_\perp)$$
$$= \int_0^{r_0{}^*} 2r^* dr^* \int_0^\infty db_z{}^* \delta[\Delta v_\perp{}^* - \Delta v_\perp{}^*(r^*, b_z{}^*)]. \tag{8.4.1}$$

We evaluated Eq. (8.4.1) numerically on the basis of the exact analysis of the function $\Delta v_\perp{}^*(r^*, b_z{}^*)$, following the procedure outlined in Section 6.6. The results are presented in Fig. 8.4. The function $\Delta v_\perp{}^*(r^*, b_z{}^*)$ can be determined analytically for half-complete collisions and for weak collisions; see Eqs. (8.2.4) and (8.2.5) respectively. One might conjecture that the function $\rho_2{}^*(\Delta v_\perp{}^*)$ is determined by weak interactions for most of the $\Delta v_\perp{}^*$-range plotted in Fig. 8.4, while its behavior for very large $\Delta v_\perp{}^*$-values stems from half-complete collisions. We will verify this statement by evaluating Eq. (8.4.1) analytically, employing the approximations given by Eqs. (8.2.4) and (8.2.5).

Let us first consider the limit of weak interactions. Instead of using Eq. (8.2.5), we prefer to exploit the analysis of the two-particle

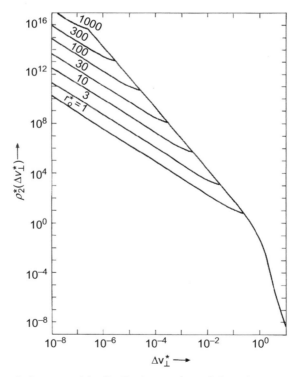

Fig. 8.4 The scaled two-particle distribution $\rho_2{}^*(\Delta v_\perp{}^*)$ for a homocentric cylindrical beam segment, plotted for different values of the scaled beam radius $r_0{}^*$.

distribution of lateral forces $p_2(F_\perp)$ presented in Section 5.8. This is justified since Eq. (8.2.5) is based on the first order perturbation approximation.

$$\Delta v_\perp \cong F_\perp T/m. \qquad (8.4.2)$$

Therefore, one may determine $p_2^*(\Delta v_\perp^*)$ from $p_2(F_\perp)$, using

$$p_2^*(\Delta v_\perp^*) = \frac{\nu r_0^{*2}}{\delta} p_2(\Delta v_\perp) = \frac{\nu r_0^{*2}}{\delta} \frac{m}{T} p_2(F_\perp = \Delta v_\perp m/T). \qquad (8.4.3)$$

Eq. (5.8.13) specifies $p_2(F_\perp)$ for an extended beam ($r_0 \to \infty$). Substitution into Eq. (8.4.3) yields

$$p_2^*(\Delta v_\perp^*) = \frac{3\sqrt{\pi}\Gamma(3/4)}{10\Gamma(1/4)} \frac{1}{\Delta v_\perp^{*5/2}}, \quad \left(\Delta v_\perp^* \ll 1, \; r_0^{*2}\Delta v_\perp^* \gg 1\right), \qquad (8.4.4)$$

in which we utilized the definitions of δ and ν given by Eq. (6.13.1). The numerical constant in the right-hand side of Eq. (8.4.4) is equal to 0.17973.

For a pencil beam ($r_0 \to 0$), one should use Eq. (5.8.18). Substitution into Eq. (8.4.3) yields

$$p_2^*(\Delta v_\perp^*) = \frac{4^{2/3} r_0^{*7/3}}{7 \Delta v_\perp^{*4/3}}, \quad \left(\Delta v_\perp^* \ll 1, r_0^{*2}\Delta v_\perp^* \ll 1\right). \qquad (8.4.5)$$

One may verify from Fig. 8.4 that the small and intermediate $\Delta v_\perp^*$-behavior of $p_2^*(\Delta v_\perp^*)$ indeed follows Eqs. (8.4.4) and (8.4.5).

In order to investigate the large $\Delta v_\perp^*$-behavior of $p_2^*(\Delta v_\perp^*)$, we start from Eq. (8.2.4). It specifies the displacement $\Delta v_\perp$ for a half-complete collision. By taking the limit $T \to \infty$, one may ignore the last term in this equation. After scaling, one obtains, with Eq. (8.4.1)

$$p_2^*(\Delta v_\perp^*) = \int_0^{r_0^*} 2r^* dr^* \int_0^\infty 2 db_z^* \delta\left(\Delta v_\perp^* - \frac{r^*}{2\left(r^{*2} + b_z^{*2}\right)^{3/4}}\right). \qquad (8.4.6)$$

By substituting

$$s = \frac{b_z^*}{r^*}, \quad t = \frac{1}{2r^{*1/2}(1+s^2)^{3/4}} \qquad (8.4.7)$$

and carrying out the t-integration, Eq. (8.4.6) transforms to

$$p_2^*(\Delta v_\perp^*) = \frac{1}{8 \Delta v_\perp^{*7}} \int_0^\infty \frac{ds}{(1+s^2)^{9/2}} \Theta\left(\Delta v_\perp^* - \frac{1}{2 r_0^{*1/2}(1+s^2)^{3/4}}\right), \qquad (8.4.8)$$

where $\Theta(x)$ is the step function defined by Eq. (3.7.8). For an extended beam ($r_0^* \to \infty$), the argument of the step function is positive for all positive values of $\Delta v_\perp^*$, and one obtains

$$\rho_2^*(\Delta v_\perp^*) = \frac{1}{8\,\Delta v_\perp^{*7}} \int_0^\infty ds\, \frac{1}{(1+s^2)^{9/2}} = \frac{2}{35}\,\frac{1}{\Delta v_\perp^{*7}}. \qquad (8.4.9)$$

This asymptote gives a good approximation of $\rho_2^*(\Delta v_\perp^*)$ for $\Delta v_\perp^* \gtrsim 1$; see Fig. 8.4.

For a pencil beam ($r_0^* \to 0$), the argument of the step function is positive for $s > s_0 \approx (2\,\Delta v_\perp^*\, r_0^{*1/2})^{-2/3}$. Thus the lower integration boundary of the integral in Eq. (8.4.8) should be replaced by s_0. As $s_0 \gg 1$, one finds in good approximation

$$\rho_2^*(\Delta v_\perp^*) = \frac{1}{8\,\Delta v_\perp^{*7}} \int_{s_0 = (2\,\Delta v_\perp^* r_0^{*1/2})^{-2/3}}^\infty ds\, 1/s^9 = \frac{r_0^{*8/3}}{2^{2/3}\,\Delta v_\perp^{*5/3}}. \qquad (8.4.10)$$

However, this asymptote is not reached for normal operating conditions ($r_0^* \gtrsim 1$) and will be disregarded in the further analysis. Eqs. (8.4.4), (8.4.5) and (8.4.9) provide a complete analysis of the behavior of the two-particle distribution $\rho_2^*(\Delta v_\perp^*)$.

The next step is to determine the scaled function $p^*(k^*)$, which follows from Eq. (8.2.7).

$$p^*(k^*) = \frac{r_0^{*2}}{\delta} p(k) = \int_0^\infty d\Delta v_\perp^*\, \rho_2^*(\Delta v_\perp^*)[1 - J_0(k^*\Delta v_\perp^*)], \qquad (8.4.11)$$

with $k^* = kv$. We evaluated the integral of Eq. (8.4.11) numerically, from the numerical data of $\rho_2^*(\Delta v_\perp^*)$. The results are shown in Fig. 8.5.

The behavior of the function $p^*(k^*)$ can be understood from the analysis of the two-particle distribution $\rho_2^*(\Delta v_\perp^*)$. Substitution of Eq. (8.4.4) into Eq. (8.4.11) yields.

$$p^*(k^*) = \frac{3\sqrt{\pi}\,\Gamma(3/4)}{10\,\Gamma(1/4)} k^{*3/2} \int_0^\infty dx\, \frac{1 - J_0(x)}{x^{5/2}} = \frac{(2\pi)^{1/2}}{15} k^{*3/2}, \qquad (8.4.12)$$

which provides an accurate approximation for $k^* \gtrsim 10$ and $k^* \lesssim 10 r_0^{*2}$. This result can also be obtained from Eq. (5.8.14), which refers to the distribution of lateral forces in an extended beam.

Substitution of Eq. (8.4.5) into Eq. (8.4.11) yields

$$p^*(k^*) = \frac{2^{4/3}}{7} r_0^{*7/3} k^{*1/3} \int_0^\infty dx\, \frac{1 - J_0(x)}{x^{4/3}} = \frac{36\,\Gamma(5/6)}{7\,\Gamma(1/6)} r_0^{*7/3} k^{*1/3}, \qquad (8.4.13)$$

which provides an accurate approximation for $k^* \gtrsim 10 r_0^{*2}$. The numerical constant in the right-hand side of Eq. (8.4.13) is equal to 1.0429. Eq. (8.4.13)

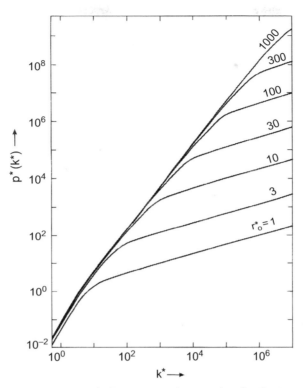

Fig. 8.5 The scaled function $p^*(k^*)$ corresponding to the distribution of transverse velocity displacements generated in a homocentric cylindrical beam segment, plotted for different values of the scaled beam radius r_0^*.

is equivalent to Eq. (5.8.19), referring to the distribution of lateral forces in a pencil beam.

For small k^*-values, the function $p^*(k^*)$ shows a quadratic k^*-dependency, which can be described as

$$\lim_{k_* \to 0} p^*(k^*) = \frac{1}{4} p_2^*(r_0^*) k^{*2}, \quad p_2^*(\infty) = 0.3026. \tag{8.4.14}$$

The small k^*-behavior of the function $p^*(k^*)$ is directly related to the second moment $\langle \Delta v_{\perp 2}^{*2} \rangle$ of the two-particle distribution $\rho_2^*(\Delta v_\perp^*)$.

$$\left\langle \Delta v_{\perp 2}^{*2} \right\rangle = p_2^*(r_0^*). \tag{8.4.15}$$

The quantity $\langle \Delta v_{\perp 2}^{*2} \rangle$ does not stem from either weak or half-complete collisions alone but is built up over the entire $\Delta v_\perp^*$-range. This prohibits

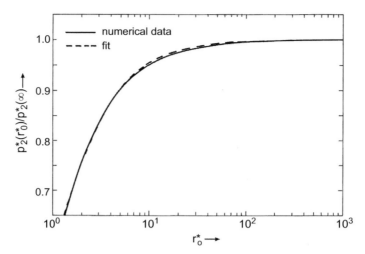

Fig. 8.6 The function $p_2^*(r_0^*)$ defined by Eq. (8.4.14) and its approximation given by Eq. (8.4.16). This function gives the r_0^*-dependency of the function $p^*(k^*)$, plotted in Fig. 8.5, for $k^* \to 0$.

an analytical estimation of its value. We evaluated the function $p_2^*(r_0^*)/p_2^*(\infty)$ numerically. The result is depicted in Fig. 8.6. The data is fitted by the function

$$p_{2a}^*(r_0^*) = \frac{p_2^*(\infty)}{\left(1 + .185/r_0^{*8/7}\right)^{7/2}}, \quad (r_0^* \gtrsim 1), \tag{8.4.16}$$

which is also plotted in Fig. 8.6. The set of Eqs. (8.4.12)–(8.4.14) provides a complete analysis of the behavior of the function $p^*(k^*)$.

The final step consists of the calculation of the displacement distribution $\rho^*(\Delta v_\perp^*)$. By scaling Eq. (8.2.8), one obtains

$$\rho^*(\Delta \mathbf{v}_\perp^*) = \nu \rho(\Delta \mathbf{v}_\perp) = \frac{1}{2\pi} \int_0^\infty k^* dk^* \; J_0(k^* \Delta v_\perp^*) e^{-4\lambda^* p^*(k^*)}, \tag{8.4.17}$$

where λ^* is the sealed linear particle density for a cylindrical beam given by, Eq. (7.4.18). The second moment of this distribution follows from the function $p^*(k^*)$, using Eq. (5.5.4) (with $m=2$).

$$\left\langle \Delta v_\perp^{*2} \right\rangle = 4 p_2^*(r_0^*) \lambda^*. \tag{8.4.18}$$

This relation can also be obtained from Eq. (8.4.15), using Eq. (5.5.10) (with $m=2$).

The FWHM of $\rho^*(\Delta v_\perp^*)$ was evaluated numerically from the numerical data of $p^*(k^*)$. The results are plotted in Fig. 8.7. A key parameter for

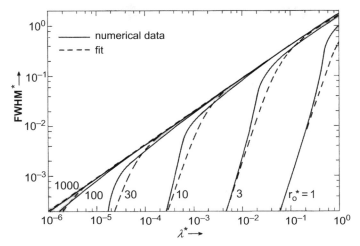

Fig. 8.7 The *FWHM* of the scaled distribution of transverse velocity displacements $\rho^*(\Delta v_\perp^*)$ generated in a homocentric cylindrical beam segment, for different values of the scaled beam radius r_0^*. The depicted fit is defined by Eqs. (8.4.25) and (8.4.26).

the interpretation of this data is the pencil beam factor for a cylindrical beam χ_p, given by Eq. (7.4.20). The following regimes should be distinguished.

(I) Gaussian regime. For $\lambda^* \gtrsim 1$ and $\chi_p \gg 1$, the distribution $\rho^*(\Delta v_\perp^*)$ is dominated by the quadratic behavior of $p^*(k^*)$ for small k^* values. This leads to a Gaussian distribution, with a *FWHM* given by

$$FWHM_G^* = 4\left[p_2^*(\infty)\ln 2\right]^{1/2} P_{PA}(r_0^*)\lambda^{*1/2} \qquad (8.4.19)$$
$$= 1.832 \; P_{PA}(r_0^*)\lambda^{*1/2},$$

where the function $P_{PA}(r_0^*)$ is defined as

$$P_{PA}(r_0^*) = \left[p_2^*(r_0^*)/p_2^*(\infty)\right]^{1/2}, \qquad (8.4.20)$$

in which $p_2^*(r_0^*)$ and $p_2^*(\infty)$ are defined by Eq. (8.4.14). The subscript *PA* indicates that the result refers to the <u>A</u>ngular deflection distribution generated in a cylindrical beam segment in which the particles run along <u>P</u>arallel trajectories.

(II) Holtsmark regime. For $\lambda^* \ll 1$ and $\chi_p \gg 1$, the distribution $\rho^*(\Delta v_\perp^*)$ becomes equal to the two-dimensional variant of the Holtsmark distribution, which is obtained by substituting Eq. (8.4.12) into Eq. (8.4.17).

$$\rho^*(\Delta \mathbf{v}_\perp^*) = \frac{1}{2\pi}\int_0^\infty k^* dk^* \; J_0(k^* \Delta v_\perp^*) e^{-\left(4\sqrt{2\pi}/15\right)\lambda^* k^{*3/2}}, \qquad (8.4.21)$$

analogous to Eq. (5.8.16), specifying the distribution of the lateral force component $F_\perp$ in an extended cylindrical beam. The shape of this type of distribution is depicted in Fig. 5.4 (curve corresponding to $\gamma = 3/2$). The *FWHM* of the distribution of Eq. (8.4.21) is equal to

$$FWHM_H^* = 2.6554 \left(4\sqrt{2\pi}/15\right)^{2/3} \lambda^{*2/3} = 2.0300 \, \lambda^{*2/3}, \qquad (8.4.22)$$

as follows with Eq. (5.9.15) and Table 5.1.

(III) Pencil beam regime. For $\chi_p \gg 1$, the distribution $\rho^*(\Delta v_\perp^*)$ is determined by the large k^*-behavior specified by Eq. (8.4.13). Substitution into Eq. (8.4.17) yields

$$\rho^*(\Delta \mathbf{v}_\perp^*) = \frac{1}{2\pi} \int_0^\infty k^* dk^* \, J_0(k^* \Delta v_\perp^*) e^{-2C_{1/3} r_0^{*7/3} \lambda^* k^{*1/3}}, \qquad (8.4.23)$$

with $C_{1/3} = 72\Gamma(5/6)/7\Gamma(1/6) = 2.0858$. Eq. (8.4.23) is equivalent to Eq. (5.8.20), specifying the distribution of the lateral force component $F_\perp$ in a pencil beam. A plot of this type of distribution is given by Fig. 5.4 (curve for $\gamma = 1/3$). The *FWHM* of the distribution of Eq. (8.4.23) follows again with Eq. (5.9.15) and Table 5.1:

$$FWHM_P^* = .013013 \left(\frac{144\Gamma(5/6)}{7\Gamma(1/6)}\right)^3 r_0^{*7} \lambda^{*3} = .94471 \, r_0^{*7} \lambda^{*3}. \qquad (8.4.24)$$

It can directly be verified that Eqs. (8.4.19), (8.4.22) and (8.4.24) provide a complete description of the different regimes shown in Fig. 8.7.

In order to interpolate Eqs. (8.4.19), (8.4.22) and (8.4.24) we use

$$FWHM^* = FWHM_H^* H_{PA}(\lambda^*, r_0^*), \qquad (8.4.25)$$

where the function $H_{PA}(\lambda^*, r_0^*)$ is defined as

$$H_{PA}(\lambda^*, r_0^*) = \left\{ \left[1 + \left(\frac{FWHM_H^*}{FWHM_G^*}\right)^6\right]^{1/7} + \left(\frac{FWHM_H^*}{FWHM_P^*}\right)^{6/7} \right\}^{-7/6},$$

which is equal to

$$H_{PA}(\lambda^*, r_0^*) = \left[\left(1 + 1.851 \frac{\lambda^*}{P_{PA}(r_0^*)^6}\right)^{1/7} + 30.82 \frac{1}{\chi_p^2}\right]^{-7/6}, \qquad (8.4.26)$$

with $\chi_p = 4\lambda^* r_0^{*3}$, as defined by Eq. (7.4.20). The function $P_{PA}(r_0^*)$ is specified by Eqs. (8.4.20) and (8.4.16), and $FWHM_H^*$ is given by Eq. (8.4.22). For comparison, the fit of Eqs. (8.4.25) and (8.4.26) is included in Fig. 8.7.

Eqs. (8.4.19) through (8.4.26) give a complete specification of the *FWHM* of the distribution of transverse velocity displacements $\rho^*(\Delta v_\perp^*)$ in terms of scaled coordinates, defined by Eq. (6.13.1). The remaining task is to remove the scaling in order to make the dependency on the experimental parameters explicit. The properties of the scaled velocity distribution $\rho^*(\Delta v_\perp^*)$ can be related to the corresponding angular deflection distribution, using

$$\Delta\alpha = \frac{\Delta v_\perp^*}{v_z^*} = \left(\frac{e}{2\pi\varepsilon_0}\frac{1}{LV}\right)^{1/3}\Delta v_\perp^*. \tag{8.4.27}$$

The general expressions of Eq. (8.4.25) transform to

$$FWHM_\alpha = C_{PHA}\,H_{PA}(\lambda^*,r_0^*)\frac{I^{2/3}L}{V^{4/3}r_0^{4/3}},$$

$$C_{PHA} = 2.6554\,\frac{m^{1/3}}{2(15\pi)^{2/3}\varepsilon_0} \tag{8.4.28}$$

In the case of electrons, one finds $C_{PHA}=1.1142$ in SI-units (note: the subscript PHA indicates that the constant applies to a cylindrical beam in which the particles run along <u>P</u>arallel trajectories and refers to the <u>H</u>oltsmark type of <u>A</u>ngular deflection distribution).

For the Holtsmark regime ($\lambda^* \ll 1$ and $\chi_p \gg 1$), one may approximate $H_{PA}(\lambda^*, r_0^*)=1$. Accordingly, Eq. (8.4.28) reduces to

$$FWHM_\alpha = C_{PHA}\frac{I^{2/3}L}{V^{4/3}r_0^{4/3}}, \quad \left(\lambda^* \ll 1, \chi_p \gg 1\right). \tag{8.4.29}$$

For the Gaussian regime ($\lambda^* \gtrsim 1$ and $\chi_p \gg 1$), Eq. (8.4.28) transforms to

$$FWHM_\alpha = C_{PGA}\,P_{PA}(r_0^*)\frac{I^{1/2}L^{2/3}}{V^{13/12}r_0}, \quad \left(\lambda^* \gtrsim 1, \chi_p \gg 1\right), \tag{8.4.30}$$

where the constant C_{PGA} is given by

$$C_{PGA} = \frac{[p_2^*(\infty)\ln 2]^{1/2}}{2^{1/12}\pi^{5/6}}\,\frac{e^{1/12}m^{1/4}}{\varepsilon_0^{5/6}},$$

in which $p_2^*(\infty)$ is specified Eq. (8.4.14). Eq. (8.4.30) can also be derived directly from Eqs. (8.4.19) and (8.4.27). In the case of electrons, one finds $C_{PGA}=0.2269$ in SI units.

For the pencil beam regime ($\chi_p \ll 1$), Eq. (8.4.28) transforms to

$$FWHM_\alpha = C_{PPA}\frac{I^3 r_0 L}{V^{5/2}}, \quad (\chi_p \ll 1), \tag{8.4.31}$$

where the constant C_{PPA} is given by

$$C_{PPA} = .013013\,\frac{\sqrt{2}}{\pi}\left(\frac{18\Gamma(5/6)}{7\Gamma(1/6)}\right)^3\frac{m^{3/2}}{\varepsilon_0 e^{7/2}} = 8.3061\times 10^{-4}\,\frac{m^{3/2}}{\varepsilon_0 e^{7/2}}, \tag{8.4.32}$$

which can also be derived directly from Eqs. (8.4.24) and (8.4.27). In the case of electrons, one finds $C_{PPA}=4.9544\times 10^{28}$ in SI-units.

It is interesting to compare the results for the distribution of transverse velocity displacements $\rho(\Delta v_\perp)$ presented in this section with the results obtained for the distribution of axial velocity displacements $\rho(\Delta v_z)$ presented in Section 7.4. These results all apply to a monochromatic homocentric cylindrical beam segment in which the particles are initially at rest in the frame of reference moving with the beam. Random velocities are generated during the flight due the conversion of potential energy into kinetic energy. On symmetry grounds, one should expect that the generated velocity distribution (in the frame of reference moving with the beam) is rotational symmetric, provided that the dimensions of the beam are large compared to the average separation of the particles. Thus, for an extended beam, the distributions $\rho(\Delta v_\perp)$ and $\rho(\Delta v_z)$ should be equivalent. However, differences will occur for pencil beams, in which the distribution of field particles around a certain test particle is nonrotational symmetric.

This reasoning is confirmed by our model, as can best be observed from the various results for the $p^*(k^*)$-transform. The small k^*-behavior is, for the separate cases, given by Eqs. (7.4.14) and (8.4.14). These equations arc identical for $r_0^* \to \infty$, which corresponds to an extended beam. As expected, different results are found for small r_0^*-values, as can be seen from the approximations for the individual $p_2^*(r_0^*)$ functions given by Eqs. (7.4.16) and (8.4.16) respectively. The intermediate k^*-behavior of $p^*(k^*)$ is again identical for both cases, as can be seen from Eqs. (7.4.12) and (8.4.12). This part of the $p^*(k^*)$ transform stems from weak interactions in an extended beam. The large k^*-behavior stems from weak interactions in a pencil beam. The corresponding Eqs. (7.4.13) and (8.4.13) are different, as was to be expected. Thus, in general, one finds that the $p^*(k^*)$ transforms of the distributions $\rho^*(\Delta v_z^*)$ and $\rho^*(\Delta v_\perp^*)$ coincide for that part of the function that represents the interactions in an extended beam segment. This can also directly be observed from Figs. 7.8 and 8.5.

It should be realized that identical $p^*(k^*)$-transforms do not necessarily lead to identical $FWHM^*$ values of the corresponding distributions $p^*(\Delta v_z^*)$ and $\rho^*(\Delta v_\perp^*)$, since they are not of the same dimension. The former is one-dimensional, while the latter is two-dimensional. A Gaussian distribution has the property that its $FWHM$ value is independent of its dimension. Accordingly, Eqs. (7.4.21) and (8.4.19) are identical, provided that $r_0^* \to \infty$. However, the $FWHM^*$-values obtained in the Holtsmark regime differ by a factor 0.9228, as can be seen from Eqs. (7.4.24) and (8.4.22). This factor follows also directly from Table 5.1.

8.5. Beam segment with a crossover of arbitrary dimensions

In the previous two sections, the distribution of transverse velocity displacements $\rho(\Delta v_\perp)$ was calculated for the extreme cases of a beam segment with a narrow crossover and a homocentric cylindrical beam segment respectively. In this section, we will consider the general case of a beam segment with arbitrary values of the geometry parameters K_1 and K_2, defined by Eq. (7.3.1). The analysis of this section is similar to that of Section 7.5, in which we considered the distribution of axial velocities $\rho(\Delta v_z)$ generated in a beam segment of arbitrary dimensions. However, in the present calculation, one cannot simply restrict the problem to the case $S_c = 0.5$ (crossover in the middle) and generalize the results afterwards. This is related to the fact that the displacement in transverse velocity $\Delta v_\perp$ experienced by the test particle in the first part of the beam segment (from the entrance plane to the crossover) can partly or entirely be cancelled out in the second part of the beam segment (from the crossover to the exit plane), depending on the location of the crossover as well as the type of collision involved. Complete cancellation occurs for a weak collision with a field particle that crosses the axis exactly in the crossover plane ($r_\perp = 0$) when his plane is located in the middle of the beam segment ($S_c = 0.5$), as can be seen from Eq. (8.2.2). In order to investigate the influence of the location of the crossover, we will perform all the numerical calculations for two different locations, corresponding to $S_c = 0.5$ and $S_c = 0.75$ respectively. The results for other S_c-values follow from the analytical analysis.

As in Section 8.3, we will employ the d_0, v_0-scaling defined by Eq. (7.3.2). The scaled two-particle distribution $\bar{p}_2(\Delta \bar{v}_\perp)$ is defined by Eq. (8.3.2). As the parameters K_1 and K_2 are not necessarily large, only a fraction of the collisions will be complete. Accordingly, one may not apply Eq. (8.3.1) for the entire integration domain of Eq. (8.3.2). The part of the integration domain that corresponds to complete collisions is indicated by the constraints of Eq. (7.5.1).

The other type of collision that can be treated by analytical means is a weak collision. The constraints of Eq. (7.5.3) indicate which part of the integration domain of Eq. (8.3.2) corresponds to weak collisions. The displacement $\Delta v_\perp$ caused by a weak collision follows from Eq. (8.2.2). Utilizing the d_0, v_0-scaling, this expression can be rewritten as

$$\Delta \bar{v}_\perp \cong \frac{1}{\bar{v}\bar{b}} \left[\left(\frac{\bar{b}}{\left[(\bar{r}_c K_1 \bar{v} + \bar{a}_0)^2 + \bar{b}^2 \right]^{1/2}} - \frac{\bar{b}}{\left[(\bar{r}_c K_2 \bar{v} - \bar{a}_0)^2 + \bar{b}^2 \right]^{1/2}} \right)^2 \right.$$

$$\left. + \left(\frac{\bar{r}_\perp \sin(\Phi)}{\bar{b}} \right)^2 \left(\frac{\bar{r}_c K_1 \bar{v} + \bar{a}_0}{\left[(\bar{r}_c K_1 \bar{v} + \bar{a}_0)^2 + \bar{b}^2 \right]^{1/2}} + \frac{\bar{r}_c K_2 \bar{v} - \bar{a}_0}{\left[(\bar{r}_c K_2 \bar{v} - \bar{a}_0)^2 + \bar{b}^2 \right]^{1/2}} \right)^2 \right]^{1/2}.$$

(8.5.1)

The computer program that was used to evaluate $\bar{p}(\Delta \bar{v}_\perp)$ from Eq. (8.3.2) checks for every two-particle collision whether all constraints of Eq. (7.5.1) are satisfied or not. If so, the collision is (nearly) complete and it determines the corresponding displacement $\Delta \bar{v}_\perp$ from Eq. (8.3.1). If not, it tests next whether one or more of the constraints of Eq. (7.5.3) are satisfied or not. If so, the collision is weak and Eq. (8.5.1) is used to evaluate the corresponding displacement $\Delta \bar{\mathbf{v}}_\perp$. For collisions that are neither weak nor complete, it determines the displacement $\Delta \bar{\mathbf{v}}_\perp$ by utilizing the numerical approach outlined in Section 6.6.

We have evaluated the distribution $\bar{p}_2(\Delta \bar{\mathbf{v}}_\perp)$ for $S_c = 0.5$ and $S_c = 0.75$, taking $K = 0.1, 0.2, 0.5, 1, 10, 100, 1000$, and $10,000$. For every S_c and K we considered the cases $\bar{r}_c = 10^0, 10^1, 10^2, 10^3, 10^4$ and 10^5. Fig. 8.8a–c show the results obtained for $S_c = 0.5$, with $K = 1, 100$ and $10,000$ respectively. Fig. 8.9b and c show the results for $S_c = 0.75$, with $K = 100$ and $K = 10,000$ respectively. The results obtained for $K = 1$ were found to be practically independent of S_c. For that reason, we did not include a separate plot for $S_c = 0.75$ and $K = 1$. All figures should be compared to Fig. 8.1, which was calculated on the basis of complete collisions. One sees that the curves are identical for large values of $\Delta \bar{v}_\perp$. However, significant differences occur for small and intermediate $\Delta v_\perp$-values, especially in combination with small $\bar{r}_c$-values. The differences become more pronounced for small K-values. Furthermore, it can be seen that results obtained for $S_c = 0.5$ and $S_c = 0.75$ show distinct differences, which become more extreme for large values of K.

Let us investigate whether these results can be understood by analytical means. The comparison with Fig. 8.1 indicates that the large $\Delta \bar{v}_\perp$-behavior stems from strong complete collisions. Their total contribution to the distribution $\bar{p}_2(\Delta \bar{\mathbf{v}}_\perp)$ depends only weakly on the parameters K and S_c. This is related to the fact that this type of collision corresponds to a small impact parameter $\bar{b}$, as is indicated by Eq. (7.5.1). The distribution of small $\bar{b}$-values is not very sensitive to the geometry parameters K and S_c. As the large $\Delta \bar{v}_\perp$-behavior stems from complete collisions, one can approximate this part of the curves by Eq. (8.3.11). The extreme case $\bar{r}_c \approx 0$ is covered by Eq. (8.3.6).

Small displacements $\Delta \bar{\mathbf{v}}_\perp$ correspond to weak, predominantly incomplete collisions with more distant field particles. Their contribution $\bar{p}(\Delta \bar{\mathbf{v}}_\perp)$ is strongly affected by the value of the beam geometry parameters K and S_c. In order to analyze the small $\Delta \bar{\mathbf{v}}_\perp$-behavior in a quantitative way, we will evaluate Eq. (8.3.2) on the basis of weak collisions, for which one may employ Eq. (8.5.1). An exact solution of this problem does not appear to be feasible, and the discussion will be restricted to the extreme cases $K \to 0$ and $K \gg 1$. For $\bar{r}_c K \ll 1$, one may approximate Eq. (8.5.1) as

$$\Delta \bar{v}_\perp \cong \frac{1}{2\bar{v}\left(\bar{r}_\perp^2 + \bar{b}_z^2\right)^{3/2}} \left[\left(-4\bar{a}_0 \bar{r}_c K \bar{v} + 4(1 - 2S_c)(\bar{r}_c K \bar{v})^2 + \cdots \right)^2 \right.$$
$$\left. + \left(\frac{\bar{r}_\perp \sin(\Phi)}{\bar{b}^2} \right)^2 \left(4\bar{b}^2 \bar{r}_c K \bar{v} - 12\bar{a}_0(1 - 2S_c)(\bar{r}_c K \bar{v})^2 + \cdots \right)^2 \right]^{1/2},$$

Statistical angular deflections

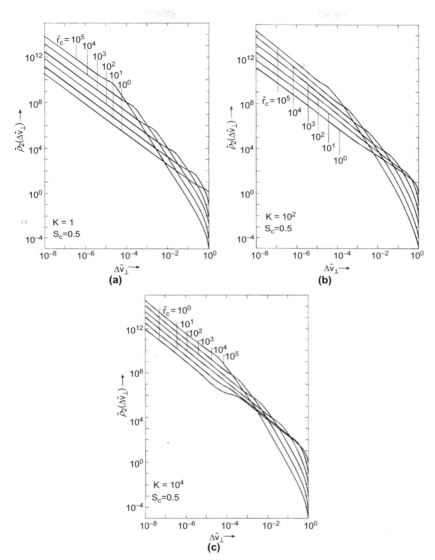

Fig. 8.8 The scaled two-particle distribution $\bar{p}_2(\Delta \bar{v}_\perp)$ for a beam segment with a crossover in the middle ($S_c = 1/2$), plotted for different values of the scaled crossover radius $\bar{r}_c$. Panels a–c pertain to different values of the beam geometry parameter $K = \alpha_0 L/2r_c$, as is indicated. The plots should be compared to those of Fig. 8.1, derived on the basis of complete collisions.

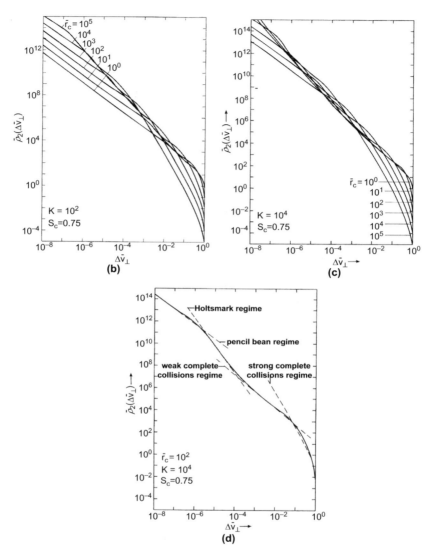

Fig. 8.9 The scaled two-particle distribution $\bar{p}_2(\Delta\bar{v}_\perp)$ for a beam segment with a crossover located at 3/4 of the segment length ($S_c = 3/4$), plotted for different values of the scaled crossover radius $\bar{r}_c$. The curves should be compared to those of Fig. 8.8b and c, which pertain to $S_c = 1/2$. The differences vanish for small K-values. Therefore, there is no panel a corresponding to $K = 1$ (which would be identical to Fig. 8.8a). Panel d explains the shape of the curves from the asymptotic behavior corresponding to the different regimes. The different asymptotes are (from left to right) specified by Eqs. (8.5.16), (8.5.10), (8.3.7) and (8.3.11).

utilizing Eqs. (8.2.3) and (7.3.1). By ignoring the second and higher order terms in $\bar{r}_c K \bar{v}$ one obtains, after some reorganization of the remaining terms,

$$\Delta \bar{v}_\perp \simeq \frac{2\bar{r}_c K \bar{r}_\perp}{\left(\bar{r}_\perp^2 + \bar{b}_z^2\right)^{3/2}}, \quad (K \to 0). \tag{8.5.2}$$

This result is identical to the scaled version of Eq. (8.2.5), which was derived for the case $v=0$. In the previous section, this expression was used to evaluate the contribution of weak interactions in a homocentric cylindrical beam segment. The corresponding results for the two-particle distribution $\rho_2(\Delta v_\perp)$ are expressed by Eqs. (8.4.4) and (8.4.5). Apparently, one may use these results to describe the limit $K \to 0$. The remaining task is therefore to transform the results from the δ, ν-scaling used in Section 8.4 to the d_0, v_0-scaling used here. By means of Eqs. (7.5.5) and (7.5.6), one obtains

$$\bar{\rho}_2(\Delta \bar{v}_\perp) = \frac{24\sqrt{2\pi}\Gamma(3/4) K^{3/2}}{5\Gamma(1/4) \bar{r}_c^{1/2} \Delta \bar{v}_\perp^{5/2}}, \quad \left[\Delta \bar{v}_\perp (\bar{r}_c K/2)^{1/3} \ll 1, \Delta \bar{v}_\perp \bar{r}_c^{2/3}/(32K)^{1/3} \gg 1\right]$$
(8.5.3)

$$\bar{\rho}_2(\Delta \bar{v}_\perp) = \frac{2^{7/3} \bar{r}_c^{2/3} K^{1/3}}{7 \Delta \bar{v}_\perp^{4/3}}, \quad \left[\Delta \bar{v}_\perp (\bar{r}_c K/2)^{1/3} \ll 1, \Delta \bar{v}_\perp \bar{r}_c^{2/3}/(32K)^{1/3} \ll 1\right]. \tag{8.5.4}$$

We note that these equations can also be derived directly from Eqs. (8.3.2) and (8.5.2). The asymptotic behavior described by Eq. (8.5.4) provides an accurate approximation of the small $\Delta \bar{v}_\perp$-behavior for small K, as can be verified from Fig. 8.8A. The behavior described by Eq. (8.5.3) becomes manifest for intermediate $\Delta \bar{v}_\perp$-values combined with small K and $\bar{r}_c$ values.

For a narrow crossover ($K_1 \gg 1$ and $K_2 \gg 1$), one may assume that $\bar{b}_z \gg \bar{r}_\perp$ for the majority of the weak collisions. Accordingly, the second term in Eq. (8.5.1) can be ignored and one finds, in good approximation, for the remaining term

$$\Delta \bar{v}_\perp \simeq \frac{1}{\bar{v}} \left| \frac{1}{\left[(\bar{r}_c K_1 \bar{v})^2 + \bar{b}_z^2\right]^{1/2}} - \frac{1}{\left[(\bar{r}_c K_2 \bar{v})^2 + \bar{b}_z^2\right]^{1/2}} \right| \quad (K \gg 1), \tag{8.5.5}$$

assuming that $\bar{v} \neq 0$. The scaled two-particle distribution follows again with Eq. (8.3.2). As Eq. (8.5.5) does not depend on $\bar{r}_\perp$ or Φ, Eq. (8.3.2) reduces to a two-dimensional integral, which can be expressed as

$$\bar{\rho}_2(\Delta \bar{v}_\perp) = \int_0^1 2\bar{v} d\bar{v} \int_0^\infty 2 d\bar{b}_z \, \delta\left(\Delta \bar{v}_\perp - \frac{F(\bar{b}_z/2\bar{r}_c K \bar{v})}{2\bar{r}_c K \bar{v}^2}\right), \tag{8.5.6}$$

in which the function $F(z)$ is given by

$$F(z) = \left| \frac{1}{[S_c^2 + z^2]^{1/2}} - \frac{1}{[(1-S_c)^2 + z^2]^{1/2}} \right|, \quad (8.5.7)$$

where the definitions of K_1 and K_2, given by Eqs. (7.3.1), were used. By substituting

$$z = \frac{\bar{b}_z}{2\bar{r}_c K \bar{v}}, \quad t = \frac{F(z)}{2\bar{r}_c K \bar{v}^2} \quad (8.5.8)$$

and carrying out the t-integration, Eq. (8.5.6) transforms to

$$\bar{p}_2(\Delta \bar{v}_\perp) = \frac{\sqrt{2}}{(\bar{r}_c K)^{1/2} \Delta \bar{v}_\perp^{5/2}} \int_0^\infty dz \, F(z)^{3/2} \Theta\left(\Delta \bar{v}_\perp - \frac{F(z)}{2\bar{r}_c K}\right), \quad (8.5.9)$$

where $\Theta(x)$ is the step function. Eq. (8.5.9) is similar to Eq. (7.5.12), specifying the two-particle distribution $\bar{p}(\Delta \bar{v}_z)$ for weak collisions in a narrow crossover. We will consider the cases $\bar{r}_c K \, \Delta \bar{v}_\perp \gg 1$ and $\bar{r}_c K \, \Delta \bar{v}_\perp \ll 1$.

For $\bar{r}_c K \, \Delta \bar{v}_\perp \gg 1$, the argument of the step function in Eq. (8.5.9) is positive for all z-values, and one obtains

$$\bar{p}_2(\Delta \bar{v}_\perp) = \frac{I_4 S_{HA}(S_c)^{3/2}}{(\bar{r}_c K)^{1/2} \Delta \bar{v}_\perp^{5/2}}, \quad (K \gg 1, \bar{r}_c K \Delta \bar{v}_\perp \gg 1), \quad (8.5.10)$$

in which

$$S_{HA}(S_c)^{3/2} = \frac{\sqrt{2}}{I_4} \int_0^\infty dz \, F(z)^{3/2} \quad (8.5.11)$$

and I_4 is some arbitrary constant that will be used to scale the function $S_{HA}(S_c)$. The function $S_{HA}(S_c)$ has the following properties:

$$\lim_{S_c \to 0} S_{HA}(S_c)(2S_c)^{1/3} = \lim_{S_c \to 1} S_{HA}(S_c)[2(1-S_c)]^{1/3} = \left(\frac{2\sqrt{\pi}\Gamma(1/4)}{I_4 \Gamma(3/4)}\right)^{2/3}$$

$$\lim_{S_c \to 1/2} S_{HA}(S_c) = \left(\frac{24\sqrt{2\pi}\Gamma(3/4)}{5I_4 \Gamma(1/4)}\right)^{2/3} |1 - 2S_c|. \quad (8.5.12)$$

As an approximation of the function $S_{HA}(S_c)$, we use

$$S_{HAa}(S_c) \cong C \left| \frac{1}{(2S_c)^{1/3}} - \frac{1}{[2(1-S_c)]^{1/3}} \right|, \quad (8.5.13)$$

which has similar properties

$$\lim_{S_c \to 0} S_{HAa}(S_c)(2S_c)^{1/3} = \lim_{S_c \to 1} S_{HAa}(S_c)[2(1-S_c)]^{1/3} = C$$
$$\lim_{S_c \to 1/2} S_{HAa}(S_c) = (2C/3)|1 - 2S_c|. \tag{8.5.14}$$

Unfortunately, one cannot choose the constants C and I_4 such that Eqs. (8.5.12) and (8.5.14) agree fully, both for $S_c \to 0$ (or $S_c \to 1$) and for $S_c \to 1/2$. Considering that the case $S_c \approx 1/2$ is the most relevant, we take

$$C = 2, \quad I_4 = \frac{3^{5/2}(2\pi)^{1/2}\Gamma(3/4)}{5\Gamma(1/4)} = 2.6413. \tag{8.5.15}$$

This choice guarantees a good approximation of the real function $S_{HA}(S_c)$ for $S_c \approx 1/2$ but leads to some overestimation of $S_{HA}(S_c)$ for $S_c \to 0$ and $S_c \to 1$. We evaluated the function $S_{HA}(S_c)$ numerically. The results are plotted in Fig. 8.10. For comparison, the approximation of Eq. (8.5.13) is included in the same figure.

For $\bar{r}_c K \Delta \bar{v}_\perp \ll 1$, the argument of the step function in Eq. (8.5.9) is only positive for large z-values. Accordingly, one may use the following asymptotic expansion of the function $F(z)$

$$\lim_{z \to \infty} F(z) = \frac{|1 - 2S_c|}{2z^3} + O(z^{-5})$$

in which $O(z^{-5})$ stands for terms of the order z^{-5}. Substitution into Eq. (8.5.9) yields

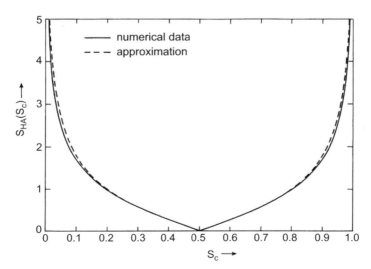

Fig. 8.10 The function $S_{HA}(S_c)$, defined by Eqs. (8.5.11), (8.5.7) and (8.5.15), and its approximation given by Eqs. (8.5.13) and (8.5.15). This function specifies the dependency of the width of the angular displacement distribution in the Holtsmark regime.

$$\bar{p}_2(\Delta \bar{v}_\perp) = \frac{2^{7/3}}{7} \frac{S_{PA}(S_c)^{1/3}(\bar{r}_c K)^{2/3}}{\Delta \bar{v}_\perp^{4/3}}, \quad (K \gg 1, \bar{r}_c K \Delta \bar{v}_\perp \ll 1), \quad (8.5.16)$$

in which the function $S_{PA}(S_c)$ represents

$$S_{PA}(S_c) = |1 - 2S_c|. \quad (8.5.17)$$

Eqs. (8.5.10) and (8.5.16) provide an accurate description of the small $\Delta \bar{v}_\perp$-behavior of the distribution $\bar{p}_2(\Delta \bar{v}_\perp)$ for large K-values and $S_c \neq 0.5$, as can be verified from Fig. 8.9b and c. In order to demonstrate the significance of the various asymptotes derived for weak and complete collisions, the curve in Fig. 8.9c corresponding to $\bar{r}_c = 10^2$ is replotted separately in Fig. 8.9d. For very small $\Delta \bar{v}_\perp$-values, the curve follows the asymptote described by Eq. (8.5.16). This result can be associated with the pencil beam regime, as will be shown later on in this section. For larger $\Delta \bar{v}_\perp$-values, the curve follows the asymptote described by Eq. (8.5.10), which is associated with the Holtsmark regime. It should be recalled that both Eqs. (8.5.16) and (8.5.10) are based on weak collisions. With further increasing $\Delta \bar{v}_\perp$, the complete collision regime is reached. The curve now follows the asymptote described by Eq. (8.3.7), which corresponds to weak complete collisions. Finally, for very large $\Delta \bar{v}_\perp$-values, the curve is dominated by strong, predominantly complete collisions and can be approximated by Eq. (8.3.11).

The previous analysis shows that the distribution $\bar{p}_2(\Delta \bar{v}_\perp)$ follows different sets of asymptotes for $K \to 0$ and $K \gg 1$. We wish to simplify the results somewhat by introducing some general equations that yield correct results for both $K \to 0$ and $K \gg 1$. By comparing Eqs. (8.5.3) and (8.5.10), it is found that one may, in general, describe the small $\Delta \bar{v}_\perp$-behavior of $\bar{p}_2(\Delta \bar{v}_\perp)$ by Eq. (8.5.10), provided that the function $S_{HA}(S_c)$ is replaced by some generalized function $S_{HA}(S_c, K)$, which has the following properties:

$$\lim_{K \to \infty} S_{HA}(S_c, K) \simeq 2 \left| \frac{1}{(2S_c)^{1/3}} - \frac{1}{[2(1-S_c)^{1/3}]} \right| \quad (8.5.18)$$

$$\lim_{K \to 0} S_{HA}(S_c, K) = \frac{4}{3} K^{4/3}$$

in which Eqs. (8.5.13) and (8.5.15) were used. Similarly, one may combine Eqs. (8.5.4) and (8.5.16) into a single expression that describes the intermediate $\Delta \bar{v}_\perp$-behavior of $\bar{p}_2(\Delta \bar{v}_\perp)$ for arbitrary K-values. This expression is the same as given by Eq. (8.5.16), provided that one replaces the function $S_{PA}(S_c)$ by some generalized function $S_{PA}(S_c, K)$, which has the following properties:

$$\lim_{K \to \infty} S_{PA}(S_c, K) = |1 - 2S_c|$$
$$\lim_{K \to 0} S_{PA}(S_c, K) = \frac{1}{K} \qquad (8.5.19)$$

in which Eq. (8.5.17) was used. We mention that the slice method (introduced in Section 5.11) can be exploited to derive some suitable expressions to approximate the required generalized functions $S_{HA}(S_c, K)$ and $S_{PA}(S_c, K)$. This analysis is postponed to Section 8.6. Here we suffice to present the results:

$$S_{HAa}(S_c, K) \cong 2 \left| \frac{1}{(1/K + 2S_c)^{1/3}} - \frac{1}{[1/K + 2(1-S_c)^{1/3}]} \right| \qquad (8.5.20)$$

$$S_{PAa}(S_c, K) \cong |1 - 2S_c| \left(\frac{1}{K} + 1 \right). \qquad (8.5.21)$$

The reader might verify that the functions $S_{HAa}(S_c, K)$ and $S_{PAa}(S_c, K)$ indeed fulfill the conditions of Eqs. (8.5.18) and (8.5.19) respectively, provided that one takes $S_c = 0$ or $S_c = 1$ in case $K \to 0$. This condition implies that a nearly cylindrical beam is represented as a beam segment with a crossover that is located either at the start or at the end of the beam segment. The results will be further discussed in Section 8.6. In the remaining analysis of this section, it is assumed that Eqs. (8.5.10) and (8.5.16), combined with the generalized functions (8.5.20) and (8.5.21), provide an accurate description of the intermediate and small $\Delta \bar{v}_\perp$-behavior of $\bar{p}_2(\Delta \bar{v}_\perp)$ respectively, which is valid for arbitrary values of K.

We now proceed with the calculation of the function $\bar{p}(\bar{k})$, which is defined by Eq. (8.3.12). We evaluated $\bar{p}(\bar{k})$ numerically from the data obtained for $\bar{p}_2(\Delta \bar{v}_\perp)$. The results for $S_c = 0.5$, with $K = 1$, 100 and 10,000 are plotted in Fig. 8.11a–c respectively. The results for $S_c = 0.75$, with $K = 100$ and 10,000 are plotted in Fig. 8.12b and c. We recall that the results become independent of S_c for small K. Thus Fig. 8.11a applies also to other S_c-values than $S_c = 0.5$. The plots should be compared with Fig. 8.2, which was derived on the basis of complete collisions. The most significant differences with the curves of Fig. 8.2 occur for intermediate and large $\bar{k}$-values. Fig. 8.2 shows a square-root dependency on $\bar{k}$ for large $\bar{k}$, as described by Eq. (8.3.16). On the other hand, the curves in Fig. 8.9d; Figs. 8.11a–c; 8.12b and 8.12c become for large $\bar{k}$ ultimately proportional to $\bar{k}^{-1/3}$. This can be understood from the analysis of the two-particle distribution $\bar{p}_2(\Delta \bar{v}_\perp)$. Substitution of Eq. (8.5.16), with the generalized function $S_{PA}(S_c, K)$, into Eq. (8.3.12) yields, after integration

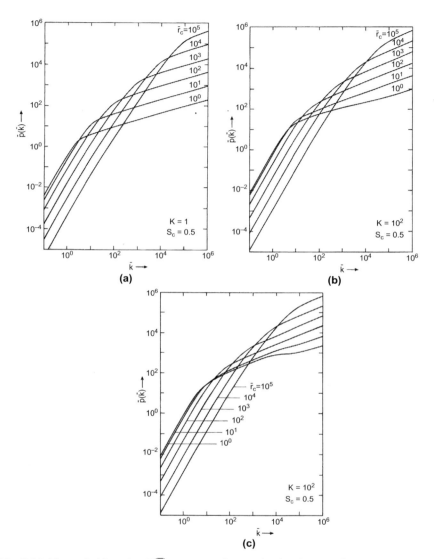

Fig. 8.11 The scaled function $\bar{p}(\bar{k})$ corresponding to the distribution of transverse velocity displacements generated in a beam segment with a crossover in the middle ($S_c = 1/2$), plotted for different values of the scaled crossover radius $\bar{r}_c$. Panels a–c pertain to different values of the beam geometry parameter $K = \alpha_0 L/2r_c$, as is indicated. The plots should be compared to those of Fig. 8.2, derived on the basis of complete collisions.

Statistical angular deflections

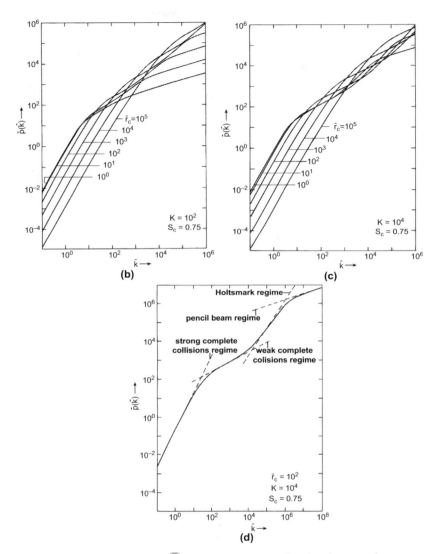

Fig. 8.12 The scaled function $\bar{p}(\bar{k})$ corresponding to the distribution of transverse velocity displacements generated in a beam segment with a crossover located at 3/4 of the segment length ($S_c = 3/4$), plotted for different values of the scaled crossover radius $\bar{r}_c$. The curves should be compared to those of Fig. 8.11b and c, which pertain to $S_c = 1/2$. The differences vanish for small K-values. Therefore, there is no panel a corresponding to $K = 1$ (which would be identical to Fig. 8.11a). Panel d explains the shape of the curves from the asymptotic behavior corresponding to the different regimes. The different asymptotes are (from left to right) specified by Eqs. (8.5.26), (8.3.16), (8.5.24) and (8.5.22).

$$\bar{p}(\bar{k}) = \frac{72\Gamma(5/6)}{7\Gamma(1/6)} S_{PA}(S_c, K)^{1/3} (\bar{r}_c K)^{2/3} \bar{k}^{1/3}, \quad (\bar{k} \to \infty), \qquad (8.5.22)$$

similar to derivation of Eq. (5.8.19). The behavior of $S_{PA}(S_c, K)$ for $K \to 0$ and $K \gg 1$ is given by Eq. (8.5.19). For $K \to 0$, Eq. (8.5.22) transforms to

$$\bar{p}(\bar{k}) = \frac{72\Gamma(5/6)}{7\Gamma(1/6)} \bar{r}_c^{2/3} K^{1/3} \bar{k}^{1/3}, \quad (K \to 0, \bar{k} \to \infty), \qquad (8.5.23)$$

which follows also directly by substitution of Eq. (8.5.4) into Eq. (8.3.12) and integration. Eq. (8.5.23) provides an accurate approximation of the large $\bar{k}$ behavior depicted in Fig. 8.11A.

The behavior described by Eq. (8.5.10) leads to a 3/2 power dependency on $\bar{k}$, which becomes manifest for intermediate $\bar{k}$ values. Substitution of Eq. (8.5.10), with the generalized function $S_{HA}(S_c, K)$, into Eq. (8.3.12) yields, after integration,

$$\bar{p}(\bar{k}) = \frac{4(3\pi)^{1/2}}{5} \frac{S_{HA}(S_c, K)^{3/2}}{(\bar{r}_c K)^{1/2}} \bar{k}^{3/2}, \qquad (8.5.24)$$

similar to the derivation of Eq. (5.8.14). In Eq. (8.5.24), we used the expression for the numerical constant I_4, given by Eq. (8.5.15). The behavior of $S_{HA}(S_c, K)$ for $K \to 0$ and $K \gg 1$ is given by Eq. (8.5.18). For $K \to 0$, Eq. (8.5.24) transforms to

$$\bar{p}(\bar{k}) = \frac{32\sqrt{\pi}}{15} \frac{K^{3/2}}{\bar{r}_c^{1/2}} \bar{k}^{3/2}, \quad (K \to 0), \qquad (8.5.25)$$

which follows also directly by substitution of Eq. (8.5.3) into Eq. (8.3.12) and integration.

For small $\bar{k}$-values, all curves in Figs. 8.11 and 8.12 become quadratic in $\bar{k}$. In order to describe this behavior, we define the function $\bar{p}_2(\bar{r}_c, K)$ as

$$\bar{p}(\bar{k}) = \frac{1}{2} \bar{p}_2(\bar{r}_c, K) \bar{k}^2, \qquad (8.5.26)$$

The function $\bar{p}_2(\bar{r}_c, K)$ is the generalization of the function $\bar{p}_2(\bar{r}_c)$, defined by Eq. (8.3.14). The dependency of the small $\bar{k}$-behavior of $\bar{p}(\bar{k})$ on S_c is very weak and will be discussed later on. We evaluated the function $\bar{p}_2(\bar{r}_c, K)$ numerically for $S_c = 0.5$. The results are plotted in Fig. 8.13.

In the limit $K \to \infty$, the function $\bar{p}_2(\bar{r}_c, K)$ should become identical to the function $\bar{p}_2(\bar{r}_c)$, as is formally expressed by Eq. (7.5.18). In the limit $K \to 0$, one should retrieve the result obtained for a homocentric cylindrical beam, which is given by Eq. (8.4.14). This condition leads to Eq. (7.5.19). Notice, however, that the saturation value $p_2^*(\infty) = 0.3026$ differs by a factor 2 from the result corresponding to the distribution of axial velocity

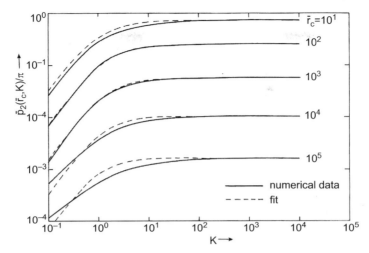

Fig. 8.13 The function $\bar{p}_2(\bar{r}_c, K)$ defined by Eq. (8.5.26) and its approximation given by Eq. (8.5.27). This function gives the dependency on $\bar{r}_c$ and K of the function $\bar{p}(\bar{k})$, plotted in Fig. 8.11a–c, for $\bar{k} \to 0$.

displacements $\rho^*(\Delta v_z^*)$. The corresponding approximations $p_{2a}(r_0^*)$ are different, too, as can be seen from Eqs. (7.4.16) and (8.4.16). As an approximation of the general function $\bar{p}_2(\bar{r}_c, K)$, we use in the present case.

$$\bar{p}_{2a}(\bar{r}_c, K) = \frac{\pi}{.894 + .3\left(\dfrac{\bar{r}_c}{K^2}\right)^{2/3}\left[1 + .5\left(\dfrac{K^2}{\bar{r}_c}\right)^{4/9}\right]^{3/2} + \dfrac{\pi \bar{r}_c}{\{2\ln[.8673(114.6 + \bar{r}_c)]\}^2}},$$

(8.5.27)

which shows the asymptotic behavior described by Eqs. (7.5.18) and (7.5.19), provided that $\bar{r}_c \gg 16K$. For comparison, the approximation of Eq. (8.5.27) is included in Fig. 8.13.

In order to demonstrate the significance of the various asymptotes, the curve in Fig. 8.12c corresponding to $\bar{r}_c = 10^2$ is replotted separately in Fig. 8.12d. The features of the plot are comparable to those of Fig. 8.9d, showing the asymptotic behavior of the two-particle distribution $\bar{p}(\Delta \bar{v}_\perp)$. For small $\bar{k}$-values the curve follows the asymptotic behavior given by Eq. (8.5.26), which is associated with the Gaussian regime. For intermediate $\bar{k}$-values the curve levels off to the asymptote predicted by Eq. (8.3.16), which corresponds to weak complete collisions. The large $\bar{k}$-behavior is determined by weak incomplete collisions. The curve follows successively the asymptotes described by Eqs. (8.5.24) and (8.5.22), which are associated with the Holtsmark regime and the pencil beam regime respectively.

The scaled displacement distribution $\bar{p}(\Delta \bar{v}_\perp)$ is defined by Eq. (8.3.17). We evaluated $\bar{p}(\Delta \bar{v}_\perp)$ numerically from the data obtained for $\bar{p}(\bar{k})$. The results for $S_c = 0.5$, with $K = 1$, 100 and 10,000 are plotted in Fig. 8.14a–c respectively. The results for $S_c = 0.75$, with $K = 100$ and 10,000 are plotted in Fig. 8.15b and 8.15c. Fig. 8.14a applies also to S_c-values other than $S_c = 0.5$, since the results become for small K-values independent of S_c, as was mentioned previously. The plots should be compared with Fig. 8.3, which was derived on the basis of complete collisions. The following regimes can be distinguished, as is explicitly shown for the case $K = 10^4$ and $S_c = 0.75$ in Fig. 8.15d.

(I) Gaussian regime. For large values of $\bar{\lambda}$, the distribution of $\bar{p}(\Delta \bar{v}_\perp)$ is determined by the small $\bar{k}$-behavior of $\bar{p}(\bar{k})$, which is specified by Eq. (8.5.26). The quadratic $\bar{k}$-dependency implies that the corresponding distribution $\bar{p}(\Delta \bar{v}_\perp)$ is Gaussian. The FWHM of this distribution is given by Eq. (8.3.19), where the function P_{CA} is now generalized to a function that depends on $\bar{r}_c, K$ and S_c,

$$P_{CA}(\bar{r}_c, K, S_c) = \frac{\bar{p}_2(\bar{r}_c, 2S_c K)^{1/2}}{2\pi^{1/2}} + \frac{\bar{p}_2[\bar{r}_c, 2(1-S_c)K]^{1/2}}{2\pi^{1/2}}, \qquad (8.5.28)$$

where $\bar{p}_2(\bar{r}_c, K)$ can be approximated by Eq. (8.5.27). Eq. (8.5.28) is based on the assumption that, in the Gaussian regime, the contributions of the first and the second part of the beam segment should be added linearly. The form of Eq. (8.5.28) is similar to that of Eq. (7.5.37). It should be noted that the dependency of the function $P_{CA}(\bar{r}_c, K, S_c)$ on S_c is, in general, very weak. In terms of the experimental parameters, the result for the Gaussian regime is expressed by Eq. (8.3.21).

(II) Weak complete collision regime. With decreasing $\bar{\lambda}$ and large K-values, the distribution $\bar{p}(\Delta \bar{v}_\perp)$ becomes dominated by the part of $\bar{p}(\bar{k})$ that is proportional to $\bar{k}^{1/2}$, as described by Eq. (8.3.16). The corresponding FWHM is given by Eq. (8.3.23). In terms of experimental parameters it leads to Eq. (8.3.24).

(III) Holtsmark regime. For an extended beam segment with a small or intermediate $\bar{\lambda}$-value, the distribution $\bar{p}(\Delta \bar{v}_\perp)$ is dominated by the 3/2 power dependency of $\bar{p}(\bar{k})$, as expressed by Eq. (8.5.24). A beam segment is called extended when $\chi_c \gg 1$. The quantity χ_c is the pencil beam factor for a beam segment with a crossover and is defined by Eq. (7.5.28). Substitution of Eq. (8.5.24) into Eq. (8.3.17) yields

$$\bar{p}(\Delta \bar{\mathbf{v}}_\perp) = \frac{1}{2\pi} \int_0^\infty \bar{k} d\bar{k} J_0(\bar{k}\Delta \bar{v}_\perp) e^{-(4/5)(3\pi/\bar{r}_c K)^{1/2} \bar{\lambda} [S_{HA}(S_c, K)\bar{k}]^{3/2}}, \qquad (8.5.29)$$

Statistical angular deflections

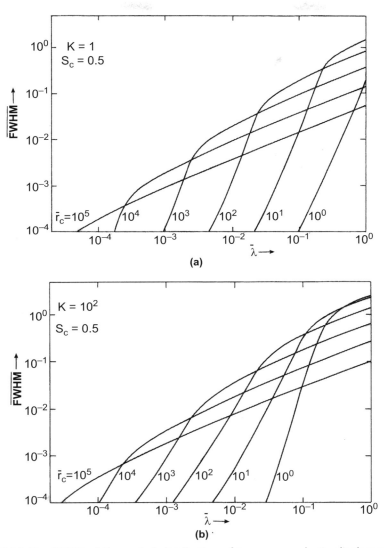

Fig. 8.14 The *FWHM* of the scaled distribution of transverse velocity displacements $\bar{\rho}(\Delta\bar{v}_\perp)$ generated in a beam segment with a crossover in the middle ($S_c = 1/2$), plotted for different values of the scaled crossover radius $\bar{r}_c$. Panels a–c pertain to different values of the beam geometry parameter $K = \alpha_0 L/2r_c$, as is indicated. The figures should be compared to Fig. 8.3, derived on the basis of complete collisions.

(Continued)

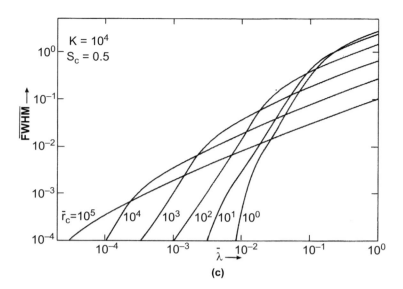

Fig. 8.14—Cont'd

which is the two-dimensional variant of the Holtsmark distribution plotted in Fig. 5.4 (curve corresponding to $\gamma = 3/2$). The FWHM of this distribution is given by

$$\overline{FWHM}_H = 2.6554 \frac{(48\pi)^{1/3}}{5^{2/3}} \frac{S_{HA}(S_c, K)\overline{\lambda}^{2/3}}{(\overline{r}_c K)^{1/3}}, \qquad (8.5.30)$$

as follows with Eq. (5.9.15) and Table 5.1. The FWHM of the corresponding angular deflection distribution is given by (using $FWHM_\alpha = \alpha_0 \overline{FWHM}$)

$$FWHM_\alpha = C_{CHA} S_{HA}(S_c, K) \frac{I^{2/3}}{V^{4/3} \alpha_0^{4/3} L^{1/3}}, \qquad (8.5.31)$$

where the constant C_{CHA} is given by

$$C_{CHA} = 2.6554 \frac{3^{1/3}}{2^{5/3}(5\pi)^{2/3}} \frac{m^{1/3}}{\varepsilon_0} = 0.19233 \frac{m^{1/3}}{\varepsilon_0}.$$

In the case of electrons, one finds $C_{CHA} = 2.1057$ in SI-units.

For $K \to 0$, one may approximate $S_{HA}(S_c, K) \approx (4/3)K^{4/3} = (2^{2/3}/3)(\alpha_0 L/r_c)^{4/3}$, as is stated by Eq. (8.5.18). Substitution into Eq. (8.5.31) leads back to Eq. (8.4.29), which was derived for a homocentric cylindrical beam segment. Notice that the constants C_{PHA} and C_{CHA} are related as

$$C_{PHA} = \frac{2^{2/3}}{3} C_{CHA}, \qquad (8.5.32)$$

as can be verified from their definitions.

Statistical angular deflections

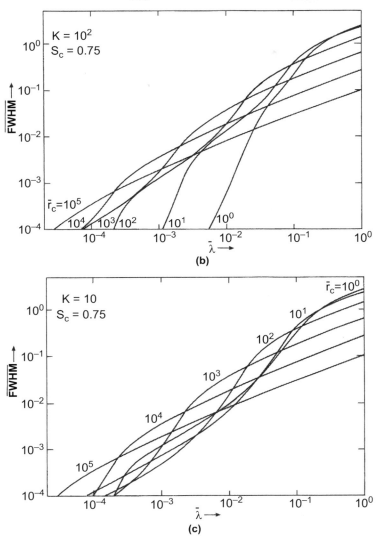

Fig. 8.15 The *FWHM* of the scaled distribution of transverse velocity displacements $\bar{p}(\Delta \bar{v}_\perp)$ generated in a beam segment with a crossover located at 3/4 of the segment length ($S_c = 3/4$), plotted for different values of the scaled crossover radius $\bar{r}_c$. The curves should be compared to those of Fig. 8.14b and c which pertain to $S_c = 1/2$. The differences vanish for small *K*-values. Therefore, there is no panel a corresponding to $K=1$ (which would be identical to Fig. 8.14a). Panel d explains the shape of the curves from the asymptotic behavior corresponding to the different regimes. The different asymptotes are (from left to right) specified by Eqs. (8.5.34), (8.5.30), (8.3.23) and (8.3.19). Panel (E) is indicative of the quality of the fit given by Eqs. (8.5.37) and (8.5.38).

(Continued)

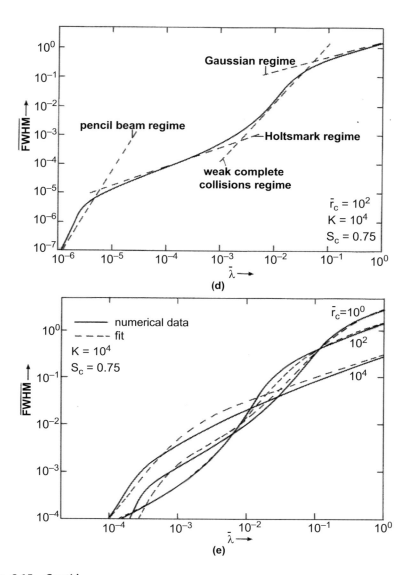

Fig. 8.15—Cont'd

(IV) Pencil beam regime. In pencil beams ($\chi_c \ll 1$), the distribution $\bar{p}(\Delta \bar{v}_\perp)$ is dominated by the $\bar{k}$-behavior given by Eq. (8.5.22). Substitution into Eq. (8.3.17) yields

$$\bar{p}(\Delta \bar{v}_\perp) = \frac{1}{2\pi} \int_0^\infty \bar{k} d\bar{k} J_0(\bar{k} \Delta \bar{v}_\perp) e^{-C_{1/3}(\bar{r}_c K)^{2/3} \bar{\lambda}[S_{PA}(S_c,K)\bar{k}]^{1/3}}, \quad (8.5.33)$$

where $C_{1/3} = 72\Gamma(5/6)/7\Gamma(1/6) = 2.0858$. This type of distribution is referred to as the two-dimensional pencil beam distribution; see Fig. 5.4 (curve corresponding to $\gamma = 1/3$). The FWHM of the distribution of Eq. (8.5.33) is given by

$$\overline{FWHM}_P = .013013 \left(\frac{72\Gamma(5/6)}{7\Gamma(1/6)} \right)^3 S_{PA}(S_c, K)(\bar{r}_c K)^2 \bar{\lambda}^3, \quad (8.5.34)$$

as follows with Eq. (5.9.15) and Table 5.1. The numerical constant in the right-hand side of Eq. (8.5.34) is equal to 0.11809. By removing the scaling, one obtains

$$FWHM_\alpha = C_{CPA} S_{PA}(S_c, K) \frac{I^3 \alpha_0 L^2}{V^{5/2}}, \quad (8.5.35)$$

where the constant C_{CPA} is given by

$$C_{CPA} = .013013 \left(\frac{18\Gamma(5/6)}{7\Gamma(1/6)} \right)^3 \frac{1}{\pi\sqrt{2}} \frac{m^{3/2}}{\varepsilon_0 e^{7/2}} = 4.1531 \times 10^{-4} \frac{m^{3/2}}{\varepsilon_0 e^{7/2}}.$$

In the case of electrons, one finds $C_{CPA} = 2.4772 \times 10^{28}$ in SI-units.

For $K \to 0$ one may approximate $S_{PA}(S_c, K) \approx 1/K = 2r_c/\alpha_0 L$, as is stated by Eq. (8.5.19). Substitution into Eq. (8.5.35) leads back to Eq. (8.4.31), which was derived for a homocentric cylindrical beam segment. Notice that the constants C_{PPA} and C_{CPA} are related as

$$C_{PPA} = 2C_{CPA}, \quad (8.5.36)$$

as can be verified from their definitions.

In order to interpolate the results obtained for the different regimes, we approximate

$$\overline{FWHM} = \overline{FWHM}_0 \, G_{CA}(\bar{\lambda}, \bar{r}_c, K, S_c),$$
$$\overline{FWHM}_0 = 2(\pi \ln 2)^{1/2} \bar{\lambda}^{1/2}, \quad (8.5.37)$$

equivalent to Eq. (7.5.30). The function G_{CA} is taken equal to

$$G_{CA}(\bar{\lambda}, \bar{r}_c, K, S_c) = \left[\left(\frac{\overline{FWHM}_0}{\overline{FWHM}_G} + \frac{\overline{FWHM}_0}{\overline{FWHM}_W} \right)^{-1} + \left(\frac{\overline{FWHM}_0}{\overline{FWHM}_H} + \frac{\overline{FWHM}_0}{\overline{FWHM}_P} \right)^{-1} \right]^{-1}.$$

By substitution of Eqs. (8.3.19), (8.3.23), (8.5.30) and (8.5.34) one obtains

$$G_{CA}(\bar{\lambda}, \bar{r}_c, K, S_c) = \left[\left(\frac{1}{P_{CA}(\bar{r}_c, K, S_c)} + \frac{A}{\bar{\lambda}^{3/2} f_\infty(\bar{r}_c)^2} \right)^{-1} \right.$$

$$\left. + \left(\frac{B(\bar{r}_c K)^{1/3}}{\bar{\lambda}^{1/6} S_{HA}(S_c, K)} + \frac{C}{\bar{\lambda}^{5/2} (\bar{r}_c K)^2 S_{PA}(S_c, K)} \right)^{-1} \right]^{-1}, \quad (8.5.38)$$

in which the constants A, B and C are given by

$$A = 7.009 \times 10^{-2}, \quad B = .6106, \quad C = 24.99. \quad (8.5.39)$$

The function $P_{CA}(\bar{r}_c, K, S_c)$ is defined by Eqs. (8.5.28) and (8.5.27). The function $f_\infty(\bar{r}_c)$ is given by Eq. (8.3.8) and can be approximated by Eq. (8.3.10). The functions $S_{HA}(S_c, K)$ and $S_{PA}(S_c, K)$ can be approximated by the functions $S_{HAa}(S_c, K)$ and $S_{PAa}(S_c, K)$, given by Eqs. (8.5.20) and (8.5.21) respectively. Fig. 8.15e compares the approximation of Eqs. (8.5.37) and (8.5.38) with the numerical data obtained for $S_c = 0.75$ and $K = 10^4$, with $\bar{r}_c = 1, 10^2$ and 10^4. One sees that the fit is rather poor in the transition areas between the different regimes.

In terms of the experimental parameters, Eq. (8.5.37) can be rewritten as

$$FWHM_a = C_{CGA} \, G_{CA}(\bar{\lambda}, \bar{r}_c, K, S_c) \sqrt{\frac{I}{V^{3/2}}}, \quad (8.5.40)$$

similar to Eq. (8.3.21), which also specifies the constant C_{CGA}. Eqs. (8.5.40) and (8.5.38) provide the most general description of the $FWHM$ of the angular deflection distribution. It applies to a beam segment of arbitrary geometry and provides accurate results as long as the theoretical parameters fulfil the conditions expressed by Eq. (7.5.33).

Similar results can be derived for the Full Width median (FW_{50}) of the angular deflection distribution. For a two-dimensional Gaussian distribution, the $FWHM$ and FW_{50} are identical. Accordingly, one may write

$$\overline{FW}_{50} = \overline{FWHM}_0 \, G_{CA}(\bar{\lambda}, \bar{r}_c, K, S_c), \quad (8.5.41)$$

similar to Eq. (8.5.37). The function $G_{CA}(\bar{\lambda}, \bar{r}_c, K, S_c)$ is again given by Eq. (8.5.38), now taking the constants A, B and C equal to

$$A = 2.657 \times 10^{-3}, \quad B = .4883, \quad C = 4.776 \times 10^{-2} \quad (8.5.42)$$

instead of the values given by Eq. (8.5.39). The results for the individual regimes can be obtained from the expressions for the $FWHM$-values, using

$$FW_{50G} = 1.0000 \ FWHM_G$$
$$FW_{50W} = 26.381 \ FWHM_W$$
$$FW_{50H} = 1.2505 \ FWHM_H$$
$$FW_{50P} = 523.29 \ FWHM_P.$$
(8.5.43)

as follows from the data of Table 5.1.

8.6. Application of the slice method

In this section, we will verify the approximations of the $S_{HA}(S_c, K)$ and $S_{PA}(S_c, K)$, given by Eqs. (8.5.20) and (8.5.21). We will show that these results can be established by means of the slice method, which is described in Section 5.11. The slice method is used to relate the results obtained for a homocentric cylindrical beam segment to those obtained for a beam segment with a crossover. The essence of the procedure is expressed by Eq. (5.11.1). This equation states that one may calculate the total displacement generated in a beam segment by adding linearly the displacements generated in a large number of thin cylindrical slices. The basic assumption is that the contributions of the individual slices to the total displacement experienced by the test particle are entirely correlated. Such is to be expected when the deviations from the unperturbed trajectories are small. Accordingly, the slice method is only applicable when the distribution $\rho(\Delta v_\perp)$ is dominated by weak collisions. This type of analysis is therefore restricted to the Holtsmark regime and the pencil beam regime.

Let us first consider the Holtsmark regime. According to Eq. (8.4.29), the contribution of a single slice at axial position z of length Δz to the total distribution of angular deflections $\rho(\Delta\alpha)$ can be expressed as

$$\Delta FWHM_\alpha = C_{PHA} \frac{J^{2/3}}{V^{4/3} r_c^{4/3}} = \frac{\Delta z}{[1 + \alpha_0|z - z_c|/r_c]^{4/3}}, \quad (8.6.1)$$

in which z_c is the z-coordinate of the crossover. For the calculation of the distribution $\rho(\Delta\alpha)$ in a beam segment with a narrow crossover, one should use a modified version of Eq. (5.11.1),

$$FWHM_\alpha = \left| \int_{z_0}^{z_1} dz \ \text{sign}(z - z_c) \frac{\Delta FWHM_\alpha}{\Delta z} \right|. \quad (8.6.2)$$

The sign function is included to account for the fact that the angular deflection experienced by the test particle during a weak collision with a single field particle changes sign when the field particle crosses the beam axis (or in fact the reference trajectory, which is here identical to the beam axis). Eq. (8.6.2) presupposes that the field particle crosses the axis at the location of the crossover. This approximation is only justified when the crossover is narrow ($K \gg 1$). Eq. (8.6.2) is therefore not suited to describe a nearly cylindrical beam segment. However, the extreme case of a cylindrical segment ($K=0$) is covered by Eq. (8.6.2) if one takes $z_c = z_0$ or $z_c = z_1$. By this choice, one accounts for the fact that in a homocentric cylindrical beam segment the particle trajectories do not cross each other.

Substitution of Eq. (8.6.1) into Eq. (8.6.2) and integration yields Eq. (8.5.31), where the function $S_{HA}(S_c, K)$ is now given by

$$S_{HA}(S_c, K) = \frac{2^{1/3} 3 C_{PHA}}{C_{CHA}} \left| \frac{1}{(1/K + 2S_c)^{1/3}} - \frac{1}{[1/K + 2(1-S_c)]^{1/3}} \right|.$$

Substitution of Eq. (8.5.32) yields the approximating function $S_{HAa}(S_c, K)$, introduced by Eq. (8.5.20). From this derivation it becomes clear that Eq. (8.5.20) is only valid for $K \gg 1$, while correct results for $K \to 0$ are obtained if one takes $S_c = 0$ or $S_c = 1$.

Let us now consider the pencil beam regime. According to Eq. (8.4.31), the contribution of a single slice at axial position z of length Δz to the total distribution of angular deflections $\rho(\Delta \alpha)$ can be expressed as

$$\Delta FWHM_\alpha = C_{PPA} \frac{I^3 r_c}{V^{5/2}} [1 + \alpha_0 |z - z_c|/r_c] \, \Delta z, \qquad (8.6.3)$$

in which z_c is again the z-coordinate of the crossover. Substitution of Eq. (8.6.3) into Eq. (8.6.2) and integration yields Eq. (8.5.35), where the function $S_{PA}(S_c, K)$ is now given by

$$S_{PA}(S_c, K) = \frac{2 C_{PPA}}{C_{CPA}} |1 - 2S_c| \left(\frac{1}{K} + 1 \right).$$

Substitution of Eq. (8.5.36) yields the approximating function $S_{PAa}(S_c, K)$, introduced by Eq. (8.5.21). As for the Holtsmark regime, it should be noticed that this equation is only valid for $K \gg 1$, while correct results for $K \to 0$ are obtained by talcing $S_c = 0$ or $S_c = 1$.

8.7. Results for Gaussian angular and spatial distributions

The results of the previous sections apply to the case of a uniform spatial and a uniform angular distribution in the crossover, see Eq. (8.2.6). We will now investigate the required modifications in case either the angular or the spatial distribution or both distributions are Gaussian. The analysis is similar to that of Section 7.6, pertaining to the Boersch effect.

We repeated the program outlined in Section 8.3, now using the modified versions of Eq. (8.2.6) in which either the spatial or the angular distribution or both distributions are taken Gaussian, following the specifications of Eq. (7.6.1) and/or Eq. (7.6.2). The principle aspects of the calculation remain the same, and the discussion here is restricted to a presentation of the main results. We evaluated Eqs. (8.2.6) and (8.2.7) on the basis of complete collisions, using Eq. (8.2.1). Analogous to Eq. (8.3.13), we found the following expressions for the scaled function $\bar{p}(\bar{k})$:

- For a Gaussian angular distribution and a Gaussian spatial distribution

$$\bar{p}_{gg}(\bar{k}) = \bar{k}\frac{16}{\sqrt{\pi}}\int_0^\infty d\bar{v}\, e^{-\bar{v}^2}\int_0^\infty dy\, e^{-y^2}\int_0^1 du\left(\frac{1}{u}-1\right)^{1/2} J_1\left(\frac{u\bar{k}\bar{v}}{R(\bar{v},y)}\right). \qquad (8.7.1)$$

- For a Gaussian angular distribution and a uniform spatial distribution

$$\bar{p}_{gu}(\bar{k}) = \bar{k}\frac{32}{\pi}\int_0^\infty d\bar{v}\, e^{-\bar{v}^2}\int_0^1 dy\sqrt{1-y^2}\int_0^1 du\left(\frac{1}{u}-1\right)^{1/2} J_1\left(\frac{u\bar{k}\bar{v}}{R(\bar{v},y)}\right). \qquad (8.7.2)$$

- For a uniform angular distribution and a Gaussian spatial distribution

$$\bar{p}_{ug}(\bar{k}) = \bar{k}\frac{16}{\sqrt{\pi}}\int_0^1 d\bar{v}\int_0^\infty dy\, e^{-y^2}\int_0^1 du\left(\frac{1}{u}-1\right)^{1/2} J_1\left(\frac{u\bar{k}\bar{v}}{R(\bar{v},y)}\right). \qquad (8.7.3)$$

Eqs. (8.7.1) through (8.7.3) are similar to Eqs. (7.6.3) through (7.6.5). As in Section 8.3, we will investigate the behavior for $\bar{k} \to 0$ and for $\bar{k} \to \infty$.

For $\bar{k} \to 0$, Eqs. (8.7.1)–(8.7.3) show a quadratic dependency on $\bar{k}$, as described by the first equation of (8.3.14). The functions $\bar{p}_{2gg}(\bar{r}_c)$, $\bar{p}_{2gu}(\bar{r}_c)$ and $\bar{p}_{2ug}(\bar{r}_c)$, which should be used in Eq. (8.3.14) for the different cases, are the same as obtained for the axial velocity distribution $\rho(\Delta v_z)$; see Eqs. (7.6.6), (7.6.7) and (7.6.8) respectively. A similar result was found in Section 8.3 for the case of a uniform angular and spatial distribution; see the second equation of (8.3.14). All functions can be approximated by Eq. (7.6.11), using different constants E and F for the individual cases, as prescribed by Eq. (7.6.12).

For $\bar{k} \to \infty$, Eqs. (8.7.1)–(8.7.3) show a square root dependency on $\bar{k}$, as described by Eq. (8.3.16). The function $f_\infty(\bar{r}_c)$ varies for the different angular and spatial distributions. Analogous to Eq. (8.3.8), one finds

$$f_{\infty gg}(\bar{r}_c) = \frac{1}{\sqrt{\pi}}\int_0^1 d\bar{v}\,\frac{\exp(-\bar{v}^2)}{\bar{v}^{1/2}}\int_0^\infty dy\exp(-y^2)R(\bar{v},y)^{1/2} \qquad (8.7.4)$$

$$f_{\infty gu}(\bar{r}_c) = \frac{2}{\pi} \int_0^1 d\bar{v} \frac{\exp(-\bar{v}^2)}{\bar{v}^{1/2}} \int_0^1 dy (1-y^2)^{1/2} R(\bar{v},y)^{1/2} \tag{8.7.5}$$

$$f_{\infty ug}(\bar{r}_c) = \frac{1}{\sqrt{\pi}} \int_0^1 \frac{d\bar{v}}{\bar{v}^{1/2}} \int_0^\infty dy \exp(-y^2) R(\bar{v},y)^{1/2}, \tag{8.7.6}$$

where the function $R(\bar{v},y)$ is given by Eq. (8.3.4). These functions have the following properties

$$\lim_{r_c \to \infty} f_{\infty gg}(\bar{r}_c) = \frac{[\Gamma(3/4)]^2}{4(2\pi)]^{1/2}} \bar{r}_c^{1/2} = .14976 \, \bar{r}_c^{1/2}$$

$$\lim_{r_c \to \infty} f_{\infty gu}(\bar{r}_c) = \frac{2}{5}\sqrt{\frac{2}{\pi}} \frac{[\Gamma(3/4)]^2}{\Gamma(1/4)} \bar{r}_c^{1/2} = .13219 \, \bar{r}_c^{1/2} \tag{8.7.7}$$

$$\lim_{r_c \to \infty} f_{\infty ug}(\bar{r}_c) = \frac{\Gamma(3/4)}{3(2\pi)]^{1/2}} \bar{r}_c^{1/2} = .16296 \, \bar{r}_c^{1/2}$$

$$\lim_{r_c \to 0} f_{\infty gg}(\bar{r}_c) = \lim_{r_c \to 0} f_{\infty gu}(\bar{r}_c) = \frac{\sqrt{\pi}}{2}, \quad \lim_{r_c \to 0} f_{\infty ug}(\bar{r}_c) = 1. \tag{8.7.8}$$

Similar to Eq. (8.3.10), one can approximate the functions $f_{\infty gg}(\bar{r}_c)$, $f_{\infty gu}(\bar{r}_c)$ and $f_{\infty ug}(\bar{r}_c)$ using

$$f_{\infty a}(\bar{r}_c) = (P + Q\bar{r}_c)^{1/2}, \tag{8.7.9}$$

in which P and Q are numerical constants, which differ for the individual cases

$$\begin{array}{ll} P_{gg} = P_{gu} = \pi/4, & P_{ug}(=P_{uu}) = 1 \\ Q_{gg} = .02243, & Q_{gu} = .01747 \\ Q_{ug} = .02655, & (Q_{uu} = .02069). \end{array} \tag{8.7.10}$$

The approximations for the functions $\bar{p}_2(\bar{r}_c)$ and $f_\infty(\bar{r}_c)$, given by Eqs. (7.6.11) and (8.7.9) respectively, contain the essential results of the analysis based on complete collisions for different types of spatial and angular distributions. The functions $\bar{p}_2(\bar{r}_c)$ and $f_\infty(\bar{r}_c)$ specify the $\bar{r}_c$-dependency of the FWHM of the angular deflection distribution in the Gaussian regime and the weak complete collision regime respectively, as can be seen from Eq. (8.3.21) with Eqs. (8.3.20) and (8.3.24).

It is often convenient to express the results obtained for non-uniform distributions in terms of the effective width measures α_{eff} and r_{eff}, which are defined as the widths for which the expressions obtained for uniform distributions yield the same results as those obtained within the full calculation, taking the proper distribution(s) into account. The resulting expressions for the angular deflections in the Gaussian regime show the same functional dependency as was found for the Boersch effect in this regime. Accordingly, the effective widths α_{eff} and r_{eff} are the same as specified by Eq. (7.6.15). For the weak complete collision regime, one should define α_{eff} and r_{eff} such that the $FWHM_\alpha$-values follow Eqs. (8.3.24) and (8.7.9). For $\bar{r}_c = 0$, the result depends only on the type of angular distribution, and one finds

$$\alpha_{eff} = \left(\frac{4}{\pi}\right)^{1/3} \alpha_c = \frac{2^{7/6}}{\pi^{1/3}} \sigma_\alpha = 1.533 \; \sigma_\alpha, \quad (\bar{r}_c = 0) \qquad (8.7.11)$$

(Weak complete coll. regime).

The effective width r_{eff} follows by considering the case $\bar{r}_c \to \infty$. The reader might verify that the resulting values for r_{eff} varies slightly for the different cases covered by Eq. (8.7.10). In general, a reasonable approximation is obtained for all $\bar{r}_c$-values by taking

$$\alpha_{eff} \cong 1.6 \; \sigma_\alpha, \quad r_{eff} \cong 1.8 \; \sigma_r \qquad (8.7.12)$$

(Weak complete coll. regime).

The exact values for α_{eff} and r_{eff} obtained for the individual cases differ less than 5% from the values given by Eq. (8.7.12).

The remaining results to be investigated pertain to the Holtsmark regime and the pencil beam regime. For those regimes, the distribution $\rho(\Delta v_\perp)$ can be determined from the distribution of the lateral interaction force $\rho(F_\perp)$ in a cylindrical beam. Accordingly, it is sufficient to determine in what way $\rho(F_\perp)$ is affected by the type of density distribution. The calculation of $\rho(F_\perp)$ for a cylindrical beam with a uniform density distribution was carried out in Section 5.8. The results are expressed by Eqs. (5.8.16) and (5.8.20). Let us reconsider this calculation, now starting from a Gaussian radial density distribution. Using Eq. (7.6.2), one may modify Eq. (5.8.9) as

$$\rho_2(F_\perp) = \int_0^\infty 2 \, dz \int_0^\infty \frac{2r \, dr}{r_0^2} e^{-(r/r_0)^2} \delta\left(F_\perp - \frac{C_0}{r^2} \frac{1}{\left(1 + (z/r)^2\right)^{3/2}}\right). \qquad (8.7.13)$$

Analogous to Eq. (5.8.12), this transforms to

$$\rho_2(F_\perp) = \frac{2C_0^{3/2}}{r_0^2 F_\perp^{5/2}} \int_0^\infty \frac{ds}{(1+s^2)^{9/4}} \exp\left(-\frac{C_0}{F_\perp r_0^2} \frac{1}{(1+s^2)^{3/2}}\right). \qquad (8.7.14)$$

As before, we will investigate the extreme cases $r_0 \to \infty$ and $r_0 \to 0$, which correspond to an extreme extended beam and a pencil beam respectively.

For $r_0 \to \infty$, one may replace the exponential function in Eq. (8.7.14) by unity, and one retrieves Eq. (5.8.13). This expression leads to the two-dimensional variant of the Holtsmark distribution given by Eq. (5.8.16). Thus, one obtains the same equation as for a uniform density distribution.

However, r_0 now stands for $\sqrt{2} \times \sigma_r$, instead of the outer beam radius. The FWHM of the corresponding angular deflection distribution is, for a homocentric cylindrical beam, given by Eq. (8.4.29) and, for a beam segment with a crossover, by Eq. (8.5.31). In general, the same equations as for a uniform distribution are obtained if one takes

$$\alpha_{eff} = \alpha_0 = \sqrt{2}\sigma_\alpha, \quad r_{eff} = r_c = \sqrt{2}\sigma_r, \quad \text{(Holtsmark regime)}. \quad (8.7.15)$$

Notice that the same effective width measures were found for the Boersch effect, as expressed by Eq. (7.6.18).

For $r_0 \to 0$, one may approximate the integral of (8.7.14) by taking $1+s^2 \simeq s^2$. This is justified since the contribution to the integral comes from large s-values. This gives

$$p_2(F_\perp) = \frac{2C_0^{3/2}}{r_0^2 F^{5/2}} \int_0^\infty \frac{ds}{s^{9/2}} \exp\left(-\frac{C_0}{F_\perp r_0^2} \frac{1}{s^3}\right) = \frac{\Gamma(1/6)}{9} \frac{C_0^{1/3} r_0^{1/3}}{F_\perp^{4/3}}, \quad (8.7.16)$$

which should be compared to Eq. (5.8.18) obtained for a uniform distribution. The FWHM of the corresponding angular deflection distribution is, for a homocentric cylindrical beam, given by Eq. (8.4.31) and, for a beam segment with a crossover, by Eq. (8.5.35). By comparison of Eqs. (8.7.16) and (5.8.18), it is found that one may use the equations for a uniform distribution if one takes

$$\alpha_{eff} = \left(\frac{7\Gamma(1/6)}{36}\right)^3 \sqrt{2}\sigma_\alpha = 1.793\ \sigma_\alpha, \quad r_{eff} = 1.793\ \sigma_r \quad (8.7.17)$$

(pencil beam regime),

which completes the analysis.

Eqs. (7.6.15), (8.7.12), (8.7.15) and (8.7.17) specify, for the different regimes, the values of the effective beam semi-angle α_{eff} of a Gaussian angular distribution and the effective crossover radius r_{eff} of a Gaussian spatial distribution. The values of α_{eff} and r_{eff} differ for the different regimes, varying between $\sqrt{2} \times \sigma$ and $1.8 \times \sigma$.

CHAPTER NINE

Trajectory displacement effect

Contents

9.1. Introduction	267
9.2. General aspects	268
9.3. Homocentric beam segment with a crossover	272
9.4. Homocentric cylindrical beam segment	286
9.5. Beam segment with a crossover of arbitrary dimensions	288
9.6. Trajectory displacement and angular deflection distribution	293
9.7. Results for Gaussian angular and spatial distributions	296

9.1. Introduction

In this chapter, the extended two-particle approach will be used to calculate the trajectory displacement effect generated in a rotational symmetric beam segment in drift space. The objective is to determine the distribution of virtual displacements $\rho(\Delta r)$ observed in a certain reference plane. The virtual displacement Δr of a test particle is determined by its spatial displacement at the end of the beam segment Δr_f, its velocity displacement $\Delta v_\perp$, and the location of the reference plane. The distribution of the transverse velocity displacements $\rho(\Delta v_\perp)$ was computed in Chapter 8. The results of that chapter are exploited to explain the mechanism of the trajectory displacement effect and to check the results in some limiting cases.

The chapter starts with a summary of the relevant results of Chapters 5 and 6, covering the basics of the statistical part of the model and the calculation of the displacement Δr caused by a two-particle collision, respectively. This material is then used to perform the calculation of $\rho(\Delta r)$ for different beam geometries. The specific cases of a homocentric beam segment with a crossover ($r_c = 0$) and a homocentric cylindrical beam segment ($\alpha_0 = 0$) are studied successively. The slice method, described in Section 5.10, is exploited to treat the general case of a beam segment with a crossover of arbitrary dimensions. For all geometries, the full trajectory displacement distribution $\rho(\Delta r)$ is computed. Explicit expressions are presented for the Full Width at Half Maximum (FWHM) of this distribution, as well as the Full Width median (FW_{50}) and the root mean square value (rms). The angular and spatial distribution in the crossover are assumed to be uniform in most calculations. Gaussian distributions are covered by introducing effective width measures for which the results obtained for uniform distributions can be used. Scaling is applied to reduce the number of independent parameters and to simplify the notation of the intermediate results. In the final results the scaling is removed in order to make the dependency on the experimental parameters explicit.

A computer program has been developed to perform the various steps of the calculation numerically. As for the angular deflection distribution, one should, in general, distinguish four regimes, each corresponding to a different type of distribution: the Gaussian regime, the weak complete collision regime, the Holtsmark regime, and the pencil beam regime. The weak complete collision regime becomes only manifest for a beam segment with a narrow crossover in which the reference plane does not coincide with the crossover plane. The widths of the various distributions show different dependencies on the experimental parameters.

The calculation of the trajectory displacement distribution $\rho(\Delta r)$ is, in general, more complex than the calculation of the distribution of the axial and transverse velocity displacements, $\rho(\Delta v_z)$ and $\rho(\Delta v_\perp)$, considered in Chapters 7 and 8 respectively. One reason is that an additional independent parameter is involved, namely the location of the reference plane. For a beam segment of relatively low particle density, with a narrow crossover in the middle, the dependency on this parameter is rather weak, since the angular deflections generated in such a beam are negligible. However, in all other cases the dependency on the location of the reference plane cannot be disregarded. Another complication is that it is not allowable to use complete collision dynamics to evaluate the displacement Δr caused by a two-particle encounter. One may use *nearly* complete collision dynamics, but the complexity of the resulting equation for the displacement Δr prohibits an exact analytical treatment of the statistical part of the problem. For the Gaussian regime, this implies that one has to rely entirely on numerical calculations, which makes a complete analysis rather involved. For all other regimes, it is possible to verify the numerical results by analytical means. A direct calculation of the behavior of the distribution in the transition areas between the regimes seems not possible. The numerical data for the $FWHM$ and the FW_{50} of the trajectory displacement distribution is approximated by analytical expressions, which cover the entire range of operating conditions.

9.2. General aspects

The trajectory displacement effect corresponds to the generation of lateral displacements Δr (observed in a certain reference plane) due to the effect of statistical Coulomb interactions. These displacements are related to the stochastic fluctuations in the charge density, which occur due to the particle nature of the beam. We recall that the trajectory displacement effect should be distinguished from the space charge effect, which is related to the average (smoothed-out) charge density in the beam. Both phenomena are caused by the transverse component of the Coulomb interaction force, but their impacts on the properties of the beam differ essentially. To isolate the trajectory displacement effect, we consider the central reference trajectory in a rotational symmetric beam. The test particles running along this trajectory experience no systematic space charge force, and their deviations are purely stochastic. A discussion of the impact of the space charge effect on the properties of the beam will be presented in Chapter 11.

Our objective is to determine the distribution of virtual displacements $\rho(\Delta r)$, observed in a certain reference plane. The virtual displacement Δr of the test particle is found by extrapolating its final perturbed position along its final perturbed velocity toward the reference plane. This procedure combines the lateral shift at the end of the beam segment Δr_f and the change in lateral velocity $\Delta v_\perp$ into the virtual shift Δr.

The location of the reference plane is specified by the parameter S_i, which is defined by Eq. (3.2.1). The reference plane is imaged down to the target. The virtual broadening in the reference plane is therefore representative for the broadening of the final probe. The reference plane will often coincide with the location of the crossover, which is specified by the parameter S_c. This parameter is also defined by Eq. (3.2.1). It should be emphasized that the reference plane and the crossover plane do not coincide in all cases of interest. For example, consider a shaped-beam lithography system employing Koehler illumination, as described by Pfeiffer (1979) or Moore et al. (1981). The shaping apertures, located in the upper part of the column, are imaged down to the target. The appropriate reference plane is therefore optically conjugated to these planes. However, the crossover between the shaping apertures is optically conjugated to the source crossover and is imaged by the second condenser and the demagnification lenses on the principal plane of the projector lens (and not on the target). Thus, in the shaping section of the column, the crossover plane and the reference plane do not coincide. This example illustrates that the quantities S_c and S_i should, in general, be treated as independent parameters.

The distribution $\rho(\Delta r)$ is in some extreme cases fully determined by the distribution of transverse velocity displacements $\rho(\Delta v_\perp)$, which was calculated in Chapter 8. One case is a beam segment with a crossover in which the reference plane is far removed from the crossover. In this situation, the main contribution to the (virtual) displacement of a test particle Δr stems from its angular deflection $\Delta \alpha = \Delta v_\perp / v_z$. The displacements Δr and $\Delta v_\perp$ are therefore related by

$$\lim_{S_i \to \pm\infty} \Delta r = |S_i T \Delta v_\perp|, \qquad (9.2.1)$$

in which the time of flight $T=L/v_z$. Another case is a cylindrical beam segment succeeded by a lens that focusses the beam in its focal plane. The transverse spatial shift Δr_{foc} in the focal plane of the lens is entirely determined by the transverse velocity shift $\Delta v_\perp$ of the test particle experienced in the cylindrical beam segment

$$\Delta r_{foc} = f \Delta v_\perp / v_z, \qquad (9.2.2)$$

where f is the focal length of the lens. Eqs. (9.2.1) and (9.2.2) will be exploited to relate the distributions $\rho(\Delta r)$ and $\rho(\Delta v_\perp)$.

As in Chapters 7 and 8, we will utilize the extended two-particle approach, which was outlined in Chapter 5. The dynamical part of the problem consists of the calculation of the virtual radial displacement Δr experienced by the test particle due to the interaction with a single field particle. This problem was studied in Chapter 6. The shift Δr can be expressed as a function of the geometrical variables $\xi = (r_\perp, b_z, v, \Phi)$, the time of flight T, the initial time $t_i = -S_c T$, and the location of the reference plane specified by the parameter S_i; see Eqs. (6.2.1)–(6.2.3) respectively and Fig. 5.1 (note: $\Phi = \psi - \varphi$). Explicit analytical equations can be determined for nearly complete collisions and for weak collisions. The virtual displacement Δr caused by a nearly complete collision ($T \gg v/b_z$) is given by

$$\Delta r \cong \frac{1}{1+q_c} \left| (S_i - S_c)Tv - \frac{b_z(1-q_c)}{2\sqrt{q_c}} \left[2\ln\left(\sqrt{\frac{q_c}{q_c+1}} \frac{2vT}{b_z} \sqrt{S_c(1-S_c)}\right) \right. \right.$$
$$\left. \left. -2 - \frac{S_i}{S_c} - \frac{1-S_i}{1-S_c} \right] + b_z\sqrt{q_c}\left[1 - \frac{2}{1+q_c}\left(\frac{S_i}{S_c} - 1\right)\right] \right|, \qquad (r_\perp = 0),$$
$$(9.2.3)$$

with

$$q_c = \left(\frac{mv^2 b_z}{2C_0}\right)^2, \qquad (r_\perp = 0), \qquad (9.2.4)$$

as follows from Eqs. (6.12.1), (6.12.2), (6.7.14) and (6.4.16). Eq. (9.2.3) refers to the case $r_\perp = 0$, which is relevant for a point crossover ($r_c = 0$). It is of importance to note that Eq. (9.2.3) diverges for $T \to \infty$, even when $S_c = S_i$. Accordingly, one cannot perform the calculation of $\rho(\Delta r)$ on the basis of complete collisions but is forced to account for the fact that T is finite. Unfortunately, this prohibits an exact analytical calculation of the statistical part of the problem. Therefore, one has to rely on numerical calculations when the contribution of nearly complete collisions to the distribution $\rho(\Delta r)$ becomes significant.

This observation also implies that an analysis of $\rho(\Delta v_\perp)$ based on complete collisions is insufficient to predict the behavior of $\rho(\Delta r)$ for $S_i \to \pm \infty$. To illustrate this point, consider Eq. (9.2.3) for $S_i \to \infty$

$$\lim_{S_i \to \infty} \frac{\Delta r}{S_i T} \simeq \frac{v}{1+q_c} \left| 1 + \frac{2C_0}{mv^3 T} \left[\frac{1-q_c}{2} \left(\frac{1}{S_c} - \frac{1}{1-S_c} \right) - \frac{2q_c}{1+q_c} \frac{1}{S_c} \right] \right| \quad (9.2.5)$$

in which we utilized Eq. (9.2.4). Eq. (9.2.5) is identical to the expression for $\Delta v_\perp$ given by Eq. (6.11.2), substituting $r_\perp = 0$. Thus, Eqs. (9.2.3) and (6.11.2) fulfill the relation given by Eq. (9.2.1). It is here essential that Eq. (6.11.2) includes the terms of the order $1/T$. However, these terms were ignored in the analysis of the distribution $\rho(\Delta v_\perp)$, assuming that the collisions can be regarded as complete ($T \to \infty$). This assumption was introduced in order to permit an analytical calculation of the problem; see Section 8.3. Unfortunately, one should, at this point, conclude that the results for $\rho(\Delta v_\perp)$ based on complete collisions are not suited to predict the behavior of $\rho(\Delta r)$ for $S_i \to \pm\infty$ by means of Eq. (9.2.1).

The virtual displacement Δr, caused by a weak collision, is for $r_\perp = 0$ equal to (reproducing Eq. 6.12.6)

$$\Delta r \simeq \frac{C_0}{mv^2} \left| \ln\left(\frac{(S_c vT + r_i)[(1-S_c)vT + r_f]}{b_z^2} \right) - vT \left(\frac{S_i}{r_i} + \frac{1-S_i}{r_f} \right) \right|, \quad (r_\perp = 0), \quad (9.2.6)$$

where r_i and r_f are given by

$$r_i = \left[(S_c vT)^2 + b_z^2 \right]^{1/2}, \qquad r_f = \left\{ [(1-S_c)vT]^2 + b_z^2 \right\}^{1/2}. \quad (9.2.7)$$

Notice that Eqs. (9.2.6) and (8.2.2) (for $r_\perp = 0$) do satisfy Eq. (9.2.1). Thus, the analysis of $\rho(\Delta v_\perp)$ based on weak collisions can be used to predict the behavior of $\rho(\Delta r)$ for $S_i \to \pm\infty$.

Eqs. (9.2.3) and (9.2.6) apply to the case that the initial relative velocity of the particles is nonzero ($v \neq 0$). In case $v = 0$, we found for a half-complete collision (reproducing Eq. 6.12.8)

$$\Delta r \simeq \left| \frac{(C_0/m)^{1/2} r_\perp T S_i}{\left(b_z^2 + r_\perp^2\right)^{3/4}} - \frac{r_\perp}{4} \left[S_i + \ln\left(\frac{8(C_0/m)^{1/2} T}{\left(b_z^2 + r_\perp^2\right)^{3/4}} \right) \right] \right|, \quad (v=0), \quad (9.2.8)$$

and for a weak collision (reproducing Eq. 6.12.9)

$$\Delta r \simeq \left| \frac{C_0}{m} \frac{r_\perp T^2 (S_i - 1/2)}{\left(b_z^2 + r_\perp^2\right)^{3/2}} \right|, \quad (v=0). \quad (9.2.9)$$

Notice that Eqs. (9.2.8) and (8.2.4) satisfy Eq. (9.2.1). The same holds true for Eqs. (9.2.9) and (8.2.5). We recall that the case $v=0$ is relevant for a monochromatic homocentric cylindrical beam segment. Eq. (9.2.2) gives the displacement Δr_{foc} in the back focal plane of a lens located at the end of a cylindrical beam segment. This displacement is entirely determined by the velocity deviation $\Delta v_\perp$ only and does not depend on the shift Δr_f of the test particle at the end of the segment. Instead of calculating Δr_{foc} from $\Delta v_\perp$ by means of Eq. (9.2.2), one can also determine this quantity directly from Eqs. (9.2.8) and (9.2.9) using

$$\Delta r_{foc} = \lim_{S_i \to \infty} \Delta r \frac{f}{S_i L}. \tag{9.2.10}$$

This relation is in agreement with Eqs. (9.2.1) and (9.2.2).

The statistical part of the problem consists of the evaluation of Eqs. (5.7.7), (5.7.8) and (5.7.10), in which $\Delta \eta$ now represents Δr,

$$\rho_2(\Delta r) = \int_0^{v_0} \frac{2v\, dv}{v_0^2} \int_0^{2\pi} \frac{d\Phi}{2\pi} \int_0^{r_c} \frac{2r_\perp\, dr_\perp}{r_c^2} \int_{-S_c L}^{(1-S_c)L} db_z\, \delta[\Delta r - \Delta r(v, \Phi, r_\perp, b_z)] \tag{9.2.11}$$

$$p(k) = \int_0^\infty d\Delta r\, \rho_2(\Delta r)[1 - J_0(k\Delta r)] \tag{9.2.12}$$

$$\rho(\Delta r) = \frac{1}{2\pi} \int_0^\infty k\, dk\, J_0(k\Delta r)\, e^{-\lambda p(k)}, \tag{9.2.13}$$

similar to Eqs. (8.2.6)–(8.2.8). We expressed $p(\xi)\, d\xi$ directly in terms of v, Φ, $r_\perp$ and b_z, using Eqs. (5.4.6) and (5.3.2). The distribution in v and $r_\perp$ are taken uniform with a cutoff at $v_0 = \alpha_0 v_z$ and r_c respectively, as prescribed by Eq. (5.3.3). The displacement distribution $\rho(\Delta r)$ depends only on the magnitude of the displacement Δr and not on its direction, which is a consequence of the rotational symmetry of the beam and the choice of the central reference trajectory. The probability of a displacement of size Δr is equal to $\rho(\Delta r) = 2\pi \Delta r^2 \rho(\Delta r)$. The reader is referred to Section 7.2 for a summary of the assumptions underlying our model for statistical interactions.

9.3. Homocentric beam segment with a crossover

In this section we will calculate the distribution of transverse displacements $\rho(\Delta r)$ for a homocentric beam segment with a crossover ($r_c = 0$). We will use the δ,ν-scaling defined by Eq. (6.13.1). The scaled two-particle distribution $\rho_2^*(\Delta r^*)$ follows from Eq. (9.2.11)

$$\rho_2^*(\Delta r^*) \doteq v_0^{*2} \rho_2(\Delta r)$$
$$= \int_0^{v_0^*} 2v^*\, dv^* \int_0^\infty 2\, db_z^*\, \delta\left[\Delta r^* - \Delta r^*(v^*, b_z^*)\right]. \tag{9.3.1}$$

We evaluated this integral for different values of v_0^* on the basis of the exact solution for the function $\Delta r^*(v^*, b_z^*)$, following the analysis of Section 6.6. The results for the case $S_c = S_i, = 1/2$ are plotted in Fig. 9.1.

Trajectory displacement effect

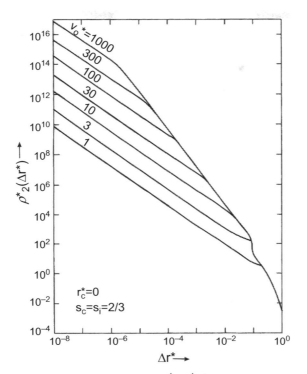

Fig. 9.1 The scaled two-particle distribution $\rho_2^*(\Delta r^*)$ for a homocentric beam segment with a crossover in the middle ($S_c = 1/2$), plotted for different values of the scaled maximum transverse velocity v_0^*. The results represent the virtual displacements observed in the crossover plane ($S_i = S_c$).

Analytical expressions for the function $\Delta r^*(v^*, b_z^*)$ can be obtained for nearly complete collisions and for weak collisions; see Eqs. (9.2.3) and (9.2.6) respectively. Unfortunately, the complexity of Eq. (9.2.3) prohibits an exact analytical treatment based on nearly complete collisions and in the analysis here we will only consider the contribution of weak collisions. This approach is appropriate to explain the data of Fig. 9.1 for small and medium Δr^*. By scaling Eq. (9.2.6) one obtains

$$\Delta r^*(v^*, b_z^*) = \frac{1}{4v^{*2}} \left| \ln\left(\frac{[S_c v^* + r_i^*][(1 - S_c)v^* + r_f^*]}{b_z^{*2}}\right) - v^* \left(\frac{S_i}{r_i^*} + \frac{1 - S_i}{r_f^*}\right) \right|, \quad (9.3.2)$$

with $r_i^* = [(S_c v^*)^2 + b_z^{*2}]^{1/2}$ and $r_f^* = \{[(1 - S_c)v^*]^2 + b_z^{*2}\}^{1/2}$. Similar to the derivation of Eq. (8.5.9), one finds from Eqs. (9.3.1) and (9.3.2)

$$\rho_2^*(\Delta r^*) = \frac{2}{\Delta r^{*5/2}} \int_0^\infty dz\, F(z)^{3/2} \Theta(\Delta r^* - F(z)/v_0^{*2}), \qquad (9.3.3)$$

in which the function $F(z)$ now stands for

$$F(z) = \frac{1}{4}\left| \ln\left(\frac{\{S_c + [S_c^2 + z^2]^{1/2}\}\{1 - S_c + [(1-S_c)^2 + z^2]^{1/2}\}}{z^2} \right) \right.$$

$$\left. - \frac{S_i}{[S_c^2 + z^2]^{1/2}} - \frac{1 - S_i}{[(1-S_c)^2 + z^2]^{1/2}} \right|.$$

$$(9.3.4)$$

We will consider the cases $v_0^* \to \infty$ and $v_0^* \to 0$ separately.

For $v_0^* \to \infty$, the argument of the step function in Eq. (9.3.3) is positive for all z-values and one obtains

$$\rho_2^*(\Delta r^*) = I_5 \frac{S_{HT}(S_c, S_i)^{3/2}}{\Delta r^{*5/2}},$$

$$S_{HT}(S_c, S_i)^{3/2} = \frac{2}{I_5} \int_0^\infty dz\, F(z)^{3/2} \qquad (v_0^* \to \infty). \qquad (9.3.5)$$

We choose the constant I_5 such that $S_{HT}(S_c = 1/2, S_i = 1/2) = 1$. This implies

$$I_5 = \frac{1}{2^{3/2}} \int_0^\infty dx \left[\ln\left(\frac{1 + (1+x^2)^{1/2}}{x} \right) - \frac{1}{(1+x^2)^{1/2}} \right]^{3/2} = 0.39507, \qquad (9.3.6)$$

as follows by numerical integration. The integral for $S_{HT}(S_c, S_i)$ was evaluated numerically for different S_c and S_i and is plotted in Fig. 9.2. A reasonable approximation of the numerical data is given by the function $S_{HTa}(S_c, S_i)$, defined as

$$S_{HTa}(S_c, S_i) = |S_{HTa}^*(S_c, S_i)|$$

$$S_{HTa}^*(S_c, S_i) \simeq \frac{3S_c - 2S_i}{[2S_c]^{1/3}} + \frac{3(1-S_c) - 2(1-S_i)}{[2(1-S_c)]^{1/3}}. \qquad (9.3.7)$$

This equation can be derived by means of the slice method as will be demonstrated in Section 9.5. By fitting the numerical data it was found that a somewhat better approximation is given by

$$S_{HTb}(S_c, S_i) = H(S_c) + |S_{HTa}^*(S_c, S_i) - H(S_c)|, \qquad (9.3.8)$$

where $S_{HTa}^*(S_c, S_i)$ is defined by Eq. (9.3.7) and the function $H(S_c)$ is equal to

Trajectory displacement effect

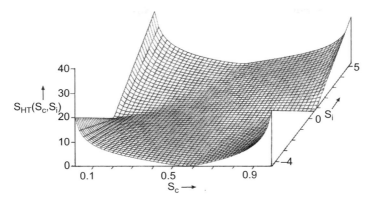

Fig. 9.2 The function $S_{HT}(S_c, S_i)$ defined by Eqs. (9.3.4)–(9.3.6). This function specifies the dependency of the width of trajectory displacement distribution on the crossover position parameter S_c and image plane position parameter S_i in the Holtsmark regime and pertains to a stigmatic crossover ($r_c^* = 0$).

$$H(S_c) = \ln\left\{1.1 + [3S_c(1 - S_c)]^4\right\}. \tag{9.3.9}$$

However, the approximation $S_{HTa}(S_c, S_i)$ given by Eq. (9.3.7) is, in general, sufficiently accurate and is preferred for its simplicity.

For the particular case that the reference plane and the crossover plane coincide ($S_c = S_i$), Eq. (9.3.7) reduces to

$$S_{HTa}(S_c) = \frac{1}{2}\left([2S_c]^{2/3} + [2(1 - S_c)]^{2/3}\right), \qquad (S_c = S_i). \tag{9.3.10}$$

This function has a maximum for $S_c = 1/2$ ($S_{HTa}(1/2) = 1$). The total variation within the full S_c-range ($0 < S_c < 1$) is approximately 20%. ($S_{HTa}(0) = S_{HTa}(1) = 2^{-1/3} = 0.7937$). In general, the function $S_{HTa}(S_c, S_i)$ increases when the crossover and the reference plane do not coincide ($S_c \neq S_i$), as can be seen from Fig. 9.2. The contribution of angular deflections becomes dominant when the crossover plane and the reference plane are separated by a large distance ($S_i \ll S_c$ or $S_i \gg S_c$).

For $v_0^* \to 0$, the argument of the step function in Eq. (9.3.3) is only positive for large z-values, and one may use the following asymptotic expansion of the function $F(z)$

$$\lim_{z \to \infty} F(z) = \frac{S_{PT}(S_c, S_i)}{48z^3} + O(z^{-5}), \tag{9.3.11}$$

in which the function $S_{PT}(S_c, S_i)$ is given by

$$S_{PT}(S_c, S_i) = |4 - 6(S_c + S_i) + 12S_c S_i|. \tag{9.3.12}$$

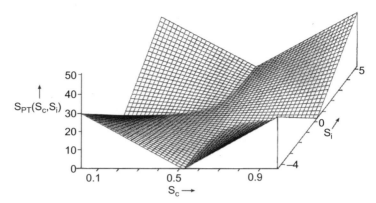

Fig. 9.3 The function $S_{PT}(S_c, S_i)$ given by Eq. (9.3.12). This function specifies the dependency of the width of trajectory displacement distribution on the crossover position parameter S_c and image plane position parameter S_i in the pencil beam regime and pertains to a stigmatic crossover ($r_c^* = 0$).

The function $S_{PT}(S_c, S_i)$ was scaled such that $S_{PT}(1/2, 1/2) = 1$. It is plotted in Fig. 9.3. For the particular case that the reference plane and the crossover plane coincide ($S_c = S_i$), one obtains

$$S_{PT}(S_c) = 4|1 + 3S_c(S_c - 1)|, \qquad (S_c = S_i). \tag{9.3.13}$$

This function has a minimum for $S_c = 1/2$ ($S_{PT}(1/2) = 1$). It is interesting to note that the behavior of the function $S_{PT}(S_c)$ is, in this respect, opposite to that of $S_{HTa}(S_c)$, given by Eq. (9.3.10). The key observation here is that the magnitude of the trajectory displacement effect decreases with the beam radius in case of an extended beam, while it increases with the beam radius for a pencil beam. The behavior of $S_{HTa}(S_c)$ and $S_{PT}(S_c)$ can now be understood from the fact that the maximum beam radius in the section becomes minimum for $S_c = 1/2$. The total variation of the function $S_{PT}(S_c)$, within the full S_c-range ($0 < S_c < 1$), is a factor 4 ($S_{PT}(0) = S_{PT}(1) = 4$), which is considerably larger than the corresponding variation of the function $S_{HTa}(S_c)$.

Substitution of Eq. (9.3.11) into Eq. (9.3.3) yields

$$\rho_2^*(\Delta r^*) = \frac{2}{6^{1/3} 7} \frac{S_{PT}(S_c, S_i)^{1/3} v_0^{*7/3}}{\Delta r^{*4/3}}, \qquad (v_0^* \to 0). \tag{9.3.14}$$

The reader might verify that the numerical data of $\rho_2^*(\Delta r^*)$ for the case $S_c = S_i = 1/2$, which is plotted in Fig. 9.1, agrees with Eqs. (9.3.5) and (9.3.14).

The scaled function $p^*(k^*)$ follows from Eq. (9.2.12),

Trajectory displacement effect

$$p^*(k^*) = \frac{v_0^{*2}}{\delta} p(k) = \int_0^\infty d\Delta r^* \rho_2^*(\Delta r^*)\left[1 - J_0(k^*\Delta r^*)\right], \quad (9.3.15)$$

with $k^* = k\delta$. This integral was evaluated numerically for the case $S_c = S_i = 1/2$, starting from the numerical data for $\rho_2^*(\Delta r^*)$. The results are plotted in Fig. 9.4.

The behavior of the function $p^*(k^*)$ can be understood from the analysis of the two particle distribution $\rho_2^*(\Delta r^*)$. For $v_0^* \to \infty$, one may use Eq. (9.3.5). By substitution into Eq. (9.3.15), one finds, after integration

$$p^*(k^*) = \frac{2\sqrt{2}\Gamma(1/4)I_5}{9\Gamma(3/4)} S_{HT}(S_c, S_i)^{3/2} k^{*3/2}, \quad (9.3.16)$$

similar to the derivation of Eq. (5.8.14). The numerical factor in the right-hand side of Eq. (9.3.16) is equal to 0.36734, as follows with Eq. (9.3.6).

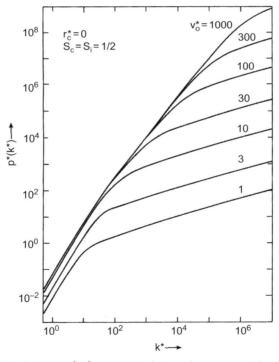

Fig. 9.4 The scaled function $p^*(k^*)$ corresponding to the trajectory displacement distribution generated in a homocentric beam segment with a crossover in the middle ($S_c = 1/2$), plotted for different values of the maximum transverse velocity v_0^*. The results pertain to the virtual displacements observed in the crossover plane ($S_i = S_c$).

Eq. (9.3.16) constitutes a good approximation of $p^*(k^*)$ for intermediate k^*-values, especially in combination with large v_0^*.

Substitution of Eq. (9.3.14) into Eq. (9.3.15) yields, after integration,

$$p^*(k^*) = \frac{2^{1/3} 18 \Gamma(5/6)}{3^{1/3} 7 \Gamma(1/6)} S_{PT}(S_c, S_i)^{1/3} v_0^{*7/3} k^{*1/3}, \quad (9.3.17)$$

similar to the derivation of Eq. (5.8.19). The numerical factor in the right-hand side of Eq. (9.3.17) is equal to 0.45553. Eq. (9.3.17) constitutes a good approximation of $p^*(k^*)$ for large k^*-values, especially in combination with small v_0^*.

For small k^*-values the function $p^*(k^*)$ becomes quadratic in k^*. We define

$$\lim_{k^* \to 0} p^*(k^*) = \frac{1}{4} p_2^*(v_0^*, S_c, S_i)^{1/2} k^{*2}. \quad (9.3.18)$$

The small k^*-behavior can directly be related to the second moment of the two-particle distribution $\rho_2^*(\Delta r^*)$

$$\langle \Delta r_2^{*2} \rangle = p_2^*(v_0^*, S_c, S_i), \quad (9.3.19)$$

as can be seen from Eq. (9.3.15) by expanding the Bessel function for small values of its argument ($J_0(z) \approx 1 - z^2/4 + \cdots$). We evaluated the quantity $\langle \Delta r_2^{*2} \rangle$ numerically as function of v_0^*, S_c and S_i. For $v_0^* \to \infty$ and $S_c = S_i = 1/2$, it was found that

$$p_2^*(\infty, 1/2, 1/2) = 0.273. \quad (9.3.20)$$

The results obtained for $S_c = S_i = 1/2$ and different v_0^*-values are plotted in Fig. 9.5. The data can be fitted using

$$p_{2_a}^*(v_0^*, 1/2, 1/2) = \frac{p_2^*(\infty, 1/2, 1/2)}{\left(1 + 1.40/v_0^{*8/7}\right)^{7/2}}. \quad (9.3.21)$$

For comparison, this approximation is included in Fig. 9.5. In order to evaluate the general behavior of the function $p_2^*(v_0^*, S_c, S_i)$, we computed the quantity $\langle \Delta r_2^{*2} \rangle$ for all combinations of v_0^*, S_c and S_i with $S_c = 0.1$, 0.2, ..., 0.9; $S_i = -0.5, -0.4, \ldots, 1.6$ and $v_0^* = 10$, 100 and 1000. The results obtained for $v_0^* \geq 100$ and $S_c = S_i$ can be approximated by

$$p_{2_a}^*(v_0, S_c, S_c) = p_2^*(\infty, 1/2, 1/2) \left[1 + 0.682(S_c - 0.5) - 0.739\left(S_c - 0.5\right)^2\right]$$

$$(v_0^* \geq 100), \quad (9.3.22)$$

which is a monotonically increasing function of S_c ($0 < S_c < 1$), which levels off for $S_c \to 1$. The quantity $\langle \Delta r_2^{*2} \rangle$ increases strongly when $S_c \neq S_i$. For all considered v_0^*-values, it was found that the results for $S_c \ll S_i$ or $S_c \gg S_i$ can be approximated as

Trajectory displacement effect

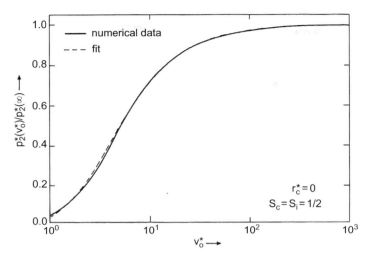

Fig. 9.5 The function $p_2^*(v_0^*)$ defined by Eq. (9.3.18) for $S_c = S_i = 1/2$ and its approximation given by Eq. (9.3.21). This function gives the v_0^*-dependency of the function $p^*(k^*)$, plotted in Fig. 9.4, for $k^* \to 0$.

$$p_{2_a}^*(v_0^*, S_c, S_i) = p_2^*(\infty, 1/2, 1/2) 113 v_0^{*1/2}(S_c - S_i)^2$$
$$(S_c \ll S_i \text{ or } S_c \gg S_i). \qquad (9.3.23)$$

On first sight, the square root dependency on v_0^* is somewhat surprising. The reader might verify with the scaling relations of Eq. (7.5.6) that the result for $\langle \Delta v_\perp^2 \rangle$ given by Eq. (8.3.18) leads, according to the relation between Δr and $\Delta v_\perp$ for $S_i \to \pm\infty$ given by Eq. (9.2.1), to a quadratic dependency on v_0^*. However, it should be recalled that Eq. (8.3.18) is based on complete collisions. This renders Eq. (9.2.1) inapplicable, as was discussed in Section 9.2.

For further use, it is convenient to define the function $P_{CT}(v_0^*, S_c, S_i)$ as

$$P_{CT}(v_0^*, S_c, S_i) = [p_2^*(v_0^*, S_c, S_i)/p_2^*(\infty, 1/2, 1/2)]^{1/2}. \qquad (9.3.24)$$

Combining the analysis of Eqs. (9.3.21)–(9.3.23) we postulate that the function $P_{CT}(v_0^*, S_c, S_i)$ can, in general, be approximated as

$$P_{CTa}(v_0^*, S_c, S_i) = \left[P_{CT_1}(v_0^*, S_c)^2 + 113 v_0^{*1/2}(S_c - S_i)^2 \right]^{1/2}, \qquad (9.3.25)$$

with

$$P_{CT1}(v_0^*, S_c) = \frac{\left[1 + 0.682(S_c - 0.5) - 0.739(S_c - 0.5)^2\right]^{1/2}}{\left(1 + 1.40/v_0^{*8/7}\right)^{7/4}} \qquad (9.3.26)$$

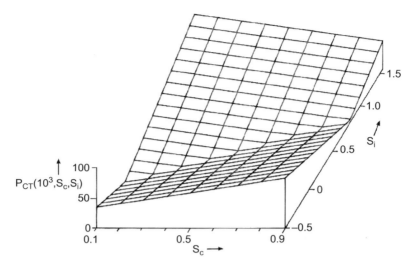

Fig. 9.6 The function $P_{CT}(v_0^*, S_c, S_i)$ defined by Eqs. (9.3.24), (9.3.19) and (9.3.20) for $v_0^* = 1000$. This function specifies the dependency of the width of trajectory displacement distribution on the crossover position parameter S_c and image plane position parameter S_i in the Gaussian regime and pertains to a stigmatic crossover ($r_c^* = 0$).

Eqs. (9.3.19), (9.3.20), (9.3.24), (9.3.25) and (9.3.26) constitute a good approximation for the numerical data of $\langle \Delta r_2^{*2} \rangle$ for all considered values of the parameters v_0^*, S_c and S_i. To illustrate the dependency on the parameters S_c and S_i, the exact numerical data obtained for the function $P_{CT}(v_0^* = 1000, S_c, S_i)$ is plotted in Fig. 9.6.

The scaled displacement distribution $\rho^*(\Delta r^*)$ follows from Eq. (9.2.13)

$$\rho^*(\Delta r^*) = \delta^2 \rho(\Delta r) = \frac{1}{2\pi} \int_0^\infty k^* dk^* J_0(k^* \Delta r^*) e^{-4\bar{\lambda} p^*(k^*)}. \tag{9.3.27}$$

The second moment of this distribution is determined by the small k^* behavior of the function $p^*(k^*)$. Using Eqs. (5.5.4) (with $m=2$) and (9.3.18) one may write

$$\langle \Delta r^{*2} \rangle = 4 p_2^*(v_0^*, S_c, S_i) \bar{\lambda}. \tag{9.3.28}$$

The relation can also be obtained from Eq. (9.3.19), using Eq. (5.5.10) (with $m=2$). We evaluated the FWHM of the distribution $\rho^*(\Delta r^*)$ numerically for $S_c = S_i = 1/2$ and various v_0^*-values, starting from the numerical data obtained for $p^*(k^*)$. The results are plotted in Fig. 9.7. Fig. 9.8 shows the results for $v_0^* = 100$, $S_c = 1/2$ and various S_i values. Different regimes should be distinguished by the type of the generated trajectory displacement distribution. The occurrence of the different regimes depends on the value of the scaled linear particle density $\bar{\lambda}$ and the pencil beam factor χ_c, which are defined by Eqs. (7.3.22) and (7.5.28) respectively.

Trajectory displacement effect 281

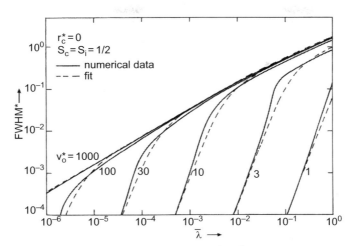

Fig. 9.7 The *FWHM* of the trajectory distribution $\rho^*(\Delta r^*)$ generated in a homocentric beam segment with a crossover in the middle ($S_c = 1/2$), plotted for different values of the maximum transverse velocity v_0^*. The results represent the virtual displacements observed in the crossover plane ($S_i = S_c$). The depicted fit is defined Eqs. (9.3.37), (9.3.40) and (9.3.39).

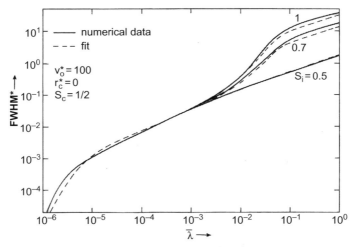

Fig. 9.8 The *FWHM* of the trajectory distribution $\rho^*(\Delta r^*)$ generated in a homocentric beam segment with a crossover in the middle ($S_c = 1/2$), plotted for different locations of the image plane, which is specified by the parameter S_i. The results pertain to a beam for which the maximum transverse velocity $v_0^* = 100$. The depicted fit is defined Eqs. (9.3.37)–(9.3.39).

(I) Gaussian regime. For $\bar{\lambda} \gg 10^{-3}$ and $\chi_c \gg 1$, the distribution $\rho^*(\Delta r^*)$ is dominated by the small k^*-behavior described by Eq. (9.3.18). This leads to a two-dimensional Gaussian distribution with a *FWHM* that is equal to

$$FWHM_G^* = 4\left[p_2^*(\infty, 1/2, 1/2)\ln 2\right]^{1/2} P_{CT}(v_0^*, S_c, S_i)\bar{\lambda}^{1/2}$$
$$= 1.740\, P_{CT}(v_0^*, S_c, S_i)\bar{\lambda}^{1/2}, \qquad (9.3.29)$$

where the function $P_{CT}(v_0^*, S_c, S_i)$ can be approximated by the function $P_{CTa}(v_0^*, S_c, S_i)$, which is defined by Eq. (9.3.25). For $S_c = S_i$ one may approximate $FWHM_G^* = FWHM_{G1}^*$, with

$$FWHM_{G1}^* = 1.740\, P_{CT1}(v_0^*, S_c)\bar{\lambda}^{1/2}, \quad (S_c = S_i), \qquad (9.3.30)$$

where the function $P_{CT1}(v_0^*, S_c)$ is given by (9.3.26). For $S_c \ll S_i$ or $S_c \gg S_i$, one may approximate $FWHM_G^* = FWHM_{G2}^*$, with

$$FWHM_{G2}^* = 18.5\,|S_c - S_i|\,v_0^{*1/4}\lambda^{1/2}, \quad (S_c \ll S_i \text{ or } S_c \gg S_i), \qquad (9.3.31)$$

as follows from Eqs. (9.3.29) and (9.3.25).

(II) Weak complete collision regime. This regime becomes only manifest for $S_c \neq S_i$. It can directly be related to the corresponding regime for the distribution of angular deflections; see Eqs. (8.3.22) and (8.3.23). For the present case of a homocentric beam segment with a crossover ($r_c = 0$), it was found from the numerical data for *FWHM** that

$$FWHM_W^* = 42.105\,|S_c - S_i|\,v_0^*\bar{\lambda}^2, \qquad (9.3.32)$$

which is in agreement with Eq. (8.3.23), utilizing Eq. (9.2.1) and the scale relations of Eqs. (7.5.5) and (7.5.6).

(III) Holtsmark regime. For $\bar{\lambda} \ll 10^{-3}$ and $\chi_c \gg 1$, the distribution $\rho^*(\Delta r^*)$ is dominated by the large k^*-behavior described by Eq. (9.3.16). Substitution into Eq. (9.3.27) yields

$$\rho^*(\Delta r^*) = \frac{1}{2\pi}\int_0^\infty k^*dk^* J_0(k^*\Delta r^*)e^{-C_{3/2}\bar{\lambda}[S_{HT}(S_c,S_i)k^*]^{3/2}}, \qquad (9.3.33)$$

with $C_{3/2} = 8\sqrt{2}\Gamma(1/4)I_5/9\Gamma(3/4) = 1.4694$. This distribution is the two-dimensional variant of the Holtsmark distribution plotted in Fig. 5.4 (curve corresponding to $\gamma = 3/2$). The *FWHM* of this distribution is given by

$$FWHM_H^* = 2.6554\,\frac{2^{7/3}\Gamma(1/4)^{2/3}I_5^{2/3}}{3^{4/3}\Gamma(3/4)^{2/3}}\,S_{HT}(S_c, S_i)\bar{\lambda}^{2/3}, \qquad (9.3.34)$$

as follows with Eq. (5.9.15) and Table 5.1. The numerical constant in the right-hand side of Eq. (9.3.34) is equal to 3.4320, using the value of I_5 given by Eq. (9.3.6).

(IV) Pencil beam regime. For $\chi_c \ll 1$, the distribution $\rho^*(\Delta r^*)$ is dominated by the k^*-behavior given by Eq. (9.3.17)

$$\rho^*(\Delta r^*) = \frac{1}{2\pi} \int_0^\infty k^* dk^* J_0(k^* \Delta r^*) e^{-C_{1/3} \bar{\lambda}[S_{PT}(S_c, S_i) v_0^{*7} k^*]^{1/3}}, \quad (9.3.35)$$

with $C_{1/3} = (2/3)^{1/3} 72\Gamma(5/6)/7\Gamma(1/6) = 1.8221$. This distribution is referred to as the two-dimensional pencil beam distribution; see Fig. 5.4 (curve corresponding to $\gamma = 1/3$). The FWHM of this distribution is given by

$$FWHM_P^* = .013013 \frac{2}{3} \left(\frac{72\Gamma(5/6)}{7\Gamma(1/6)} \right)^3 S_{PT}(S_c, S_i) v_0^{*7} \bar{\lambda}^3, \quad (9.3.36)$$

as follows with Eq. (5.9.15) and Table 5.1. The constant in the right-hand side of Eq. (9.3.36) is equal to 0.078729.

In order to interpolate Eqs. (9.3.29), (9.3.32), (9.3.34) and (9.3.36) we use

$$FWHM^* = FWHM_0^* H_{CT}(\bar{\lambda}, v_0^*, S_c, S_i),$$
$$FWHM_0^* = 3.4320 \bar{\lambda}^{2/3}, \quad (9.3.37)$$

where the function $H_{CT}(\bar{\lambda}, v_0^*, S_c, S_i)$ is defined as

$$H_{CT}(\bar{\lambda}, v_0^*, S_c, S_i)$$
$$= \left\{ \left[\left(\frac{FWHM_0^*}{FWHM_{G1}^*} \right)^6 + \left(\frac{FWHM_0^*}{FWHM_H^*} \right)^6 \right]^{1/7} + \left(\frac{FWHM_0^*}{FWHM_P^*} \right)^{6/7} \right\}^{-7/6}$$
$$+ \left[\left(\frac{FWHM_0^*}{FWHM_{G2}^*} \right)^3 + \left(\frac{FWHM_0^*}{FWHM_W^*} \right)^3 \right]^{-1/3},$$

which is identical to

$$H_{CT}(\bar{\lambda}, v_0^*, S_c, S_i) =$$

$$\left[\left(\frac{1}{S_{HT}(S_c, S_i)^6} + \frac{A\bar{\lambda}}{P_{CT1}(v_0^*, S_c)^6} \right)^{-1/7} + \frac{B}{v_0^{*6} \bar{\lambda}^2 S_{PT}(S_c, S_i)^{6/7}} \right]^{-7/6}$$

$$+ |S_c - S_i| \left(\frac{C \bar{\lambda}^{-1/2}}{v_0^{*3/4}} + \frac{D}{v_0^{*3} \bar{\lambda}^4} \right)^{-1/3},$$

(9.3.38)

in which the constants A, B, C and D are given by

$$A = 58.88, \quad B = 25.42, \quad C = 6.38 \times 10^{-3}, \quad D = 5.415 \times 10^{-4}. \tag{9.3.39}$$

For the case $S_c = S_i$, Eq. (9.3.38) reduces to

$$H_{CT}(\bar{\lambda}, v_0^*, S_c) = \left[\left(\frac{1}{S_{HT}(S_c)^6} + \frac{A\bar{\lambda}}{P_{CT1}(v_0^*, S_c)^6} \right)^{1/7} \right. \tag{9.3.40}$$

$$\left. + \frac{16B}{\chi_c^2 S_{PT}(S_c)^{6/7}} \right]^{-7/6} \quad (S_c = S_i),$$

where the pencil beam factor χ_c is defined by Eq. (7.5.28). The function $S_{HT}(S_c)$ can be approximated by the function $S_{HTa}(S_c)$ of Eq. (9.3.10). The function $S_{PT}(S_c)$ is given by Eq. (9.3.13). For comparison, the fit of Eqs. (9.3.37) and (9.3.38) is included in Figs. 9.7 and 9.8.

Similar results can be derived for the Full Width median (FW_{50}) of the trajectory displacement distribution $\rho(\Delta r)$. The results for the individual regimes can be determined from the FWHM-values by means of Eq. (8.5.43). The general expression for the entire range of operating conditions becomes

$$FW_{50}^* = 1.2505 \; FWHM_0^* H_{CT}(\bar{\lambda}, v_0^*, S_c, S_i), \tag{9.3.41}$$

similar to Eq. (9.3.37). The function $H_{CT}(\bar{\lambda}, v_0^*, S_c, S_i)$ is again given by Eq. (9.3.38), now taking the constants A, B, C and D equal to

$$A = 225.2, \quad B = .1439, \quad C = 1.25 \times 10^{-2}, \quad D = 5.768 \times 10^{-8} \tag{9.3.42}$$

instead of the values given by Eq. (9.3.39).

By removing the scaling in Eq. (9.3.37), one obtains for the FWHM of the trajectory displacement distribution (using $FWHM_r = \delta FWHM^*$)

$$FWHM_r = C_{CHT} H_{CT}(\bar{\lambda}, v_0^*, S_c, S_i) \frac{I^{2/3} L^{2/3}}{V^{4/3} \alpha_0^{4/3}}, \tag{9.3.43}$$

where the constant C_{CHT} is given by

$$C_{CHT} = 2.6554 \frac{\Gamma(1/4)^{2/3} I_5^{2/3}}{2^{1/3} 3^{4/3} \pi \Gamma(3/4)^{2/3}} \frac{m^{1/3}}{\varepsilon_0} = 0.17205 \frac{m^{1/3}}{\varepsilon_0}$$

using the numerical value of I_5 given by Eq. (9.3.6). In the case of electrons, one finds $C_{CHT} = 1.8837$ in SI-units. Eq. (9.3.43), in combination with Eq. (9.3.38), constitutes the main result of this section. It gives the FWHM of the trajectory displacement distribution $\rho(\Delta r)$, generated in a homocentric beam segment with a crossover ($r_c = 0$), for the entire range of operating conditions.

We conclude this section by investigating the dependencies on the experimental parameters in the main regimes covered by Eqs. (9.3.43) and (9.3.38). For $\bar{\lambda} \lesssim 0.01$ and $\chi_c \gg 1$, one may approximate $H_{CT} \approx S_{HT}(S_c, S_i)$, which yields the result for the Holtsmark regime

$$FWHM_r = C_{CHT} S_{HT}(S_c, S_i) \frac{I^{2/3} L^{2/3}}{V^{4/3} \alpha_0^{4/3}}, \qquad (9.3.44)$$

which can also be found directly from Eq. (9.3.34).

For $\bar{\lambda} \gtrsim 0.05$ and $\chi_c \gg 1$, one obtains the result for the Gaussian regime

$$FWHM_r = C_{CGT} P_{CT}(v_0^*, S_c, S_i) \frac{I^{1/2} L^{2/3}}{V^{13/12} \alpha_0}, \qquad (9.3.45)$$

where the constant C_{CGT} is given by

$$C_{CGT} = \frac{[p_2^*(\infty, 1/2, 1/2) \ln 2]^{1/2}}{2^{1/12} \pi^{5/6}} \frac{e^{1/12} m^{1/4}}{\varepsilon_0^{5/6}},$$

in which the constant $p_2^*(\infty, 1/2, 1/2)$ is specified by Eq. (9.3.20). The numerical constant in the right-hand side of the equation for C_{CGT} is equal to 0.158. In the case of electrons, one finds $C_{CGT} = 0.215$ in SI-units.

For $S_c \neq S_i$, $\chi_c \gg 1$ and intermediate $\bar{\lambda}$-values, one obtains the result for the weak complete collision regime. By removing the scaling in Eq. (9.3.32), it follows that

$$FWHM_r = C_{CWA} |S_c - S_i| \frac{I^2 L}{\alpha_0^3 V^3}, \qquad (9.3.46)$$

where the constant C_{CWA} is defined in Eq. (8.3.24). Eqs. (9.3.46) and (8.3.24) fulfill Eq. (9.2.1), using that $\Delta \alpha = \Delta v_\perp / v_z$. This implies that the spatial broadening described by Eq. (9.3.46) is the result of the angular deflections produced by weak complete collisions.

For $\chi_c \ll 1$, Eq. (9.3.38) gives $H_{CT} \approx 9.033 \times 10^{-4} S_{PT}(S_c, S_i) \chi_c^{7/3}$. Substitution in Eq. (9.3.43) yields

$$FWHM_r = C_{CPT} S_{PT}(S_c, S_i) \frac{I^3 \alpha_0 L^3}{V^{5/2}}, \qquad (9.3.47)$$

where the constant C_{CPT} is given by

$$C_{CPT} = .013013 \left(\frac{18 \Gamma(5/6)}{7 \Gamma(1/6)}\right)^3 \frac{1}{6\pi \sqrt{2}} \frac{m^{3/2}}{\varepsilon_0 e^{7/2}} = 6.9218 \times 10^{-5} \frac{m^{3/2}}{\varepsilon_0 e^{7/2}}.$$

Eq. (9.3.47) can also be derived directly from Eq. (9.3.36). In the case of electrons, one finds $C_{CPT} = 4.1287 \times 10^{27}$ in SI-units.

9.4. Homocentric cylindrical beam segment

In this section, we will consider the case of a monochromatic homocentric cylindrical beam, which is succeeded by a lens. The angular displacement $\Delta \alpha = \Delta v_\perp / v_z$ experienced by a particle in the cylindrical beam segment results in a spatial shift Δr_{foc} in the backfocal plane of the lens, which is given by Eq. (9.2.2). It should be emphasized that Δr_{foc} is entirely determined by $\Delta \alpha$ only and does not depend on the spatial displacement experienced by the particle in the cylindrical beam segment. Accordingly, the trajectory displacement distribution $p(\Delta r_{foc})$, in the backfocal plane of the lens, is fully determined by the distribution $p(\Delta v_\perp)$, generated in the cylindrical beam segment. This distribution was calculated in Section 8.4.

Combining Eqs. (9.2.2) and (8.4.28), the FWHM of the distribution $p(\Delta r_{foc})$ can, in general, be expressed as

$$FWHM_r = FWHM_\alpha f = C_{PHA} H_{PA}(\bar{\lambda}^*, v_0^*) \frac{I^{2/3} Lf}{V^{4/3} r_0^{4/3}}, \qquad (9.4.1)$$

in which f denotes the focal distance of the lens. The constant C_{PHA} and the function $H_{PA}(\bar{\lambda}^*, v_0^*)$ are given by Eqs. (8.4.28) and (8.4.26) respectively. Eq. (9.4.1) covers the entire range of operating conditions. As described in Section 8.4, three regimes should be distinguished by the type of distribution which is generated: the Gaussian regime, the Holtsmark regime and the pencil beam regime. The corresponding results for the individual regimes can directly be obtained from Eqs. (8.4.29)–(8.4.31) respectively, again using that $FWHM_r = FWHM_\alpha f$.

It is interesting to compare the effects generated in a homocentric cylindrical beam with radius r_0 to those generated in a homocentric beam segment with a crossover in the middle and beam semi-angle $\alpha_0 = 2 r_0/L$. The length L is the same for both geometries. This choice of parameters implies that the beam radius at the entrance and the exit plane of the beam segment is the same for both geometries; see Fig. 9.9. Assume that a lens focusses the beam in a spot at a distance $L/2$ from the end of the beam segment. For the crossover, this implies that the transverse magnification from the crossover to this image plane equals unity. For the cylindrical beam, it implies that $f = L/2$. We now define the ratio R as

$$R = \frac{FWHM_{r,\text{crossover}}}{FWHM_{r,\text{cylinder}}} = \frac{FWHM_{r,\text{crossover}}}{f FWHM_{\alpha,\text{cylinder}}}. \qquad (9.4.2)$$

Substitution of Eqs. (9.3.43) and (9.4.1) yields

$$R = \frac{C_{CHT} H_{CT}(\bar{\lambda}, v_0^*, 1/2, 1/2)}{2^{1/3} C_{PHA} H_{PA}(4\bar{\lambda}, v_0^*/2)} = 1.3418 \frac{H_{CT}(\bar{\lambda}, v_0^*, 1/2, 1/2)}{H_{PA}(4\bar{\lambda}, v_0^*/2)}, \qquad (9.4.3)$$

Trajectory displacement effect

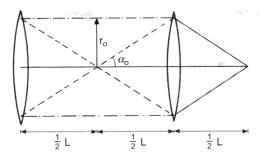

Fig. 9.9 Schematic view of two beam geometries in a segment of length L between two lenses. One is a homocentric cylindrical beam ($\alpha_0 = 0$) and the other is a homocentric beam segment with a crossover in the middle ($r_c = 0$, $S_c = 1/2$). The corresponding contributions to the trajectory displacement effect observed in the succeeding crossover (located at a distance $L/2$ after the second lens) are compared; see Eqs. (9.4.4) and (9.4.6).

where it was used that $r_0 = \alpha_0 L/2$ and $f = L/2$, which implies $r_0^* = v_0^*/2$ and $\lambda^* = 4\bar{\lambda}$. The last relation follows from the definitions of $\bar{\lambda}$ and λ^*, given by Eq. (7.3.22) and (7.4.18) respectively.

The results for the individual regimes can be determined from Eq. (9.4.3) or directly from Eq. (9.4.2) by substituting the corresponding results for the *FWHM*-values. For the Gaussian regime, one finds

$$R_G = .948 \frac{P_{CT}(v_0^*, 1/2, 1/2)}{P_{PA}(v_0^*/2)} \quad \text{(Gaussian regime)}. \tag{9.4.4}$$

Utilizing Eqs. (9.3.25), (9.3.26), (8.4.20) and (8.4.16), this gives

$$\lim_{v_0^* \to \infty} R_G = 0.948, \quad \lim_{v_0^* \to 0} R_G = 0.116. \tag{9.4.5}$$

Practical conditions correspond to $v_0^* \gg 1$, for which $R_G \approx 0.948$. Similarly, one finds for the Holtsmark regime and the pencil beam regime

$$\begin{aligned} R_H &= 1.34 \quad \text{(Holtsmark regime)} \\ R_P &= 0.333 \quad \text{(Pencil beam regime)}. \end{aligned} \tag{9.4.6}$$

Eqs. (9.4.5) and (9.4.6) show that the ratio R of the magnitude of the trajectory displacement effect generated in the two different geometries is, in general, a simple numerical constant, which differs slightly for the various regimes. However, for all regimes, the ratio R is close to unity, which implies that the size of the trajectory displacement effect generated in the beam segment with a crossover is of the same order of magnitude as the effect generated in the cylindrical beam.

On first sight, this result is perhaps somewhat surprising. Clearly, the particle density becomes much higher in the case of a crossover than for a cylindrical beam. For that reason one might conjecture that the trajectory displacement effect generated in the crossover is considerably larger, too. Indeed, the deviations from the unperturbed trajectories will be larger in the case of a crossover than for a cylindrical beam. However, what counts is the virtual broadening in the reference plane. Here, one should realize that the final lateral spatial shift and velocity shift generated in the segment with a crossover are partly canceled out by the extrapolation to a virtual position in the crossover plane. Our calculation shows that the resulting virtual broadening is of the same order of magnitude as the broadening generated in a cylindrical beam. It should be noted that the situation is quite different for the Boersch effect, as the reader might verify from the results presented in Chapter 7. The Boersch effect, which is related to the generated axial velocity spread, will be much larger for the crossover case than for the cylindrical beam.

9.5. Beam segment with a crossover of arbitrary dimensions

So far, we considered the trajectory displacement effect generated in the extreme cases of a homocentric beam segment with a crossover and a homocentric cylindrical beam segment respectively. We now wish to consider the general case of a beam segment with a crossover that has some finite nonzero radius r_c. For a homocentric crossover ($r_c=0$), it was found that the trajectory displacement effect is, in general, a function of four theoretical parameters, namely $\bar{\lambda}$, v_0^*, S_c and S_i, as is formally expressed by Eq. (9.3.37). For a crossover of arbitrary dimensions, the results will, in addition, depend on the scaled crossover radius $r_c^* = r_c/\delta$, and one ends up with five independent parameters. We note that it is sometimes more convenient to use the d_0, v_0-scaling defined by Eq. (7.3.2) instead of the δ, ν-scaling given by Eq. (6.13.1). In this representation, the parameters v_0^* and r_c^* are replaced by the parameters K and $\bar{r}_c$. The relation between both sets of parameters is given by Eq. (7.5.6). Clearly, in both representations, one requires five independent theoretical parameters to describe the general case.

A straightforward analysis of the dependency on all five parameters is rather involved. In principle, one could extend the approach outlined in Section 9.3 to include the dependency on r_c^*. However, such an approach would require an extreme large number of numerical evaluations of the two-particle distribution $\rho_2^*(\Delta r^*)$, while each evaluation now requires the solution of a four-dimensional integral instead of a two-dimensional integral, as can be seen from Eq. (9.2.11). We therefore did not attempt a systematic approach to this problem. Instead, we will use an approximate method starting from the results obtained for the extreme cases of a homocentric beam segment with a crossover and a homocentric cylindrical beam segment.

In order to get some indication of the dependency on the scaled crossover radius r_c^*, we evaluated the case $S_c = S_i = 1/2$ and $v_0^* = 200$ for various values of r_c^*, employing our numerical model. The results for the scaled FWHM of the trajectory displacement distribution $\rho^*(\Delta r^*)$ are plotted in Fig. 9.10. We recall that for practical operating conditions the scaling length δ is of the order 10^{-5} m (for $V = 10$ kV and $L = 0.1$ m, one finds $\delta = 14.2 \mu m$). Accordingly, the data of Fig. 9.10 indicates that the dependency on r_c^* is rather weak for the investigated circumstances. Thus, contrary to the Boersch effect, the trajectory displacement effect generated in a narrow crossover is not very sensitive to the crossover radius.

We will now exploit the slice method, which is outlined in Section 5.11, to investigate the dependency on the parameters v_0^* and r_c^* (or K and $\bar{r}_c$) in a more systematic way. The analysis presented here is similar to that of Section 8.6, concerning the distribution $\rho(\Delta v_\perp)$. The essence of the slice method is expressed by Eq. (5.11.1). This equation states that one may calculate the total displacement generated in a beam segment by adding linearly the displacements generated in a large number of thin cylindrical slices. The basic assumption is that the contributions of the individual slices

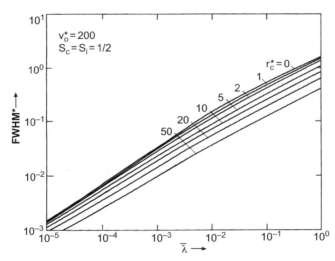

Fig. 9.10 The FWHM of the trajectory distribution $\rho^*(\Delta r^*)$ generated in a beam segment with a crossover in the middle ($S_c = 1/2$), plotted for different values of the scaled crossover radius r_c^*. The results represent the virtual displacements observed in the crossover plane ($S_i = S_c$) and pertain to a beam for which the maximum scaled transverse velocity $v_0^* = 200$.

to the total displacement experienced by the test particle are entirely correlated. Such is to be expected when the deviations from the unperturbed trajectories are small. Accordingly, one should only apply the slice method when the distribution $\rho(\Delta r)$ is dominated by weak collisions. This type of analysis is therefore restricted to the Holtsmark regime and the pencil beam regime.

Let us first consider the Holtsmark regime. According to Eq. (8.4.29), the contribution of a single slice at axial position z of length Δz to the *FWHM* of the total trajectory displacement distribution $\rho(\Delta r)$ can be expressed as

$$\Delta FWHM_r = C_{PHA} \frac{I^{2/3}}{V^{4/3} r_c^{4/3}} \frac{(z - z_i) \Delta z}{[1 + \alpha_0 |z - z_c|/r_c]^{4/3}}, \quad (9.5.1)$$

in which z_c and z_i denote the z-coordinates of the crossover and the image plane respectively. For the calculation of the distribution $\rho(\Delta r)$ in a beam segment with a narrow crossover, we will use

$$FWHM_r = \left| \int_{z_0}^{z_1} dz \, \text{sign}(z - z_c) \frac{\Delta FWHM_r}{\Delta z} \right|. \quad (9.5.2)$$

The sign function is included to account for the fact that the angular deflection experienced by the test particle during a weak collision with a single field particle changes sign when the field particle crosses the beam axis (or in fact the reference trajectory that is here identical to the beam axis). The quantity $\Delta FWHM_r$ given by Eq. (9.5.1) can become negative depending on the sign of the quantity $z - z_i$. Notice that the sign changes of the quantities $z - z_c$ and $z - z_i$ cancel out when $z_c = z_i$. Eq. (9.5.2) presupposes that the field particle crosses the axis exactly in the crossover. This approximation is only justified when the crossover is narrow ($K \gg 1$). Eq. (9.5.2) is therefore not suited to describe a nearly cylindrical beam segment. However, the extreme case of a cylindrical segment ($K = 0$) is covered, too, by taking $z_c = z_0$ or $z_c = z_1$.

Substitution of Eq. (9.5.1) into Eq. (9.5.2) and integration yields Eq. (9.3.44), where the function $S_{HT}(S_c, S_i)$ is now replaced by

$$S_{HTa}(S_c, S_i, K) = \left| \frac{3S_c - 2S_i + 3/2K}{(1/K + 2S_c)^{1/3}} + \frac{3(1 - S_c) - 2(1 - S_i) + 3/2K}{[1/K + 2(1 - S_c)]^{1/3}} - \frac{3}{K^{2/3}} \right|, \quad (9.5.3)$$

in which we approximated

$$\frac{C_{PHA}}{C_{CHT}} \frac{3}{2^{2/3}} = 1.1179 \approx 1. \quad (9.5.4)$$

The constants C_{PHA} and C_{CHT} are defined in Eqs. (8.4.28) and (9.3.43) respectively. The function $S_{HTa}(S_c, S_i, K)$ is the generalization of the function $S_{HTa}(S_c, S_i)$ given by Eq. (9.3.7). Both equations become identical for $K \to \infty$. As was mentioned previously, one should realize that the derivation of Eq. (9.5.3) presupposes $K \gg 1$. However, correct results for $K \to 0$ are obtained if one takes $S_c = 0$ or $S_c = 1$.

Let us now consider the pencil beam regime. According to Eq. (8.4.31), the contribution of a single slice at axial position z of length Δz to the total distribution of trajectory displacements $\rho(\Delta r)$ can be expressed as

$$\Delta FWHM_r = C_{PPA} \frac{l^3 r_c}{V^{5/2}} \left[1 + \alpha_0 |z - z_c|/r_c\right](z - z_i) \Delta z. \quad (9.5.5)$$

Substitution into Eq. (9.5.2) and integration yields Eq. (9.3.47), where the function $S_{PT}(Sc, S_i)$ is now replaced by

$$S_{PTa}(S_c, S_i, K) = \left| 4 - 6(S_c + S_i) + 12 S_c S_i + \frac{3 - 6 S_c^2 - 6 S_i (1 - 2 S_c)}{K} \right| \quad (9.5.6)$$

in which we used that

$$C_{PPA} = 12 C_{CPT}, \quad (9.5.7)$$

as follows from the definitions of the constants C_{CPA} and C_{CPT}. The function $S_{PTa}(S_c, S_i, K)$ is the generalization of the function $S_{PTa}(S_c, S_i)$ given by Eq. (9.3.12). These equations become identical for $K \to \infty$. As in the case of the Holtsmark regime, it should be noted that Eq. (9.5.6) is only valid for $K \gg 1$, while correct results for $K \to 0$ are obtained if one takes $S_c = 0$ or $S_c = 1$.

The remaining task is to extend the results for the Gaussian regime and the weak complete collision regime. For the latter it was found that the scaled FWHW follows Eq. (9.3.32). This result is directly related to the corresponding FWHM of the angular deflection distribution, which is specified by Eq. (8.3.24). Accordingly, it is reasonable to extend Eq. (9.3.32) for the case $r_c \neq 0$ as

$$FWHM_W^* = 42.105 |S_c - S_i| \left(1 + .08275 v_0^{*2} r_c^* \right) v_0^* \bar{\lambda}^2, \quad (9.5.8)$$

using the approximation of $f_\infty(\bar{r}_c)$ given by Eq. (8.3.10) and the relation $\bar{r}_c = 4 v_0^{*2} r_c^*$.

In order to obtain a systematic analysis of the dependency of the FWHM in the Gaussian regime on the parameters v_0^*, r_c^*, S_c and S_i, one should extend the numerical calculations of the mean square value $\langle \Delta r^{*2} \rangle$ to include nonzero r_c^*-values. As we did not perform such a systematic analysis, we will just present some provisional results to cover this case. The data of Fig. 9.10 shows that the FWHM of the scaled trajectory displacement distribution falls off with r_c^* as

$$FWHM_{G1}^* = \frac{FWHM_{G1}^*(r_c^* = 0)}{(1 + 0.30 r_c^*)^{1/2}}, \qquad (S_c = S_i, \ v_0^* = 200). \qquad (9.5.9)$$

Therefore, we generalize the function P_{CT1} specified by Eq. (9.3.26) as

$$P_{CT1}(v_0^*, r_c^*, S_c) = \frac{\left[1 + 0.682(S_c - 0.5) - 0.739(S_c - 0.5)^2\right]^{1/2}}{\left(1 + 1.40/v_0^{*8/7}\right)^{7/4} (1 + 0.30 r_c^*)^{1/2}}, \qquad (9.5.10)$$

which should yield accurate results for most conditions but may overestimate the exact result for small v_0^*-values in combination with large r_c^*-values. These conditions pertain to a nearly cylindrical extended beam. Eq. (9.3.26) is relevant for the case $S_c = S_i$. We did not further investigate the case $S_c \neq S_i$, and we will use Eq. (9.3.31) also for nonzero r_c^*-values.

The modifications outlined above have to be introduced in the general expression for the FWHM of the trajectory displacement distribution given by Eq. (9.3.43). This equation remains valid for a beam segment of arbitrary dimensions, provided that the original function $H_{CT}(\bar{\lambda}, v_0^*, S_c, S_i)$ is replaced by a function $H_{CT}(\bar{\lambda}, v_0^*, r_c^*, S_c, S_i)$ that depends also on r_c^*. Similar to the derivation of Eq. (9.3.38), it was found that

$$H_{CT}(\bar{\lambda}, v_0^*, r_c^*, S_c, S_i) = \left[\left(\frac{1}{S_{HT}(S_c, S_i, K)^6} + \frac{A\bar{\lambda}}{P_{CT1}(v_0^*, r_c^*, S_c)^6} \right)^{1/7} \right.$$

$$+ \frac{B}{v_0^{*6} \bar{\lambda}^2 S_{PT}(S_c, S_i, K)^{6/7}} \Bigg]^{-7/6}$$

$$\left. + |S_c - S_i| \left(\frac{C\bar{\lambda}^{1/2}}{v_0^{*3/4}} + \frac{D}{\left(1 + .08272 v_0^{*2} r_c^*\right)^3 v_0^{*3} \bar{\lambda}^4} \right)^{-1/3} \right..$$

$$(9.5.11)$$

The constants A, B, C and D remain the same and are specified by Eq. (9.3.39). The functions $S_{HT}(S_c, S_i, K)$ and $S_{PT}(S_c, S_i, K)$ can be approximated by the expressions given by Eqs. (9.5.3) and (9.5.6) respectively, while the function $P_{CT1}(v_0^*, r_c^*, S_c)$ is defined by Eq. (9.5.10). Notice that $K = v_0^*/2 r_c^*$, as expressed by Eq. (7.5.6). Correct results for $K \to 0$ are obtained by substituting $S_c = 0$ or $S_c = 1$, as was mentioned in the derivation of the functions $S_{HTa}(S_c, S_i, K)$ and $S_{PTa}(S_c, S_i, K)$. Eqs. (9.3.43) and (9.5.11) constitute the main result of this chapter. The modifications required to obtain the Full Width median value FW_{50} are indicated by Eqs. (9.3.41) and (9.3.42).

9.6. Trajectory displacement and angular deflection distribution

In this section, we will investigate the results of the previous section for the limiting cases $S_i \to \infty$ and $K \to 0$. Our objective is to clarify the relation with the equations for the $FWHM$ of the distribution of angular deflections presented in Chapter 8. The results for the different regimes involved will be investigated separately.

According to Eq. (9.2.1), the relation between $FWHM_r$ and $FWHM_\alpha$ can, in general, be expressed as

$$FWHM_\alpha = \lim_{S_i \to \pm\infty} \frac{FWHM_r}{|S_i L|}, \qquad (9.6.1)$$

utilizing that $T = L/v_z$ and $\Delta\alpha \approx \Delta v_\perp / v_z$. Eq. (9.6.1) expresses that the trajectory displacement effect observed in a far removed reference plane is entirely the result of angular deflections.

For the Holtsmark regime, the $FWHM$ of the trajectory displacement distribution is given by Eq. (9.3.44). For the general case, the function S_{HT} can be approximated by the function S_{HTa}, which is given by Eq. (9.5.3). Substitution into Eq. (9.6.1) yields

$$FWHM_\alpha = C_{CHT} \left| \frac{2}{(1/K + 2S_c)^{1/3}} - \frac{2}{[1/K + 2(1 - S_c)]^{1/3}} \right| \frac{I^{2/3}}{V^{4/3} \alpha_0^{4/3} L^{1/3}}, \qquad (9.6.2)$$

which is in agreement with Eqs. (8.5.31) and (8.5.20), approximating that

$$\frac{C_{CHA}}{C_{CHT}} = \left(\frac{3^5 \pi \Gamma(3/4)}{2^4 5^2 I_5^2 \Gamma(1/4)^2} \right)^{1/3} = 1.1179 \approx 1, \qquad (9.6.3)$$

equivalent to Eq. (9.5.4). The constant I_5 is given by Eq. (9.3.6).

For the pencil beam regime, the $FWHM$ of the trajectory displacement distribution is given by Eq. (9.3.47). The function S_{PT} can be approximated by the function S_{PTa}, which is given by Eq. (9.5.6). Substitution into Eq. (9.6.1) yields

$$FWHM_\alpha = 6C_{CPT} \left| 1 - 2S_c \right| \left(\frac{1}{K} + 1 \right) \frac{I^3 \alpha_0 L^2}{V^{5/2}}, \qquad (9.6.4)$$

which is in agreement with Eqs. (8.5.35) and (8.5.21), using that

$$6C_{CPT} = C_{CPA}, \qquad (9.6.5)$$

as follows from the definition of the constants C_{CPT} and C_{CPA}.

For the weak complete collision regime, one may verify that substitution of Eq. (9.3.46) into Eq. (9.6.1) directly leads to the $FWHM_\alpha$ given by Eq. (8.3.24). Notice that Eq. (9.3.46) pertains to the case $r_c = 0$. For the general case, the function $f_\infty(\bar{r}_c)$ was included, using that $\bar{r}_c = 4v_0 {*2} r_c^*$, as expressed by Eq. (9.5.8). For the Gaussian regime, one cannot simply relate the results obtained for $FWHM_r$ to those obtained for $FWHM_\alpha$, as was discussed in Section 9.2.

Another extreme case that is of interest is that of a nearly cylindrical beam. In terms of the parameter K, which is defined by Eq. (7.3.1), this case corresponds to $K \to 0$. We will demonstrate that the following relation between $FWHM_\alpha$ and $FWHM_r$ should exist when weak collisions are dominant

$$FWHM_r = |S_i - 1/2| L\ FWHM_\alpha, \quad (K \to 0, \text{ weak collisions}). \quad (9.6.6)$$

This equation can be understood from the observation that a weak collision in a homocentric cylindrical beam segment leads to a virtual spatial displacement Δr that is proportional to the distance from the reference plane to the middle of the beam segment ($=|S_i - 1/2|L$) times the total angular displacement $\Delta \alpha$ caused by the collision. This observation can be understood as follows. Assume that a field particle exerts a lateral force $F_\perp$ on the test particle during the flight through the beam segment from z_0 to z_1. As the particles are initially running along parallel trajectories and only experience a small deviation from these trajectories, the force $F_\perp$ will approximately be constant. The angular displacement $\Delta \alpha$ of the test particle can therefore be expressed as

$$\Delta \alpha = \int_{z_0}^{z_1} dz \frac{F_\perp}{mv_z^2} = \frac{F_\perp L}{mv_z^2}, \quad (9.6.7)$$

with $L = z_1 - z_0$. Similarly, the virtual displacement Δr in a reference plane at $z = z_r$ can be expressed as

$$\Delta r = \int_{z_0}^{z_1} dz |z_r - z| \frac{F_\perp}{mv_z^2} = |(z_0 - z_r)^2 - (z_1 - z_r)^2| \frac{F_\perp}{2mv_z^2}$$
$$= |S_i - 1/2| \frac{F_\perp L^2}{mv_z^2}, \quad (9.6.8)$$

Using that $z_r - z_0 = S_i L$. The derivation of Eq. (9.6.8) is equivalent to that of Eq. (5.3.7). From Eqs. (9.6.7) and (9.6.8), one finds

$$\Delta r = |S_i - 1/2| L\ \Delta \alpha. \quad (9.6.9)$$

By assuming that this result applies to all interactions of the test particle with the individual field particles, one obtains Eq. (9.6.6).

Let us now verify whether our results based on weak interactions fulfill Eq. (9.6.6). The analysis of the previous sections showed that weak incomplete interactions in an extended beam generate a Holtsmark type of distribution with a *FWHM* that is given by Eq. (9.3.44). The *FWHM* of the distribution generated by weak incomplete interactions in a pencil beam is given by Eq. (9.3.47). Weak *complete* collisions are absent for $K \to 0$ and the corresponding regime becomes only manifest for $K \gg 1$. The Gaussian regime is dominated by strong collisions, and the corresponding results will not fulfill Eq. (9.6.6). Accordingly, the analysis here should be restricted to the Holtsmark regime and the pencil beam regime.

The dependency of $FWHM_r$ on S_c, S_i and K is for the Holtsmark regime described by the function $S_{HTa}(S_c, S_i, K)$ and specified by Eq. (9.5.3). The derivation of this function presupposes that $S_c=0$ or $S_c=1$ for $K \to 0$. By power expansion one finds

$$\lim_{K \to 0} S_{HTa}(S_c, S_i, K) = \frac{4}{3} K^{4/3} \left| \frac{1}{2} - S_c^2 - S_i + 2S_c S_i \right|.$$

Taking $S_c=0$ or $S_c=1$, one obtains with Eq. (9.3.44)

$$\lim_{K \to 0} FWHM_r = C_{CHT} \frac{2^{2/3}}{3} \left| \frac{1}{2} - S_i \right| \frac{l^{2/3} L^2}{V^{4/3} r_c^{4/3}}. \qquad (9.6.10)$$

Similarly, one obtains from Eq. (8.5.20)

$$\lim_{K \to 0} S_{HAa}(S_c, K) = \frac{4}{3} K^{4/3} |1 - 2S_c|.$$

Taking $S_c=0$ or $S_c=1$, this gives with Eq. (8.5.31)

$$\lim_{K \to 0} FWHM_a = C_{CHA} \frac{2^{2/3}}{3} \frac{l^{2/3} L^2}{V^{4/3} r_c^{4/3}}. \qquad (9.6.11)$$

Indeed, Eqs. (9.6.10) and (9.6.11) satisfy Eq. (9.6.6) within the approximation of Eq. (9.6.3).

The dependency of $FWHM_r$ on S_c, S_i and K is for the pencil beam regime described by the function $S_{PTa}(S_c, S_i, K)$ specified by Eq. (9.5.6). By power expansion, one finds

$$\lim_{K \to 0} S_{PTa}(S_c, S_i, K) = \frac{6}{K} \left| \frac{1}{2} - S_c^2 - S_i(1 - 2S_c) \right|$$

Taking $S_c=0$ or $S_c=1$, one obtains with Eq. (9.3.47)

$$\lim_{K \to 0} FWHM_r = 12 C_{CPT} \left| \frac{1}{2} - S_i \right| \frac{l^3 r_c L^2}{V^{5/2}}. \qquad (9.6.12)$$

From Eq. (8.5.21), one obtains

$$\lim_{K \to 0} S_{PAa}(S_c, K) = \frac{1}{K} |1 - 2S_c|.$$

Taking $S_c=0$ or $S_c=1$, this gives with Eq. (8.5.35)

$$\lim_{K \to 0} FWHM_\alpha = 2C_{CPA} \frac{I^3 r_c L}{V^{5/2}}. \qquad (9.6.13)$$

Utilizing the relation between the constants C_{CPT} and C_{CPA} given by Eq. (9.6.5), one sees that Eqs. (9.6.12) and (9.6.13) satisfy Eq. (9.6.6).

9.7. Results for Gaussian angular and spatial distributions

The results of the previous sections apply to the case of a uniform spatial and a uniform angular distribution. Similar to the analysis of Sections 7.6 and 8.7, we will now determine the effective width measures, r_{eff} and α_{eff}, of a Gaussian spatial and angular distribution respectively, for which the equations obtained for uniform distributions yield accurate results. We repeated the calculations outlined in this chapter, now starting from the modified versions of Eq. (9.2.11) that are obtained by replacing either the uniform distribution in v or the uniform distribution in $r_\perp$ or both distributions by their Gaussian counterparts given by Eqs. (7.6.1) and (7.6.2) respectively. As the results are quite similar to those of Section 8.6, we will not go into detail.

For the Gaussian regime, the numerical calculations of the mean square broadening $\langle \Delta r^{*2} \rangle$ were reconsidered. It was found that the equations obtained for the case of a uniform angular and a uniform spatial distribution can be retrieved by using

$$\alpha_{eff} = \sqrt{2}\sigma_\alpha, \quad r_{eff} = \sqrt{2}\sigma_r, \quad \text{(Gaussian regime)}. \qquad (9.7.1)$$

The weak complete collision regime is entirely determined by angular deflections, as was discussed previously. Accordingly, one should use Eq. (8.7.12) for the calculation of the effective width measures of the Gaussian spatial and angular distribution.

For the Holtsmark regime and the pencil beam regime, the trajectory displacement effect is directly related to the distribution of the lateral interaction force $\rho(\mathbf{F}_\perp)$ occurring in the beam. As the same holds true to for the corresponding regimes of the angular deflection distribution, we may lake over the analysis of Section 8.7. For the Holtsmark regime, the effective width measures should therefore be calculated from Eq. (8.7.15). For the pencil beam regime, one should use Eq. (8.7.17).

The *FWHM* of the trajectory displacement distribution generated in a beam segment with a crossover is, in the case of a uniform angular and a uniform spatial distribution, given by Eqs. (9.3.43) and (9.5.11). The different regimes are implicitly covered by these equations, and the question arises how to implement the results of this section. The best approach is to use the full equations found for each of the four combinations of a uniform or a Gaussian angular distribution and a uniform or a Gaussian spatial distribution (*uu*, *gg*, *gu* and *ug*) instead of employing the effective width measures α_{eff} and r_{eff}. This approach leads, for each of these cases, to a different version of the function $H_{CT}(\bar{\lambda}, v_0{}^*, r_c{}^*, S_c, S_i)$, which is for a uniform angular and a uniform spatial distribution (uu) given by Eq. (9.5.11). For the resulting set of equations, the reader is referred to Section 16.5, which summarizes the analytical prescriptions for the calculations of the trajectory displacement effect.

CHAPTER TEN

Further investigations on statistical interactions

Contents

10.1. Introduction	299
10.2. Exact approach for off-axis reference trajectories	300
10.3. Approximating approach for off-axis reference trajectories	307
10.4. Nonmonochromatic beams	312
10.5. Beams in an external uniform axial electrostatic field	320
10.6. Relativistic beams	324
Appendix 10.A. Distributions for the average reference trajectory	327

10.1. Introduction

This chapter reports on some further investigations on statistical Coulomb interactions in particle beams. Part of these investigations concern the verification of the validity of the assumptions underlying the calculations presented in the previous chapters. The other investigations study the feasibility of extending the model to beams located in a uniform acceleration field and to relativistic beams.

The extended two-particle model was outlined in Chapter 5 and applied to the Boersch effect, the effect of statistical angular deflections, and the trajectory displacement effect in Chapters 7–9 respectively. These calculations pertain to the displacements experienced by the test particles running along the *central (on-axis) reference trajectory* in a rotational symmetric beam segment in drift space. Furthermore, the calculations were restricted to initially *monochromatic beams* with respect to the normal energy of the particles. These simplifications were introduced to facilitate the calculations, assuming that they do not affect the results in any significant way. This will be verified here within the framework of the extended two-particle model by taking a more general approach.

First, we will treat the problem of including off-axis reference trajectories. Rather than evaluating the effect for a specific on- or off-axis reference trajectory, we will evaluate the average result for all possible reference

Advances in Imaging and Electron Physics, Volume 230
ISSN 1076-5670
https://doi.org/10.1016/B978-0-443-29784-7.00010-8

Copyright © 2024 Elsevier Inc.
All rights are reserved, including
those for text and data mining,
AI training, and similar technologies.

trajectories in the beam. In general, the exact solution of this problem requires the evaluation of an eight-dimensional integral. However, for a homocentric beam, the problem reduces to the evaluation of a four-dimensional integral, which can be handled by numerical means. The cases of a homocentric cylindrical beam and a homocentric beam segment with a crossover are studied. The expectation is that if the central reference trajectory approximation is justified for those extreme beam geometries, it will be justified for any other beam geometry as well. For the Boersch effect produced in a beam segment with a crossover, one cannot rely on this approach, since the case of a homocentric crossover is not representative. As an alternative, an approximating method will be used, which is in fact extremely accurate for paraxial beams. The analysis shows that the average results for all reference trajectories are, in general, 10–20% smaller than those obtained for the central reference trajectory.

Next, the interaction effects generated in nonmonochromatic beams will be studied. The calculations arc restricted to the extreme cases of a homocentric cylindrical beam and a homocentric beam segment with a crossover. The analysis shows that the impact of the initial energy spread on the statistical interaction effects is negligible for practical operating conditions.

Finally, some preliminary results of an analysis will be presented to extend the model for statistical interactions to beams that are accelerated or retarded and to high energy beams. It will be shown that the extended two particle model can straightforwardly be modified to handle beams that are located in a uniform axial electrostatic field. The modifications concern the dynamical part of the problem only. The essential point is to treat the two-particle dynamics at the entrance and the exit of the field in a proper way. The chapter concludes with a brief discussion of the relativistic corrections required to treat high energy beams, in which the particle velocities observed in the laboratory system are comparable to the speed of light. For a paraxial beam in drift space, one may assume that the *relative* velocities of the particles are small compared to the speed of light, and the required modifications only concern the coordinate transformation between the laboratory system and the frame of reference moving with the beam. The analysis shows that it is sufficient to replace the beam potential V appearing in the final equations for the various interaction effects by the well-known relativistic potential V^*.

10.2. Exact approach for off-axis reference trajectories

In Section 5.6, we studied the required modifications of the basic equations of the extended two-particle model to cover off-axis reference trajectories. It was shown that the function $p(k)$, defined by Eq. (5.4.8), will have a nonvanishing imaginary part $p_i(\mathbf{k})$ when the reference trajectory is off-axis. It was found that the impact of the imaginary part $p_i(\mathbf{k})$ on the displacement distribution $\rho(\Delta\eta)$ is twofold. The first order term in $\mathbf{k}$ of the imaginary part $p_i(\mathbf{k})$ leads to a uniform shift of size $\langle\Delta\eta\rangle$, which represents the space charge effect. The third and higher order terms in $\mathbf{k}$ of the imaginary part $p_i(\mathbf{k})$ cause the shifted distribution $\rho(\Delta\eta-\langle\Delta\eta\rangle)$ to become asymmetric, which means that $\rho(\Delta\eta-\langle\Delta\eta\rangle)\neq\rho(-\Delta\eta+\langle\Delta\eta\rangle)$.

The symmetry in the problem can be restored by taking the average over all possible reference trajectories. The odd moments of the resulting displacement distribution $\rho(\Delta\eta)$ will be zero, just as for the on-axis reference trajectory, and the function $\rho(\mathbf{k})$ will only have a real part. In this section, we will perform this type of calculation for some extreme beam geometries and compare the results with those obtained previously for the on-axis reference trajectory. The objective of this analysis is to show that the differences with the results obtained for the on-axis reference trajectory are minor. This would, *a posteriori*, justify the approach taken in the previous chapters. The space charge shift $\langle\Delta\eta\rangle$ occurring for the individual off-axis reference trajectories will be calculated separately in Chapter 11.

Fig. 10.1 shows the unperturbed trajectories of a test and a field particle at the moment that the field particle passes the x, y-plane of the laboratory system. The z-axis of this system coincides with the beam axis, while the x, y-plane coincides with the crossover plane (for a cylindrical beam the choice of the x, y-plane is arbitrary). The x-axis is chosen parallel to $\mathbf{r}_{\perp t}$ (thus $\varphi_t = 0$). The set of coordinates defining the unperturbed positions and velocities of the test and field (or colliding) particle is equal to

$$\xi = (\xi_t, \xi_c) \quad \text{with} \quad \begin{aligned} \xi_c &= (r_{\perp c}, \varphi_c, v_c, \psi_c, b_z) \\ \xi_t &= (r_{\perp t}, v_t, \psi_t) \end{aligned} \quad (10.2.1)$$

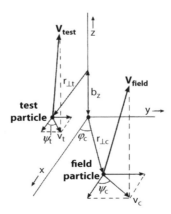

Fig. 10.1 Unperturbed coordinates of an off-axis test particle and a field particle at the moment that the field particle passes the x, y-plane of the laboratory system.

Eq. (10.2.1) should be compared with Eq. (5.2.1), which applies to the on-axis reference trajectory ($r_{\perp t} = 0$, $v_t = 0$).

For a specific (on- or off-axis) reference trajectory, the integral over the distribution of field particles appearing in the definition of the two-particle distribution $p_2(\Delta \eta)$ given by Eq. (5.4.10) can be expressed as

$$\int d\xi p(\xi) = \int d\xi_c p(\xi_c)$$
$$= \int_0^\infty dv_c f_v(v_c) \int_0^{2\pi} \frac{d\psi_c}{2\pi} \int_0^\infty dr_{\perp c} f_r(r_{\perp c}) \int_0^{2\pi} \frac{d\varphi_c}{2\pi} \int_{-S_c L}^{(1-S_c)L} db_z \quad (10.2.2)$$

in which the distributions f_v and f_r correspond to the distribution of transverse velocities and the density distribution in the crossover respectively. Eq. (5.3.3) specifies these functions in case of uniform distributions and Eq. (5.3.4) in case of Gaussian distributions. Notice in Eq. (10.2.2) that the integration over the angles ψ_c and φ_c are kept separate. For the on-axis reference trajectory, they can be combined into a single integration over the relative angle $\Phi = \psi_c - \varphi_c$ due to the rotational symmetry of the problem. In that case, Eq. (10.2.2) leads back to Eqs. (5.3.2) and (5.4.6).

In order to determine the average over all reference trajectories, one should use

$$\int d\xi p(\xi) = \int d\xi_t p(\xi_t) \int d\xi_c p(\xi_c)$$
$$= \int_0^\infty dv_t f_v(v_t) \int_0^{2\pi} \frac{d\psi_t}{2\pi} \int_0^\infty dr_{\perp t} f_r(r_{\perp t}) \int d\xi_c p(\xi_c) \quad (10.2.3)$$

in which the integral over ξ_c represents the five-dimensional integral given by Eq. (10.2.2). Eq. (5.4.10) can now be expressed as

$$p_2(\Delta \eta) = \int d\xi_t p(\xi_t) \int d\xi_c p(\xi_c) \delta[\Delta \eta - \Delta \eta_2(\xi_t, \xi_c)], \quad (10.2.4)$$

which represents an eight-dimensional integral. For short we will refer to the results based on Eq. (10.2.4) as the results for the "average reference trajectory," as opposed to the results for the "on-axis reference trajectory" presented previously.

In order to facilitate the numerical calculations, we will concentrate on two extreme cases, for which Eq. (10.2.4) reduces to a four-dimensional integral. One case is that of a homocentric beam segment with a crossover with a uniform angular distribution, for which one finds

$$\int d\xi_t p(\xi_t) \int d\xi_c p(\xi_c) = \int_0^{v_0} \frac{2v_r \, dv_r}{v_0^2} \int_0^{2\pi} \frac{d\Phi_v}{2\pi} \int_0^{v_0} \frac{2v_c \, dv_c}{v_0^2} \int_{-S_c L}^{(1-S_c)L} db_z \quad (10.2.5)$$

with $\Phi_v = \psi_c - \psi_t$ and $v_0 = \alpha_0 v_z$. The other case is that of a cylindrical homocentric beam segment with a uniform current density distribution, which leads to

$$\int d\xi_t p(\xi_t) \int d\xi_c p(\xi_c) = \int_0^{r_0} \frac{2r_{\perp r} dr_{\perp r}}{r_0^2} \int_0^{2\pi} \frac{d\Phi_r}{2\pi} \int_0^{r_0} \frac{2r_{\perp c} dr_{\perp c}}{r_0^2} \int_{-S_c L}^{(1-S_c)L} db_z \quad (10.2.6)$$

with $\Phi_r = \psi_c - \psi_t$ and r_0 the beam radius.

Let us first consider the case of a homocentric cylindrical beam segment. The distribution of axial velocity displacements $\rho(\Delta v_z)$ for the on-axis reference trajectory was calculated in Section 7.4. We repeated the calculation of the scaled two-particle distribution $\rho_2^*(\Delta v_z^*)$, now using the distribution of field and test particles given by Eq. (10.2.6). Fig. 10.2a shows the results for different values of the scaled beam radius r_0^*. For comparison, we included, in the same figure, the results obtained previously for the on-axis reference trajectory. The second step is the calculation of the $p^*(k^*)$-transform, defined by Eq. (7.4.11). The results of this step are shown in Fig. 10.2b. The third and final step is the evaluation of the FWHM of the scaled displacement distribution $\rho^*(\Delta v_z^*)$, defined by Eq. (7.4.17), as a function of the scaled linear particle density λ^*. The results of this step are given in Fig. 10.2c.

Similar calculations were carried out for the distribution of transverse velocities $\rho(\Delta v_\perp)$ generated in a homocentric cylindrical beam segment. The results for the scaled two-particle distribution $\rho_2^*(\Delta v_\perp^*)$, the $p^*(k^*)$-transform and the FWHM of the scaled displacement distribution $\rho^*(\Delta v_\perp^*)$ are shown in Fig. 10.3a-c respectively. The functions $p^*(k^*)$ and $\rho^*(\Delta v_\perp^*)$ are defined by Eqs. (8.4.11) and (8.4.17) respectively. For comparison, we included in the same figures the results obtained for the on-axis reference trajectory, which were presented previously in Section 8.4.

Figs. 10.2c and 10.3c, pertaining to the distributions $\rho^*(\Delta v_z^*)$ and $\rho^*(\Delta v_\perp^*)$ respectively, show similar features. For very large values of the scaled beam radius r_0^*, the two types of calculations lead to identical results. This reflects that in a very wide beam most of the reference trajectories are sufficiently far removed from the edge of the beam to be treated as a central trajectory in an infinite wide beam. For less wide beams, which are nevertheless extended ($\chi_p = 4\lambda^* r_0^{*3} = \lambda r_0 \gg 1$), the calculation for an average reference trajectory leads to smaller results than that for the on-axis reference trajectory. This is due to the fact that the impact on a test particle moving along the edge of the beam is smaller than on a central test particle. For pencil beams ($\chi_p \ll 1$), one finds different results for $\rho^*(\Delta v_z^*)$ and $\rho^*(\Delta v_\perp^*)$. The displacements in axial velocity Δv_z are now fully determined by axial separation of the particles, which is the same for on- and off-axis test particles. Accordingly, Fig. 10.2c shows, in the pencil beam regime, equal results for the average and the on-axis reference trajectory.

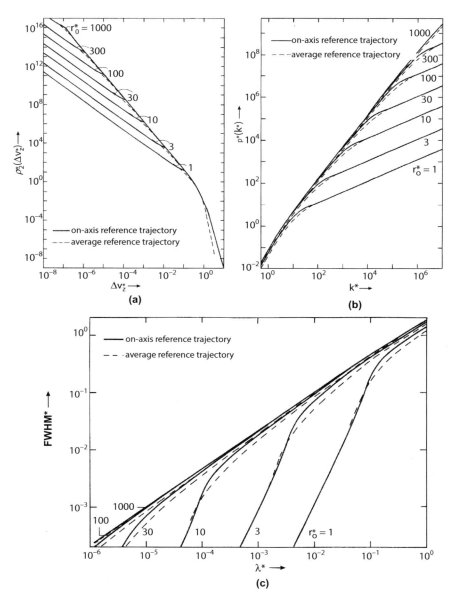

Fig. 10.2 Comparison of the results for the on-axis and the average reference trajectory for the axial velocity displacements generated in a homocentric cylindrical beam segment. Panels a, b and c show the scaled two-particle distribution $\rho_2^*(\Delta v_z^*)$, its transform $p^*(k^*)$ and the FWHM of the scaled displacement distribution $\rho^*(\Delta v_z^*)$ respectively. The results for the on-axis reference trajectory were plotted previously in Figs. 7.7, 7.8 and 7.10.

Further investigations on statistical interactions

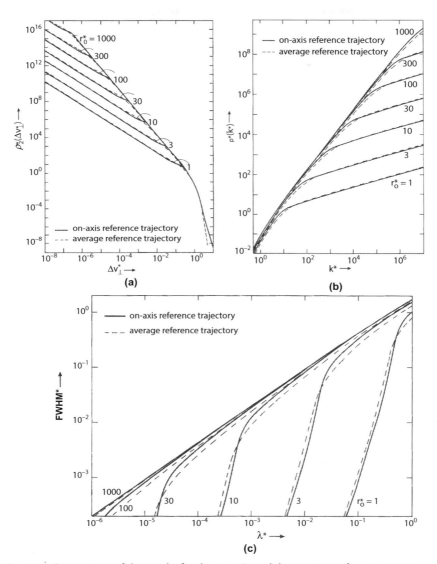

Fig. 10.3 Comparison of the results for the on-axis and the average reference trajectory for the transverse velocity displacements generated in a homocentric cylindrical beam segment. Panels a, b and c show the scaled two-particle distribution $\rho_2^*(\Delta v_\perp^*)$, its transform $p^*(k^*)$ and the FWHM of the scaled displacement distribution $\rho^*(\Delta v_\perp^*)$ respectively. The results for the on-axis reference trajectory were plotted previously in Figs. 8.4, 8.5 and 8.7.

However, the transverse velocity displacements $\Delta v_\perp$ become zero when all particles are on a row and increase proportional to r_0 for small but non-zero r_0. Therefore, one should expect that the impact on a particle moving along the edge of a pencil beam is larger than for a central particle. Accordingly, Fig. 10.3c shows, in the pencil beam regime, larger results for the average reference trajectory than for the on-axis reference trajectory.

Let us now consider the case of a homocentric beam segment with a crossover. The scaled trajectory displacement distribution $\rho^*(\Delta r^*)$ was calculated in Section 9.3 for the on-axis reference trajectory. We repeated this calculation, now using the distribution of field and test particles given by Eq. (10.2.5). The final results for the FWHM of $\rho^*(\Delta r^*)$ are presented in Fig. 10.4. The reader might verify that the characteristics of this plot are similar to that of Fig. 10.3c, pertaining to the distribution of transverse velocity displacements generated in a homocentric cylindrical beam. The plots of the two-particle distributions $\rho^*(\Delta r^*)$ and the $p^*(k^*)$ are omitted, since they are also quite similar to the corresponding plots obtained for a homocentric cylindrical beam, which are shown in Figs. 10.3a and 10.3b respectively.

The Boersch effect, occurring in a beam segment with a crossover, could, in principle, be handled in the same way, by evaluating the distribution of axial velocities $\rho(\Delta v_z)$ for a homocentric beam. However, since

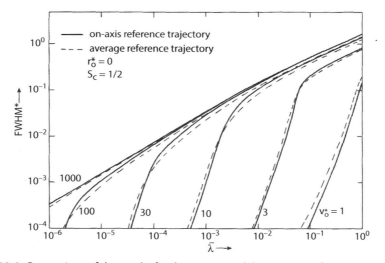

Fig. 10.4 Comparison of the results for the on-axis and the average reference trajectory for the FWHM of the scaled trajectory displacement distribution $\rho^*(\Delta r^*)$ generated in a homocentric beam segment with a crossover in the middle ($S_c = 1/2$). The results for the on-axis reference trajectory were plotted previously in Fig. 9.7.

the Boersch effect depends strongly on the size of the crossover, the case of a homocentric beam ($r_c = 0$) is not representative. In order to study the more practical case of an extended crossover, we will follow an alternative method, which is the subject of the next section.

10.3. Approximating approach for off-axis reference trajectories

In this section, a method will be outlined to reduce the dimension of the integral for the two-particle distribution $\rho(\Delta \eta)$, given by Eq. (10.2.4). For paraxial beams, one may in good approximation replace the exact solution for the displacement of an off-axis test particle $\Delta \eta = \Delta \eta_2(\xi_c, \xi_t)$ by the solution for an on-axis test particle $\Delta \eta = \Delta \eta_2(\xi')$, where $\xi' = \{r_\perp', \varphi', v', \psi', b_z'\}$ and

$$\begin{aligned} r_\perp' &\cong \left[r_{\perp c}^2 + r_{\perp t}^2 - 2 r_{\perp c} r_{\perp t} \cos(\varphi_c) \right]^{1/2} \\ \varphi' &\cong \arctan^* \left[\frac{r_{\perp c} \sin \varphi_c}{r_{\perp c} \cos \varphi_c - r_{\perp t}} \right] - \psi_t \\ v' &= \left[v_c^2 + v_t^2 - 2 v_c v_t \cos(\psi_c - \psi_t) \right]^{1/2} \\ \psi' &\cong \arctan^* \left[\frac{v_c \sin(\psi_c - \psi_t)}{v_c \cos(\psi_c - \psi_t) - v_t} \right] \\ b_z' &\cong b_z \end{aligned} \qquad (10.3.1)$$

in which

$$\arctan^*\left(\frac{y}{x}\right) = \left(1 - \frac{x}{|x|}\right) \frac{\pi}{2} + \arctan\left(\frac{y}{x}\right). \qquad (10.3.2)$$

Eq. (10.3.1) represent the coordinate transformation from the original coordinate system $R = (x, y, z)$, which is drawn in Fig. 10.1, to the coordinate system $R' = (x', y', z')$ in which the z'-axis is directed along the velocity of the test particle $\mathbf{v}_{\text{test}}$ and the x', z'-plane coincides with the plane through the original z-axis and the vector $\mathbf{v}_{\text{test}}$. In the expressions for $r_\perp'$, φ' and b_z, we ignored the second and higher order terms in the angle $\alpha_t \approx v_t / v_z$, which is allowed for paraxial beams.

As the integrand in Eq. (10.2.4) now depends on ξ_t and ξ_c only through ξ', one may replace the probability $p(\xi_t) p(\xi_c)$ by the probability $p(\xi')$, defined as

$$p(\xi') = \int d\xi_t p(\xi_t) \int d\xi_c p(\xi_c) \, \delta[\xi' - \xi'(\xi_t, \xi_c)], \qquad (10.3.3)$$

where $\xi'(\xi_t, \xi_c)$ represents the coordinate transformation of Eq. (10.3.1). The two-particle distribution can now be expressed as

$$\rho_2(\Delta\eta) = \int d\xi' p(\xi')\delta[\Delta\eta - \Delta\eta_2(\xi')], \qquad (10.3.4)$$

as follows with Eq. (10.2.4).

The problem of determining $\rho_2(\Delta\eta)$ is now reduced to the evaluation of the five-dimensional integral of Eq. (10.3.4), provided that the function $p(\xi')$ can be evaluated from Eq. (10.3.3). Let us consider this problem first. Due to the rotational symmetry of the beam, one may conjecture that for any given value of $r_\perp'$ and v' the distribution over the angles φ' and ψ' is uniform over 2π. The main problem is therefore to determine the distributions over $r_\perp'$ and v'. Eqs. (10.3.1) and (10.3.3) show that the calculation of both distributions requires the solution of the same mathematical problem. Given the distribution $f(r)$, one should evaluate the function $p(r)$, using

$$p(r) = \int_0^\infty dr_1 f(r_1) \int_0^\infty dr_2 f(r_2) \int_0^{2\pi} \frac{d\phi}{2\pi} \delta\left[r - \{r_1^2 - r_2^2 - 2r_1 r_2 \cos\phi\}^{1/2}\right]. \qquad (10.3.5)$$

The function $p(r)$ can be visualized as the distribution of the distance $r = |\mathbf{r}_1 - \mathbf{r}_2|$ between two points, $\mathbf{r}_1$ and $\mathbf{r}_2$, in a plane when both points are distributed as $f(|\mathbf{r}|)$ with respect to the origin and uniform over an angle 2π. In Appendix 10.A it is demonstrated that when $f(r)$ is a two-dimensional uniform distribution, i.e., $f(r) = f_u(r)$, with

$$f_u(r) = \begin{cases} 2r/R^2 & \text{for } 0 \leq r \leq R \\ 0 & \text{for } r > R, \end{cases} \qquad (10.3.6)$$

Eq. (10.3.5) yields $p(r) = p_u(r)$, where

$$p_u(r) = \begin{cases} \dfrac{4r}{\pi R^2}\left[\arccos\left(\dfrac{r}{2R}\right) - \dfrac{r}{2R}\sqrt{1 - \left(\dfrac{r}{2R}\right)^2}\right] & \text{for } 0 \leq r \leq 2R \\ 0 & \text{for } r > 2R. \end{cases} \qquad (10.3.7)$$

When $f(r)$ is a two-dimensional Gaussian distribution, i.e., $f(r) = f_g(r)$, with

$$f_g(r) = \frac{2r}{R^2} e^{-(r/R)^2}, \qquad (10.3.8)$$

one finds, from Eq. (10.3.5), that $p(r) = p_g(r)$, where

$$p_g(r) = \frac{r}{R^2} e^{-(r/R)^2/2}, \qquad (10.3.9)$$

which is also demonstrated in Appendix 10.A. Notice that the resulting distribution $p_g(r)$ is again Gaussian, but its width is a factor $\sqrt{2}$ larger than the original distribution $f_g(r)$.

The probability distribution $p(\xi')$ occurring in Eq. (10.3.4) can now be expressed as

$$\int d\xi' p(\xi') = \int_0^\infty dv' p_v(v') \int_0^{2\pi} \frac{d\Phi'}{2\pi} \int_0^\infty dr_\perp' p_r(r_\perp') \int_{-S_c L}^{(1-S_c)L} db_z', \qquad (10.3.10)$$

with $\Phi' = \psi' - \varphi'$. For the functions p_v and p_r, one should use Eq. (10.3.7) when the corresponding distributions f_v and f_r are uniform and use Eq. (10.3.9) when they are Gaussian, taking the proper measure for the width R. With Eqs. (10.3.4) and (10.3.10), the calculation of the average two-particle distribution $\rho_2(\Delta\eta)$ for all possible reference trajectories is reduced to the same form as the calculation of $\rho_2(\Delta\eta)$ for the on-axis reference trajectory. For the dynamical part, one may use the same function $\Delta\eta_2(\xi)$ as used for the on-axis reference trajectory. The differences between both calculations concern the distributions of v' and $r_\perp'$ only.

We will concentrate on the case that the distributions in v and r are uniform, leading to distributions in r' and v' of the type given by Eq. (10.3.7). The case of a Gaussian distribution in v or r does not require any separate calculation. Eq. (10.3.9) shows that, when the distribution in v or r is Gaussian, the distribution in v' or r' is Gaussian, too, with a width that is larger by a factor $\sqrt{2}$. Accordingly, for this case, one should expect that a calculation based on Eqs. (10.3.4) and (10.3.10) leads to the same final equations as obtained for the on-axis reference trajectory but with the crossover radius r_c and the beam semi-angle α_0 replaced by some effective quantities, which are $\sqrt{2}$ times larger than the original ones.

The correctness of the approximating method outlined above can be verified by applying it to the extreme cases of a homocentric cylindrical beam segment and a homocentric beam segment with a crossover, which were studied in the previous section using the exact approach. For a homocentric beam, Eq. (10.3.4), defining the two-particle distribution $\rho_2(\Delta\eta)$, reduces to a two-dimensional integral. We evaluated this integral for each of the cases studied in the previous section. The results were found to be in very close agreement with those presented in Figs. 10.2–10.4, justifying the use of the approximating method.

The approximating method to treat off-axis reference trajectories was utilized to evaluate the distribution of axial velocity displacements $\rho(\Delta v_z)$, generated in a beam segment with an extended but narrow crossover, assuming a uniform angular and spatial distribution. The corresponding calculation for the on-axis reference trajectory was presented in Section 7.3. The calculation for the average reference trajectory is quite similar but starts from the modified distribution of field particles defined by Eqs. (10.3.10) and (10.3.7). As the principle aspects involved are the same, we will restrict the discussion to the presentation of the main results.

Fig. 10.5a shows the results for the scaled two-particle distributions $\bar{p}_2(\Delta\bar{v}_z)$, for different values of the scaled beam radius $\bar{r}_c$. For comparison, the data obtained previously for the on-axis reference trajectory is included in the same figure. The next step is the calculation of the $\bar{p}(\bar{k})$ transform, defined by Eq. (7.3.14). The results are shown in Fig. 10.5b. Finally, the FWHM of the scaled displacement distribution $\bar{p}(\Delta\bar{v}_z)$ was evaluated, which is defined by Eq. (7.3.21). The results of this step are presented in Fig. 10.5c.

Following the calculation scheme of Section 7.3, the reader might verify that the scaled $\bar{p}(\bar{k})$ transform for the average reference trajectory, shown in Fig. 10.5B, is given by

$$\bar{p}(\bar{k}) = 16\bar{k} \int_0^2 d\bar{v} F(\bar{v}) \int_0^{2\pi} \frac{d\Phi}{2\pi} \int_0^2 t\, dt\, F(t)$$
$$\times \int_0^1 du \frac{(1-u^2)^{1/2}}{u} \sin\left(\frac{\bar{k}\bar{v}u}{(4 + \bar{v}^4 t^2 \bar{r}_c^2 \sin^2\Phi)^{1/2}}\right), \quad (10.3.11)$$

with

$$F(x) = \frac{2}{\pi}\left[\arccos(x/2) - (x/2)\sqrt{1-(x/2)^2}\right]. \quad (10.3.12)$$

We will investigate the behavior of $\bar{p}(\bar{k})$ for $\bar{k} \to 0$ and for $\bar{k} \to \infty$.

For $\bar{k} \to 0$, Eq. (10.3.11) shows a quadratic dependency on $\bar{k}$, which can be described by the first equation of (7.3.16), taking the function $\bar{p}_2(\bar{r}_c)$ equal to

$$\bar{p}_2(\bar{r}_c) = 2\int_0^2 \bar{v}\, d\bar{v} F(\bar{v}) \int_0^{2\pi} d\Phi \int_0^2 t\, dt\, F(t) [1 + \bar{v}^4 t^2 \bar{r}_c^2 \sin^2\Phi/4]^{-1/2}$$
$$= 8\int_0^2 \bar{v}\, d\bar{v} F(\bar{v}) \int_0^2 t\, dt\, F(t) \frac{1}{[1+\bar{v}^4 t^2 \bar{r}_c^2/4]^{1/2}} K\left(\frac{\bar{v}^2 t \bar{r}_c/2}{[1+\bar{v}^4 t^2 \bar{r}_c^2/4]^{1/2}}\right),$$
$$(10.3.13)$$

where K is the complete elliptic integral of the first kind. The function $\bar{p}_2(\bar{r}_c)$, defined by Eq. (10.3.13), is plotted in Fig. 10.6. For comparison, we also plotted the result for the on-axis reference trajectory, given by the second equation of (7.3.16). The function $\bar{p}_2(\bar{r}_c)$ given by Eq. (10.3.13) can be approximated by Eq. (7.6.11), taking

$$E = 37.38, \quad F = 1.471, \quad (10.3.14)$$

which should be compared to the values E_{uu} and F_{uu} given by Eq. (7.6.12).

Further investigations on statistical interactions 311

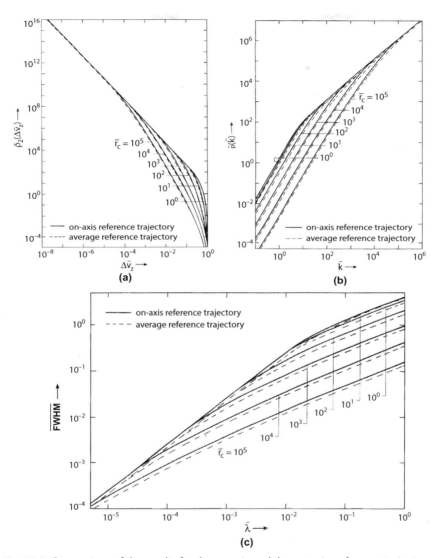

Fig. 10.5 Comparison of the results for the on-axis and the average reference trajectory for the axial velocity displacements generated by complete collisions in a beam segment with a crossover. Panels a, b and c show the scaled two-particle distribution $\bar{p}_2(\Delta \bar{v}_z)$, its transform $\bar{p}(\bar{k})$ and the FWHM of the scaled displacement distribution $\bar{p}(\Delta \bar{v}_z)$ respectively. The results for the on-axis reference trajectory were plotted previously in Figs. 7.1, 7.2 and 7.5.

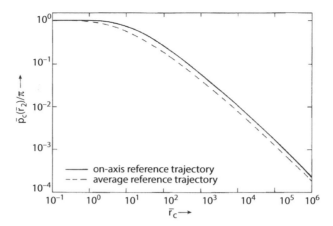

Fig. 10.6 Comparison of the functions $\bar{p}_2(\bar{r}_c)$ for the on-axis and the average reference trajectory, defined by Eqs. (7.3.16) and (10.3.13) respectively. The function for the on-axis reference trajectory was plotted previously in Fig. 7.3.

For $\bar{k} \to \infty$, Eq. (10.3.11) shows a linear dependency on $\bar{k}$, which is given by

$$\bar{p}(\bar{k}) = \bar{p}_\infty \bar{k}, \quad \bar{p}_\infty = 32/3 \quad (\bar{k} \to \infty). \tag{10.3.15}$$

This result should be compared to Eq. (7.3.19), pertaining to the on-axis reference trajectory.

The analysis of Section 10.2 and the present section shows that the results obtained for the average reference trajectory are for most conditions somewhat smaller than those obtained for the on-axis reference trajectory. However, in general, the differences are within 20% and in many cases even within 10%. The analysis justifies the assumption that the results obtained for the on-axis reference trajectory are representative for other trajectories as well.

10.4. Nonmonochromatic beams

The calculations carried out in Chapters 7–9 rely on the assumption that the beam is initially monochromatic with respect to the normal energy. This means that it is assumed that all particles initially have the same axial velocity. In this section, we will study the more realistic case of a beam in which the particles start with a certain axial velocity spread. The objective of this analysis is to show that the differences with the results obtained for an initially monochromatic beam are minor for practical conditions.

Let us assume that the initial axial velocity spread of the field particles with respect to the test particle is given by the Gaussian distribution

$$f(\Delta v_z) = \frac{1}{\sqrt{\pi}\Delta v_{z0}} e^{-(\Delta_z/\Delta v_{z0})^2}, \qquad (10.4.1)$$

which has a *FWHM* equal to

$$FWHM_v = 2(\ln 2)^{1/2}\Delta v_{z0}. \qquad (10.4.2)$$

For the on-axis reference trajectory, the set of coordinates defining the unperturbed position and velocity of the field particle with respect to the test particle is now given by

$$\xi = (r_\perp, \Phi, v_\perp, \Delta v_z, b_z), \qquad (10.4.3)$$

similar to Eq. (5.2.1). The angles φ and ψ were combined as $\Phi = \psi - \varphi$, using that the size of the displacement $|\Delta\eta|$ depends only on φ and ψ through the relative angle Φ.

For the integral over the distribution of field particles in Eq. (5.4.10), which defines the two-particle distribution $p_2(\Delta\eta)$, one should now use

$$\int d\xi p(\xi) = \int_0^{v_0} \frac{2v_\perp \, dv_\perp}{v_0^2} \int_{-\infty}^{\infty} \frac{d\Delta v_z}{\sqrt{\pi}\Delta v_{z0}} e^{-(\Delta v_z/\Delta v_{z0})^2}$$
$$\times \int_0^{2\pi} \frac{d\Phi}{2\pi} \int_0^{r_c} \frac{2r_\perp \, dr_\perp}{r_c^2} \int_{-S_c L}^{(1-S_c)L} db_z. \qquad (10.4.4)$$

We will restrict the analysis to the extreme cases of a homocentric beam segment with a crossover ($r_c = 0$) and a homocentric cylindrical beam segment ($v_0 = 0$). For the former, one may omit the integration over $r_\perp$ and Φ. For the latter, the integration over v and Φ can be left out. In both cases the remaining integral is three-dimensional.

We will again employ the δ, ν-scaling, defined by Eq. (6.13.1). The scaled axial velocity displacement Δv_z^* is related to the relative energy displacement $\Delta E/E$ as

$$\Delta v_z^* = \left(\frac{\pi\varepsilon_0}{4e}\right)^{1/3} V^{1/3} L^{1/3} \frac{\Delta E}{E}, \qquad (10.4.5)$$

using that $\Delta E \cong mv_z\Delta v_z$. For electrons, the physical constant in front of the right-hand side of Eq. (10.4.5) is equal to 351.43 in *SI*-units. Taking as a numerical example an electron beam with $V = 10\,\text{kV}$, $L = 0.1\,\text{m}$ and $\Delta E/E = 10^{-4}$ (which implies $\Delta E = 1\,\text{eV}$), one finds $\Delta v_z^* = 0.35$. Thus, for practical beams, the scaled *FWHM* of the initial distribution of axial velocities $FWHM_v^*$ ($= FWHM_v/\nu$) is of the order 1.

Let us first consider the case of a homocentric cylindrical beam segment. We repeated the calculations described in Sections 7.4 and 8.4, now including the integration over Δv_z prescribed by Eq. (10.4.4).

For the scaled beam radius r_0^*, we considered the values 1000, 100, 10 and 1, and for each r_0^* we took $FWHM_v^* = 0, 0.2, 0.4, 1, 2, 4, 10$ and 20.

Fig. 10.7a, b and c show part of the results obtained for the scaled two-particle distribution $\rho_2^*(\Delta v_z^*)$, the corresponding $p^*(k^*)$ transform, and the $FWHM$ of the scaled displacement distribution $\rho^*(\Delta v_z^*)$ respectively. One sees that the magnitude of the Boersch effect in a cylindrical beam decreases with increasing $FWHM_v^*$. The effect is the strongest in the Gaussian regime ($\lambda^* \gtrsim 1$ and $\chi_p \gg 1$) and becomes larger for small r_0^*-values. In the Gaussian regime, the $FWHM$ of $\rho^*(\Delta v_z^*)$ can be expressed as

$$FWHM_G^* = 1.832 P_{PE}(r_0^*, FWHM_v^*) \lambda^{*1/2}, \qquad (10.4.6)$$

similar to Eq. (7.4.21). In Fig. 10.7D, the function $P_{PE}(r_0^*, FWHM_v^*)$ is plotted as function of $FWHM_v^*$ for different values of the scaled beam radius r_0^*.

Fig. 10.8a, b and c show part of the results obtained for the scaled two-particle distribution $\rho_2^*(\Delta v_\perp^*)$, the corresponding $p^*(k^*)$ transform and the $FWHM$ of the scaled displacement distribution $\rho^*(\Delta v_\perp^*)$ respectively. The function $P_{PA}(r_0^*, FWHM_v^*)$, which is plotted in Fig. 10.8d, is the generalization of the function $P_{PA}(r_0^*)$, defined by Eq. (8.4.20). It specifies the dependency of the $FWHM$ of $\rho^*(\Delta v_\perp^*)$ on r_0^* and $FWHM_v^*$ in the Gaussian regime, as can be seen from Eq. (8.4.19). All figures show the same characteristics as the equivalent figures obtained for the axial velocity displacements, but the decrease of the $FWHM$ of $\rho^*(\Delta v_\perp^*)$ with increasing $FWHM_v^*$ is stronger.

We also studied the impact of the initial energy spread on the trajectory displacement distribution $\rho^*(\Delta r^*)$ generated in a homocentric beam segment with a crossover in the middle ($S_c = 0.5$). The displacements refer to the crossover plane ($S_i = 0.5$). For the scaled maximum transverse velocity v_0^*, we considered the values 1000, 100, 10 and 1, and for each v_0^*, we took $FWHM_v^* = 0, 0.2, 0.4, 1, 2, 4, 10$ and 20. Part of the results obtained for the two-particle distribution $\rho_2^*(\Delta r^*)$ the $p^*(k^*)$ transform, and the $FWHM$ of the displacement distribution $\rho^*(\Delta r^*)$ are plotted in the Fig. 10.9a, b and c respectively. In general, the results are not very sensitive to $FWHM_v^*$. Fig. 10.9c shows that the trajectory displacement effect generated in a homocentric beam segment with a crossover increases slightly with $FWHM_v^*$. The strongest increase is found for small v_0^*-values. It also shows that the deviations from the monochromatic case ($FWHM_v^* = 0$) are the largest for those operating conditions that correspond to the transition area between the pencil beam and the extended beam regimes (that is, the Holtsmark and the Gaussian regime). For the Gaussian regime the dependencies on v_0^* and $FWHM_v^*$ are represented by the function $P_{CT}(v_0^*, FWHM_v^*)$, which is shown in Fig. 10.9d. This function is the equivalent of $P_{CT}(v_0^*, S_c, S_i)$ for $S_c = S_i = 0.5$ and nonzero $FWHM_v^*$.

Further investigations on statistical interactions 315

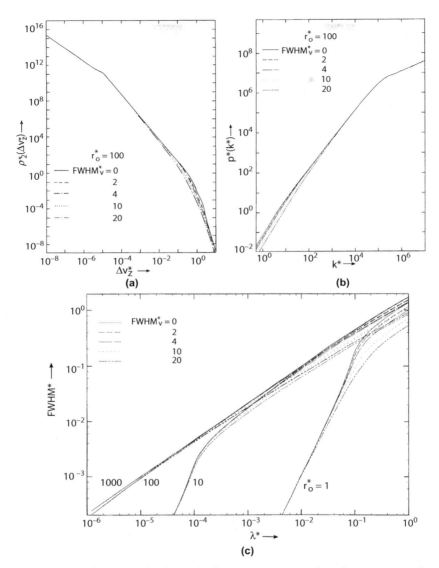

Fig. 10.7 Results for the axial velocity displacements generated in a homocentric cylindrical beam segment, for different values of the initial axial velocity spread $FWHM_v^*$. Panels a, b and c show the scaled two-particle distribution $\rho_2^*(\Delta v_z^*)$, its transform $p^*(k^*)$ and the FWHM of the scaled displacement distribution $\rho^*(\Delta v_z^*)$ respectively. The curves for the monochromatic case ($FWHM_v^* = 0$) were plotted previously in Figs. 7.7, 7.8 and 7.10. Panel d gives the results for the function $P_{PE}(r_0^*, FWHM_v^*)$, which is defined by Eq. (10.4.6). This function specifies the dependency on r_0^* and $FWHM_v^*$ of the function $p^*(k^*)$, plotted in panel b, for $k^* \to 0$. This part of the $p^*(k^*)$ transform determines the large λ^*-behavior of the $FWHM^*$ plotted in panel c.

(Continued)

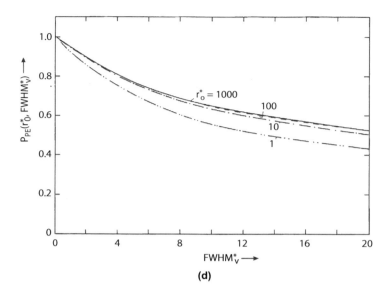

Fig. 10.7—Cont'd

The function $P_{CT}(v_0^*, S_c, S_i)$ is defined by Eq. (9.3.24) and is used in Eq. (9.3.29), specifying the FWHM of $p^*(\Delta r^*)$ in the Gaussian regime.

Some physical appreciation for the results of this section can be gained by studying the contributions of the different types of collisions involved. Consider Figs. 10.7a and 10.8b, showing the scaled two-particle distributions $\rho_2^*(\Delta v_z^*)$ and $\rho_2^*(\Delta v_\perp^*)$ respectively, for a cylindrical beam, with $r_0^* = 100$. One sees that an increase of $FWHM_v^*$ causes a decrease in the distributions for large displacements. For $FWHM_v^* = 0$, the large displacements stem from *half-complete collisions* as is shown by Eqs. (7.4.9) and (8.4.9). For such a collision, the initial kinetic energy in the center of mass system is zero, and the initial potential energy is fully converted into kinetic energy during the flight. With increasing $FWHM_v^*$, the initial kinetic energy in the center of mass system will in general be nonzero. For a collision in which the initial potential energy is fully converted into kinetic energy, the change in relative velocity Δv is determined by

$$4C_0/mr_i = (v_i + \Delta v)^2 - \Delta v^2 = 2v_i \Delta v + \Delta v^2, \qquad (10.4.7)$$

where r_i is the initial distance between the particles and v_i is their initial relative velocity. One sees that the velocity change Δv decreases with increasing v_i, which is a direct consequence of the quadratic dependency of kinetic energy on velocity.

Further investigations on statistical interactions 317

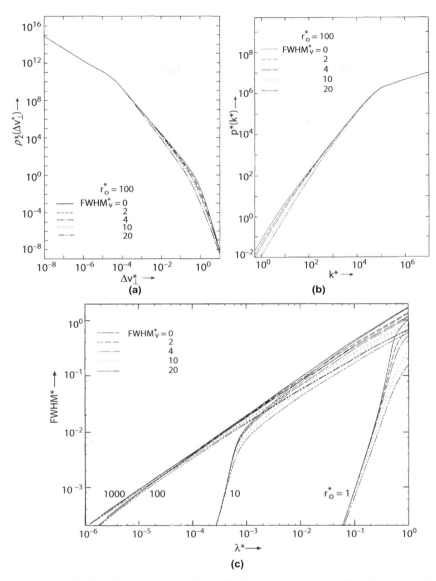

Fig. 10.8 Results for the transverse velocity displacements generated in a homocentric cylindrical beam segment, for different values of the initial axial velocity spread $FWHM_v^*$. Panels a, b and c show the scaled two-particle distribution $\rho_2^*(\Delta v_\perp^*)$, its transform $p^*(k^*)$ and the $FWHM$ of the scaled displacement distribution $\rho^*(\Delta v_\perp^*)$ respectively. The curves for the monochromatic case ($FWHM_v^* = 0$) were plotted previously in Figs. 8.4, 8.5 and 8.7. Panel d gives the results for the function $P_{PA}(r_0^*, FWHM_v^*)$, which is the generalization of the function $P_{PA}(r_0^*)$, defined by Eq. (8.4.20). It specifies the dependency on r_0^* and $FWHM_v^*$ of the function $p^*(k^*)$, plotted in panel b, for $k^* \to 0$. This part of the $p^*(k^*)$ transform determines the large λ^*-behavior of the $FWHM^*$ plotted in panel c.

(Continued)

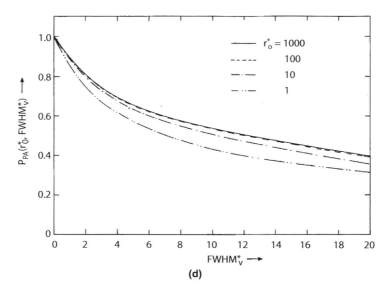

(d)

Fig. 10.8—Cont'd

Small displacements stem from *weak uncomplete collisions*. Using first order perturbation dynamics, one finds for the relative velocity displacement Δv caused by such a collision

$$\Delta v \cong 2C_0 T/mr_i^2, \qquad (10.4.8)$$

in which T is the time of flight. This result does not depend on the initial relative velocity v_i, which explains why the distribution of small displacements in Figs. 10.7a and 10.8b is hardly affected by the magnitude of initial axial velocity spread, represented by $FWHM_v{}^*$.

Fig. 10.9a gives the scaled two-particle distributions $\rho_2{}^*(\Delta r^*)$ in a homocentric beam segment with a crossover with $v_0{}^* = 10$. The curves show substantially less variation with $FWHM_v{}^*$ than the corresponding curves obtained for a cylindrical beam segment. The physical reason is that, even for $FWHM_v{}^* = 0$, almost all collisions start with a large initial relative velocity v_i, due to the lateral motion of the particles. Therefore, the additional axial velocity components Δv_z occurring for nonzero $FWHM_v{}^*$ hardly change the initial conditions of the various collisions unless the relative transverse velocity $v_\perp$ and/or the impact parameter b are very small. The increase of $FWHM_v{}^*$ has some effect for large displacements, which stem predominantly from collisions with small b and/or $v_\perp$.

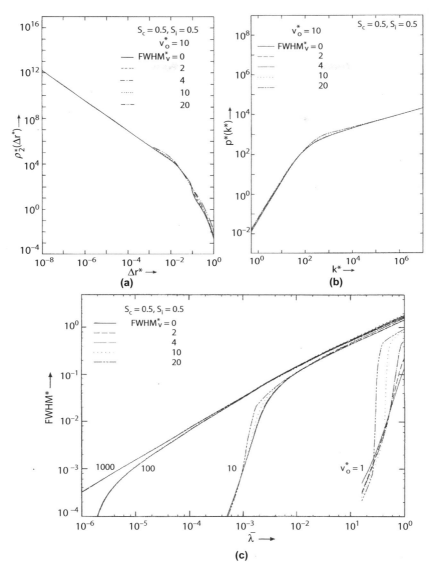

Fig. 10.9 Results for the trajectory displacement effect generated in a homocentric beam segment with a crossover in the middle ($S_c = 1/2$), for different values of the initial axial velocity spread $FWHM_v^*$. The results pertain to the virtual displacements observed in the crossover plane ($S_i = S_c$) and pertain to a beam for which the maximum transverse velocity $v_0^* = 10$. Panels a, b and c show the scaled two-particle distribution $\rho_2^*(\Delta r^*)$, its transform $p^*(k^*)$ and the FWHM of the scaled displacement distribution $\rho^*(\Delta r^*)$ respectively. The curves for the monochromatic case ($FWHM_v^* = 0$) were plotted previously in Figs. 9.1, 9.4 and 9.7. Panel d gives the results for the function $P_{CT}(r_0^*, FWHM_v^*)$, which is the equivalent of the function $P_{CT}(r_0^*, S_c, S_i)$, defined by Eq. (9.3.24), and pertains to $S_c = S_i = 1/2$ and nonzero $FWHM_v^*$. It specifies the dependency on r_0^* and $FWHM_v^*$ of the function $p^*(k^*)$, plotted in Panel b, for $k^* \to 0$. This part of the $p^*(k^*)$ transform determines the large $\bar{\lambda}$-behavior of the $FWHM^*$ plotted in panel c.

(Continued)

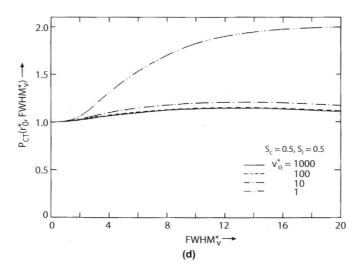

Fig. 10.9—Cont'd

From Figs. 10.7c, 10.8c and 10.9c, one sees that, for values of $FWHM_v^*$ of the order one or less and practical particle densities ($\bar{\lambda}$ or λ^* smaller than 10^{-2}), the differences with the results obtained for the monochromatic case ($FWHM_v^* = 0$) are negligible. This data supports the assumption that, as far as the calculation of statistical interactions is concerned, practical beams can be treated as monochromatic.

10.5. Beams in an external uniform axial electrostatic field

In this section we will investigate how the model for statistical interactions can be extended to cover beams that are accelerated or retarded by a uniform electrostatic field. Consider the system schematically drawn in Fig. 10.10. In the area between the two grids located at $z = z_0$ and $z = z_1$ the beam is accelerated from a potential V_0 to a potential V_1. The electrostatic field produced by the grids is assumed to be perfectly uniform. Furthermore, it is assumed that the beam is initially monochromatic with respect to the normal energy, which means that all particles have the same initial axial velocity v_{z0}.

One way to evaluate the effect of statistical interactions in presence of an acceleration field is by means of the slice method described in Section 5.11. The function $FWHM_P(I, V, r_0, \Delta z)$, appearing in the integrand of Eq. (5.11.1), depends on the beam voltage. Instead of using a constant V, one now takes $V = V(z)$, where

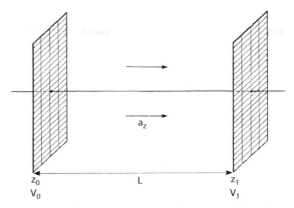

Fig. 10.10 Schematic view of a system of two grids separated by a distance L. The grids produce a uniform axial electrostatic field, which causes an acceleration a_z of the particles in the beam.

$$V(z) = V_0 + \frac{V_1 - V_0}{L}(z - z_0). \qquad (10.5.1)$$

Most theories on statistical interactions that treat the case of an accelerated beam utilize this type of approach; see for instance Knauer (1981), Massey et al. (1981) or Sasaki (1986). It should be emphasized that the slice method presupposes that the displacements from the unperturbed trajectories are small, which limits its application to beams with a relatively low particle density. In the following, we will outline a more general approach, utilizing the features of the extended two-particle model. For the basics of this approach, the reader is referred to Section 5.4.

The extended two-particle model evaluates the distribution of displacements of a set of test particles running along a certain trajectory in the beam. Each test particle is surrounded by a set of field particles representing the beam. The initial coordinates of the $N-1$ field particles with respect to the test particle are specified by the set $\xi_1, \xi_2, \ldots, \xi_{N-1}$, in which ξ is defined by Eq. (5.2.1). Here we will assume that these coordinates refer to the unperturbed situation (without interaction) *before* the particles reach the acceleration zone. The dynamical part of the problem is to evaluate the displacement of the test particle $\Delta\eta = \Delta\eta(\xi_1, \xi_2, \ldots, \xi_{N-1})$ caused by the configuration of field particles $\xi_1, \xi_2, \ldots, \xi_{N-1}$, taking both the mutual interaction and the presence of the external field into account. The statistical part of the problem is the same as before and is expressed by Eq. (5.2.3).

As the modifications concern the dynamical part of the problem only, we will concentrate on that part. First, we will introduce some additional physical quantities, which are required in the discussion. Next, we will consider the acceleration process from a macroscopic point of view. Finally, we will investigate how the detailed trajectories should be evaluated taking the mutual Coulomb repulsion as well as the acceleration into account.

The initial velocity v_{z0} and the axial acceleration a_z experienced by a particle with mass m and charge e are equal to

$$v_{z0} = \sqrt{\frac{2eV_0}{m}}, \quad a_z = \frac{e(V_1 - V_0)}{mL}, \qquad (10.5.2)$$

where $L = z_1 - z_0$, the distance between the grids. In case of deceleration, $V_1 < V_0$ and a_z is negative. The time of flight T in the presence of acceleration ($V_1 \neq V_0$) and the time of flight T_0 in absence of acceleration ($V_1 = V_0$) are given by

$$T = \frac{v_{z0}}{a_z}\left(\sqrt{1 + \frac{2a_z L}{v_{z0}^2}} - 1\right), \quad T_0 = \frac{L}{v_{z0}}. \qquad (10.5.3)$$

It is convenient to introduce the acceleration coefficient κ, defined as

$$\kappa = (V_1/V_0)^{1/2}. \qquad (10.5.4)$$

The relation between the flight times T and T_0 can now be expressed as

$$\frac{T}{T_0} = \frac{2}{\kappa + 1}, \qquad (10.5.5)$$

which can directly be verified from Eqs. (10.5.3).

The impact of the acceleration on the macroscopic properties of the beam can be summarized as follows:

- The time of flight T is reduced by a factor $(\kappa+1)/2$, as shown by Eq. (10.5.5).
- The linear particle density λ is decreased by a factor κ, as can be seen from Eq. (3.2.4).
- The angular distribution of the particles shrinks with a factor κ, since the axial velocities increase with a factor κ, while the transverse velocities are not affected by the acceleration field. Notice also that a transverse velocity displacement $\Delta v_\perp$ occurring after acceleration corresponds to a κ times smaller angular displacement as the same velocity displacement occurring before acceleration.
- Any initial axial velocity spread is reduced by a factor κ, as follows from Eq. (3.7.14). Notice also that an axial velocity displacement Δv_z occurring after acceleration corresponds to a κ times larger energy displacement as the same velocity displacement occurring before acceleration.

The description of the detailed particle trajectories should be consistent with these macroscopic observations.

The model for statistical interactions is based on a reduction of the full N-particle dynamical problem into $N-1$ two particle problems, as described in Section 5.4. The main problem of this section is, therefore, to describe a two-particle collision in the presence of a uniform external field. In Chapter 6, we considered the dynamical problem of two free particles interacting through the Coulomb force. We will show that this analysis requires only some minor modifications to treat the present case. The essential observation is that once both particles have entered the acceleration zone the motion in their center of mass system is the same as for two free particles, due to the uniformity of the external field. Accordingly, one can employ the same set of equations, now using the adjusted time of flight T specified by Eq. (10.5.5). The remaining problem is to treat the dynamics of the particles at the entrance and the exit plane in a proper way.

Consider two particles axially separated by a distance b_{z0} that enter the acceleration zone with an axial velocity v_{z0}. The particle in front will enter the acceleration field earlier than the other particle. This time difference Δt_0 is equal to b_{z0}/v_{z0}. At the moment that the second particle enters the acceleration zone, the first particle has already increased its axial velocity by an amount δv_{z0}, while the axial separation of the two particles b_{z0} has increased by an amount δb_{z0}, where

$$\delta b_{z0} = \frac{1}{2} a_z \left(\frac{b_{z0}}{v_{z0}}\right)^2, \quad \delta v_{z0} = a_z \frac{b_{z0}}{v_{z0}}. \tag{10.5.6}$$

Using Eqs. (10.5.2) and (10.5.4), these quantities can be expressed as

$$\frac{\delta b_{z0}}{b_{z0}} = (\kappa^2 - 1) \frac{b_{z0}}{4L}, \quad \frac{\delta v_{z0}}{v_{z0}} = (\kappa^2 - 1) \frac{b_{z0}}{2L}. \tag{10.5.7}$$

Due to the velocity difference δv_{z0}, the axial separation of the particles will increase linearly with time during the period that both particles are in the acceleration zone. The particle in front will leave the acceleration zone earlier than the other particle. This time difference Δt_1 is equal to Δt_0 and the corresponding reduction in the axial separation and relative axial velocity compensates the differences δb_{z0} and δv_{z0} produced at the entrance plane. The net increase in axial separation δb_z is, therefore, equal to $T \times \delta v_{z0}$ and the final axial separation b_{z1} is given by

$$b_{z1} = b_{z0} + T\,\delta v_{z0} = \kappa b_{z0}, \tag{10.5.8}$$

as follows with Eqs. (10.5.5) and (10.5.7). This result is consistent with the observation that the linear particle density λ decreases with a factor κ due to the acceleration.

We will assume that the average axial separation $\langle s \rangle = 1/\lambda$ is very small compared to the length L of the acceleration zone. This implies that the time intervals Δt_0 and Δt_1 defined above are, in general, small compared to the total flight time T. Accordingly, during the major part of the flight, both particles are in the acceleration region. The calculation of the two-particle displacement $\Delta \eta(\xi)$ now proceeds as follows:

- The relative initial coordinates $\xi = (r_\perp, \varphi, b_{z0}, v_\perp, \psi)$ are adjusted to account for the acceleration of the foremost particle during the time interval Δt_0 that the other particle has not reached the acceleration zone yet. In this step, the coordinates ξ are replaced by ξ', with

$$\xi' = (r_\perp, \varphi, b_{z0} + \delta b_{z0}, v_\perp, \delta v_{z0}, \psi), \tag{10.5.9}$$

where δb_{z0} and δv_{z0} are given by Eq. (10.5.7). Contrary to ξ the coordinates ξ' include a relative axial velocity.
- The displacement $\Delta \eta(\xi')$ is evaluated using the analysis for two interacting particles in drift space presented in Chapter 6. For the flight time T, one should now use Eq. (10.5.5) instead of Eq. (6.2.2). Special care should be taken with the calculation of the virtual spatial displacement Δr in the reference plane. The final coordinates should be extrapolated (backwards) to the reference plane along a trajectory that is curved, due to the presence of the acceleration field.
- The axial velocity displacement Δv_z is corrected for the fact that the interaction causes the actual flight time through the acceleration zone to be shorter than T for the particle ahead and longer than T for the particle behind. If Δz is the relative axial displacement caused by the interaction, the correction term involved is equal to

$$\delta \Delta v_z = -v_{z1}\left(\sqrt{1 + \frac{2a_z \Delta z}{v_{z1}^2}} - 1\right) = -v_{z0}\kappa\left(\sqrt{1 + \frac{2\Delta z(\kappa - 1)}{Tv_{z0}\kappa^2}} - 1\right), \tag{10.5.10}$$

using Eq. (10.5.4).
The function $\Delta \eta(\xi)$ now specifies the two-particle displacement in presence of a uniform acceleration field, which is characterized by the coefficient κ. As for the case without an external field, the two-particle distribution $p_2(\Delta \eta)$ follows from Eq. (5.4.10). The next step is to evaluate the $p(\mathbf{k})$ transform from Eq. (5.4.11). Finally, the required N-particle displacement distribution $\rho(\Delta \eta)$ follows with Eq. (5.4.9).

10.6. Relativistic beams

In this section we will investigate, in general terms, what modifications are required to extend the applicability of the models for statistical and space charge interactions to high energy beams, in which the average velocity of the particles is comparable to the speed of light. It is assumed that the reader is familiar with the basic notions of the theory of special relativity.

The analysis will be restricted to paraxial beams in drift space. In accordance with the paraxial condition, it is justifiable to assume that the *relative* velocities of the particles are small compared to their average velocity in the laboratory system. This implies that it is permitted to use non-relativistic equations to describe the relative motion of the particles, and the required modifications only concern the coordinate transformation between the laboratory system S and the frame of reference S' moving with beam. For beams in drift space, the system S' moves with a constant axial velocity v_z relative to S. Accordingly, the problem is covered by the theory of special relativity. For high energy beams, the Galileo transformation, which was used so far to relate the coordinates in the systems S and S', has to be replaced by the Lorentz transformation. We will investigate the corresponding changes of the basic equations of our model.

For a relativistic beam, Eq. (3.2.2) is not suited for evaluating the average axial velocity of the particles observed in the laboratory system S. Instead, one should use the relativistic relation between momentum p and energy E, which follows from the covariance of the energy momentum four vector

$$p^2 = E^2/c^2 - m_0^2 c^2, \qquad (10.6.1)$$

where c is the speed of light, m_0 is the rest mass of the particles, and E their average total energy, which is equal to

$$E = m_0 c^2 + eV. \qquad (10.6.2)$$

Substitution into Eq. (10.6.1) yields

$$p^2 = 2m_0 eV \left(1 + \frac{eV}{2m_0 c^2}\right), \qquad (10.6.3)$$

which is the relativistic equivalent of Eq. (3.2.2). For paraxial beams, the lateral momentum $p_\perp$ is extremely small compared to the axial momentum p_z and in the remaining analysis we will identify p_z with p.

In particle optics one usually accounts for relativistic effects by introducing the so-called relativistic beam potential V^*, which is defined such that the nonrelativistic relation $p^2 = 2m_0 eV^*$ yields the same result as the relativistic Eq. (10.6.3). Accordingly, it is given by

$$V^* = V\left(1 + \frac{eV}{2m_0 c^2}\right). \qquad (10.6.4)$$

We will come back to the use of the relativistic potential V^* at the end of this section.

The time of flight T (or T_f) is in the nonrelativistic model given by $T = L/v_z$, as expressed by Eq. (3.2.3) (where v_z is denoted as $\langle v_\parallel \rangle$). In the relativistic model, the time of flight differs for the observers in the systems S and S',

$$T = \frac{\gamma_r m_0 L}{p}, \quad T' = \frac{m_0 L}{p}, \qquad (10.6.5)$$

where γ_r can be expressed as

$$\gamma_r = \frac{E}{m_0 c^2} = 1 + \frac{eV}{m_0 c^2}, \qquad (10.6.6)$$

using the celebrated relation $E = mc^2$. Eq. (10.6.2) was exploited in the second step. The proper measure for the duration of the interaction, to be used in the dynamical part of the calculation, is T', measured by the observer in S'. It is specified by the second equation of (10.6.5). This equation provides the relativistic equivalent of Eq. (3.2.3). Notice that L is the length of the beam segment measured in the laboratory system S.

The linear particle densities λ and λ', observed in S and S' respectively, are equal to

$$\lambda = \frac{\gamma_r m_0 I}{ep}, \quad \lambda' = \frac{m_0 I}{ep}, \qquad (10.6.7)$$

where the beam current I is observed in the laboratory system S. The measure for the linear particle density to be used in the statistical part of the calculation is λ', which is observed in S'. The second equation of (10.6.7) is the relativistic equivalent of Eq. (3.2.4).

Finally, it should be realized that the lateral velocities of the particles differ for the observers in S and S'. However, the transverse momentum $p_\perp$ of a particle is the same for both observers. The maximum transverse momentum p_0 is, in paraxial approximation, given by

$$p_0 \approx \alpha_0 p, \qquad (10.6.8)$$

where α_0 is the beam semi-angle observed in the laboratory system S. Eq. (10.6.8) replaces the nonrelativistic equation $v_0 \approx \alpha_0 v_z$, which is used throughout the model. Notice that the lateral dimensions of the beam are the same for the observers in S and S', thus the beam radius r_0 and the crossover radius r_c can be used in the same way as in the nonrelativistic model.

The macroscopic condition of the beam, observed in the reference system S', is now specified in terms of the experimental parameters, measured in the laboratory system S. The calculations can be carried out as before, using the appropriate values for the time of flight T', the linear particle density λ', and the maximum transverse momentum p_0, specified by Eqs. (16.6.5), (16.6.7) and (16.6.8) respectively. The results of the model should be expressed in terms of the generated distributions of axial and transverse momentum, $\rho(\Delta p_z')$ and $\rho(\Delta p_\perp')$, and the trajectory displacement distribution $\rho(\Delta r')$, all observed in S'. The transformation to the desired energy distribution $\rho(\Delta E)$, the angular deflection distribution

$p(\Delta a)$, and the trajectory displacement distribution $p(\Delta r)$ observed in the laboratory system S can be performed with

$$\Delta E \approx p \Delta p_z'/m_0$$
$$\Delta a \approx \Delta p_\perp'/p \qquad (10.6.9)$$
$$\Delta r = \Delta r',$$

which completes the analysis.

The discussion given above shows that the essential equations of the model can simply be adjusted to the relativistic case by expressing them in terms of the components of the momentum vectors p and p' instead of the velocity components of the particles. The value of p should be evaluated from the relativistic Eq. (10.6.3). As an alternative, one may use the original equations and account for the relativistic effects by means of the relativistic potential V^* specified by Eq. (10.6.4). Accordingly, we conclude that for a paraxial beam in drift space, it is sufficient to replace the beam potential V, appearing in the final equations for the various interaction effects, by its relativistic counterpart V^*.

Appendix 10.A. Distributions for the average reference trajectory

This appendix discusses the derivation of the distributions $p_u(r)$ and $p_g(r)$ given by Eqs. (10.3.7) and (10.3.9) respectively. The evaluation of Eq. (10.3.5) can be divided into two steps by introducing the auxiliary function $h(r, r_1)$:

$$h(r, r_1) = \int_0^\infty dr_2\, f(r_2) \int_0^{2\pi} \frac{d\varphi}{2\pi} \delta\left[r - \{r_1^2 - r_2^2 - 2r_1 r_2 \cos \varphi\}^{1/2}\right]. \quad (10.A.1)$$

This function will be evaluated first. The function $p(r)$ then follows from $h(r, r_1)$ as

$$p(r) = \int_0^\infty dr_1\, f(r_1) h(r, r_1). \qquad (10.A.2)$$

As was mentioned in Section 10.3, the function $p(r)$ represents the distribution of the distance $r = |\mathbf{r}_1 - \mathbf{r}_2|$ between two points, $\mathbf{r}_2$ and $\mathbf{r}_2$, in a plane when the distribution of the distance between the individual points and the origin is, for both points, given by the distribution $f(r)$. The function $h(r, r_1)$ represents the distribution of the distance $r = |\mathbf{r}_1 - \mathbf{r}_2|$ for a given value of r_1. We will successively treat the cases of a uniform $f(r)$ and a Gaussian $f(r)$.

Let $f(r) = f_u(r)$, where $f_u(r)$ is the uniform distribution given by Eq. (10.3.6). Substitution into Eq. (10.A.1) and integration yields

$$h_u(r, r_1) = \begin{cases} 2r/R^2 & \text{for } 0 \leq r \leq R - r_1 \\ \dfrac{2r}{\pi R^2} \arccos\left(\dfrac{r_1^2 + r^2 - R^2}{2r_1 r}\right) & \text{for } R - r_1 < r \leq R + r_1 \\ 0 & \text{for } r > R + r_1. \end{cases}$$
(10.A.3)

The reader might verify that this result can straightforwardly be understood from geometrical arguments. By substitution of Eq. (10.A.3) into Eq. (10.A.2) and integration, one finds Eq. (10.3.7). We notice that an alternative way to derive Eq. (10.3.7) is to prove that the quantity $p_u(r)R^2/2r$ is equal to the ratio A_1/A_2, where A_1 is the mutual surface of two circles, both with radius R, whose centers are separated by a distance r and $A_2 = \pi R^2$, the surface of one circle.

Let $f(r) = f_g(r)$, where $f_g(r)$ is the Gaussian distribution given by Eq. (10.3.8). Substitution into Eq. (10.A.1) yields

$$h_g(r, r_1) = \frac{2r}{R^2} e^{-(r/R)^2 - (r_i/R)^2} I_0\left(\frac{2rr_1}{R^2}\right),$$
(10.A.4)

using the integral

$$\int_a^b dx \frac{e^{-qx}}{[(x-a)(b-x)]^{1/2}} = \pi e^{-(a+b)q/2} I_0[(b-a)q/2],$$

where I_0 is the modified Bessel function of the first kind and zero order. From Eqs. (10.A.4) and (10.A.2), one obtains Eq. (10.3.9), using the Laplace transform,

$$\int_0^\infty dx\, I_0\left[2(ax)^{1/2}\right] e^{-qx} = \frac{1}{q} e^{a/q},$$

which is given by Erdélyi et al. (1954).

CHAPTER ELEVEN

Space charge effect in low density particle beams

Contents

11.1.	Introduction	329
11.2.	General aspects	330
11.3.	Beams with laminar flow	338
11.4.	First order perturbation theory	341
11.5.	First order optical properties of the space charge lens	343
11.6.	Third order geometrical aberrations of the space charge lens	346
11.7.	Beam segment with a narrow crossover	350
11.8.	Homocentric cylindrical beam segment	352
11.9.	Addition of the effects generated in individual beam segments	355

11.1. Introduction

This chapter discusses an analytical model for the space charge effect in a beam segment in drift space. The model is based on first order perturbation theory. This approximation should be accurate for beams of relatively low particle density in which the particles experience only small deviations from their unperturbed trajectories. The theory applies to the same types of beams as the model for statistical interactions, discussed in the previous chapters.

It will be demonstrated that the impact of the space charge effect on the lateral properties of the beam corresponds to the action of a negative lens. This space charge lens is ideal if the current density distribution is uniform in every cross section of the beam. Accordingly, it causes a defocussing of the image as well as some (de)magnification. These effects can be compensated by proper lens adjustment and do not influence the resolution of the system. For non-uniform current density distributions, the space charge lens is non-ideal. In addition to a defocussing and a (de)magnification, it produces aberrations causing a nonrefocusable blurring of the image. We will describe the space charge lens in terms of its first and third order properties, analogous to the description of conventional lenses used in geometrical optics.

11.2. General aspects

The space charge effect in charged particles beams has been studied by many authors during the past 60 years. The reader is referred to Kirstein et al. (1967), Nagy and Szilagyi (1974) or Pierce (1954) for a general discussion of the subjects involved. The majority of the contributions deal with the beam spreading observed in high intensity beams or study the space charge effect occurring in the vicinity of the cathode surface of a thermionic gun. In this chapter, we will restrict ourselves to the impact of space charge on the lateral properties of low and medium density beams in drift space.

Both the (deterministic) space charge effect and the effect of statistical interactions are generated by the mutual Coulomb repulsion of the charged particles constituting the beam. In order to emphasize this common basis, we will start from a general description, which covers both phenomena, using the terminology and notation introduced in Chapter 5. The specific aspects of an analytical model for the space charge effect will be considered next.

The displacement distribution $\rho(\Delta\eta)$ contains all information regarding the different manifestations of Coulomb particle-particle interactions. It specifies the probability that a test particle experiences a displacement $\Delta\eta$ due to the Coulomb interaction with the other particles in the beam, referred to as field particles. Equivalently, one may describe $\rho(\Delta\eta)$ as the distribution of displacements $\Delta\eta$ experienced by a large set of test particles, which would, in the absence of Coulomb interactions, all run along the same trajectory in the beam. The unperturbed trajectory of the test particles is called the reference trajectory. In this chapter, we are primarily interested in the lateral properties of the beam; thus we consider the displacements

$$\Delta\eta = \Delta\mathbf{v}_\perp \quad \text{or} \quad \Delta\eta = \Delta\mathbf{r}_\perp. \tag{11.2.1}$$

The quantities $\Delta\mathbf{v}_\perp$ and $\Delta\mathbf{r}_\perp$ specify the lateral displacement in velocity and position of a test particle at the end of the beam segment.

In Chapter 5, the following representation of the distribution $\rho(\Delta\eta)$ was introduced (reproducing Eq. (5.2.3))

$$\rho(\Delta\eta) = \int d\boldsymbol{\xi}_1 d\boldsymbol{\xi}_2 \cdots d\boldsymbol{\xi}_{N-1} P_N(\boldsymbol{\xi}_1, \boldsymbol{\xi}_2, \cdots, \boldsymbol{\xi}_{N-1}) \delta[\Delta\eta - \Delta\eta(\boldsymbol{\xi}_1, \boldsymbol{\xi}_2, \cdots, \boldsymbol{\xi}_{N-1})], \tag{11.2.2}$$

in which the set of generalized coordinates $\boldsymbol{\xi}_1, \boldsymbol{\xi}_2, \ldots, \boldsymbol{\xi}_{N-1}$ specifies the initial configuration of field particles relative to the test particle; see Eq. (5.2.1) and Fig. 5.1. Given this set of coordinates, one can in principle determine the deviation of the test particle $\Delta\eta = \Delta\eta(\xi_1, \xi_2, ..., \xi_{N-1})$, using Coulomb's law and classical mechanics. This evaluation is referred to as the dynamical part of the problem. The statistical part of the problem is the determination of the probability $P_N(\xi_1, \xi_2, ..., \xi_{N-1})$ of the configuration $\xi_1, \xi_2, ..., \xi_{N-1}$ and the evaluation of the distribution of displacements $\rho(\Delta\eta)$ from Eq. (11.2.2). We note that Eq. (11.2.2) expresses the distribution $\rho(\Delta\eta)$ as an ensemble average. The ensemble average is assumed to be equivalent to a time average, which corresponds to the distribution of displacements of a large set of test particles successively arriving at the end of the beam segment. The reader should be alert that we will often describe a quantity in terms of a time average, while the corresponding equation represents an ensemble average.

Assuming that one is somehow able to determine the distribution $\rho(\Delta\eta)$, the average shift of the test particles $\langle \Delta\eta \rangle$ can be calculated as

$$\langle \Delta\eta \rangle = \int d\Delta\eta \, \rho(\Delta\eta) \, \Delta\eta$$
$$= \int d\xi_1 \, d\xi_2 \cdots d\xi_{N-1} \, P_N(\xi_1, \xi_2, ..., \xi_{N-1}) \, \Delta\eta(\xi_1, \xi_2, ..., \xi_{N-1}). \quad (11.2.3)$$

This shift depends strongly on the choice of the reference trajectory. For the central reference trajectory in a rotational symmetric beam, one expects that $\langle \Delta\eta \rangle = 0$, since there is no average space charge force acting on a particle that is located on the axis. In the previous chapters, this phenomenon was exploited to isolate the effect of statistical interactions. However, this chapter focusses on the calculation of the average shift $\langle \Delta\eta \rangle$, and we should consider off-axis reference trajectories in order to obtain nonvanishing results.

The average shift $\langle \Delta\eta \rangle$ is directly related to the average electrostatic force acting on the test particles during their flight. In order to make this relation explicit, we express the displacement $\Delta\eta(\xi_1, \xi_2, ..., \xi_{N-1})$ of the test particle caused by a given set of field particles $\xi_1, \xi_2, ..., \xi_{N-1}$ as an integral over time (reproducing Eq. (5.3.5)),

$$\Delta\eta(\xi_1, \xi_2, ..., \xi_{N-1}) = \int_{t_i}^{t_f} dt \, \mathbf{G_N}(\xi_1, \xi_2, ..., \xi_{N-1}, t), \quad (11.2.4)$$

in which t_i and t_f are the initial and final time of interaction, defined by the moments that the test particle enters and leaves the beam segment respectively. For the displacement in transverse velocity ($\Delta\eta = \Delta\mathbf{v}_\perp$), the function $\mathbf{G_N}(\xi_1, \xi_2, ..., \xi_{N-1}, t)$ is given by

$$\mathbf{G_N}(\xi_1, \xi_2, ..., \xi_{N-1}, t) = \mathbf{F}_\perp(\xi_1, \xi_2, ..., \xi_{N-1}, t)/m, \quad (11.2.5)$$

in which $\mathbf{F}_\perp(\boldsymbol{\xi}_1, \boldsymbol{\xi}_2, \ldots, \boldsymbol{\xi}_{N-1}, t)$ represents the transverse component of the total interaction force acting on the test particle at time t. For the displacement in transverse position ($\Delta \boldsymbol{\eta} = \Delta \mathbf{r}_\perp$), one finds (reproducing Eq. (5.3.7))

$$\mathbf{G}_N(\boldsymbol{\xi}_1, \boldsymbol{\xi}_2, \ldots, \boldsymbol{\xi}_{N-1}, t) = (t_f - t) \mathbf{F}_\perp(\boldsymbol{\xi}_1, \boldsymbol{\xi}_2, \ldots, \boldsymbol{\xi}_{N-1}, t)/m. \tag{11.2.6}$$

Eqs. (11.2.5) and (11.2.6) follow directly from the definition of the function $\mathbf{G}_N(\boldsymbol{\xi}_1, \boldsymbol{\xi}_2, \ldots, \boldsymbol{\xi}_{N-1}, t)$, given by Eq. (11.2.4).

We will now consider the case $\Delta \boldsymbol{\eta} = \Delta \mathbf{v}_\perp$ and $\Delta \boldsymbol{\eta} = \Delta \mathbf{r}_\perp$ separately. By substituting Eqs. (11.2.4) and (11.2.5) into Eq. (11.2.3) and interchanging the integration over t and the integration over $\boldsymbol{\xi}_1, \boldsymbol{\xi}_2, \ldots, \boldsymbol{\xi}_{N-1}$, one obtains

$$\langle \Delta \mathbf{v}_\perp \rangle = \int_{t_i}^{t_f} dt \frac{\langle \mathbf{F}_\perp(t) \rangle}{m}. \tag{11.2.7}$$

Similarly, one finds with Eq. (11.2.6)

$$\langle \Delta \mathbf{r}_\perp \rangle = \int_{t_i}^{t_f} dt (t_f - t) \frac{\langle \mathbf{F}_\perp(t) \rangle}{m}. \tag{11.2.8}$$

The quantity $\langle \mathbf{F}_\perp(t) \rangle$ denotes the average transverse component of the interaction force experienced by the test particles at time t,

$$\langle \mathbf{F}_\perp(t) \rangle = \int d\boldsymbol{\xi}_1 d\boldsymbol{\xi}_2 \cdots d\boldsymbol{\xi}_{N-1} P_N(\boldsymbol{\xi}_1, \boldsymbol{\xi}_2, \ldots, \boldsymbol{\xi}_{N-1}) \mathbf{F}_\perp(\boldsymbol{\xi}_1, \boldsymbol{\xi}_2, \ldots, \boldsymbol{\xi}_{N-1}, t). \tag{11.2.9}$$

Eqs. (11.2.7)–(11.2.9) constitute the basic equations for the space charge effect within the formalism of Chapter 5. An important difference with the theory for the statistical effects is that one may determine the average over all configurations $\boldsymbol{\xi}_1, \boldsymbol{\xi}_2, \ldots, \boldsymbol{\xi}_{N-1}$ first and carry out the time integration next, without loss of generality. This is simply due to the fact that the characteristic quantities for the space charge effect, $\langle \Delta \mathbf{r}_\perp \rangle$ and $\langle \Delta \mathbf{v}_\perp \rangle$, are linear in the interaction force, while the characteristic quantities for the statistical effects, $\langle \Delta v_z^2 \rangle, \langle \Delta v_\perp^2 \rangle$ and $\langle \Delta r_\perp^2 \rangle$, are quadratic in the interaction force.

A straightforward analytical evaluation of Eqs. (11.2.7)–(11.2.9) does not appear to be feasible. In order to determine the function $\mathbf{F}_\perp(\boldsymbol{\xi}_1, \boldsymbol{\xi}_2, \ldots, \boldsymbol{\xi}_{N-1}, t)$, one needs to know the actual positions of all particles at time t. However, the computation of these positions requires the knowledge of the particle trajectories under the influence of the interaction force $\mathbf{F}_\perp$ up to that time. In order to disentangle this many-body problem by analytical means, one is forced to introduce some fundamental assumptions.

The essential observation is that one is not really interested in the value of $\mathbf{F}_\perp(\boldsymbol{\xi}_1, \boldsymbol{\xi}_2, \ldots, \boldsymbol{\xi}_{N-1}, t)$ for any specific configuration but rather in the ensemble average $\langle \mathbf{F}_\perp(t) \rangle$. One may conjecture that the calculation of $\langle \mathbf{F}_\perp(t) \rangle$ does not require any detailed knowledge of the coordinates of the individual field particles in the separate configurations constituting

Space charge effect in low density particle beams

the ensemble but only involves some smoothed-out distribution of charge, which can directly be expressed in terms of the macroscopic properties of the beam. Let us investigate this idea more closely.

Consider a single configuration of field particles. Let $\rho(\mathbf{r}, t)$ be the microscopic density distribution at time t

$$\rho(\mathbf{r},t) = \sum_{i=1}^{N-1} e\delta[\mathbf{r} - \mathbf{r}_i(t)] \tag{11.2.10}$$

where $\delta(\mathbf{r})$ is the three-dimensional delta-Dirac function. We recall that the quantity e denotes the particle charge (not the unit of elementary charge). The charge density distribution $\rho(\mathbf{r},t)$ contains all microscopic information of the individual particle positions as functions of time. If one knows this distribution, one can directly calculate the force $\mathbf{F}(\mathbf{r},t)$, acting at time t on the test particle located at position $\mathbf{r}$, using

$$\mathbf{F}(\mathbf{r},t) = \frac{e}{4\pi\varepsilon_0} \int d\mathbf{r}_1 \rho(\mathbf{r}_1, t) \frac{\mathbf{r} - \mathbf{r}_1}{|\mathbf{r} - \mathbf{r}_1|^3}. \tag{11.2.11}$$

It is sometimes convenient to express this equation in an alternative form. By exploiting the vector relation

$$\frac{\partial}{\partial \mathbf{r}} \frac{\mathbf{r} - \mathbf{r}_1}{|\mathbf{r} - \mathbf{r}_1|^3} = 4\pi\delta(\mathbf{r} - \mathbf{r}_1), \tag{11.2.12}$$

Eq. (11.2.11) can be transformed to

$$\int_S \mathbf{F}(\mathbf{r},t) \cdot d\mathbf{S} = \frac{e}{\varepsilon_0} \int_V \rho(\mathbf{r},t) d\mathbf{r}, \tag{11.2.13}$$

in which S is a closed surface surrounding the volume V. Eq. (11.2.13) is known as Gauss's theorem for the electrostatic force produced by a configuration of charges. Gauss's theorem states that the total flux of $\mathbf{F}$ through the surface S depends only on the total charge within volume V, whatever the exact location of the individual charges within this volume.

Eqs. (11.2.11) and (11.2.13) are true for every configuration of particles in the ensemble. Accordingly, one should expect that the same relations do apply to the corresponding ensemble averaged quantities. By taking the ensemble average of both sides of Eq. (11.2.11), one directly obtains

$$\langle \mathbf{F}(\mathbf{r},t) \rangle = \frac{e}{4\pi\varepsilon_0} \int d\mathbf{r}_1 \langle \rho(\mathbf{r}_1, t) \rangle \frac{\mathbf{r} - \mathbf{r}_1}{|\mathbf{r} - \mathbf{r}_1|^3}, \tag{11.2.14}$$

in which $\langle \rho(\mathbf{r}_1,t) \rangle$ is called the smoothed-out distribution of charge. Employing Eq. (11.2.12), this result can be transformed to

$$\int_S \langle \mathbf{F}(\mathbf{r},t) \rangle \cdot d\mathbf{S} = \frac{e}{\varepsilon_0} \int_V \langle \rho(\mathbf{r},t) \rangle d\mathbf{r}, \tag{11.2.15}$$

which can also be obtained from Eq. (11.2.13) by taking the ensemble average of both sides. Eq. (11.2.15) relates the average force $\langle \mathbf{F}(\mathbf{r}, t) \rangle$ directly to the smoothed-out distribution of charge $\langle \rho(\mathbf{r}, t) \rangle$. Eq. (11.2.15) is often identified with Gauss's theorem. However, strictly speaking Gauss's theorem pertains to a specific configuration of particles and not to the ensemble average.

From now on we will restrict ourselves to the case of a rotational symmetric beam. Let $r_0(z)$ be a measure for the beam radius at axial position z and let $r_m(z)$ be the corresponding outer beam radius. For a uniform distribution, it is sufficient to specify the outer beam radius $r_m(z)$, taking $r_0(z) = r_m(z)$. For a Gaussian distribution, one should take $r_m(z) = \infty$ and relate $r_0(z)$ to some characteristic width measure, for instance the rms-value of the spatial distribution at position z. In general, we will express the smoothed-out distribution of charge in a rotational symmetric beam as

$$\langle \rho(r,z) \rangle = \begin{cases} \dfrac{e\lambda}{\pi r_0(z)^2} \left[a(z) + b(z) \left(\dfrac{r}{r_0(z)} \right)^2 + c(z) \left(\dfrac{r}{r_0(z)} \right)^4 + \cdots \right] & \text{for } r \leq r_m(z) \\ 0 & \text{for } r > r_m(z), \end{cases}$$

(11.2.16)

in which λ is the linear particle density defined by Eq. (3.2.4). The type of distribution is specified by the quantities $a(z), b(z), c(z), \ldots$, which may still depend on the axial coordinate z. For all z, these quantities are related through the normalization of $\langle \rho(r, z) \rangle$:

$$1 = \frac{1}{e\lambda} \int_0^{r_m(z)} \langle \rho(r,z) \rangle 2\pi r \, dr.$$

(11.2.17)

In the representation of Eq. (11.2.16), a uniform current density distribution corresponds to

$$a(z) = 1, \quad b(z) = c(z) = \cdots = 0, \quad r_m(z) = r_0(z),$$

(11.2.18)

while a Gaussian current density distribution is represented as

$$a(z) = 1, \quad b(z) = \frac{-1}{2!}, \quad c(z) = \frac{1}{3!}, \ldots, \quad r_m(z) = \infty, \quad r_0(z) = \sqrt{2}\sigma(z),$$

(11.2.19)

with $\sigma(z) = [\langle r(z)^2 \rangle / 2]^{1/2}$, the well-known sigma-value of the Gaussian distribution at axial position z. Finally, a parabolic current density distribution is specified by

$$a(z) = 2, \quad b(z) = -2, \quad c(z) = \cdots = 0, \quad r_m(z) = r_0(z).$$

(11.2.20)

The advantage of the form of Eq. (11.2.16) is that one can study the space charge effect on a general basis without making any *a priori* assumptions on the exact distribution of charge in the beam. We will study the impact of each term in $\langle\rho(r,z)\rangle$ separately. For a uniform distribution, one only has to consider the first term. For nonuniform distributions one has to consider higher order terms as well.

We will now evaluate Eq. (11.2.15) using the representation of $\langle\rho(r,z)\rangle$ given by Eq. (11.2.16). Consider a cylinder of length Δz and radius r centered around the beam axis of a rotational symmetric beam. The average lateral force component $\langle\mathbf{F}_\perp\rangle$ is directed perpendicular to the cylinder surface. We will ignore any variation in the axial component of the average force (thus assuming $\langle F_\parallel\rangle=0$ or $\langle F_\parallel\rangle=$ constant). Accordingly, the left-hand side of Eq. (11.2.15) is equal to $\langle\mathbf{F}_\perp\rangle\times 2\pi r\,\Delta z$, which gives

$$\langle\mathbf{F}_\perp(r,z)\rangle = \frac{e}{\varepsilon_0 r}\int_0^r \langle\rho(r_1,z)\rangle r_1 dr_1. \tag{11.2.21}$$

By substitution of Eq. (11.2.16) into Eq. (11.2.21), one obtains for the average value of the transverse component of the force acting on a test particle in the beam ($r < r_m(z)$)

$$\langle\mathbf{F}_\perp(r,z)\rangle = \frac{e^2\lambda}{2\pi\varepsilon_0}\left[a(z)\frac{r}{r_0(z)^2} + b(z)\frac{r^3}{2r_0(z)^4} + c(z)\frac{r^5}{3r_0(z)^6} + \cdots\right]. \tag{11.2.22}$$

The first term between brackets in Eq. (11.2.22) is proportional to the distance to the axis r. For a uniform current density distribution this is the only nonvanishing term, as follows with Eq. (11.2.18). Notice that an ideal lens has the property that it deflects the particles by an angle that is proportional to the radial distance r. Hence, for a uniform current density distribution the space charge force $\langle F_\perp(r,z)\rangle$ acts as an ideal lens. For non-uniform distributions the remaining terms in Eq. (11.2.22) do not all vanish, and the space charge force $\langle F_\perp(r,z)\rangle$ acts as a non-ideal lens. Besides its imaging action, it introduces aberrations as well. The second term between brackets in Eq. (11.2.22) corresponds to third order aberrations, similar to those generated by a conventional electrostatic or magnetic lens.

The radial equation of motion of a test particle moving under influence of a lateral force $\langle F_\perp(r,z)\rangle$ is given by

$$\frac{d^2 r(t)}{dt^2} = \frac{\langle F_\perp[r(t),z(t)]\rangle}{m}. \tag{11.2.23}$$

In general, one may write

$$\frac{d^2 r}{dt^2} = \frac{d^2 r}{dz^2}\left(\frac{dz}{dt}\right)^2 + \frac{dr}{dz}\frac{d^2 z}{dt^2}.$$

As we restrict ourselves to beam segments in drift space, the second term can be disregarded (since z is then linear in t: $z = z_0 + v(t - t_0)$). With Eqs. (11.2.22) and (11.2.23), one now obtains

$$\frac{d^2r}{dz^2} = k\left[a(z)\frac{r}{r_0(z)^2} + b(z)\frac{r^3}{2r_0(z)^4} + c(z)\frac{r^5}{3r_0(z)^6} + \cdots\right], \quad (11.2.24)$$

where the quantity k is given by

$$k = \frac{e\lambda}{4\pi\varepsilon_0 V} = \frac{1}{4\pi\varepsilon_0}\left(\frac{m}{2e}\right)^{1/2}\frac{I}{V^{3/2}}. \quad (11.2.25)$$

Eq. (11.2.24) is the ray equation for a rotational symmetric beam in drift space, which takes the interaction force corresponding to the smoothed-out distribution of charge into account. It should be emphasized that this ray equation pertains to the *average trajectory* of a large set of test particles, starting with the same initial condition.

The ray Eq. (11.2.24) relies on an essential assumption that might not be obvious from its derivation. The average trajectory of the test particle is calculated as if it is moving in some *external* electrostatic field. This "external" field is produced by the smoothed-out distribution of charge of the field particles. In this approach, one considers only the perturbation of the test particle and ignores the influence of the test particle on the field particles surrounding it. Due to this lack of feedback (from the coordinates of the test particle to the coordinates of the field particles), the model becomes unrealistic in case the space charge effect is produced by strong interactions between a limited number of particles.

It is instructive to consider the space charge effect occurring in a beam with a narrow waist of radius r_c (we use the term "waist" instead of cross-over, since the particles do not necessarily cross the axis). Let α_0 be the beam semi-angle and assume that the current density distribution is approximately uniform in every cross section of the beam. Consider a particle moving along the edge of the beam toward the waist of the beam. Thus, $r = r_0(z)$ and $\alpha = \alpha_0$. From Eqs. (11.2.22) and (11.2.18), one finds for the space charge force acting on this particle

$$\langle F_\perp \rangle = \frac{e^2\lambda}{2\pi\varepsilon_0 r_0(z)}. \quad (11.2.26)$$

According to this equation, the space charge force $\langle F_\perp \rangle$ acting on the test particle is expected to diverge for a waist radius that approaches zero (thus $r_0(z) = r_c \to 0$). At this point, one should notice that $r_0(z)$ represents the beam radius in presence of the space charge repulsion, which will, in general, be larger than the unperturbed beam radius. When studying the particle motion in the vicinity of the waist under influence of the space charge force $\langle F_\perp \rangle$, it is useful to consider the extreme cases of a beam of very high particle density and a beam of low particle density. Let us study the implications of Eq. (11.2.26) for these cases separately.

In a beam of very high particle density, the particles running along the edge of the beam are expected to remain at the edge of the beam (without crossing the axis), due to the repulsion of the total space charge present in the inner part of the beam. The corresponding waist radius r_c will be non-zero, even for a homocentric beam. Accordingly, the space charge force $\langle F_\perp \rangle$ computed from Eq. (11.2.26) will be finite provided that one accounts for the impact of the space charge repulsion on the distribution of field particles. If the stream of particles is such that none of the particle trajectories cross each other, one speaks of "laminar flow." This condition implies that all particles located at a certain position have the same velocity. This type of flow will be further investigated in the next section.

In a beam of low particle density, a particle running along the edge of the beam is expected to cross the axis unless it happens to collide with some other particle traveling with approximately the same axial coordinate but diametrically opposite. In this type of beam, the space charge effect may cause an axial shift and a slight (de)magnification of the crossover, but it will not change the characteristics of the beam drastically. Accordingly, a homocentric beam is expected to have a crossover of zero radius, and a straightforward application of Eq. (11.2.26) to compute the space charge force $\langle F_\perp \rangle$ will lead to divergent results at the location of this crossover. Clearly, in reality the space charge force acting on the test particle is always finite, due to the fact that the minimum distance between the test and a field particle is limited by their mutual interaction. The divergence of Eq. (11.2.26) for this case is a consequence of the decoupling of the coordinates of the test particle from the coordinates of the field particles. However, anticipating the results of Section 11.7, we mention that this decoupling does not represent any problem in the actual calculation of the space charge effect in beams of low particle density with a narrow crossover.

In order to specify the distinction between beams of low and beams of high particle density, let us consider the number of particles simultaneously present in a cylindrical volume in the beam with some characteristic length Δz and some characteristic radius r. A beam of high particle density is defined as a beam for which this number is large compared to unity. Taking $r = r_m(z)$ and Δz of the order $r_m(z)$, this condition can be expressed as

$$\frac{r_m(z)}{e} \int_0^{r_m(z)} \langle \rho(r,z) \rangle 2\pi r \, dr \gg 1. \tag{11.2.27}$$

We will refer to Eq. (11.2.27) as the continuum condition. It is postulated that the continuum condition should be fulfilled in order to obtain laminar flow. However, we add that this condition does not guarantee laminar flow.

For a beam with a uniform current density distribution, Eq. (11.2.27) transforms to

$$\lambda r_0(z) \gg 1, \qquad (11.2.28)$$

as follows with Eqs. (11.2.16) and (11.2.18). This equation expresses that the beam radius $r_0(z)$ should be large compared to the average axial separation of the particles $\langle s \rangle = 1/\lambda$. It should be fulfilled for all z, including the location of the waist ($r_0(z) = r_c$). In order to verify Eq. (11.2.28), one needs to know the actual dimensions of the beam, which occur under influence of the space charge force.

The classic approach to the space charge effect is to assume laminar flow. We will briefly review this approach in Section 11.3. However, as it turns out, the continuum condition of Eq. (11.2.27) is, in general, not fulfilled for the particle beams used in lithography and microscopy instruments. Therefore, the laminar flow model is not suited to describe the space charge effect occurring in the drift sections of such systems. The remaining part of the chapter concerns an alternative approach that is appropriate for this case. In this model, the ray Eq. (11.2.24) is solved in first order perturbation approximation. The basic philosophy of the model is described in Section 11.4. The actual calculations are presented in Sections 11.5 and 11.6, treating the first order optical properties of the space charge lens and its third order aberrations respectively. In Sections 11.7 and 11.8, the results of the model are examined for the specific cases of a beam segment with a narrow crossover and a homocentric cylindrical beam segment respectively. Finally, Section 11.9 considers the problem of adding the contributions of the individual beam segments to determine the total effect generated in the entire beam.

11.3. Beams with laminar flow

The classic approach to the problem of calculating the spreading of a beam under influence of its own space charge is to assume laminar flow. The first contributions to the subject were given by McGreggor-Morris and Mines (1925) and Watson (1927). Detailed discussions and extensive references can be found in the standard textbooks on electron optics, for instance see Glaser (1952), Klemperer (1953), El-Kareh and El-Kareh (1970) or Hutter (1967). We will only briefly investigate the subject here. Our main interest is to determine the range of validity of the laminar-flow approach.

Consider a rotational symmetric homocentric converging beam with initial radius r_i and slope r_i'. Assume that the current density distribution in the initial plane ($z = z_i$) is uniform. The reader might verify that the assumption of laminar flow in a homocentric beam implies that the distribution will be uniform in other planes as well. The space charge force, which is for a uniform distribution directly proportional to the distance to the axis, may change the size of the beam, but it does not alter its shape. This type of laminar flow is referred to as congruent flow. In case of congruent flow, it is sufficient to determine the trajectory of a particle moving along the edge of the beam, thus taking $r = r_0(z)$.

The appropriate ray equation for the case of a uniform beam follows from Eqs. (11.2.24) and (11.2.18). Multiplication of both sides of Eq. (11.2.24) with (dr/dz) and integration results in

$$\frac{1}{2}\left(\frac{dr_0}{dz}\right)^2 = k\ln\left(\frac{r_0}{r_i}\right) + \frac{1}{2}r_i'^2. \tag{11.3.1}$$

The minimum beam radius r_c follows by taking $(dr_0/dz) = 0$. Eq. (11.3.1) now yields

$$r_c = r_i \exp\left(-r_i'^2/2k\right) \cong r_i \exp(-1/4\bar{\lambda}), \tag{11.3.2}$$

where $\bar{\lambda}$ is the scaled linear particle density given by

$$\bar{\lambda} = \frac{k}{2\alpha_0^2} = \frac{m^{1/2}}{2^{7/2}\pi\varepsilon_0 e^{1/2}}\frac{I}{\alpha_0^2 V^{3/2}}, \tag{11.3.3}$$

as was previously defined by Eq. (7.3.22). In Eq. (11.3.2), we approximated $r_i' \approx \alpha_0$, where α_0 is the beam semi-angle at the initial plane. This is justified since we consider only paraxial beams. The z-coordinate for which the beam radius reaches its minimum value follows by integrating Eq. (11.3.1)

$$z_c - z_i = \int_{r_i}^{r_c}\left[2k\ln\left(\frac{r_0}{r_i}\right) + r_i'^2\right]^{-1/2} dr_0.$$

By substituting $r_0 = r_i\exp(\zeta^2 - r_i'^2/2k)$, this transforms to

$$z_c - z_i = \sqrt{\frac{2}{k}}r_c\int_0^{|r_i'|/\sqrt{2k}}\exp(\zeta^2)\,d\zeta \cong \frac{r_c}{\alpha_0\bar{\lambda}^{1/2}}\int_0^{1/2\bar{\lambda}}\exp(\zeta^2)\,d\zeta, \tag{11.3.4}$$

where r_c is given by Eq. (11.3.2) and the linear particle density $\bar{\lambda}$ by Eq. (11.3.3). Eqs. (11.3.2) and (11.3.4) specify the size and axial position of the minimum cross section of a uniform homocentric beam, assuming laminar flow.

In the design of oscilloscope tubes, one likes to determine the radius r_s of the smallest spot that can be obtained at the screen, given a certain beam current I, beam voltage V, and distance to the screen $L = z_s - z_i$. Given these parameters, the spot size at the screen can be minimized by varying r_i'. Schwartz (1957) presented a universal curve from which the ratio r_s/r_i can be determined as function of $\bar{\lambda}^{1/2}$ (our notation). He pointed out that the minimum spot at the screen is obtained by focussing the beam in such a manner that its minimum cross section occurs somewhat ahead of the screen. For high $\bar{\lambda}$-values, the spot radius r_s can be appreciably smaller than the spot radius r_c obtained when the minimum cross section is made to occur at the screen. The reader is referred to the original publication of Schwartz (1957) or to El-Kareh and El-Kareh (1970) for a more detailed discussion of this phenomenon. One should keep in mind that the so-called universal beam spreading curve is, in fact, not universal at all. It relies on the continuum approximation, which breaks down for low particle densities, as was pointed out earlier. Van den Broek (1984) investigated the influence of the current density distribution on the spot growth. The results are presented in graphs similar to those of Schwartz, showing that the spot growth for non-uniform beams is larger than for uniform beams.

We conclude this section by examining the conditions for which the laminar flow model is justified. Substitution of Eq. (11.3.2) into the continuum condition of Eq. (11.2.28) yields

$$\bar{r}_i \gg \frac{\exp(1/4\bar{\lambda})}{\bar{\lambda}}, \qquad (11.3.5)$$

in which $\bar{r}_i$ is given by

$$\bar{r}_i = \frac{r_i}{d_0} = \frac{8\pi\varepsilon_0}{e}\alpha_0^2 V r_i, \qquad (11.3.6)$$

where the scaling quantity d_0 is defined by Eq. (7.3.2). The right-hand side of Eq. (11.3.5) increases exponentially with decreasing $\bar{\lambda}$. Given a certain value of $\bar{r}_i$, Eq. (11.3.5) specifies a lower limit for $\bar{\lambda}$. For an electron beam with $V=10$ kV, $\alpha_0=10$ mR and $r_i=100$ μm, one finds from Eq. (11.3.6) $\bar{r}_i = 1.39 \times 10^5$. According to Eq. (11.3.5), one should demand $\bar{\lambda} \gtrsim 5.0 \times 10^{-2}$, which implies $I > 0.66$ mA, as follows with Eq. (11.3.3). This is a typical value of the minimum beam current necessary to fulfill the continuum condition at the smallest cross section. Clearly, this type of current is so large that it is beyond the scope of this monograph; see Section 1.1. It is interesting to note that a transition region may occur in which the continuum approximation is justified for the particles at the edge of the beam but not for the inner beam particles. In that case the outer beam particles still form a waist, while the inner beam particles already cross over.

It should be emphasized that our model for statistical interactions described in Chapter 5 breaks down in case Eq. (11.3.5) would be satisfied. The assumption that the field particles can be considered as statistically independent, as expressed by Eq. (5.3.1), would clearly not be justified. Similarly, it would be incorrect to represent the total deviation of the test particle as the sum of independent two particle effects, as expressed by Eq. (5.3.12). We recall that our model relies on the assumption that the test particle is not involved in more than one strong interaction during its flight. The majority of all interactions are assumed to generate only small deviations from the unperturbed trajectory. In the next section, we will present a method to calculate the space charge effect in beams to which our model of statistical interactions applies, that is, beams of low or medium current density.

11.4. First order perturbation theory

Consider a rotational symmetric paraxial beam segment with a crossover. Let the beam radius at axial position z be characterized by $r_0(z)$, which is, in absence of particle interactions, given by

$$\tilde{r}_0(z) = r_c + \alpha_0 |z - z_c|. \tag{11.4.1}$$

The tilde ($\tilde{\ }$) refers to the unperturbed condition of the beam. The axial coordinate z_c specifies the location of the crossover plane. The unperturbed trajectory of a test particle can be represented as

$$\tilde{r}(z) = \left[r_\perp^2 + \alpha^2(z - z_c)^2 + 2r_\perp \alpha(z - z_c)\cos(\psi - \varphi) \right]^{1/2}, \tag{11.4.2}$$

as can be understood from Fig. 5.1. A meridian trajectory (which is a trajectory that intersects the beam axis) corresponds to $\psi = \varphi$ or $\psi = \varphi + \pi$. For these trajectories Eq. (11.4.2) transforms to

$$\tilde{r}(z) = |r_\perp \pm \alpha(z - z_c)|,$$

with the $+$ sign referring to the case $\psi = \varphi$ and the $-$ sign to $\psi = \varphi + \pi$. By allowing negative values for $\tilde{r}(z)$, $r_\perp$, and α, one can represent a meridian trajectory as

$$\tilde{r}(z) = r_\perp - \alpha(z - z_c), \tag{11.4.3}$$

where α is defined such that a positive value corresponds to a trajectory that runs from large $r_\perp$ to small $r_\perp$ values with increasing z; see Fig. 11.1. Notice that a meridian trajectory in a rotational symmetric beam remains meridian also in presence of the space charge repulsion. From now on, we will restrict ourselves to trajectories described by Eq. (11.4.3) and assume that the results apply to other trajectories as well.

The model presented in this section starts from the ray Eq. (11.2.24). In the first order perturbation approximation, the total space charge force acting on the test particle is evaluated from the unperturbed dimensions of the beam and the unperturbed trajectory of the test particle, specified by $\tilde{r}_0(z)$ and $\tilde{r}(z)$ respectively. By replacing r and $r_0(z)$ in the right-hand side of the ray Eq. (11.2.24) by $\tilde{r}_0(z)$ and $\tilde{r}(z)$ respectively, one finds for the angular and spatial displacements of the test particle

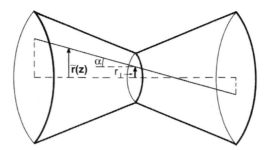

Fig. 11.1 Rotational symmetric beam segment with a crossover. The radial coordinate $\tilde{r}(z)$ of an (unperturbed) meridian trajectory is specified by the angle α and the radial coordinate $r_\perp$ in the crossover plane.

$$\Delta\alpha_f \approx \frac{\Delta v_{\perp f}}{v_z} = k\int_{z_i}^{z_f} dz \left[a\frac{\tilde{r}(z)}{\tilde{r}_0(z)^2} + b\frac{\tilde{r}(z)^3}{2\tilde{r}_0(z)^4} + c\frac{\tilde{r}(z)^5}{3\tilde{r}_0(z)^6} + \cdots \right]$$

$$\Delta r_{\perp f} = k\int_{z_i}^{z_f} dz (z_f - z) \left[a\frac{\tilde{r}(z)}{\tilde{r}_0(z)^2} + b\frac{\tilde{r}(z)^3}{2\tilde{r}_0(z)^4} + c\frac{\tilde{r}(z)^5}{3\tilde{r}_0(z)^6} + \cdots \right], \quad (11.4.4)$$

similar to Eqs. (11.2.7) and (11.2.8). The axial coordinates z_i and z_f refer to the initial and final plane of the beam segment. The coefficients $a, b, c, \ldots$ are from now on assumed to be independent of z, which means that the shape of the current density distribution is assumed to be the same in every cross section of the beam. Notice that the quantities $\Delta\alpha_f$ and $\Delta r_{\perp f}$ may become negative.

The virtual displacement $\Delta r_\perp$ in a reference plane with axial coordinate z_r follows from

$$\Delta r_\perp = \Delta r_{\perp f} - (z_f - z_r)\Delta\alpha_f. \quad (11.4.5)$$

By substituting Eq. (11.4.4) into Eq. (11.4.5), one obtains

$$\Delta r_\perp = k\int_{z_i}^{z_f} dz(z_r - z)\left[a\frac{\tilde{r}(z)}{\tilde{r}_0(z)^2} + b\frac{\tilde{r}(z)^3}{2\tilde{r}_0(z)^4} + c\frac{\tilde{r}(z)^5}{3\tilde{r}_0(z)^6} + \cdots \right], \quad (11.4.6)$$

similar to the second equation of (11.4.4). The difference with this equation is that z_f is replaced by z_r. The integral in Eq. (11.4.6) can be evaluated term by term. Accordingly, one may write

$$\Delta r_\perp = a\Delta r_{\perp 1} + b\Delta r_{\perp 2} + \cdots, \quad (11.4.7)$$

in which

$$\Delta r_{\perp 1} = k\int_{z_i}^{z_f} dz\,(z_r - z)\frac{\tilde{r}(z)}{\tilde{r}_0(z)^2} \quad (11.4.8)$$

$$\Delta r_{\perp 2} = k \int_{z_i}^{z_f} dz \, (z_r - z) \frac{\tilde{r}(z)^3}{2\tilde{r}_0(z)^4}. \tag{11.4.9}$$

In the next section, we will evaluate the quantity $\Delta r_{\perp 1}$ given by Eq. (11.4.8). It specifies the first order optical properties of the space charge lens. These properties can be expressed in terms of a defocussing distance and a magnification. The quantity $\Delta r_{\perp 2}$ given by Eq. (11.4.9) specifies the third order aberration properties of the space charge lens. It will be evaluated in Section 11.6.

11.5. First order optical properties of the space charge lens

In this section, we will evaluate the quantity $\Delta r_{\perp 1}$ given by Eq. (11.4.8). Let us define $z_c = 0$, which implies that the x,y-plane of the laboratory system coincides with the crossover plane. Substitution of Eqs. (11.4.1) and (11.4.3) into Eq. (11.4.8) yields

$$\frac{\Delta r_{\perp 1}}{r_c} = 2\bar{\lambda}\left(R\frac{r_{\perp}}{r_c} + A\frac{\alpha}{\alpha_0}\right), \tag{11.5.1}$$

in which R and A represent

$$R = \int_{x_i}^{x_f} \frac{(x_r - x)}{[1 + |x|]^2} dx, \quad A = \int_{x_i}^{x_f} \frac{x(x - x_r)}{[1 + |x|]^2} dx, \tag{11.5.2}$$

where x is defined as

$$x = \frac{\alpha_0 z}{r_c}. \tag{11.5.3}$$

The scaled linear particle density $\bar{\lambda}$ is given by Eq. (11.3.3). By carrying out the integrals in Eq. (11.5.2), one finds

$$R = \ln\left(\frac{1 + K_1}{1 + K_2}\right) + \frac{1}{1 + K_1} - \frac{1}{1 + K_2} + x_r\left[2 - \frac{1}{1 + K_1} + \frac{1}{1 + K_2}\right] \tag{11.5.4}$$

$$A = 2K + 2 - \frac{1}{1 + K_1} - \frac{1}{1 + K_2} - 2\ln[(1 + K_1)(1 + K_2)]$$
$$+ x_r\left[\ln\left(\frac{1 + K_1}{1 + K_2}\right) + \frac{1}{1 + K_1} - \frac{1}{1 + K_2}\right], \tag{11.5.5}$$

in which the quantities K, K_1 and K_2 are defined by (as in Eq. (7.3.1))

$$K_1 = \frac{\alpha_0 L_1}{r_c}, \quad K_2 = \frac{\alpha_0 L_2}{r_c}, \quad K = \frac{K_1 + K_2}{2} = \frac{\alpha_0 L}{2r_c}, \tag{11.5.6}$$

where $L_i = -z_i$ is the distance between the initial plane and the crossover, $L_2 = z_f$ is the distance between the crossover and the final plane, and $L = L_1 + L_2$ is the total beam length.

We will now assume that the reference plane is located in the vicinity of the crossover. This implies that $x_r \ll 1$, and one may neglect the terms in Eqs. (11.5.4) and (11.5.5) that are proportional to x_r. The resulting expressions can be written as

$$R \approx R_1(K_1) - R_1(K_2), \quad R_1(K) = \ln(1 + K) - \frac{K}{1 + K} \tag{11.5.7}$$

$$A \approx A_1(K_1) - A_1(K_2), \quad A_1(K) = K + \frac{K}{1 + K} - 2\ln(1 + K). \tag{11.5.8}$$

Notice that one may add an arbitrary constant to $R_1(K)$ without affecting the outcome of R. The functions $R_1(K)$ and $A_1(K)$ are plotted in Fig. 11.2. For the case of a crossover located in the middle of the beam segment (thus $K_1 = K_2 = K$), one finds $R = 0$ and $A = 2A_1(K)$.

We like to express the results given by Eqs. (11.5.1), (11.5.7) and (11.5.8) in terms of a defocussing distance Δz_f and a magnification M, both referring to the crossover. Therefore we have to determine the axial position z_r of the plane in which the crossover is imaged by the space charge lens. The radial coordinate of the perturbed trajectory of the test particle can be expressed as

$$r(z) = \tilde{r}(z) + \Delta r_\perp = r_\perp - \alpha z + a\Delta r_{\perp 1} + b\Delta r_{\perp 2} + \cdots, \tag{11.5.9}$$

as follows with Eqs. (11.4.3) and (11.4.7) using that $z_c = 0$. The axial coordinate z_r of the image plane follows by solving

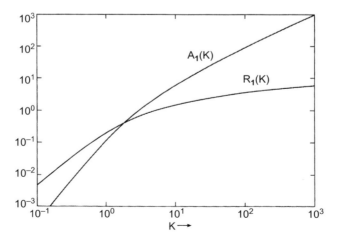

Fig. 11.2 The first order optical functions $R_1(K)$ and $A_1(K)$ of the space charge lens, which are specified by Eqs. (11.5.7) and (11.5.8) respectively.

$$\left[\frac{\partial r(z)}{\partial \alpha}\right]_{z=z_r} = 0. \tag{11.5.10}$$

With Eqs. (11.5.9), (11.5.1) and (11.5.8) one obtains (note that $x_r = \alpha_0 z_r / r_c$)

$$x_r \approx a2\bar{\lambda}A \approx a2\bar{\lambda}[A_1(K_1) + A_1(K_2)], \tag{11.5.11}$$

ignoring the terms $b \ \Delta r_{\perp 2}+\cdots$ in Eq. (11.5.9). With Eq. (11.5.3), one finds for the defocussing distance $\Delta z_f \ (= z_r - z_c = z_r)$

$$\Delta z_f \approx a2\bar{\lambda}L\frac{1}{2K}[K_1 Z(K_1) + K_2 Z(K_2)], \tag{11.5.12}$$

in which the function $Z(K)$ is defined as

$$Z(K) = \frac{A_1(K)}{K} = 1 + \frac{1}{1+K} - \frac{2}{K}\ln(1+K). \tag{11.5.13}$$

This function is plotted in Fig. 11.3. For $K \gtrsim 100$, one may approximate $Z(K) \approx 1$ within 10% accuracy. From Eq. (11.5.12), it now follows that for $K_1 \gtrsim 100$ and $K_2 \gtrsim 100$ the defocussing distance Δz_f is practically independent of the position of the crossover. Notice that Δz_f is always positive. Thus, the virtual crossover, which is imaged by the lens at the end of the beam segment, is always located after the crossover that would occur in the unperturbed beam.

Utilizing Eq. (11.3.3), one can express Eq. (11.5.12) in terms of the experimental parameters

$$\Delta z_f \approx C_{SC}\frac{1}{2K}(K_1 Z(K_1) + K_2 Z(K_2))\frac{aIL}{\alpha_0^2 V^{3/2}}, \tag{11.5.14}$$

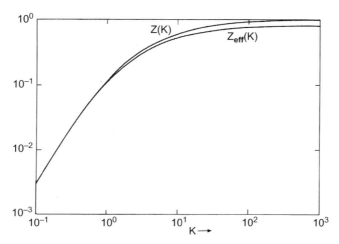

Fig. 11.3 The defocussing functions $Z(K)$ and $Z_{eff}(K)$ of the space charge lens, which are specified by Eqs. (11.5.13) and (11.6.13) respectively.

where the constant C_{SC} is given by

$$C_{SC} = \frac{m^{1/2}}{2^{5/2}\pi\varepsilon_0 e^{1/2}}. \qquad (11.5.15)$$

In the case of electrons, one finds $C_{SC} = 1.5154 \times 10^4$ in SI-units.

Eq. (11.5.14) shows that the defocussing distance increases linearly with the beam current. If one does not refocus the spot while changing the beam current, the space charge effect causes a blurring of the image. The point-spread function representing the blurring at the location of the original (unperturbed) crossover is given by

$$\rho(\Delta r_\perp) = \frac{1}{\Delta z_f} f_\alpha(\Delta r_\perp / \Delta z_f), \qquad (11.5.16)$$

where $f_\alpha(\alpha)$ is the angular distribution at the crossover. The FWHM of this distribution is equal to

$$FWHM_r = \Delta z_f FWHM_\alpha, \qquad (11.5.17)$$

where $FWHM_\alpha$ is the FWHM of the angular distribution $f_\alpha(\alpha)$. The corresponding blurring observed at the target follows by multiplying $FWHM_r$ with the magnification of the system from the crossover to the target.

The magnification M of the crossover, due to action of the space charge lens, follows with Eq. (11.5.9):

$$M = \frac{r(z_r)}{r_\perp} \approx 1 - \alpha \frac{z_r}{r_\perp} + a \frac{\Delta r_{\perp 1}(z_r)}{r_\perp} \qquad (11.5.18)$$

(ignoring the terms $b\ \Delta r_{\perp 2} + \cdots$). By substituting Eqs. (11.5.1), (11.5.7) and (11.5.11), one obtains

$$M \approx 1 + a2\bar{\lambda}[R_1(K_1) - R_1(K_2)], \qquad (11.5.19)$$

in which the function $R_1(K)$ is given by Eq. (11.5.7). Eq. (11.5.9) shows that in a beam segment with a crossover halfway ($K_1 = K_2$) the magnification generated by the space charge effect in the first half of the beam is exactly compensated in the second half. Thus, for this particular geometry, one finds $M = 1$. Furthermore, one sees that $M > 1$ when $K_1 > K_2$ and $M < 1$ when $K_1 < K_2$.

11.6. Third order geometrical aberrations of the space charge lens

In this section, we will evaluate the quantity $\Delta r_{\perp 2}$, given by Eq. (11.4.9). We define again $z_c = 0$. Substitution of Eqs. (11.4.1) and (11.4.3) into Eq. (11.4.9) yields

$$\frac{\Delta r_{\perp 2}}{r_c} = 2\bar{\lambda}\left[D\left(\frac{r_\perp}{r_c}\right)^3 + F\frac{\alpha}{\alpha_0}\left(\frac{r_\perp}{r_c}\right)^2 + C\left(\frac{\alpha}{\alpha_0}\right)^2\frac{r_\perp}{r_c} + S\left(\frac{\alpha}{\alpha_0}\right)^3\right], \qquad (11.6.1)$$

Space charge effect in low density particle beams

in which D, F, C and S represent

$$D = \int_{x_i}^{x_f} \frac{(x_r - x)}{2[1 + |x|]^4} dx, \quad F = \int_{x_i}^{x_f} \frac{3x(x - x_r)}{2[1 + |x|]^4} dx$$

$$C = \int_{x_i}^{x_f} \frac{3x^2(x_r - x)}{2[1 + |x|]^4} dx, \quad S = \int_{x_i}^{x_f} \frac{x^3(x - x_r)}{2[1 + |x|]^4} dx, \quad (11.6.2)$$

where x is defined by Eq. (11.5.3) and the scaled linear particle density $\bar{\lambda}$ is given by Eq. (11.3.3). By carrying out the integrals in Eq. (11.6.2), one finds

$$D = \frac{1}{6(1 + K_1)^3} - \frac{1}{6(1 + K_2)^3} - \frac{1}{4(1 + K_1)^2} + \frac{1}{4(1 + K_2)^2}$$

$$+ \frac{x_r}{6}\left[2 - \frac{1}{(1 + K_1)^3} - \frac{1}{(1 + K_2)^3}\right]$$

$$F = \frac{-1}{2(1 + K_1)^3} - \frac{1}{2(1 + K_2)^3} + \frac{3}{2(1 + K_1)^2} + \frac{1}{2(1 + K_2)^2}$$

$$- \frac{3}{2(1 + K_1)} - \frac{3}{2(1 + K_2)} + 1$$

$$+ x_r\left[\frac{1}{2(1 + K_1)^3} - \frac{1}{2(1 + K_2)^3} - \frac{3}{4(1 + K_1)^2} + \frac{3}{4(1 + K_2)^2}\right]$$

$$C = \frac{1}{2(1 + K_1)^3} - \frac{1}{2(1 + K_2)^3} - \frac{9}{4(1 + K_1)^2} + \frac{9}{4(1 + K_2)^2}$$

$$+ \frac{9}{2(1 + K_1)} - \frac{9}{2(1 + K_2)} + \frac{3}{2}\ln\left(\frac{1 + K_1}{1 + K_2}\right)$$

$$- x_r\left[\frac{1}{2(1 + K_1)^3} + \frac{1}{2(1 + K_2)^3} - \frac{3}{2(1 + K_1)^2}\right.$$

$$\left. - \frac{3}{2(1 + K_2)^2} + \frac{3}{2(1 + K_1)} + \frac{3}{2(1 + K_2)} - 1\right]$$

$$S = -\frac{1}{6(1 + K_1)^3} - \frac{1}{6(1 + K_2)^3} + \frac{1}{(1 + K_1)^2} + \frac{1}{(1 + K_2)^2}$$

$$- \frac{3}{1 + K_1} - \frac{3}{1 + K_2} + \frac{13}{3} - 2\ln[(1 + K_1)(1 + K_2)] + K$$

$$+ \frac{x_r}{2}\left[\ln\left(\frac{1 + K_1}{1 + K_2}\right) + \frac{1}{3(1 + K_1)^3} - \frac{1}{3(1 + K_2)^3}\right.$$

$$\left. - \frac{3}{2(1 + K_1)^2} + \frac{3}{2(1 + K_2)^2} + \frac{1}{1 + K_1} - \frac{3}{1 + K_2}\right],$$

in which the quantities K, K_1 and K_2 are defined by Eq. (11.5.6). We will now assume that $x_r \ll 1$ and ignore the terms proportional to x_r. This way, one obtains

$$D \approx D_1(K_1) - D_1(K_2), \quad F \approx F_1(K_1) + F_1(K_2)$$
$$C \approx C_1(K_1) - C_1(K_2), \quad S \approx S_1(K_1) + S_1(K_2), \qquad (11.6.3)$$

in which the functions $D_1(K)$, $F_1(K)$, $C_1(K)$ and $S_1(K)$ can be expressed as

$$D_1(K) = \frac{3K^2 + K^3}{12(1+K)^3} \qquad (11.6.4)$$

$$F_1(K) = \frac{K^3}{2(1+K)^3} \qquad (11.6.5)$$

$$C_1(K) = \frac{3}{2}\ln(1+K) - \frac{6K + 15K^2 + 11K^3}{4(1+K)^3} \qquad (11.6.6)$$

$$S_1(K) = \frac{1}{2}K - 2\ln(1+K) + \frac{9K + 21K^2 + 13K^3}{6(1+K)^3}. \qquad (11.6.7)$$

These functions are plotted in Fig. 11.4. For the case of a crossover located in the middle of the beam segment (thus $=K_1=K_2=K$), one finds from Eq. (11.6.3) that $D=C=0$, $F=2F_1(K)$ and $S=2S_1(K)$.

Fig. 11.4 shows that the term proportional to S dominates Eq. (11.6.1) for $K \gg 1$. This term corresponds to the spherical aberration of the space charge lens. Let us investigate the properties of this term in more detail. The coefficient of spherical aberration C_s of a lens is usually defined as

$$\Delta r_\perp = -C_s \alpha^3. \qquad (11.6.8)$$

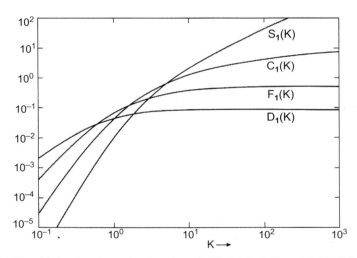

Fig. 11.4 The third order aberration functions $D_1(K)$, $F_1(K)$, $C_1(K)$ and $S_1(K)$ of the space charge lens, which are specified by Eqs. (11.6.4), (11.6.5), (11.6.6) and (11.6.7) respectively.

The minus sign implies that a lens with a positive C_s deflects the outer rays by a larger angle than required for ideal imaging. The corresponding deviation $\Delta r_\perp$ is negative.

With Eqs. (11.4.7), (11.6.1) and (11.6.3), one obtains from Eq. (11.6.8) for the coefficient of spherical aberration of the space charge lens

$$C_s = -b\frac{2\bar{\lambda}L}{\alpha_0^2}\frac{1}{2K}[S_1(K_1) + S_1(K_2)], \qquad (11.6.9)$$

where $2K = K_1 + K_2$ and the function $S_1(K)$ is given by Eq. (11.6.7). Substitution of the expression for $\bar{\lambda}$ given by Eq. (11.3.3) yields

$$C_s = C_{SC}\frac{1}{2K}[S_1(K_1) + S_1(K_2)]\frac{-bIL}{\alpha_0^4 V^{3/2}}, \qquad (11.6.10)$$

where the constant C_{SC} is given by Eq. (11.5.15). We recall that the parameter b is a measure for the non-uniformity of the current density distribution; see Eq. (11.2.16). A Gaussian distribution has the properties $a = 1$, $b = -1/2$, as is specified by Eq. (11.2.19). Eq. (11.6.10) shows that the coefficient of spherical aberration C_s is positive when b is negative. A negative value of b implies that the current density at the edge of the beam is smaller than in the center of the beam, as can be seen from Eq. (11.2.16). This is usually the case in practical beams. In order to exploit the space charge lens to compensate the spherical aberration of the system components (which have a positive C_s), one should produce a beam that has the property $b > 0$. This means that the current density at the edge of the beam should be larger than in the center of the beam.

As is well known from geometrical optics, the disk of least confusion of a beam that is dominated by spherical aberrations is located at a distance $(3/4)C_s\alpha_0^2$ ahead of the Gaussian image plane and has a diameter $d_s \approx (1/2)C_s\alpha_0^3$. The effective defocusing distance Δz_{eff} of the space charge lens is now defined as

$$\Delta z_{eff} = \Delta z_f - \frac{3}{4}C_s\alpha_0^2. \qquad (11.6.11)$$

By substitution of Eqs. (11.6.10) and (11.5.14), one finds an expression for Δz_{eff} that has the same form as Eq. (11.5.14) but with the function $Z(K)$ replaced by a function $Z_{eff}(K)$, which is given by

$$\begin{aligned}Z_{eff}(K) &= Z(K) + \frac{3b}{4a}\frac{S_1(K)}{K} \\ &= 1 + \frac{3b}{8a} + \frac{1}{1+K} + \frac{1}{a}\frac{b\,9 + 21K + 13K^2}{8(1+K)^3} - \left(2 + \frac{3b}{2a}\right)\frac{\ln(1+K)}{K}.\end{aligned}$$
$$(11.6.12)$$

For a Gaussian current density distribution, one obtains with Eq. (11.2.19)

$$Z_{eff}(K) = \frac{13}{16} + \frac{7 + 11K + 3K^2}{16(1+K)^3} - \frac{5\ln(1+K)}{4K}, \quad (11.6.13)$$

which is plotted in Fig. 11.3. The diameter of the disk of least confusion d_s is equal to

$$d_s = \frac{1}{2}C_s\alpha_0^3 = C_{SC}\frac{1}{2K}[S_1(K_1) + S_1(K_2)]\frac{-bIL}{2\alpha_0 V^{3/2}}, \quad (11.6.14)$$

as follows with Eq. (11.6.10).

11.7. Beam segment with a narrow crossover

In this section, we will study the results of the previous two sections for the case of a beam segment with a narrow crossover. This geometry corresponds to $r_c \to 0$, thus $K \to \infty$. We will start by investigating the first order optical properties of the space charge lens and will discuss its third order properties next. For $K \to \infty$, one may expand Eqs. (11.5.4), (11.5.13) and (11.6.12) in the following power series

$$R_1(K) = \ln(K) - 1 + \frac{2}{K} - \frac{3}{2K^2} + \frac{4}{3K^3} + O(K^{-4}) \quad (K \to \infty) \quad (11.7.1)$$

$$Z(K) = 1 - \frac{2\ln(K)}{K} + \frac{1}{K} - \frac{3}{K^2} + \frac{2}{K^3} + O(K^{-4}) \quad (K \to \infty) \quad (11.7.2)$$

$$Z_{eff}(K) = 1 + \frac{3b}{8a} - \left(2 + \frac{3b}{2a}\right)\frac{\ln(K)}{K} + \left(1 + \frac{13b}{8a}\right)\frac{1}{K} - \left(3 + \frac{15b}{4a}\right)\frac{1}{K^2}$$
$$+ \left(2 + \frac{15b}{4a}\right)\frac{1}{K^3} + O(K^{-4}) \quad (K \to \infty),$$

$$(11.7.3)$$

in which $O(K^{-4})$ represents terms of the order K^{-4}. From Eqs. (11.7.2), (11.5.14) and (11.5.17), one obtains for the extreme case of a homocentric beam with a crossover ($r_c = 0$, $K = \infty$) with a uniform current density distribution in every cross section of the beam ($a = 1$)

$$\Delta z_f = C_{SC}\frac{IL}{\alpha_0^2 V^{3/2}}, \quad FWHM_r = C_{SC}\frac{2IL}{\alpha_0 V^{3/2}} \quad (a = 1, K \to \infty), \quad (11.7.4)$$

where the constant C_{SC} is defined by Eq. (11.5.15). For a Gaussian current density distribution ($a = 1$, $b = -1/2$), one finds an effective defocussing distance that is smaller by a factor 13/16

$$\Delta z_{eff} = C_{SC}\frac{13IL}{16\alpha_0^2 V^{3/2}} \quad (a = 1, b = -1/2, K \to \infty), \quad (11.7.5)$$

as follows from Eqs. (11.5.14) and (11.7.3). For an electron beam with $I = 1\,\mu A$, $L = 0.1$ m, $V = 10$ mR, Eqs. (11.7.4) and (11.7.5) yield $\Delta z_f = 15\,\mu m$, $\Delta z_{eff} = 12\,\mu m$ and $FWHM_r = 0.30\,\mu m$.

The magnification of a narrow crossover M can be determined from Eqs. (11.5.19) and (11.7.1). For a converging beam with a uniform current density distribution ($a=1$, $K_1 \to \infty$ and $K_2=0$), one obtains

$$(M-1)r_c \approx C_{SC} \frac{IL}{\alpha_0 V^{3/2}} \frac{\ln(K_1)}{K_1} \approx \Delta z_f \alpha_0 \frac{\ln(K_1)}{K_1}, \qquad (11.7.6)$$

utilizing Eqs. (11.3.3) and (11.7.4). Eq. (11.7.6) shows that the lateral displacement $(M-1)r_c$ associated with the magnification effect is, for the considered geometry ($K \to \infty$), typically much smaller than the displacement $\Delta z_f \alpha_0$ associated with the defocussing effect. This can also directly be observed from Fig. 11.2, employing Eqs. (11.5.1), (11.5.7) and (11.5.8). The ratio $\Delta z_f \alpha_0 / Mr_c$ becomes even larger for a beam segment with a narrow crossover that is located somewhere in the middle ($K_1 \approx K_2$), due to the cancellation of the contributions to M of the first and the second part of the beam segment; see Eq. (11.5.7).

The third order aberrations of the space charge lens are described by Eq. (11.6.1). The functions D, F, C, and S are defined by Eqs. (11.6.3)–(11.6.7). For $K \to \infty$, these functions can be expanded as

$$D_1(K) = \frac{1}{12} - \frac{4}{K^2} - \frac{2}{3K^3} + O(K^{-4}) \qquad (K \to \infty) \quad (11.7.7)$$

$$F_1(K) = \frac{1}{2} - \frac{3}{2K} + \frac{3}{K^2} - \frac{5}{K^3} + O(K^{-4}) \qquad (K \to \infty) \quad (11.7.8)$$

$$C_1(K) = \frac{3}{2}\ln(K) - \frac{11}{4} + \frac{6}{K} - \frac{15}{2K^2} + \frac{10}{K^3} + O(K^{-4}) \qquad (K \to \infty) \quad (11.7.9)$$

$$S_1(K) = \frac{1}{2}K - 2\ln(K) + \frac{13}{6} - \frac{5}{K} + \frac{5}{K^2} - \frac{35}{6K^3} + O(K^{-4}) \quad (K \to \infty). \quad (11.7.10)$$

These equations show that, for large K-values, the dominating third order term is spherical aberration, as was mentioned previously. This can also be seen from Fig. 11.4. For the extreme case of a homocentric beam segment with a crossover ($K_1 \to \infty$ and $K_2 \to \infty$), with a Gaussian current density distribution ($a=1$, $b=-1/2$), one obtains from Eqs. (11.6.10), (11.6.14) and (11.7.10)

$$C_s = S_{SC} \frac{IL}{4\alpha_0^4 V^{3/2}}, \quad d_s = C_{SC} \frac{IL}{8\alpha_0 V^{3/2}} \quad (a=1, b=-1/2, K \to \infty). \quad (11.7.11)$$

We recall that d_s represents the diameter of the disk of least confusion. The corresponding axial location is specified by Eq. (11.7.5). For an electron beam with $I=1\,\mu A$, $L=0.01\,m$, $V=10\,kV$ and $\alpha_0=10\,mR$, these equations yield $C_s = 38\,mm$ and $d_s = 19\,nm$.

11.8. Homocentric cylindrical beam segment

We will now consider the case of a homocentric cylindrical beam segment with a narrow crossover. This geometry corresponds to $\alpha_0=0$, thus $K=0$. As in the previous section, we will start by investigating the first order optical properties of the space charge lens and will discuss its third order properties next. For $K \to 0$, one may expand Eqs. (11.5.4), (11.5.13) and (11.6.12) in the following power series

$$R_1(K) = \frac{1}{2}K^2 - \frac{2}{3}K^3 + \frac{3}{4}K^4 - \frac{4}{5}K^5 + O(K^6) \quad (K \to 0) \tag{11.8.1}$$

$$Z(K) = \frac{1}{3}K^2 - \frac{1}{2}K^3 + \frac{3}{5}K^4 - \frac{2}{3}K^5 + O(K^6) \quad (K \to 0) \tag{11.8.2}$$

$$Z_{eff}(K) = \frac{1}{3}K^2 - \frac{1}{2}K^3 + \left(\frac{3}{5} + \frac{3b}{40a}\right)K^4 - \left(\frac{2}{3} + \frac{b}{8a}\right)K^5 + O(K^6) \quad (K \to 0). \tag{11.8.3}$$

Eqs. (11.8.2) and (11.8.3) show that, for $K \to 0$, one may approximate $Z_{eff}(K) \approx Z(K)$, which is correct up to fourth order terms in K. This behavior is confirmed by Fig. 11.3. For the extreme case that $\alpha_0=0$ (thus $K=0$), one obtains from Eqs. (11.5.1), (11.5.7), (11.5.8), (11.5.13), (11.8.1), (11.8.2) and (11.5.6)

$$\Delta r_{\perp 1} = 2\lambda^*(S_c - 1/2) r_\perp \quad (\alpha_0 = 0), \tag{11.8.4}$$

in which the scaled linear particle density for a cylindrical beam λ^* is defined as (as stated previously by Eq. (7.4.18))

$$\lambda^* = 4\bar{\lambda}K^2 = \frac{kL^2}{2r_0^2} = \frac{m^{1/2}}{2^{7/2}\pi\varepsilon_0 e^{1/2}} \frac{IL^2}{r_0^2 V^{3/2}}, \tag{11.8.5}$$

in which r_0 denotes the radius of the beam. As the crossover disappears for $\alpha_0 \to 0$, the parameter S_c in Eq. (11.8.4) merely defines the location of the image plane. In fact, S_c replaces here the parameter S_i, defined by Eq. (3.2.1). Eq. (11.8.4) can also be obtained directly by performing the integral in Eq. (11.4.8), taking $\tilde{r}(z) = r_\perp$ and $\tilde{r}_0(z) = r_0$ independent of z.

The third order aberration functions D_1, F_1, C_1 and S_1 given by Eqs. (11.6.4)–(11.6.7) can, for $K \to 0$, be expanded as

$$D_1(K) = \frac{1}{4}K^2 - \frac{2}{3}K^3 + \frac{5}{4}K^4 - 2K^5 + O(K^6) \quad (K \to 0) \tag{11.8.6}$$

$$F_1(K) = \frac{1}{2}K^3 - \frac{3}{2}K^4 + 3K^5 + O(K^6) \quad (K \to 0) \tag{11.8.7}$$

$$C_1(K) = \frac{3}{8}K^4 - \frac{6}{5}K^5 + O(K^6) \quad (K \to 0) \tag{11.8.8}$$

$$S_1(K) = \frac{1}{10}K^5 + O(K^6) \quad (K \to 0) \tag{11.8.9}$$

Substitution into Eq. (11.6.1), using Eq. (11.6.3), yields for the extreme case $\alpha_0 = 0$

$$\Delta r_\perp^2 = \lambda^*(S_c - 1/2)\frac{r_\perp^3}{r_0^2} \quad (\alpha_0 = 0), \tag{11.8.10}$$

utilizing Eq. (11.8.5). Eq. (11.8.10) can also be obtained directly by performing the integral in Eq. (11.4.9), taking $\tilde{r}(z) = r_\perp$ and $\tilde{r}_0(z) = r_0$.

Eqs. (11.8.4) and (11.8.10) specify the first and third order optical properties of the space charge lens associated with a cylindrical beam segment in terms of the displacements Δr_1 and Δr_2 in a certain plane determined by the parameter S_c. We will now translate these quantities in terms of the corresponding properties of the beam in the back-focal plane of the lens, succeeding the beam segment. Fig. 11.5 shows a cylindrical beam segment of length L between two lenses. The beam radius is r_0. The focal distance of the second lens is f. The displacement Δr_f in the back-focal plane of the lens can be expressed as

$$\Delta r_f = \lim_{S_c \to -\infty} \left|\frac{f}{S_c L}\right| \Delta r, \tag{11.8.11}$$

in which Δr is the displacement in the plane defined by S_c. Eq. (11.8.11) is equivalent to Eq. (9.2.10) (We recall that the parameter S_c is, for the present case, identical to S_i). It expresses that the displacement Δr_f is entirely the result of the angular displacement generated in the cylindrical beam segment.

With Eq. (11.8.11), one obtains from Eqs. (11.8.4) and (11.8.10)

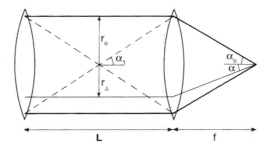

Fig. 11.5 Schematic view of a homocentric cylindrical beam segment (solid lines) and a homocentric beam segment with a crossover in the middle (dashed lines). The corresponding space charge effect, observed in the crossover after the second lens, is the same for both geometries.

$$\Delta r_{1,f} = 2\lambda^* \frac{fr_\perp}{L} \qquad (11.8.12)$$

$$\Delta r_{2,f} = 2\lambda^* \frac{fr_\perp^3}{Lr_0^2}. \qquad (11.8.13)$$

The displacement $\Delta r_{1,f}$ is directly proportional to $r_\perp$ and can, therefore, be expressed in terms of a defocussing distance Δz_f. We define

$$\Delta z_f = a \, \Delta r_{1,f} \frac{f}{r_\perp}, \qquad (11.8.14)$$

utilizing Eq. (11.4.7). The constant a depends on the type of current density distribution in the cross section of the beam, as can be seen from Eq. (11.2.16). Substitution of Eqs. (11.8.12) and (11.8.5) into Eq. (11.8.14) yields

$$\Delta z_f = a2\lambda^* \frac{f^2}{L} = C_{SC} \frac{aILf^2}{r_0^2 V^{3/2}}, \qquad (11.8.15)$$

where the constant C_{SC} is defined by Eq. (11.5.15). Similarly, one can express the third order aberration specified by Eq. (11.8.13) as

$$\Delta r = C_{SC} \frac{-bILfr_\perp^3}{2r_0^4 V^{3/2}} = C_{SC} \frac{-bILf^4\alpha^3}{2r_0^4 V^{3/2}}, \qquad (11.8.16)$$

using that for a paraxial beam $r_\perp \approx \alpha f$, as can be seen from Fig. 11.5. The minus sign was included to express that a positive value of b leads to an aberration that corresponds to a too strong focussing of the outer rays. Utilizing Eq. (11.6.8), one can express this result in terms of a coefficient of spherical aberration C_s,

$$C_s = C_{SC} \frac{-bILf^4}{2r_0^4 V^{3/2}}. \qquad (11.8.17)$$

We recall that a Gaussian current density distribution corresponds to $a=1$ and $b=-1/2$, as specified by Eq. (11.2.19).

Eqs. (11.8.15) and (11.8.17) constitute the main results of this section. It is interesting to compare these results to the corresponding results of the previous section. Assume that the cylindrical beam segment in Fig. 11.5 is replaced by a homocentric beam segment with a crossover, as indicated in the figure by the dashed lines. One may now use Eqs. (11.7.4) and (11.7.11) to calculate the defocussing distance $\Delta z_{f,c}$ and the coefficient of spherical aberration $C_{s,c}$, both referring to the crossover. In order to compare the outcome with the results of this section, one has to translate these quantities to the image plane of the second lens, using

$$\Delta z_f = \left(\frac{2f}{L}\right)^2 \Delta z_{f,c}$$

$$C_s = \frac{2f}{L} \frac{\alpha_1^3}{\alpha_0^3} C_{s,c} = \left(\frac{2f}{L}\right)^4 C_{s,c}. \qquad (11.8.18)$$

The quantity $2f/L$ is equal to the transverse magnification from the crossover to image plane of the second lens. Substitution of Eqs. (11.7.4) and (11.7.11) into Eq. (11.8.18) yields

$$\Delta z_f = C_{SC} \frac{IL}{\alpha_1{}^2 V^{3/2}} \left(\frac{2f}{L}\right)^2 = C_{SC} \frac{IL}{\alpha_0{}^2 V^{3/2}}$$
$$C_s = C_{SC} \frac{IL}{4\alpha_1{}^4 V^{3/2}} \left(\frac{2f}{L}\right)^4 = C_{SC} \frac{IL}{4\alpha_0{}^4 V^{3/2}},$$
$(a=1, b=-1/2)$ (11.8.19)

using that $\alpha_1 = 2f\alpha_0/L$. Since $r_0 = \alpha_0 f$, one finds that Eq. (11.8.19) are identical to Eqs. (11.8.15) and (11.8.17), taking $a=1$ and $b=-1/2$. Thus, the space charge effect generated in the homocentric beam segment with a crossover is identical to that generated in the homocentric cylindrical beam segment as far as the impact on the spot in the image plane of the second lens is concerned.

11.9. Addition of the effects generated in individual beam segments

We conclude this chapter by investigating how to add the space charge effects generated in the different segments of a beam. We have demonstrated that the space charge effect generated in a single beam segment corresponds to the action of a negative lens. Accordingly, the total effect generated in the entire beam can be determined by combining the different space charge lenses associated with the individual beam segments, employing the laws of geometrical optics. As the solution of this problem is straightforward, we will restrict the discussion to some general remarks.

The defocusing distance Δz_f of a crossover is directly proportional to the factor $2\bar{\lambda}L$ and a function that depends on the geometry parameters K_1 and K_2; see Eq. (11.5.12). In order to obtain the defocusing of the final probe caused by the space charge effect in segment i, one has to multiply the defocusing of the corresponding crossover $\Delta z_{f,i}$ with M_i^2, where M_i denotes the transverse magnification of the crossover to the final probe (using that the axial magnification is the square of the transverse magnification). The quantity $\bar{\lambda}$ is proportional to $I/\alpha_0{}^2 V^{3/2}$; see Eq. (11.3.3). In absence of acceleration areas and apertures, one may assume that I and V are constant. The semi-angle α_0 is related to the semi-angle at the final probe α_p as $\alpha_0 = M\alpha_p$. This implies that the scaled linear particle densities in segment i and in the final probe, $\bar{\lambda}_i$ and $\bar{\lambda}_p$, are related as $\bar{\lambda}_i = \bar{\lambda}_p/M_i^2$. Accordingly, one finds that the total defocusing of the final probe for a beam consisting of N_b beam segments is proportional to

$$\sum_{i=1}^{N_b} 2\bar{\lambda}_i L_i M_i^2 = C_{SC} \frac{I}{V^{3/2}\alpha_p{}^2} \sum_{i=1}^{N_b} L_i, \qquad (11.9.1)$$

where L_i is the length of segment number i. For a narrow crossover ($K \gg 1$), the geometry-dependent factor in Eq. (11.5.12) is equal to unity. Eq. (11.9.1) indicates that if the condition $K \gg 1$ is fulfilled for all crossovers, the total defocussing is given by the same equation as obtained for a single beam segment, provided that one substitutes for L the length of the total beam and for α_0 the probe semi-angle α_p. The same holds true when one or more beam segments are nearly cylindrical ($K \rightarrow 0$), as follows from the analysis of the previous section. However, for intermediate K values, one has to account for the geometry dependent factors of each individual beam segment. In general, one should calculate the effective length of the beam $L_{\textit{eff}}$ as

$$L_{\textit{eff}} = \sum_{i=1}^{N_b} L_i \frac{1}{2K_i}[L_{1i}Z(K_{1i}) + L_{2i}Z(K_{2i})], \qquad (11.9.2)$$

in which K_i, K_{1i} and K_{2i} represent the parameters K, K_1 and K_2 for beam segment number i. The function $Z(K)$ is specified by Eq. (11.5.13).

The third order aberration properties associated with the space charge effect in a single beam segment are described by Eq. (11.6.1). When the displacement $\Delta r_{\perp 2}$ is magnified with a factor M to the target, one can express the corresponding displacement $\Delta r_{\perp 2, p}$ in the final probe as

$$\frac{\Delta r_{\perp 2,p}}{r_p} = 2\bar{\lambda}M^2 \left[\frac{D}{M^2}\left(\frac{r_\perp}{r_c}\right)^3 + \frac{F}{M^2}\frac{\alpha}{\alpha_0}\left(\frac{r_\perp}{r_c}\right)^2 + \frac{C}{M^2}\left(\frac{\alpha}{\alpha_0}\right)^2 \frac{r_\perp}{r_c} + \frac{S}{M^2}\left(\frac{\alpha}{\alpha_0}\right)^3 \right], \qquad (11.9.3)$$

where $r_p = Mr_c$. The quantity $\bar{\lambda}M^2$ as well as the ratios $r_\perp/r_c$ and α/α_0 are independent of M. Accordingly, one should calculate the total aberration X ($= D$, F, C or S) as

$$X_p = \sum_{i=1}^{N_b} \frac{X_i}{M_i^2} \qquad (11.9.4)$$

where X_i denotes the aberration constant of segment number i, while M_i is the magnification of crossover number i to the target.

CHAPTER TWELVE

Calculation of different spot- and edge-width measures

Contents

12.1.	Introduction	357
12.2.	Spot-width obtained by knife-edge scans	358
12.3.	Edge-width of a shaped spot	361
12.4.	Trajectory displacement effect	365
12.5.	Chromatic aberration	370
12.6.	Spherical aberration	384
12.7.	Space charge effect	388
12.8.	Results for a truncated Gaussian angular distribution	390

12.1. Introduction

This chapter is concerned with the calculation of various experimental measures used to characterize the resolution of probe forming instruments. In Gaussian beam systems, in which the source is imaged on the target, the spot width is usually evaluated by scanning the beam across a knife edge. The transmitted current is measured as function of the position of the beam. The spot width follows as the distance between two characteristic current levels, for instance those corresponding to 10% and 90% of the total beam current.

In shaped beam systems, in which an aperture is imaged on the target, one is primarily interested in the edge width of the spot. As a measure for the edge width, one usually considers the distance between the points corresponding to two characteristic intensity levels, for instance those corresponding to 10% and 90% of the intensity level at the center of the spot. It will be demonstrated that this type of edge width is equivalent to the spot width of a Gaussian spot determined by knife edge scans. The corresponding profiles can both be interpreted as the convolution of a step function with the point-spread function, representing the total aberrations of the system. In general, we will denote this type of width as $d_{p,1-p}$, where p refers to the choice of the characteristic intensity level ($0 < p < 1/2$).

Advances in Imaging and Electron Physics, Volume 230
ISSN 1076-5670
https://doi.org/10.1016/B978-0-443-29784-7.00012-1

In some applications, one likes to know the Full Width of the spot that contains a certain fraction f of the total beam current ($0<f<1$). This width is here denoted as FW_f. The Full Width median FW_{50} value corresponds to $f=0.5$.

The material of this chapter serves to express the results of the previous chapters in terms of the corresponding $d_{p,1-p}$ and FW_f values. The main part of the chapter is concerned with the trajectory displacement effect and the effect of axial chromatic aberration related to the energy spread produced by the Boersch effect. The contribution of spherical aberration and (uncompensated) space charge defocussing is also studied. The type of angular distribution at the target should be specified for the evaluation of the $d_{p,1-p}$ and FW_f values corresponding to chromatic aberration, spherical aberration, and the blurring due to uncompensated space charge defocussing. The cases of a uniform, a Gaussian, and a truncated Gaussian distribution are considered. The latter corresponds to a Gaussian angular distribution, which is limited by an aperture.

12.2. Spot-width obtained by knife-edge scans

Consider a spot, characterized by a current-density distribution $j(x, y)$ like the one depicted in Fig. 12.1. Assume that $j(x, y)$ is rotational symmetric with respect to the origin ($x=0$, $y=0$) and has a maximum at the origin. Let $j(x, y)$ be normalized to unity

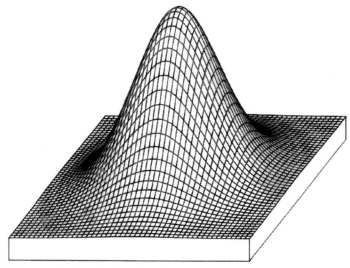

Fig. 12.1 Current density distribution of a Gaussian spot.

Calculation of different spot- and edge-width measures

$$\int_{-\infty}^{\infty} dx \int_{-\infty}^{\infty} dy\, j(x,y) = 1. \tag{12.2.1}$$

The associated one-dimensional current density distribution $j(x)$ is defined as

$$j(x) = \int_{-\infty}^{\infty} dy\, j(x,y), \tag{12.2.2}$$

which can be considered as a projection of the two-dimensional distribution $j(x, y)$ on the x-axis. The edge integral $I_E(x)$ is defined as

$$I_E(x) = \int_{-\infty}^{x} dx'\, j(x'). \tag{12.2.3}$$

It represents the fraction of the total current that is passed along a knife edge, when the center of the spot is located at a distance x from the edge. This situation is shown in Fig. 12.2. The general behavior of $I_E(x)$ is depicted in Fig. 12.3a. It is convenient to express $I_E(x)$ in terms of the auxiliary function $I(x)$, using

$$I_E(x) = \frac{1}{2}[1 + I(x)], \tag{12.2.4}$$

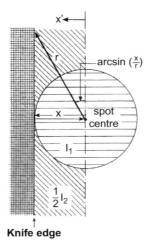

Fig. 12.2 Schematic view of a spot that is scanned over a knife-edge. The center of the spot is located at a distance x from the knife-edge. The fraction $I_E(x)$ of the total current that passes through is equal to $[1 + I_1(x) + I_2(x)]/2$, as expressed by Eqs. (12.2.4) and (12.2.9). The quantities $I_1(x)$ and $I_2(x)$ are specified by Eqs. (12.2.10) and (12.2.11) respectively.

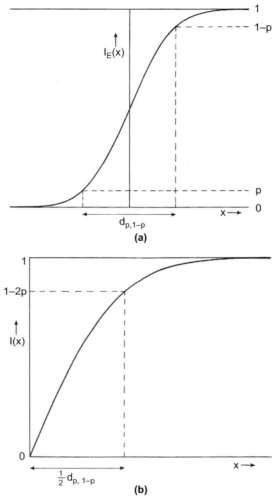

Fig. 12.3 Behavior of the functions $I_E(x)$ and $I(x)$, defined by Eqs. (12.2.4) and (12.2.5).

where

$$I(x) = 2 \int_0^x dx' j(x'). \quad (12.2.5)$$

The function $I(x)$ is antisymmetric: $I(-x) = -I(x)$. Its general behavior for positive x is shown in Fig. 12.3b.

The width measure $d_{p,1-p}$ is defined as the distance between the points for which $I_E(x)$ equals p and $1-p$ respectively, as is indicated in Fig. 12.3a. Accordingly, one finds

$$d_{p,1-p} = I_E^{-1}(1-p) - I_E^{-1}(p) = 2I^{-1}(1-2p), \quad (12.2.6)$$

where I_E^{-1} and I^{-1} denote the inverse functions of I_E and I respectively. Eqs. (12.2.6), (12.2.5) and (12.2.2) specify the width $d_{p,1-p}$, obtained by scanning the beam across a knife edge, in terms of the current density distribution $j(x, y)$.

In the actual calculations, we will employ cylindrical coordinates, which is convenient since $j(x, y)$ is assumed to be rotational symmetric. The cylindrical current-density distribution $j(r)$ is related to $j(x, y)$ as

$$j(r) = 2\pi r j(x, y), \qquad (12.2.7)$$

where $r^2 = x^2 + y^2$. When $j(x, y)$ is normalized, as expressed by Eq. (12.2.1), $j(r)$ is normalized, too:

$$\int_0^\infty dr\, j(r) = 1. \qquad (12.2.8)$$

Employing cylindrical coordinates, Eq. (12.2.5) transforms to

$$I(x) = I_1(x) + I_2(x), \qquad (12.2.9)$$

where $I_1(x)$ and $I_2(x)$ are defined as

$$I_1(x) = \int_0^x dx\, j(r) \qquad (12.2.10)$$

$$I_2(x) = \frac{2}{\pi} \int_x^\infty dr\, \arcsin(x/r) j(r), \qquad (12.2.11)$$

as can be understood from Fig. 12.2. $I_1(x)$ represents the fraction of the total current contained within the circle of radius x. $I_2(x)$ is twice the fraction of the current contained in the area outside this circle between the lines in y-direction through $x'=0$ and $x'=x$. Eqs. (12.2.6) and (12.2.9), in combination with Eqs. (12.2.10) and (12.2.11), specify the width $d_{p,1-p}$ in terms of the cylindrical current density distribution $j(r)$.

The function $I_1(x)$ will be used to calculate the full width of the current density distribution FW_f, which contains a certain fraction f of the total current

$$FW_f = 2 I_1^{-1}(f), \qquad (12.2.12)$$

in which I_1^{-1} denotes the inverse function of I_1. The case $f = 1/2$ corresponds to the full width median value FW_{50}, introduced in Section 5.9. We will use Eqs. (12.2.10) and (12.2.12) to calculate the width FW_f from the current density distribution $j(r)$.

12.3. Edge-width of a shaped spot

Fig. 12.4 shows the current density distribution $j(x, y)$ of a shaped spot. The profile can be represented as the convolution of an ideal rectangular distribution $A(x, y)$ with a point spread function $S(x, y)$, which is determined by the aberrations of the system,

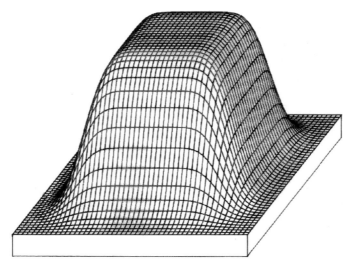

Fig. 12.4 Current density distribution of a shaped spot.

$$j(x,y) = \int_{-\infty}^{\infty} dx' \int_{-\infty}^{\infty} dy'\ A(x',y')S(x-x', y-y') \qquad (12.3.1)$$

where $A(x, y)$ is defined as

$$A(x,y) = \frac{1}{d_x d_y}\Theta(d_x/2-|x|)\Theta(d_y/2-|y|), \qquad (12.3.2)$$

in which d_x and d_y are the full width values of the ideal spot, in x- and y-direction respectively. $\Theta(x)$ is the step function defined by Eq. (3.7.8). Eq. (12.3.1) presupposes that the aberrations, represented by $S(x, y)$, are independent of the (lateral) location in the image plane. The distribution $A(x, y)$ is normalized to unity. The corresponding distribution $j(x, y)$ specified by Eq. (12.3.1) will be normalized, too, provided that $S(x, y)$ is normalized.

We will assume that the width of the point spread function $S(x, y)$ is small compared to both d_x and d_y. This implies that the rectangular shape of the ideal spot is practically maintained. The main effect of the aberrations is an increase of the edge width, which is small compared to the dimensions of the spot. Furthermore, we will assume that $S(x, y)$ is rotational symmetric. Under these circumstances, it is sufficient to study the one-dimensional equivalent of Eq. (12.3.1),

$$j(x) = \int_{-\infty}^{\infty} dy\, j(x,\ y) = \int_{-\infty}^{\infty} dx'\ A(x')S(x-x'), \qquad (12.3.3)$$

where $A(x)$ and $S(x)$ are defined as

$$A(x) = \int_{-\infty}^{\infty} dy\, A(x, y) = \frac{1}{d_x} \Theta(d_x/2 - |x|) \quad (12.3.4)$$

$$S(x) = \int_{-\infty}^{\infty} dy\, S(x, y). \quad (12.3.5)$$

The functions $j(x)$, $A(x)$ and $S(x)$ can be considered as the projection of their two-dimensional counterparts on the x-axis. Fig. 12.5 shows the general behavior of $j(x)$ and $A(x)$.

By substitution of Eqs. (12.3.4) and (12.3.5) into Eq. (12.3.3), one obtains

$$j(x) = \frac{1}{2d_x}\left[I(d_x/2 + x) + I(d_x/2 - x)\right], \quad (12.3.6)$$

in which $I(x)$ is defined as

$$I(x) = 2\int_0^x dx'\, S(x'). \quad (12.3.7)$$

Notice that Eq. (12.3.6) relies on the fact that $S(x)$ is a symmetric function, which means $S(-x) = S(x)$. This property follows from our assumption that $S(x, y)$ is rotational symmetric.

Let us consider the right edge of the profile $j(x)$ plotted in Fig. 12.5. This location corresponds to $x \approx d_x/2$. As the width of $S(x)$ is assumed to be

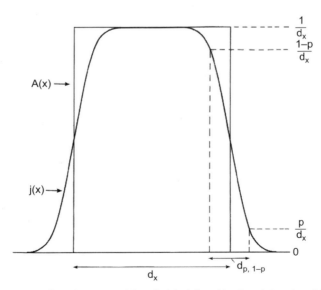

Fig. 12.5 Behavior of the functions $j(x)$ and $A(x)$, defined by Eqs. (12.3.3) and (12.3.4). The edge width $d_{p,1-p}$ is defined as the distance between the intensity levels (within the edge) that are equal to p and $1 - p$ times the intensity level at the center of the spot.

small compared to d_x, one may now approximate $I(d_x/2 + x) \approx I(d_x) \approx I(\infty) = 1$ and Eq. (12.3.6) transforms to

$$j(x) \approx \frac{1}{2d_x} [1 + I(d_x/2 - x)] \quad (x \approx d_x/2). \tag{12.3.8}$$

The $d_{p,1-p}$ edge width is defined as the distance between the points for which the current density $j(x)$ equals $p \times j(0)$ and $(1-p) \times j(0)$ respectively, where $j(0)$ is the current density at the center of the spot; see again Fig. 12.5. Utilizing Eq. (12.3.8), one can express $d_{p,1-p}$ as

$$d_{p,1-p} = j^{-1}[(1-p)/d_x] - j^{-1}[p/d_x] = 2I^{-1}(1-2p), \tag{12.3.9}$$

similar to Eq. (12.2.6). j^{-1} and I^{-1} denote the inverse functions of $j(x)$ and $I(x)$ respectively. Given the point spread function $S(x, y)$, the edge-width $d_{p,1-p}$ is fully specified by Eqs. (12.3.5), (12.3.7) and (12.3.9).

Eqs. (12.3.9) and (12.3.7) represent the same mathematical problem as Eqs. (12.2.6) and (12.2.5). Therefore, we conclude that the calculation of the spot width obtained by scanning the beam across a knife edge is mathematically equivalent to the calculation of the edge width of a shaped spot. The point spread function $S(x, y)$, used to describe the edge blurring of the shaped spot, plays the same role as the current density distribution $j(x, y)$, used for the Gaussian spot. Both quantities are determined by the brightness of the source, the aberrations of the optical components in the column, and the effect of Coulomb interactions between the beam particles.

From now on, we will employ the notation used in Eqs. (12.2.6) and (12.2.5). We emphasize that the resulting $d_{p,1-p}$ value represents the edge width of a shaped spot, as well as the spot width of a Gaussian spot determined by knife-edge scans. Eqs. (12.2.9)–(12.2.11) are exploited to perform the calculation in cylindrical coordinates. In the next sections, we will successively study the extreme cases where the distribution $j(r)$ is entirely determined by the trajectory displacement effect, chromatic aberration, spherical aberration and space-charge defocussing respectively. Thus, in each case, it is assumed that the ideal image is a point, while the blurring of this point is the result of the considered aberration only. Accordingly, one may regard $j(r)$ as the cylindrical point-spread function associated with that aberration.

In all cases, the calculation proceeds as follows. First, we will determine the distribution $j(r)$ corresponding to the considered aberration. Next, we will use $j(r)$ to evaluate the integrals $I_1(x)$ and $I(x)$ defined by Eqs. (12.2.9)–(12.2.11). Finally, the FW_f value will be determined from $I_1(x)$ by solving Eq. (12.2.12), while the $d_{p,1-p}$ value follows from $I(x)$ by solving Eq. (12.2.6).

It should be noticed that in practical systems the point-spread function $j(r)$ will not be determined by a single aberration but rather by the simultaneous action of a number of aberrations. The calculation of the corresponding $j(r)$ is more involved and will not be considered here. Some related problems in the image formation of Scanning Transmission Electron Microscopy (STEM) instruments are discussed by Mory (1985) and Mory et al. (1987). We note that, given the function $j(r)$ for certain combination of aberrations, one can again use the procedure outlined above to evaluate the corresponding FW_f and $d_{p,1-p}$ values.

12.4. Trajectory displacement effect

Let us assume that the current density distribution $j(r)$ of the spot is entirely determined by the trajectory displacement effect. Utilizing the representation of Eq. (5.7.10), one can now express $j(r)$ as

$$j(r) = \int_0^\infty dk\, kr J_0(kr) e^{-\lambda p(k)}, \tag{12.4.1}$$

where λ is the linear particle density defined by Eq. (3.2.4), and $p(k)$ is the function representing $j(r)$ in the k-domain. Following the analysis of Section 5.9, we approximate $\lambda p(k)$ as (reproducing Eq. (5.9.1))

$$\lambda p(k) = A_\gamma k^\gamma, \tag{12.4.2}$$

where γ is a numerical constant and A_γ is a physical quantity, which is determined by the experimental parameters. This representation permits a straightforward discussion of the different types of displacement distributions involved, as was shown in Section 5.9. We recall that, for the trajectory displacement effect, the parameter γ can take on the values $\gamma = 2$ (Gaussian distribution), $\gamma = 3/2$ (Holtsmark type of distribution), $\gamma = 1$ (Lorentzian distribution) and $\gamma = 1/3$ (two-dimensional pencil beam distribution).

Substitution of Eq. (12.4.1) into Eq. (12.2.10) and integration over r yields

$$I_1(x) = \int_0^\infty dk\, x J_1(kx) e^{-\lambda p(k)}, \tag{12.4.3}$$

using the identity

$$z J_0(z) = \frac{d}{dz}[z J_1(z)]. \tag{12.4.4}$$

In fact, Eq. (12.4.3) was utilized in Section 5.9 to calculate the Full Width median FW_{50} for the case of a two-dimensional distribution ($n=2$).

Substitution of Eq. (12.4.1) into Eq. (12.2.11) yields

$$I_2(x) = \int_0^\infty dk\, x A_2(kx) e^{-\lambda p(k)}, \tag{12.4.5}$$

in which the function $A_2(z)$ is defined as

$$A_2(z) = \frac{2}{\pi} \int_1^\infty dt\, zt J_0(zt)\, \arcsin(1/t).$$

Partial integration, employing Eq. (12.4.4), gives

$$A_2(z) = -J_1(z) + \frac{2}{\pi} \frac{\sin(z)}{z} \tag{12.4.6}$$

in which we utilized the integral

$$\int_1^\infty dt\, \frac{J_1(zt)}{(t^2-1)^{1/2}} = \frac{\sin(z)}{z}. \tag{12.4.7}$$

By combining Eqs. (12.4.3), (12.4.5) and (12.4.6), one finds for the function $I(x)$, defined by Eq. (12.2.9),

$$I(x) = \frac{2}{\pi} \int_0^\infty dk\, \frac{\sin(kx)}{k} e^{-\lambda p(k)}. \tag{12.4.8}$$

The function $I(x)$ will be used in the calculation of the $d_{p,1-p}$ width from Eq. (12.2.6).

We note that the derivation of Eq. (12.4.8) can be performed in a more direct way, by using rectangular coordinates. According to Eq. (5.7.6), one may express the one dimensional distribution $j(x)$ as

$$j(x) = \frac{1}{\pi} \int_0^\infty dk\, \cos(kx) e^{-\lambda p(k)}, \tag{12.4.9}$$

where $p(k)$ is the same function as appearing in Eq. (12.4.1). Substitution into Eq. (12.2.5) and integration over x directly yields Eq. (12.4.8).

We will now replace the function $\lambda p(k)$ by the expression given by Eq. (12.4.2). In order to bring the resulting expression(s) in a form that is independent of the experimental parameters (represented by the quantity A_γ), we will employ the scaled quantities

$$\bar{k} = A_\gamma^{1/\gamma} k, \qquad \bar{x} = A_\gamma^{-1/\gamma} x \tag{12.4.10}$$

first introduced by Eq. (5.9.13). Eqs. (12.4.3) and (12.4.8) now transform to

$$I_1(\bar{x}) = \int_0^\infty d\bar{k}\, \bar{x} J_1(\bar{k}\bar{x}) e^{-\bar{k}^\gamma} \tag{12.4.11}$$

$$I(\bar{x}) = \frac{2}{\pi} \int_0^\infty d\bar{k}\, \frac{\sin(\bar{k}\bar{x})}{\bar{k}} e^{-\bar{k}^\gamma}. \tag{12.4.12}$$

We evaluated Eqs. (12.4.11) and (12.4.12) numerically, as function of $\bar{x}$ for γ is 2, 3/2, 1, 1/2 and 1/3. The results are plotted in Figs. 12.6 and 12.7.

Explicit analytical expressions for $I_1(\bar{x})$ and $I(\bar{x})$ can be obtained in the cases $\gamma = 2$ (Gaussian trajectory displacement distribution) and $\gamma = 1$

Calculation of different spot- and edge-width measures

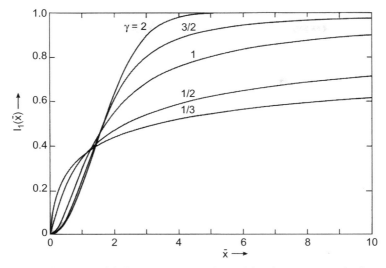

Fig. 12.6 The function $I_1(\tilde{x})$ for a spot that is limited by the trajectory displacement effect; see Eq. (12.4.11).

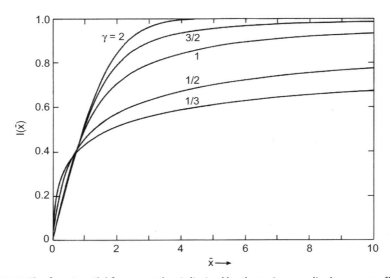

Fig. 12.7 The function $I(\tilde{x})$ for a spot that is limited by the trajectory displacement effect; see Eq. (12.4.12).

(Lorentzian trajectory displacement distribution). For $\gamma=2$, one obtains from Eqs. (12.4.11) and (12.4.12)

$$I_1(\bar{x}) = 1 - e^{-(\bar{x}/2)^2} \quad (\gamma = 2) \tag{12.4.13}$$

$$I(\bar{x}) = \mathrm{erf}(\bar{x}/2) \quad (\gamma = 2), \tag{12.4.14}$$

where $\mathrm{erf}(x)$ is the Fresnel error function, defined by the first equation of (4.5.9). Notice that Eqs. (12.4.13) and (12.4.14) can also be derived from Eqs. (5.9.3) (taking $n=2$ and $\Delta\eta=r$), (12.2.9)–(12.2.11), using the scaling of Eq. (12.4.10). For $\gamma=1$, one obtains from Eqs. (12.4.11) and (12.4.12)

$$I_1(\bar{x}) = 1 - \frac{1}{(1+\bar{x}^2)^{1/2}} \quad (\gamma = 1) \tag{12.4.15}$$

$$I(\bar{x}) = \frac{2}{\pi}\arctan(\bar{x}) \quad (\gamma = 1), \tag{12.4.16}$$

which can also be derived from Eqs. (5.9.6) (taking $\Delta\eta=r$), (12.2.9)–(12.2.11), using the scaling of Eq. (12.4.10).

The next step, is the calculation of the $d_{p,1-p}$ width from Eq. (12.2.6) and the FW_f value from Eq. (12.2.12). For $\gamma=2$, one finds with Eqs. (12.4.13) and (12.4.14)

$$FW_f = 4\sqrt{\ln\left(\frac{1}{1-f}\right)} \quad (\gamma = 2) \tag{12.4.17}$$

$$\bar{d}_{p,1-p} = 4\,\mathrm{erf}^{-1}(1-2p) \quad (\gamma = 2), \tag{12.4.18}$$

in which erf^{-1} denotes the inverse of the error function. For $\gamma=1$, one finds with Eqs. (12.4.15) and (12.4.16)

$$FW_f = 2\sqrt{\left(\frac{1}{1-f}\right)^2 - 1} \quad (\gamma = 1) \tag{12.4.19}$$

$$\bar{d}_{p,1-p} = 2\tan[\pi(1-2p)/2] \quad (\gamma = 1). \tag{12.4.20}$$

In fact, Eqs. (12.4.17) and (12.4.19) were exploited in Eqs. (5.9.4) and (5.9.8) to calculate the FW_{50} value (which corresponds to $f=1/2$) of the two-dimensional ($n=2$) Gaussian and Lorentzian distribution respectively.

In order to determine the FW_f and $d_{p,1-p}$ values for arbitrary γ-values, we followed a numerical approach. Eq. (12.2.12) was solved for γ values in the range $0.2<\gamma<2$, using the expression for $I_1(x)$ given by Eq. (12.4.11). For the parameter f, we considered the values 0.1, 0.2, ..., 0.9. The results arc plotted in Fig. 12.8. Eq. (12.2.6) was solved for the same range of γ-values, using the expression for $I(x)$ given by Eq. (12.4.12). For the parameters p, we considered the values 0.05, 0.10, ..., 0.45. The results are presented in Fig. 12.9.

Calculation of different spot- and edge-width measures

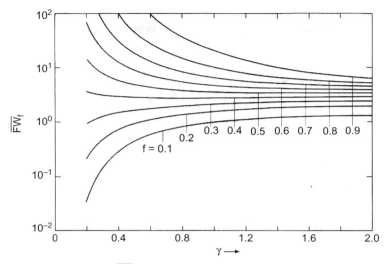

Fig. 12.8 The scaled width $\overline{FW}_f$ for a spot that is limited by the trajectory displacement effect as a function of the type of trajectory displacement distribution represented by the parameter γ. See Eqs. (12.2.12), (12.4.10) and (12.4.11).

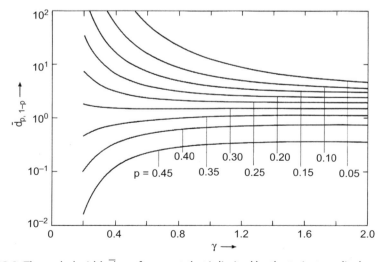

Fig. 12.9 The scaled width $\overline{d}_{p,1-p}$ for a spot that is limited by the trajectory displacement effect as a function of the type of trajectory displacement distribution represented by the parameter γ. See Eqs. (12.2.6), (12.4.10) and (12.4.12).

We note that the Full Width at Half Maximum (*FWHM*) of the distribution $j(r)$, determined by the trajectory displacement effect, can directly be obtained from the analysis presented in Section 5.9. The quantity $FWHM_y$, which is the *FWHM* of the scaled distribution given by Eq. (5.9.14), is plotted in Fig. 5.5A. Table 5.1 summarizes the numbers for some specific cases. The scaling, which is defined by Eq. (5.9.13), is the same as used in this section; see Eq. (12.4.10). This allows a direct comparison of Figs. 12.8, 12.9 and 5.5A (curve corresponding to a two-dimensional distribution). Fig. 5.5B gives the corresponding values of the scaled Full Width median value FW_{50y}. The curve in this figure given for a two-dimensional distribution ($n=2$) corresponds to the curve presented in Fig. 12.8 for $f=0.5$.

12.5. Chromatic aberration

In this section, we will study the properties of the probe in a chromatically limited system. Let us assume that the energy distribution is entirely the result of statistical particle-particle interactions (Boersch effect). According to Eq. (5.7.6), the energy distribution can then be expressed as

$$\rho(\Delta E) = \frac{1}{\pi} \int_0^\infty dk \cos(k\Delta E) e^{-\lambda p(k)}, \quad (12.5.1)$$

which is a symmetric function in ΔE. The function $p(k)$ represents the energy distribution in the k-domain.

Before embarking on the actual calculations, we will briefly consider some related subjects. In order to determine the fraction of the particles contained within a certain energy range, one has to evaluate the following integral,

$$E_1(x) = 2 \int_0^x d\Delta E \rho(\Delta E), \quad (12.5.2)$$

similar to Eq. (12.2.10). The Full Width energy spread $FW_{E,f}$, which contains a fraction f of the particles, follows by solving the equation

$$FW_{E,f} = 2E_1^{-1}(f), \quad (12.5.3)$$

similar to Eq. (12.2.12). E_1^{-1} denotes the inverse function of E_1. By substitution of Eq. (12.5.1) into Eq. (12.5.2) and integration over ΔE, one obtains

$$E_1(x) = \frac{2}{\pi} \int_0^\infty dk \frac{\sin(kx)}{k} e^{-\lambda p(k)}. \quad (12.5.4)$$

In fact, this equation was used in Section 5.9 to calculate the Full Width median FW_{50} for the general case of a one-dimensional displacement distribution ($n=1$). By comparison with Eq. (12.4.8), one sees that the function $E_1(x)$ is identical to the function $I(x)$ obtained for the trajectory displacement effect. Accordingly, the solution of Eq. (12.5.3) can be obtained from the results plotted in Fig. 12.9, using that $FW_{E,f}$ is here equivalent to $d_{p,1-p}$, where $f = 1 - 2p$.

The relation between the energy deviation ΔE of a particle and its corresponding deviation Δv_z, from the average axial velocity v_z, is given by

$$\Delta E \cong mv_z \Delta v_z, \quad (12.5.5)$$

provided that $v_z \gg \Delta v_z$. The energy distribution $\rho(\Delta E)$ and the velocity distribution $\rho_v(\Delta v_z)$ are, therefore, related as

$$\rho(\Delta E) = \frac{1}{mv_z} \rho_v(\Delta E / mv_z). \quad (12.5.6)$$

The corresponding functions in the k-domain are related by

$$p(k) = p_v(kmv_z), \quad (12.5.7)$$

as can be seen from the structure of Eq. (12.5.1). Utilizing the representation of Eq. (12.4.2), one finds

$$A_\gamma = A_{\gamma,v}(mv_z)^\gamma, \quad (12.5.8)$$

where A_γ represents the energy distribution and $A_{\gamma,v}$ the distribution of axial velocities. In general, the reader should be alert that A_γ may represent different physical quantities.

Let us now return to the main subject of this section and investigate the distribution $j(r)$ of a probe that is dominated by chromatic aberration. Consider the single thin-lens system of Fig. 12.10. The distance between the object plane and the lens is p; the distance between the lens and the image plane is q. A particle leaving the object plane from the axis with a certain angle α_1 and an energy equal to the average energy E will cross the axis in the image plane with an angle $\alpha = \alpha_1 p/q$. Another particle that starts out along the same trajectory but with an energy smaller than the average energy E will be deflected stronger and crosses the axis before reaching the image plane, as is shown by the dashed line in Fig. 12.10. In general, one finds that the radial position r of a particle in the vicinity of the image plane, which left the object from the axis, depends on its energy deviation ΔE, its angle with the axis α and the distance to the image plane Δz, as

$$r = \left| \Delta z \alpha - C_c \frac{\Delta E}{E}\left(1 + \frac{\Delta z}{q}\right)\alpha \right|, \quad (12.5.9)$$

where C_c is the constant of chromatic aberration of the lens. The radial position in the (Gaussian) image plane follows by taking $\Delta z = 0$,

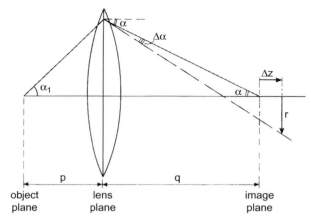

Fig. 12.10 Schematic view of a single lens system. The image is affected by the aberrations of the lens. An angular shift $\Delta\alpha$ in the lens plane translates into a spatial shift r near the image plane.

$$r = C_c \frac{|\Delta E|}{E} \alpha. \tag{12.5.10}$$

From this equation, one sees that the distribution of r in the Gaussian image plane, denoted as $j(r)$, depends on the energy distribution $\rho(\Delta E)$ and the angular distribution $f_\alpha(\alpha)$ near the image plane. Eq. (12.5.9) shows that the distribution $j(r)$ also depends on the axial position, represented by Δz. However, we will restrict the analysis to the case $\Delta z = 0$, that is, to the calculation of $j(r)$ in the Gaussian image plane.

Assuming that the energy distribution $\rho(\Delta E)$ and the angular distribution $f_z(\alpha)$ are independent, one can express $j(r)$ as

$$j(r) = \int d\Delta E \rho(\Delta E) \int d\alpha f_\alpha(\alpha) \delta\left(r - C_c \frac{|\Delta E|}{E} \alpha\right), \tag{12.5.11}$$

where $\delta(x)$ is the delta-Dirac function. For the angular distribution $f_\alpha(\alpha)$, we write

$$f_\alpha(\alpha) = \frac{2\alpha}{\alpha_0^2} F_\alpha(\alpha/\alpha_0) \quad (\alpha > 0), \tag{12.5.12}$$

in which the function F_α is for a uniform angular distribution given by

$$F_\alpha(x) = \Theta(1 - x), \tag{12.5.13}$$

where $\Theta(z)$ denotes the step function. In case of a Gaussian angular distribution, we use

$$F_\alpha(x) = \exp(-x^2). \tag{12.5.14}$$

Finally, a truncated Gaussian distribution is represented as

$$F_a(x) = \frac{\exp(-x^2)\,\Theta(1 - x\alpha_0/\alpha_m)}{1 - \exp\left[-(\alpha_m/\alpha_0)^2\right]}. \quad (12.5.15)$$

This type of distribution corresponds to a Gaussian illumination of the image, which is limited by an aperture. The angle α_m denotes the maximum angle of the rays passing through the aperture.

By substituting Eqs. (12.5.1) and (12.5.12) into Eq. (12.5.11) and eliminating the δ-function by means of its definition, one obtains

$$j(r) = \frac{4E^2 r}{\pi C_c^2 \alpha_0^2} \int_0^\infty dk\, e^{-\lambda p(k)} \int_0^\infty d\Delta E \, \frac{\cos(k\Delta E)}{\Delta E^2} F_a\left(\frac{Er}{C_c \alpha_0 \Delta E}\right). \quad (12.5.16)$$

We will exploit the representation of $p(k)$ given by Eq. (12.4.2) to rewrite Eq. (12.5.16). In addition, we introduce the following scaling

$$\bar{r} = \frac{E}{C_c \alpha_0 A_\gamma^{1/\gamma}} r, \qquad \bar{j}(\bar{r}) = \frac{C_c \alpha_0 A_\gamma^{1/\gamma}}{E} j(r)$$
$$\bar{k} = A_\gamma^{1/\gamma} k, \qquad \Delta \bar{E} = A_\gamma^{-1/\gamma} \Delta E. \quad (12.5.17)$$

Eq. (12.5.16) now transforms to

$$\bar{j}(\bar{r}) = \frac{4\bar{r}}{\pi} \int_0^\infty d\bar{k}\, e^{-\bar{k}^\gamma} \int_0^\infty d\Delta \bar{E} \, \frac{\cos(\bar{k}\Delta \bar{E})}{\Delta \bar{E}^2} F_a(\bar{r}/\Delta \bar{E}). \quad (12.5.18)$$

This equation can be rewritten as

$$\bar{j}(\bar{r}) = \frac{4}{\pi} \int_0^\infty d\bar{k}\, C(\bar{k}\bar{r}) e^{-\bar{k}^\gamma}, \quad (12.5.19)$$

where the function C depends on the type of angular distribution. For a uniform angular distribution ($C = C_u$), one finds with Eq. (12.5.13)

$$C_u(z) = \int_1^\infty dx\, \frac{\cos(xz)}{x^2} \quad \text{(uniform } f_\alpha(\alpha)\text{)}. \quad (12.5.20)$$

For a Gaussian angular distribution ($C = C_g$), one finds with Eq. (12.5.14)

$$C_g(z) = \int_0^\infty dx\, \frac{\cos(xz)}{x^2} e^{-1/x^2} \quad \text{(Gaussian } f_\alpha(\alpha)\text{)}. \quad (12.5.21)$$

The case of a truncated Gaussian distribution will be considered separately in Section 12.8.

The functions $C_u(z)$ and $C_g(z)$ were evaluated numerically and are plotted in Fig. 12.11. For further numerical calculations, we use the following approximations

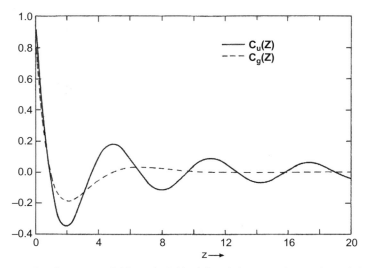

Fig. 12.11 The functions $C_u(z)$ and $C_g(z)$ defined by Eqs. (12.5.20) and (12.5.21) respectively.

$$C_{u,a}(z) = \frac{1 - 0.1751z + 2.121z \cos(z)}{1 + 3.584z + 1.116z^2 + z^3} - \frac{z \sin(z)}{0.2976 + 0.9307z + z^2}$$

$$C_{g,a}(z) = \frac{\sqrt{\pi}/2 + 1.842z - 0.5902z^2 + 6.362z^3 \sin(3.932\sqrt{z} + 5.326)e^{-z/8}}{1 + 3.877z + 4.831z^2 + 9.105z^3 + 3.155z^4 + z^5}.$$

(12.5.22)

The form of Eq. (12.5.22) was chosen such that the behavior for $z \to 0$ and $z \to \infty$ agrees with the behavior of the exact functions, given by Eq. (12.5.20) and (12.5.21). This also explains some of the numerical constants in Eq. (12.5.22). The remaining constants were determined by fitting the functions to the numerical data. The accuracy of these approximations is, in general, better than 0.5%.

Eq. (12.5.19) was evaluated numerically, using the approximations of Eq. (12.5.22). The relevant values of γ are 2 (Gaussian energy distribution), 3/2 (Holtsmark type of energy distribution), 1 (Lorentzian energy distribution) and 1/2 (energy distribution generated in a pencil beam). The corresponding distributions found for a uniform angular distribution are plotted in Fig. 12.12a, while the results for a Gaussian angular distribution are plotted in Fig. 12.12b.

For the cases $\gamma = 2$ and $\gamma = 1$, one can evaluate $\bar{j}(\bar{r})$ analytically without introducing any approximations. From Eqs. (12.5.19) and (12.5.20), one finds for the case of a uniform angular distribution

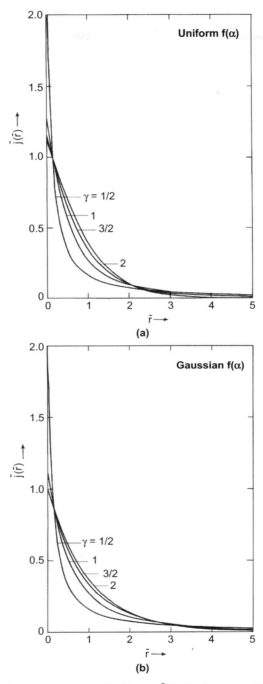

Fig. 12.12 The scaled current density distribution $\bar{j}(\bar{r})$ of a chromatically limited spot for different types of energy distributions represented by the parameter γ. Panels a and b pertain to a uniform and a Gaussian angular distribution respectively and are evaluated from Eqs. (12.5.19) and (12.5.22). Notice that the distributions are normalized in conformity with Eq. (12.2.8).

$$\bar{j}(\bar{r}) = \frac{2}{\sqrt{\pi}} e^{-(\bar{r}/2)^2} + \bar{r}[\text{erf}(\bar{r}/2) - 1] \quad (\gamma = 2, \text{ uniform } f_a(\alpha))$$

(12.5.23)

$$\bar{j}(\bar{r}) = \frac{4}{\pi}\left[1 + \bar{r}\left(\arctan(\bar{r}) - \frac{\pi}{2}\right)\right] \quad (\gamma = 1, \text{ uniform } f_a(\alpha)).$$

From Eqs. (12.5.19) and (12.5.21), one finds for the case of a Gaussian angular distribution

$$\bar{j}(\bar{r}) = e^{-\bar{r}} \qquad (\gamma = 2, \text{ Gaussian } f_a(\alpha))$$

$$\bar{j}(\bar{r}) = \frac{2}{\sqrt{\pi}} + 2\bar{r}[\text{erf}(\bar{r}) - 1]e^{\bar{r}^2} \quad (\gamma = 1, \text{ Gaussian } f_a(\alpha)).$$

(12.5.24)

The analytical results of Eqs. (12.5.23) and (12.5.24) were exploited to check the accuracy of the numerical calculations based on the approximations of Eq. (12.5.22). The numerical accuracy of the results plotted in Fig. 12.12a and b was found to be better than 1%.

The next step is the evaluation of the functions $I_1(x)$ and $I(x)$ from Eqs. (12.2.9)–(12.2.11). We will scale the quantity x in the same way as r and the quantities I and I_1 in the same way as j, see Eq. (12.5.17). From Eqs. (12.2.10) and (12.5.19), one obtains

$$\bar{I}_1(\bar{x}) = \int_0^{\bar{x}} d\bar{r} \bar{j}(\bar{r}) = \int_0^{\infty} d\bar{k} \frac{B_1(\bar{k}\bar{x})}{\bar{k}} e^{-\bar{k}^{\gamma}},$$

(12.5.25)

in which the function $B_1(z)$ is defined as

$$B_1(z) = \frac{4}{\pi} \int_0^z dt\, C(t).$$

(12.5.26)

Similarly, one finds from Eqs. (12.2.9), (12.2.10), (12.2.11) and (12.5.19)

$$\bar{I}(\bar{x}) = \bar{I}_1(\bar{x}) + \frac{2}{\pi} \int_{\bar{x}}^{\infty} d\bar{r} \arcsin(\bar{x}/\bar{r})\bar{j}(\bar{r}) = \int_0^{\infty} d\bar{k} \frac{B(\bar{k}\bar{x})}{\bar{k}} e^{-\bar{k}^{\gamma}},$$

(12.5.27)

in which the function $B(z)$ is defined as

$$B(z) = B_1(z) + \frac{8}{\pi^2} \int_z^{\infty} dt\, \arcsin(z/t) C(t).$$

(12.5.28)

By substitution of Eq. (12.5.20) into Eqs. (12.5.26) and (12.5.28), one finds, after some straightforward transformations, for a uniform angular distribution

$$B_{1u}(z) = \frac{2}{\pi}[\sin z + zC_u(z)]$$

$$B_u(z) = \frac{8}{\pi^2} \int_1^{\infty} dt\, \frac{\sin(tz)}{t^3} \sqrt{t^2 - 1} \qquad (\text{uniform } f_a(\alpha)),$$

(12.5.29)

where the function $C_u(z)$ is given by Eq. (12.5.20). Similarly, one obtains with Eq. (12.5.21) for a Gaussian angular distribution

$$B_{1g}(z) = \frac{2z}{\pi} \int_0^\infty dt \, \cos(tz)\left(1 - e^{-1/t^2}\right)$$

$$B_g(z) = \frac{4}{\pi^{3/2}} \int_0^\infty dt \, \frac{\sin(tz)}{t^2} e^{-1/t^2} \qquad \text{(Gaussian } f_\alpha(\alpha)\text{),} \qquad (12.5.30)$$

The functions $B_u(z)$, $B_{1g}(z)$ and $B_g(z)$ were evaluated numerically and are plotted in Fig. 12.13. For further numerical calculations, we use the following approximations

$$B_{1g,a}(z) = \frac{2z/\sqrt{\pi} + 1.120z + 21.09z^3 \sin\left(3.51\sqrt{z+2.91} + 0.751\right)e^{-z/8}}{1 + 2.298z + 6.979z^2 + 28.47z^3 + 8.066z^4 + z^5}$$

$$B_{u,a}(z) = \frac{1.80z^{5/6} - 1.395z + 1.617z\cos(z - 0.582)}{1 + 3.709z + 1.098z^2 + z^3}$$

$$+ \frac{z^{3/2}\cos(z + 0.857)}{3.067 + 9.538z + 2.312z^2 + z^3}$$

(12.5.31)

$$B_{g,a}(z) = \frac{1.62z^{5/6} - 0.1801z - .04237z^2 + 4.637z^3 \sin\left(3.70\sqrt{z+3.763} + 0.0869\right)e^{-z/8}}{1 + 2.345z + 2.722z^2 + 7.616z^3 + 4.250z^4 + z^5}.$$

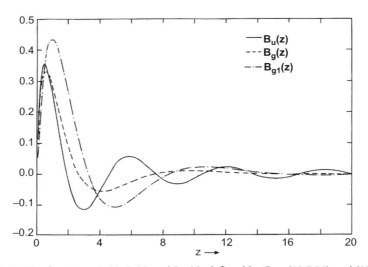

Fig. 12.13 The functions $B_u(z)$, $B_g(z)$ and $B_{1g}(z)$, defined by Eqs. (12.5.29) and (12.5.30).

The form of Eq. (12.5.31) was chosen such that the behavior for $z \to 0$ and $z \to \infty$ agrees with the behavior of the exact functions, given by Eqs. (12.5.29) and (12.5.30). This also explains some of the numerical constants in Eq. (12.5.31). The remaining constants were determined by fitting the functions to the numerical data. The accuracy of these approximations is, in general, better than 0.5%. The first equation of (12.5.29) shows that the function $B_{1u}(z)$ is directly related to the function $C_u(z)$, for which one may use the approximation given by the first equation of (12.5.22).

Eq. (12.5.25) and the first equation of (12.5.29), which apply to the calculation of $\bar{I}_1(\bar{x})$ for a uniform angular distribution, were evaluated numerically, using the approximation of $C_{u,a}(z)$ given by the first equation of (12.5.22). The results are plotted in Fig. 12.14a. Similar results were obtained for a Gaussian angular distribution, using Eq. (12.5.25) and the first equation of (12.5.31). These results are plotted in Fig. 12.14b. The function $\bar{I}(\bar{x})$ was evaluated from Eq. (12.5.27), using the approximations given by the second and third equation of (12.5.31), for a uniform and a Gaussian angular distribution respectively. The results are plotted in Fig. 12.15a and b.

The cases $\gamma = 2$ (Gaussian energy distribution) and $\gamma = 1$ (Lorentzian energy distribution) permit a more detailed analytical treatment. For $\gamma = 2$ and a uniform angular distribution, one finds

$$\bar{I}_1(\bar{x}) = \frac{\bar{x}}{\sqrt{\pi}} e^{-(\bar{x}/2)^2} + \mathrm{erf}(\bar{x}/2) + \frac{1}{2}\bar{x}^2 [\mathrm{erf}(\bar{x}/2) - 1]$$

$$(\gamma = 2, \text{ uniform } f_a(\alpha))$$

$$\bar{I}(\bar{x}) = \frac{4}{\pi} \int_1^\infty dt \frac{\sqrt{t^2-1}}{t^3} \mathrm{erf}\left(\frac{t\bar{x}}{2}\right) \qquad (12.5.32)$$

$$= 1 - \frac{2\bar{x}}{\pi^{3/2}} \int_1^\infty dt \left[\arccos(1/t) - \frac{\sqrt{t^2-1}}{t^2}\right] e^{-(t\bar{x}/2)^2}.$$

For $\gamma = 1$ and a uniform angular distribution, one finds

$$\bar{I}_1(\bar{x}) = \frac{2}{\pi} \left[\bar{x} + (\bar{x}^2 + 1) \arctan(\bar{x})\right] - \bar{x}^2$$

$$(\gamma = 1, \text{ uniform } f_a(\alpha)) \quad (12.5.33)$$

$$\bar{I}(\bar{x}) = \frac{8}{\pi^2} \int_1^\infty dt \frac{\sqrt{t^2-1}}{t^3} \arctan(t\bar{x}).$$

For $\gamma = 2$ and a Gaussian angular distribution, one finds

$$\bar{I}_1(\bar{x}) = 1 - e^{-\bar{x}}$$

$$(\gamma = 2, \text{ Gaussian } f_a(\alpha)) \quad (12.5.34)$$

$$\bar{I}(\bar{x}) = \frac{2}{\sqrt{\pi}} \int_0^\infty dt\, \mathrm{erf}(\bar{x}/2t) e^{-t^2} = 1 - \frac{2\bar{x}}{\pi} \int_1^\infty dt\, \arccos(1/t) e^{-t\bar{x}}.$$

Calculation of different spot- and edge-width measures 379

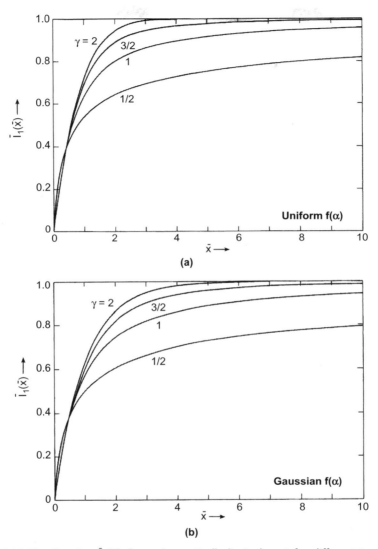

Fig. 12.14 The function $\bar{I}_1(\bar{x})$ for a chromatically limited spot for different types of energy distributions represented by the parameter γ. Panels a and b pertain to a uniform and a Gaussian angular distribution respectively. They are evaluated from Eq. (12.5.25), using the approximations for the function $B_1(z)$ given by Eqs. (12.5.29) and (12.5.22) for panel a and the first equation of (12.5.31) for panel b.

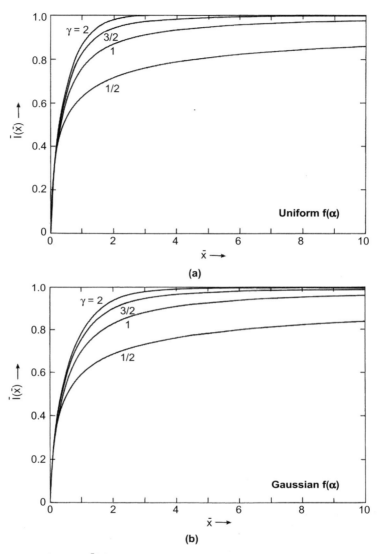

Fig. 12.15 The function $\bar{I}(\bar{x})$ for a chromatically limited spot, for different types of energy distributions, represented by the parameter γ. Panels a and b pertain to a uniform and a Gaussian angular distribution respectively. They are evaluated from Eq. (12.5.27), using the approximations for the function $B(z)$ given by the second and the third equation of (12.5.31) respectively.

Anally, for $\gamma=1$ and a Gaussian angular distribution, one finds

$$\bar{I}_1(\bar{x}) = 1 + [\text{erf}(\bar{x}) - 1]e^{\bar{x}^2}$$

$$\bar{I}(\bar{x}) = \frac{4}{\pi^{3/2}} \int_0^\infty dt \, \arctan(\bar{x}/t) e^{-t^2}. \qquad (\gamma = 1, \text{Gaussian } f_\alpha(\alpha)) \quad (12.5.35)$$

The results of Eqs. (12.5.32)–(12.5.35) were exploited to check the accuracy of the numerical calculations based on the approximations given by Eq. (12.5.31). The accuracy of the results plotted in Figs. 12.14B, 12.15A and B was found to be better than 1%.

The last step is the calculation of the $d_{p,1-p}$ width from Eq. (12.2.6) and the FW_f value from Eq. (12.2.12). The only case for which we found an explicit analytical expression is the calculation of the FW_f value of the current density distribution corresponding to a Gaussian energy distribution ($\gamma=2$) and a Gaussian angular distribution. From Eq. (12.2.12) and the first equation of (12.5.34), it follows for this particular case

$$\overline{FW}_f = 2\ln\left(\frac{1}{1-f}\right) \quad (\gamma = 2, \text{Gaussian } f_\alpha(\alpha)). \quad (12.5.36)$$

We determined the FW_f and $d_{p,1-p}$ values for arbitrary γ-values, using a numerical approach. Eq. (12.2.12) was solved for γ values in the range $0.2<\gamma<2$, using the expression for $I_1(x)$ given by Eq. (12.5.25). For the function $B_{1u}(z)$, we used the approximation given by Eqs. (12.5.29) and (12.5.22). For the function $B_{1g}(z)$, we exploited the approximation given by the first equation of (12.5.31). For the parameter f, we considered the values 0.1, 0.2, …, 0.9. The results are plotted in Fig. 12.16a and b for a uniform and a Gaussian angular distribution respectively. Eq. (12.2.6) was solved for the same range of γ-values using the expression for $I(x)$ given by Eq. (12.5.27) and the approximations of Eq. (12.5.31). For the parameters p, we considered the values 0.05, 0.10, …, 0.45. The results are presented in Fig. 12.17a and b for a uniform and a Gaussian distribution respectively.

In order to relate the $d_{p,1-p}$ and FW_f values presented in this section to the Full Width at Half Maximum of the energy distribution ($FWHM_E$), one has to remove the scaling defined by Eq. (12.5.17). In addition, one should express the quantity A_γ in terms of $FWHM_E$. According to Eq. (5.9.15), one may write

$$FWHM_E = A_\gamma^{1/\gamma} FWHM_\gamma. \quad (12.5.37)$$

The quantity $FWHM_\gamma$ is plotted in Fig. 5.5A (curve corresponding to the one-dimensional distribution). Table 5.1 summarizes the numbers for some specific cases. From Eqs. (12.5.17) and (12.5.37), one finds

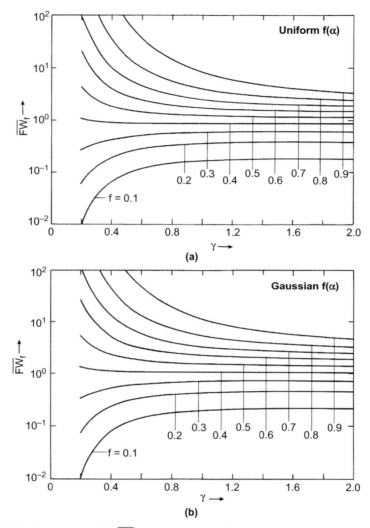

Fig. 12.16 The scaled width $\overline{FW}_f$ for a chromatically limited spot as a function of the type of energy distribution, represented by the parameter γ. Panels a and b pertain to a uniform and a Gaussian angular distribution respectively. See Eqs. (12.2.12) and (12.5.17) and the caption of Fig. 12.14.

Calculation of different spot- and edge-width measures

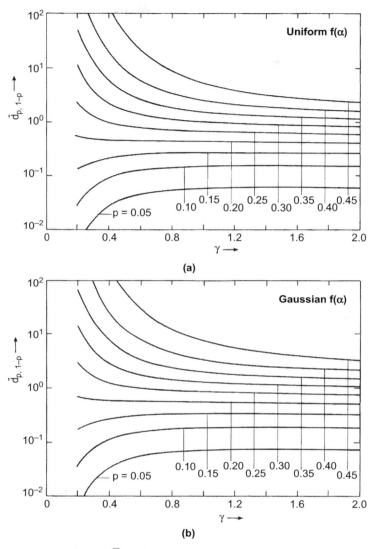

Fig. 12.17 The scaled width $\bar{d}_{p,1-p}$ for a spot that is limited by chromatic aberration as function of the type of energy distribution, represented by the parameter γ. Panels a and b pertain to a uniform and a Gaussian angular distribution respectively. See Eqs. (12.2.6) and (12.5.17) and the caption of Fig. 12.15.

$$FW_f = \frac{\overline{FW_f}}{FWHM_\gamma} C_c \alpha_0 \frac{FWHM_E}{E}$$

$$\overline{d}_{p,1-p} = \frac{\overline{d}_{p,1-p}}{FWHM_\gamma} C_c \alpha_0 \frac{FWHM_E}{E}. \tag{12.5.38}$$

The numerical factors $\overline{FW_f}/FWHM_\gamma$ and $\overline{d}_{p,1-p}/FWFIM_\gamma$ can be determined from Figs. 12.16, 12.17 and 5.5A (or Table 5.1). For example, in case of a Gaussian energy distribution ($\gamma = 2$) and $p = 0.1$ and $f = 0.5$, one finds for a uniform and a Gaussian angular distribution respectively

$$\overline{FW}_{50}/FWHM_\gamma = 0.344, \quad \overline{d}_{10,90}/FWHM_\gamma = 0.477 \quad (\gamma = 2, \text{uniform } f_\alpha(\alpha))$$

$$\overline{FW}_{50}/FWHM_\gamma = 0.419, \quad \overline{d}_{10,90}/FWHM_\gamma = 0.635 \quad (\gamma = 2, \text{Gaussian } f_\alpha(\alpha)).$$

Finally, we note that one cannot define a *FWHM* of the current density distribution corresponding to a chromatically limited spot, since the quantity $j(r)/2\pi r$ diverges for $r \to 0$, This can, for instance, be seen from Eqs. (12.5.23) and (12.5.24). We recall that $j(r)$ represents a point spread function. For a practical spot, the ideal image is not a point and the resulting distribution, which is a convolution of the ideal image and the point spread function, will be finite for $r \to 0$.

12.6. Spherical aberration

In this section we will study the properties of a probe that is limited by spherical aberration. Consider again the single thin-lens system of Fig. 12.10. Assume that all particles have the same energy, which implies that chromatic aberration is absent. For an ideal lens, a particle leaving the object plane from the axis, with a certain angle α_1, will cross the axis in the image plane, irrespective of the value of α_1. However, in presence of spherical aberration, the outer rays (with a large angle α_1) are focussed more strongly than the rays near the axis. Therefore, the outer rays will cross the axis at a location closer to the lens. In general, one finds that the radial position r of a particle in the vicinity of the image plane that left the object from the axis, depends on its angle with the axis α. and the distance to the image plane Δz, as

$$r = \left| \Delta z \alpha + C_s \left(1 + \frac{\Delta z}{q}\right) \alpha^3 \right|, \tag{12.6.1}$$

where C_s is the constant of spherical aberration of the lens (which is assumed to be positive). The radial position in the (Gaussian) image plane follows by taking $\Delta z = 0$

$$r = C_s \alpha^3. \tag{12.6.2}$$

From this equation, one sees that the distribution of r in the Gaussian image plane, denoted as $j(r)$, depends on the angular distribution $f_a(\alpha)$ near the image plane. Eq. (12.6.1) shows that the distribution $j(r)$ also depends on the axial position represented by Δz. However, we will restrict the analysis to the case $\Delta z = 0$, that is, to the calculation of $j(r)$ in the Gaussian image plane. It is well known that, when the probe is dominated by spherical aberration, the plane of best focus is located at a distance $\Delta z = -0.75 C_s \alpha_0^2$ ahead of the Gaussian image plane, where α_0 represents the beam semi-angle subtended at the image. The radius of the spot in this plane is equal to $0.25 C_s \alpha_0^3$, which is four times smaller than the radius in the Gaussian image plane. It should be noted that the shape of the current density distribution $j(r)$ is entirely different for both planes. When spherical aberration is not dominant, it is more appropriate to evaluate the contribution to the probe size in the Gaussian image plane. We will concentrate on this case and start from Eq. (12.6.2).

Similar to Eq. (12.5.11), one can express the distribution $j(r)$ associated with spherical aberration as

$$j(r) = \int d\alpha f_a(\alpha) \delta(r - C_s \alpha^3). \quad (12.6.3)$$

The angular distribution $f_a(\alpha)$ is again given by Eq. (12.5.12), in combination with either Eq. (12.5.13), for a uniform distribution, or Eq. (12.5.14), for a Gaussian distribution. We will investigate these two cases.

For convenience of notation, we introduce the following scaling

$$\bar{r} = \frac{r}{C_s \alpha_0^3}, \quad \bar{j}(\bar{r}) = C_s \alpha_0^3 j(r). \quad (12.6.4)$$

For a uniform angular distribution, one obtains from Eqs. (12.6.3), (12.5.12) and (12.5.13)

$$\bar{j}(\bar{r}) = \int_0^1 dt\, 2t\, \delta(\bar{r} - t^3) = \frac{2}{3\bar{r}^{1/3}} \Theta(1 - \bar{r}) \quad (\text{uniform } f_a(\alpha)). \quad (12.6.5)$$

For a Gaussian angular distribution, one finds from Eqs. (12.6.3), (12.5.12) and (12.5.14)

$$\bar{j}(\bar{r}) = \int_0^\infty dt\, 2t e^{-t^2} \delta(\bar{r} - t^3) = \frac{2}{3\bar{r}^{1/3}} e^{-\bar{r}^{2/3}} \quad (\text{Gaussian } f_a(\alpha)), \quad (12.6.6)$$

similar to Eq. (12.6.5).

The next step is to evaluate the functions $I_1(x)$ and $I(x)$ from Eqs. (12.2.9)–(12.2.11). We will scale the quantity x in the same way as r and the quantities I and I_1 in the same way as j; see Eq. (12.6.4). With Eq. (12.6.5), one obtains for a uniform angular distribution (with $0 \leq \bar{x} \leq 1$)

$$\bar{I}_1(\bar{x}) = \bar{x}^{2/3} \qquad \text{(uniform } f_a(\alpha)) \qquad (12.6.7)$$

$$\begin{aligned}\bar{I}(\bar{x}) &= \frac{2}{\pi}\arcsin(\bar{x}) + \frac{6}{\pi}\bar{x}^{2/3}\int_{\bar{x}^{1/3}}^{1} dt \frac{1}{(1-t^6)^{1/2}} \\ &= 1 - \frac{6}{\pi}\int_{\bar{x}^{1/3}}^{1} dt \frac{t^2 - \bar{x}^{2/3}}{(1-t^6)^{1/2}} \qquad \text{(uniform } f_a(\alpha)).\end{aligned} \qquad (12.6.8)$$

Similarly, one obtains with Eq. (12.6.6), for a Gaussian angular distribution,

$$\bar{I}_1(\bar{x}) = 1 - e^{-\bar{x}^{2/3}} \qquad \text{(Gaussian } f_a(\alpha)) \qquad (12.6.9)$$

$$\begin{aligned}\bar{I}(\bar{x}) &= 1 - \frac{3}{\pi}\int_1^\infty du \frac{e^{-u\bar{x}^{2/3}}}{u(u^3-1)^{1/2}} \\ &= 1 - \frac{6}{\pi}\int_0^1 dt \frac{t^2 e^{-\bar{x}^{2/3}/t^2}}{(1-t^6)^{1/2}} \qquad \text{(Gaussian } f_a(\alpha)).\end{aligned} \qquad (12.6.10)$$

The integrals of Eqs. (12.6.8) and (12.6.10) were evaluated numerically. In order to improve the numerical accuracy, we replaced the t-integration by an integration over z, using

$$z = (1/t^2 - 1)^{1/2} \Rightarrow t^2 = 1/(1+z^2),$$
$$dt/(1-t^6)^{1/2} = -dz/(3 + 3z^2 + z^4)^{1/2}.$$

The numerical results obtained for the functions $\bar{I}(\bar{x})$ given by Eqs. (12.6.8) and (12.6.10), are plotted in Fig. 12.18.

The last step is the calculation of the $d_{p,1-p}$ width from Eq. (12.2.6) and the FW_f value from Eq. (12.2.12). For the FW_f value, one directly finds from Eqs. (12.6.7) and (12.6.9)

$$\overline{FW}_f = 2f^{3/2} \qquad \text{(uniform } f_a(\alpha)) \qquad (12.6.11)$$

$$\overline{FW}_f = 2\left[\ln\left(\frac{1}{1-f}\right)\right]^{3/2} \qquad \text{(Gaussian} f_a(\alpha)). \qquad (12.6.12)$$

We determined the $d_{p,1-p}$ values from Eqs. (12.2.6), (12.6.8) and (12.6.10) by numerical means. The results are plotted in Fig. 12.19.

Calculation of different spot- and edge-width measures

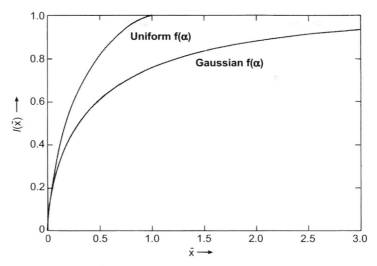

Fig. 12.18 The function $\bar{I}(\bar{x})$ for a spot that is limited by spherical aberration. See Eqs. (12.6.8) and (12.6.10) for the case of a uniform and a Gaussian angular distribution respectively.

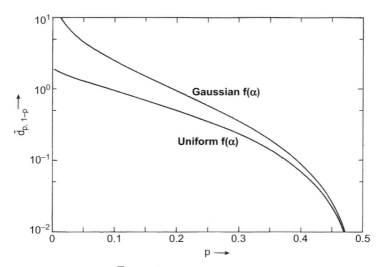

Fig. 12.19 The scaled width $\bar{d}_{p,1-p}$ of a spot that is limited by spherical aberration. It is evaluated from Eq. (12.2.6), using Eqs. (12.6.8) and (12.6.10) for the case of a uniform and a Gaussian angular distribution respectively and the scaling of Eq. (12.6.4).

12.7. Space charge effect

In Chapter 11, it was demonstrated that the space charge effect occurring in the drift space of a column can be described in terms of the properties of a negative lens. The first order properties of the space charge lens cause a defocussing of the image, as well as some (de)magnification. These effects can, in principle, be compensated by proper adjustment of the lenses in the column. It was also found that the space charge lens will be non-ideal when the lateral current density distribution in the beam is non-uniform. For those conditions, the space charge lens shows nonvanishing third order properties, which cause a nonrefocusable blurring of the image. In general, the dominant contribution to the blurring stems from spherical aberration.

The impact of spherical aberration on the FW_f and $d_{p,1-p}$ width measures is covered by the analysis of the previous section. In this section, we will study the impact of the first order properties of the space charge lens, assuming that the corresponding defocussing of the probe is not compensated. This situation is usually encountered in variable shaped spot lithography columns, in which it is difficult to adjust the projector lens on a spot-by-spot basis.

The current density distribution $j(r)$, associated to a defocussing of the spot by a distance Δz_f, is equal to

$$j(r) = \frac{1}{\Delta z_f} f_\alpha(r/\Delta z_f), \qquad (12.7.1)$$

which is also expressed by Eq. (11.5.16), however using a somewhat different notation. The function f_α specifies, as before, the angular distribution near the target. For convenience of notation, we introduce the following scaling

$$\bar{r} = \frac{r}{\Delta z_f \alpha_0}, \quad \bar{j}(\bar{r}) = \Delta z_f \alpha_0 j(r). \qquad (12.7.2)$$

For a uniform angular distribution, one obtains from Eqs. (12.7.1), (12.5.12) and (12.5.13)

$$\bar{j}(\bar{r}) = 2\bar{r}\Theta(1 - \bar{r}). \qquad (12.7.3)$$

Similarly, one obtains with Eq. (12.5.14), for a Gaussian angular distribution,

$$\bar{j}(\bar{r}) = 2\bar{r}e^{-\bar{r}^2}. \qquad (12.7.4)$$

As before, the functions $I_1(x)$ and $I(x)$ follow with Eqs. (12.2.9)–(12.2.11). We will scale the quantity x in the same way as r and the quantities I and I_1 in the same way as j; see Eq. (12.7.2). From Eq. (12.7.3), one obtains for a uniform angular distribution

$$\bar{I}_1(\bar{x}) = \bar{x}^2 \qquad \text{(uniform } f_\alpha(\alpha)) \qquad (12.7.5)$$

$$\bar{I}(\bar{x}) = \frac{2}{\pi}\left(\arcsin(\bar{x}) + \bar{x}\sqrt{1 - \bar{x}^2}\right) \qquad \text{(uniform } f_\alpha(\alpha)), \qquad (12.7.6)$$

where $0 \leq \bar{x} \leq 1$. From Eq. (12.7.4), one obtains for a Gaussian angular distribution

$$\bar{I}_1(\bar{x}) = 1 - e^{-\bar{x}^2} \quad \text{(Gaussian } f_a(\alpha)\text{)} \tag{12.7.7}$$

$$\bar{I}(\bar{x}) = \text{erf}(\bar{x}) \quad \text{(Gaussian } f_a(\alpha)\text{)}. \tag{12.7.8}$$

The final step is the calculation of the corresponding $d_{p,1-p}$ and FW_f values. From Eqs. (12.2.12), (12.7.5) and (12.7.7), one obtains for the FW_f value, in case of a uniform and a Gaussian angular distribution respectively

$$\overline{FW}_f = 2f^{1/2} \quad \text{(uniform } f_a(\alpha)\text{)} \tag{12.7.9}$$

$$\overline{FW}_f = 2\left[\ln\left(\frac{1}{1-f}\right)\right]^{1/2} \quad \text{(Gaussian } f_a(\alpha)\text{)}. \tag{12.7.10}$$

From Eqs. (12.2.6) and (12.7.8), one obtains for a Gaussian angular distribution

$$\bar{d}_{p,1-p} = 2\,\text{erf}^{-1}(1-2p) \quad \text{(Gaussian } f_a(\alpha)\text{)}. \tag{12.7.11}$$

The inverse of the function $\bar{I}(\bar{x})$ specified by Eq. (12.7.6) cannot be made explicit. In order to determine the $d_{p,1-p}$ value for a uniform angular distribution, we solved Eq. (12.2.6), combined with Eq. (12.7.6), numerically. The results are plotted in Fig. 12.20, which also shows the relation given by Eq. (12.7.11), applying to the case of a Gaussian angular distribution.

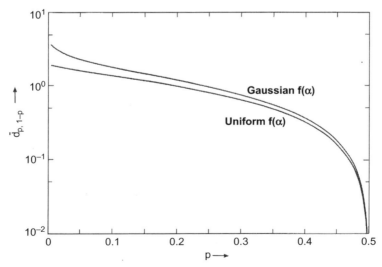

Fig. 12.20 The scaled width $\bar{d}_{p,1-p}$ for a spot that is limited by (uncompensated) space charge defocussing. The curve for a uniform angular distribution is evaluated from Eqs. (12.2.6) and (12.7.6), using the scaling of Eq. (12.7.2). The curve for a Gaussian angular distribution is given by Eq. (12.7.11).

12.8. Results for a truncated Gaussian angular distribution

So far, we restricted our calculations to a uniform and a Gaussian angular distribution. In this section, we will outline a method to approximate the intermediate case of a truncated Gaussian angular distribution. This type of distribution is defined by Eqs. (12.5.12) and (12.5.15).

Suppose that, for a uniform angular distribution, some arbitrary width measure w of the spot is given by

$$w = C\alpha_m^q, \qquad (12.8.1)$$

where α_m denotes the maximum (cut-off) beam semi-angle and C some constant. The parameter q is characteristic for the aberration considered. For example, chromatic aberration corresponds to $q=1$ and spherical aberration to $q=3$, as can be seen from Eqs. (12.5.10) and (12.6.2) respectively.

We now define the effective angle α_{eff} of an arbitrary angular distribution $f_\alpha(\alpha)$, to be used in Eq. (12.8.1), as

$$\alpha_{eff} = \left[\frac{q+2}{2} \int_0^{\alpha_m} d\alpha f_\alpha(\alpha)\alpha^q\right]^{1/q}. \qquad (12.8.2)$$

The factor in front of the integral was chosen such that $\alpha_{eff} = \alpha_m$ when the angular distribution $f_\alpha(\alpha)$ is uniform. For a truncated Gaussian angular distribution, one obtains by substitution of Eqs. (12.5.12) and (12.5.15) into Eq. (12.8.2)

$$\frac{\alpha_{eff}}{\alpha_0} = \left[\frac{q+2}{1-\exp(-\alpha_m^2/\alpha_0^2)} \int_0^{\alpha_m/\alpha_0} dt\, t^{q+1} e^{-t^2}\right]^{1/q}, \qquad (12.8.3)$$

which specifies α_{eff}/α_0 as function of α_m/α_0. We evaluated Eq. (12.8.3) numerically, for different values of q. The results are plotted in Fig. 12.21. For $\alpha_m/\alpha_0 \ll 1$ the angular distribution is nearly uniform and one finds $\alpha_{eff}/\alpha_0 \approx \alpha_m/\alpha_0$, and thus $\alpha_{eff} \approx \alpha_m$. For $\alpha_m/\alpha_0 \gg 1$ the angular distribution is nearly Gaussian and the ratio α_{eff}/α_0 saturates to a value that is characteristic for this type of distribution.

We note that the integral in Eq. (12.8.3) can be expressed in elementary functions (and the error function) when q is a positive integer ($q=1, 2, 3, \ldots$). For instance, for $q=1, 2$ and 3 one finds respectively

$$\frac{\alpha_{eff}}{\alpha_0} = \frac{3\sqrt{\pi}}{4} \frac{\operatorname{erf}(\alpha_m/\alpha_0) - (2/\sqrt{\pi})(\alpha_m/\alpha_0)\exp(-\alpha_m^2/\alpha_0^2)}{1-\exp(-\alpha_m^2/\alpha_0^2)} \qquad (q=1)$$

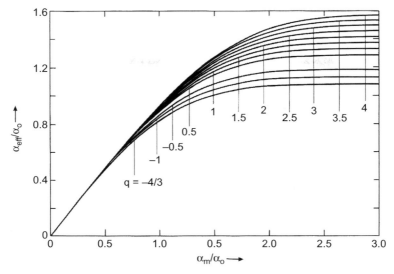

Fig. 12.21 The effective angle α_{eff} of a truncated Gaussian distribution as defined by Eq. (12.8.3).

$$\frac{\alpha_{\text{eff}}}{\alpha_0} = \sqrt{2}\left[\frac{1 - \left[1 + (\alpha_m/\alpha_0)^2\right]\exp(-\alpha_m^2/\alpha_0^2)}{1 - \exp(-\alpha_m^2/\alpha_0^2)}\right]^{1/2} \quad (q=2)$$

$$\frac{\alpha_{\text{eff}}}{\alpha_0} = \frac{(15\sqrt{\pi})^{1/3}}{2}$$
$$\times \left[\frac{\text{erf}\left(\frac{\alpha_m}{\alpha_0}\right) - \left(\frac{4}{3\sqrt{\pi}}\right)\left[\frac{3\alpha_m}{2\alpha_0} + \left(\frac{\alpha_m}{\alpha_0}\right)^3\right]\exp\left(\frac{-\alpha_m^2}{\alpha_0^2}\right)}{1 - \exp\left(\frac{-\alpha_m^2}{\alpha_0^2}\right)}\right]^{1/3} \quad (q=3),$$

(12.8.4)

as follows straightforwardly by integration.

It should be emphasized that the calculation of α_{eff} by means of Eq. (12.8.3) provides only a rough method to calculate the FW_f and $d_{p,1-p}$ values for a truncated Gaussian angular distribution. The extreme cases $\alpha_m/\alpha_0 \to 0$ (uniform distribution) and $\alpha_m/\alpha_0 \to \infty$ (Gaussian distribution) were explicitly considered in the previous sections and can therefore be used to check the accuracy of Eq. (12.8.3). As an example, let us consider the calculation of the FW_{50} (FW_f for $f=0.5$) and $d_{10,90}$ ($d_{p,1-p}$ for $p=0.1$) values for a chromatically limited spot. From the first equation of (12.8.4), one finds for $\alpha_m/\alpha_0 \to \infty$ that $\alpha_{\text{eff}}/\alpha_0 = 3\sqrt{\pi}/4 = 1.33$, which can also be observed from Fig. 12.21 (curve for $q=1$). As the width measures FW_f and $d_{p,1-p}$ increase linearly with α (since $q=1$), this analysis predicts that the results obtained for a Gaussian angular distribution are approximately a factor 1.33 larger than the results obtained for a uniform angular distribution (for the same value α_0). The results of the exact analysis carried out in Section 12.5 are plotted in Figs. 12.16a, 12.16b, 12.17a and 12.17b. Let us consider the cases $\gamma=2$ (Gaussian energy distribution) and $\gamma=1$ (Lorentzian energy distribution). For $\gamma=2$, one finds the following values for the ratio of the results obtained for a Gaussian and a uniform angular distribution

$$\frac{FW_{50;g}}{FW_{50;u}} = \frac{1.39}{1.15} = 1.21, \qquad \frac{d_{10,90;g}}{d_{10,90;u}} = \frac{2.10}{1.61} = 1.30 \qquad (\gamma = 2). \qquad (12.8.5)$$

Similarly, one obtains for $\gamma=1$

$$\frac{FW_{50;g}}{FW_{50;u}} = \frac{1.55}{1.24} = 1.25, \qquad \frac{d_{10,90;g}}{d_{10,90;u}} = \frac{3.35}{2.52} = 1.33 \qquad (\gamma = 1). \qquad (12.8.6)$$

Eqs. (12.8.5) and (12.8.6) show that the exact ratio α_{eff}/α_0 depends on the width measure considered, as well as the type of energy distribution. This particular example shows that the approximating value $\alpha_{eff}/\alpha_0=1.33$ is within 10% of the exact values. One should expect that the results for a truncated Gaussian angular distribution (α_m/α_0 finite) are at least as accurate as the results for a Gaussian angular distribution ($\alpha_m/\alpha_0 \to \infty$) presented above.

CHAPTER THIRTEEN

Monte Carlo simulation of particle beams

Contents

13.1.	Introduction	393
13.2.	Source routine	395
13.3.	Optical elements	398
13.4.	Numerical ray tracing	401
13.5.	Analytical ray tracing	406
13.6.	Simulation of large currents near the source	412
13.7.	Correction of finite-size effects	413
13.8.	Data analysis	416
13.9.	Accuracy limitations of the MC program	421
13.10.	MC simulation vs analytical modeling	423
13.11.	Program organization and examples	424
Appendix 13. A. Random number routine		430
Appendix 13. B. Polynomial fit algorithm		433

13.1. Introduction

This chapter discusses the basics of numerical Monte Carlo (MC) simulation of charged particle beams. This technique provides a direct method to determine the impact of Coulomb interactions on the properties of the beam. The essential part of a MC simulation program is a ray tracing routine, which computes the particle trajectories by numerical integration of the equations of motion, taking the mutual Coulomb repulsion into account. The solution of the N-body problem obtained this way does not rely on any fundamental physical approximations. Therefore, one application of the MC simulation method is to verify the validity of the assumptions underlying the analytical models. A comparison of the results of MC simulations with the analytical theory presented in the previous chapters is reported in Chapter 14.

In general, a MC simulation program contains the following elements: a source routine to simulate a random initial condition of an ensemble (sample/bunch) of particles; routines to simulate lenses and other optical components; ray tracing routines; routines to translate the final coordinates of the particles in terms of some characteristic quantities, such as the width of the energy distribution, the defocussing distance and the width of the spatial distribution in the plane of best focus. We will successively describe each of these elements. Various concepts presented here are based on the ideas of Groves et al. (1979) and Groves (1981a,b). The actual realization of our program was considerably facilitated by the fact that Dr. T. R. Groves has made a basic version of his program available.

The different methods to perform the ray tracing will be considered in some detail. We will discuss a numerical third order integration procedure, which uses a variable time step. A similar procedure is proposed by Meisburger (1983) and Munro (1987). As a spin-off of the analysis of the two-particle dynamical problem, which was presented in Chapter 6, we were able to produce a semi-analytical ray tracing method. This method constitutes an alternative to the conventional numerical ray tracing routine. The analytical ray tracing method was found to be extremely accurate for the operating conditions encountered in practical columns, while it increases the speed of the program with as much as two orders of magnitude. This technique is, therefore, referred to as Fast Monte Carlo simulation (FMC). Part of the material presented here was reported previously by Jansen (1987) and van der Mast and Jansen (1987).

The accuracy of the MC simulation method is limited by ray tracing errors, statistical errors and errors related to the finite size of the sample. Several provisions have been introduced in the program to estimate the impact of these errors on the accuracy the final results. A correction of the final velocities is described, which reduces the error in the generated energy spread due to finite size effects. Under normal conditions, statistical errors and finite size errors are dominant, leading to an error in the final results that is typically 10% for the median width of the energy and trajectory displacement distribution and somewhat larger for the Full Width at Half Maximum (*FWHM*) of these distributions and the edge width of shaped spots.

The main part of the chapter is dedicated to the fundamental subjects involved. The actual organization of our program is briefly outlined at the end of this chapter, where we also give some simulation examples. The column that is to be simulated can be specified by editing the so-called *system* file. A command structure has been developed for a flexible definition of the various elements, constituting the column. The first command should define a single or multisource. It can be succeeded by any combination of commands, specifying drift spaces, thin lenses, deflectors, quadrupoles and round or rectangular apertures. Other commands are available to instruct the program how to analyze the final coordinates. The so-called MC *data* file contains the variable beam properties (such as the beam current, the beam voltage and the initial energy spread) as well as a number of general program parameters. Separate output files are used for the storage of the final coordinates, the positions in the target plane and the final energy and spatial distribution. A separate graphics program is available to plot the data on screen or paper.

13.2. Source routine

The source routine generates the initial coordinates of a sample of particles, representing the beam in the vicinity of the emission surface. The number of particles in the sample N_{sam} is typically chosen between 100 and 1000. A pseudo-random number generator is used to assign a position and a velocity to every particle, in such a way that their distribution in phase space resembles the macroscopic properties of the beam.

The set of quantities used to specify the macroscopic condition of the beam depends on the type of source that is to be simulated. In our program, we have implemented two different types of single sources. The essential difference between the two algorithms concerns the formation of the velocity distribution. The formation of the spatial distribution is the same. Type 1 is a somewhat artificial source in which the particles start at full speed. The velocity distribution of the particles is specified in terms of their angular distribution and their energy distribution. Type 2 is a thermionic source in which the particles are emitted according to the half Maxwell-Boltzmann distribution. In this algorithm, the particles may start at full speed, as in the type-one source, but they can also start at zero beam potential and be accelerated afterwards in the ray tracing routine. The ray tracing is the subject of Sections 13.4 and 13.5. We note that the ray tracing allows for a uniform external acceleration field only. For that reason, we did not attempt to model field emitters, which require the incorporation of a spherical symmetric field. An array of sources, all of the same type, can be simulated by the program. This possibility was included to model columns employing multiple emitters, as proposed by van der Mast et al. (1985). In this section, we will discuss the algorithms used to model a single source. The extension to multiple sources is straightforward and will not be considered here.

For the type one source, one has to specify the initial velocity distribution in terms of the angular distribution and the energy distribution of the particles at emission. The angular distribution can be chosen circular uniform or circular Gaussian. The energy distribution can be chosen uniform, Gaussian or Lorentzian. The user has to specify the Half Width at Half Maximum (*HWHM*) of both distributions. Finally, the beam potential V defines the average energy of the particles, which is equal to eV. From this information, the program selects for each particle randomly an angle α with the axial direction, an polar angle φ_v (between 0 and 2π) and an energy deviation ΔE from the average energy eV. A description of the random number generator that serves this purpose is presented in Appendix 13.A. Once the values α, ψ_v and ΔE have been selected for a certain particle, the program calculates the corresponding total velocity v as

$$v = \sqrt{\frac{2(eV + \Delta E)}{m}}, \qquad (13.2.1)$$

while the individual velocity components follow, using

$$\begin{aligned} v_x &= v\cos(\varphi_v)\sin(\alpha) \\ v_y &= v\sin(\varphi_v)\sin(\alpha) \\ v_z &= v\cos(\alpha). \end{aligned} \qquad (13.2.2)$$

The z-axis here coincides with the beam axis. Alternatively, one can instruct the program (by setting $ICONS = 2$) to associate the energy distribution entirely with the z-component of velocity, using

$$\begin{aligned} v_x &= v\cos(\varphi_v)\tan(\alpha) \\ v_y &= v\sin(\varphi_v)\tan(\alpha) \\ v_z &= v. \end{aligned} \qquad (13.2.3)$$

This feature is primarily intended to generate a beam that is monochromatic with respect to the axial velocity of the particles. This case is considered in our analytical model.

In the thermionic source (the type 2 source) the velocity distribution is fully determined by the temperature of the cathode surface T_{ca} only, as is expressed by Eq. (3.7.7). Defining the quantities $\varepsilon_{\|}$ and $\varepsilon_{\perp}$ as

$$\varepsilon_{\|} = \frac{mv_{\|}^2}{2k_B T_{ca}}, \quad \varepsilon_{\perp} = \frac{mv_{\perp}^2}{2k_B T_{ca}}, \qquad (13.2.4)$$

one can rewrite Eq. (3.7.7) as

$$f_i(\varepsilon_{\|}, \varepsilon_{\perp}, \varphi_v)\, d\varepsilon_{\|}\, d\varepsilon_{\perp}\, d\varphi_v = \frac{1}{2\pi} e^{-(\varepsilon_{\|} + \varepsilon_{\perp})}\, d\varepsilon_{\|}\, d\varepsilon_{\perp}\, d\varphi_v, \qquad (13.2.5)$$

in which φ_v represents again the polar angle, which is uniformly distributed over 2π. Notice that both $\varepsilon_{\|}$ and $\varepsilon_{\perp}$ are distributed according to the one-dimensional exponential distribution $f_{1e}(x)$

$$f_{1e}(x) = e^{-x}. \qquad (13.2.6)$$

This distribution can easily be generated, as is discussed in Appendix 13.A. Given a certain value for $\varepsilon_{\|}$ and $\varepsilon_{\perp}$, the program determines the total velocity of the particle v and its angle α with the axial direction, using

$$v = \sqrt{\frac{2(\varepsilon_{\|} + \varepsilon_{\perp})k_B T_{ca}}{m}}, \quad \alpha = \sqrt{\frac{\varepsilon_{\perp}}{\varepsilon_{\|}}}. \qquad (13.2.7)$$

The individual velocity components follow again with Eqs. (13.2.2) or (13.2.3).

The specification of the current density distribution at the emission surface is the same for both types of sources. It can be chosen circular uniform, circular Gaussian or rectangular uniform. The last possibility is included to facilitate the simulation of shaped beam lithography systems. The source dimensions are expressed in terms of the HWHM of the current density distribution. In case of a circular current density distribution, the program selects for each particle randomly a radius r and a polar angle φ_p. The x and y component of the particle position follow from

$$x = r\cos(\varphi_p)$$
$$y = r\sin(\varphi_p). \tag{13.2.8}$$

In case of a rectangular uniform distribution, the x and y coordinates are chosen uniform in the intervals $(-HWHM_x, +HWHM_x)$ and $(-HWHM_y, +HWHM_y)$ respectively. The program allows for a shift in the x and y direction, which can be exploited to simulate some misalignment of the source.

The axial distribution of the particles is determined by the total beam current I. For particles of charge e, the average time T_{sam} required to emit N_{sam} particles is given by

$$T_{sam} = \frac{eN_{sam}}{I}. \tag{13.2.9}$$

Each particle is assigned a time T, which is randomly chosen in the interval $[-T_{sam}/2, T_{sam}/2]$, using a uniform distribution. The source routine and ray tracing routines are organized such that the particles are effectively emitted one by one from the cathode surface.

The nominal length of the sample L_{sam} is a function of the beam current I, the beam potential V and the number of particles in the sample N_{sam}

$$L_{sam} = N_{sam}\langle s \rangle = \sqrt{\frac{2e^3}{m}} \frac{N_{sam}\sqrt{V}}{I}, \tag{13.2.10}$$

where $\langle s \rangle$ denotes the average axial separation of the particles, which is specified by Eq. (3.2.6). In the case of an electron beam with $V = 10\,kV$, $I = 1\,\mu A$ and a sample size $N_{sam} = 100$, one obtains $L_{sam} = 95\,\mu m$.

The actual sample length will always differ somewhat from the nominal length L_{sam}, since the positions of the particles are chosen randomly. Moreover, the sample length may increase during the flight—first, because the energy spread of the particles causes some spatial dispersion and second, because the collective space charge force is unbalanced for particles near the edge of the bunch. The particles at the front of the bunch experience an artificial acceleration, while the particles at the back of the sample are artificially retarded. This so-called finite size effect will be further discussed in Section 13.7. In general, one should choose the sample length L_{sam} large enough to render finite size effects negligible. In this respect, problems may arise with the simulation of high current beams operating at a low beam potential. Eq. (13.2.10) shows that such beams require an extreme large N_{sam} in order to obtain a sufficient sample length L_{sam}. However, the value of N_{sam} is, in general, limited by cpu time constraints and in some cases by the available memory space. An alternative way to model a high-current beam in the vicinity of the source will be discussed in Section 13.6.

13.3. Optical elements

Lenses, deflectors and quadrupoles can be simulated by the program in thin-lens approximation. Round and rectangular apertures are incorporated, too. We will refer to these elements as optical components.

The first order imaging properties of a thin lens are fully specified by the axial position of the lens and its focal distance f. Consider a particle that reaches the lens plane with an axial velocity v_z and transverse velocity $\mathbf{v}_\perp$. Let $r_\perp$ be its radial distance to the axis when it crosses the lens plane. The first order imaging properties of the lens are simulated by adding a transverse velocity $\Delta \mathbf{v}_{\perp 1}$ to the velocity of the particle, which is chosen equal to

$$\Delta \mathbf{v}_{\perp 1} = -\frac{v_z}{f} \mathbf{r}_\perp. \tag{13.3.1}$$

A lens is called ideal when it shows only first order imaging properties. Eq. (13.3.1) expresses that such a lens can be simulated by shifting the transverse velocity of each particle opposite to the direction of $\mathbf{r}_\perp$ by an amount that is directly proportional to the length of this vector.

The performance of actual lenses is limited by third and higher order geometrical aberrations as well as chromatic aberrations. In addition, magnetic lenses may introduce rotational errors. These effects are incorporated within the thin-lens approximation. The chromatic aberration of a lens causes an additional shift in the transverse velocity $\Delta \mathbf{v}_{\perp c}$, which is equal to

$$\Delta \mathbf{v}_{\perp c} = C_c \frac{v_z}{f^2} \left(\frac{v^2}{\langle v^2 \rangle} - 1 \right) \mathbf{r}_\perp, \tag{13.3.2}$$

where $\langle v^2 \rangle$ represents the mean square velocity of all particles and C_c is the constant of chromatic aberration of the lens, for an object that is located at infinity. Notice that the term between brackets is equal to $\Delta E/E$.

Third order geometrical aberrations are taken into account by introducing the additional shift in transverse velocity $\Delta \mathbf{v}_{\perp 3}$, with

$$\Delta \mathbf{v}_{\perp 3} = -C_s \frac{v_z r_\perp^2}{f^4} \mathbf{r}_\perp, \tag{13.3.3}$$

where C_s is the constant of spherical aberration of the lens for an object that is located at infinity.

The program does not account for the rotation of the beam by a magnetic lens. However, the aberrations caused by the non-uniformity of the rotation are included within the thin lens approximation. The transverse velocity and position of a particle are rotated around the center of the lens over an angle $\Delta\varphi$, which depends on the radial distance to the axis $r_\perp$ and the total velocity v, as

$$\Delta\varphi = \frac{R_d}{f^2} r_\perp^2 - R_c \left(\frac{v^2}{\langle v^2 \rangle} - 1 \right), \qquad (13.3.4)$$

where R_d represents the rotational distortion of the lens and R_c represents the chromatic error in the rotation. The rotation over the angle $\Delta\varphi$ is incorporated by shifting the x and y coordinates of the particle with

$$\begin{aligned} \Delta x &= -y\,\Delta\varphi, & \Delta y &= x\,\Delta\varphi, \\ \Delta v_x &= -v_y\,\Delta\varphi, & \Delta v_y &= v_x\,\Delta\varphi, \end{aligned} \qquad (13.3.5)$$

assuming that $\Delta\varphi \ll 1$. This scheme, which was proposed by Barth (1988), guarantees conservation of the (canonical) angular momentum of the particle around the z-axis.

The lens routine accounts for the rotational shifts given by Eq. (13.3.5) first and computes the total velocity shift $\Delta\mathbf{v}_\perp$ specified by Eqs. (13.3.1)–(13.3.3) next, using

$$\begin{aligned} \Delta\mathbf{v}_\perp &= \Delta\mathbf{v}_{\perp 1} + \Delta\mathbf{v}_{\perp c} + \Delta\mathbf{v}_{\perp 3} \\ &= -\frac{v_z}{f}\left[1 - \frac{C_c}{f}\left(\frac{v^2}{\langle v^2 \rangle} - 1 \right) + C_s \frac{r_\perp^2}{f^3} \right] \mathbf{r}_\perp. \end{aligned} \qquad (13.3.6)$$

The user has to specify the axial position of the lens, its focal distance f and the aberration constants C_c, C_s, R_d and R_c. In addition, one can specify a shift of the center of the lens relative to the beam axis in order to simulate some misalignment.

The lens procedure described by Eqs. (13.3.5) and (13.3.6) affects the transverse velocity of a particle. Its axial velocity remains the same. This implies that the total kinetic energy of the particle is changed in this scheme. However, the user may instruct the program (by setting ICONS $=1$) to compensate this effect by multiplying each of the final x, y and z velocity components with a factor

$$F = \frac{\left(v_{x,i}^2 + v_{y,i}^2 + v_{z,i}^2 \right)^{1/2}}{\left(v_{x,f}^2 + v_{y,f}^2 + v_{z,f}^2 \right)^{1/2}}, \qquad (13.3.7)$$

where $v_{x,i}$, $v_{y,i}$ and $v_{z,i}$ are the initial velocity components of the particle (before entering the lens) and $v_{x,f}$, $v_{y,f}$ and $v_{z,f}$ are the final velocity components (when leaving the lens).

So far, we considered the action of a rotational symmetric thin lens. Similarly, one can describe the action of a thin deflector by specifying its axial position, the deflection angles $\Delta \alpha_x$ and $\Delta \alpha_y$ in x and y direction respectively and a constant D_c, which covers the dispersion of the deflector. The change in the transverse velocity of a particle caused by the deflector is computed by

$$\Delta v_x = \tan(\alpha_x) v_z \left[1 - D_c \left(\frac{v^2}{\langle v^2 \rangle} - 1 \right) \right]$$
$$\Delta v_y = \tan(\alpha_y) v_z \left[1 - D_c \left(\frac{v^2}{\langle v^2 \rangle} - 1 \right) \right]. \tag{13.3.8}$$

As for the lens, the user may instruct the program to multiply the final velocity by the factor F, defined by Eq. (13.3.7).

Quadrupoles are also included in the program, in thin-lens approximation. A quadrupole is specified by its axial position, the focal distances f_x and f_y, the constants of chromatic aberration C_{cx} and C_{cy} and the spherical aberration constants C_{30}, C_{03}, C_{21} and C_{12}. The change in transverse velocity of a particle caused by the quadrupole is computed by

$$\Delta v_x = -\frac{v_z}{f_x} \left[1 - \frac{C_{cx}}{f_x} \left(\frac{v^2}{\langle v^2 \rangle} - 1 \right) + \frac{1}{f_x^3} (C_{30} x^2 + C_{12} y^2) \right] x$$
$$\Delta v_y = -\frac{v_z}{f_y} \left[1 - \frac{C_{cy}}{f_y} \left(\frac{v^2}{\langle v^2 \rangle} - 1 \right) + \frac{1}{f_y^3} (C_{30} y^2 + C_{21} x^2) \right] y, \tag{13.3.9}$$

similar to Eq. (13.3.6), describing the action of a thin rotational symmetric lens. For a weak lens, one may assume that $C_{30} = C_{03}$ and $C_{21} = C_{12}$. If specified, the final velocity of the particle is multiplied with the factor F, defined by Eq. (13.3.7). The center of the quadrupole can be shifted in x and y direction.

Apertures are modeled by means of an index table, which contains the indices of the particles present in the sample. The aperture routine checks for every particle, whether its perturbed trajectory passes through the aperture or not. The parameter N_{sam} is then set to the actual number of particles that have passed the aperture, and the indices of these particles are stored in the first N_{sam} positions of the index table. The routine that is called next by the program retrieves the coordinates of the particles by reading the indices from the first N_{sam} positions in the index table. The remaining coordinates are ignored.

The user has to specify the axial location of the aperture, its shape and its size. The shape can be round or rectangular. The center of the aperture can be shifted in x and y direction. An option is included to "invert" the aperture, which means that the program keeps track of the particles that normally do *not* pass through the aperture and omits the others. This feature can, for example, be used to simulate a ring source by combining a round source and a somewhat smaller round inverted aperture, both located at the same axial position.

13.4. Numerical ray tracing

The ray tracing routine calculates the trajectories of the particles during the passage from one optical component to the next, taking the mutual Coulomb repulsion of the particles into account. A uniform axial accelerating or retarding external field can be included. Two different ray tracing methods are implemented in the program. The numerical method (in the program referred to as DRIFT1) is described here, while the so called analytical ray tracing method (in the program referred to as DRIFT2) is the subject of Section 13.5.

The total force exerted on particle i in a sample consisting of N_{sam} particles is equal to

$$\mathbf{F}_i = C_0 \sum_{j \neq i}^{N_{sam}} \frac{\mathbf{r}_{ij}}{r_{ij}^3} + \mathbf{F}_{i,e}, \qquad (13.4.1)$$

where $\mathbf{r}_{ij} = \mathbf{r}_i - \mathbf{r}_j$, the relative position vector of particles i and j and $C_0 = e^2/4\pi\varepsilon_0$. $\mathbf{F}_{i,e}$ denotes the external field force acting on particle i. For the moment, we will disregard the external field and concentrate on the first term in Eq. (13.4.1).

The equation of motion of particle number i is, in absence of external fields, given by

$$\mathbf{r}_i^{(II)} = \frac{C_0}{m} \sum_{j \neq i}^{N_{sam}} \frac{\mathbf{r}_{ij}}{r_{ij}^3}, \qquad (13.4.2)$$

in which $\mathbf{r}_i^{(II)}$ denotes the second derivative of $\mathbf{r}_i$ with respect to time. By differentiating Eq. (13.4.2) once, one obtains

$$\mathbf{r}_i^{(III)} = \frac{C_0}{m} \sum_{j \neq i}^{N_{sam}} \left(\frac{\mathbf{v}_{ij}}{r_{ij}^3} - \frac{3(\mathbf{r}_{ij} \cdot \mathbf{v}_{ij})\mathbf{r}_{ij}}{r_{ij}^5} \right), \qquad (13.4.3)$$

where $\mathbf{v}_{ij} = \dot{\mathbf{r}}_{ij} = \mathbf{v}_i - \mathbf{v}_j$, the relative velocity vector of particles i and j. Given the positions $\mathbf{r}_i$ and velocities $\mathbf{v}_i$ of all particles at time t, the second and third time derivatives $\mathbf{r}_i^{(II)}$ and $\mathbf{r}_i^{(III)}$ can be calculated for each particle from Eqs. (13.4.2) and (13.4.3) respectively. The new positions and velocities of each particle at time $t + \Delta t$ are determined next, using the following approximations:

$$\mathbf{r}_i(t+\Delta t) \cong \mathbf{r}_i(t) + \mathbf{v}_i(t)\Delta t + \frac{1}{2}\mathbf{r}_i^{(\mathrm{II})}(t)\Delta t^2 + \frac{1}{6}\mathbf{r}_i^{(\mathrm{III})}(t)\Delta t^3$$
$$\mathbf{v}_i(t+\Delta t) \cong \mathbf{v}_i(t) + \mathbf{r}_i^{(\mathrm{II})}(t)\Delta t + \frac{1}{2}\mathbf{r}_i^{(\mathrm{III})}(t)\Delta t^2. \tag{13.4.4}$$

Thus, a third order expansion is used for updating the positions and a second order expansion for updating the velocities.

The time step size Δt is adjusted after every step, in order to maintain a certain local integration error. Two different types of step size algorithms are implemented. In the first type, the program evaluates the average absolute value of the second and third time derivatives $\mathbf{r}_i^{(\mathrm{II})}$ and $\mathbf{r}_i^{(\mathrm{III})}$ from

$$\langle r^{(\mathrm{II})} \rangle = \frac{1}{N_{sam}} \sum_{i=1}^{N_{sam}} |\mathbf{r}_i^{(\mathrm{II})}|, \quad \langle r^{(\mathrm{III})} \rangle = \frac{1}{N_{sam}} \sum_{i=1}^{N_{sam}} |\mathbf{r}_i^{(\mathrm{III})}|. \tag{13.4.5}$$

The nominal time steps Δt_2 and Δt_3 are evaluated next, defined as the time steps that lead to a contribution of the second and third order term in the first expansion of Eq. (13.4.4), which are equal to the predefined values ε_2 and ε_3 respectively

$$\Delta t_2 = \left(\frac{2\varepsilon_2}{\langle r^{(\mathrm{II})}\rangle}\right)^{1/2}, \quad \Delta t_3 = \left(\frac{6\varepsilon_3}{\langle r^{(\mathrm{III})}\rangle}\right)^{1/3}. \tag{13.4.6}$$

The next time step Δt is taken equal to the minimum of Δt_2 and Δt_3.

The second type of time step algorithm evaluates the maximum absolute value of the second and third time derivatives, denoted as $r^{(\mathrm{II})}_{max}$ and $r^{(\mathrm{III})}_{max}$ respectively. The nominal step sizes Δt_2 and Δt_3 follow again by Eq. (13.4.6), in which $\langle r^{(\mathrm{II})}\rangle$ and $\langle r^{(\mathrm{III})}\rangle$ are now replaced by $r^{(\mathrm{II})}_{max}$ and $r^{(\mathrm{III})}_{max}$ respectively. This algorithm guarantees that the perturbations, corresponding to the second and third order terms in the expansion for $\mathbf{r}_i(t+\Delta t)$, do not exceed the predefined values ε_1 and ε_2 for any of the particles. This algorithm was proposed by Meisburger (1983) and reported by Munro (1987). We note that in most applications, one is not really interested in the exact coordinates of a few strongly deflected particles, and the first algorithm is therefore preferred.

The values for the parameters ε_2 and ε_3 are estimated by the program from the input parameter N_{step}, which specifies the nominal number of steps per meter, which should on the average be maintained during the ray tracing. This quantity can directly be related to the run time of the program, as will be discussed later. For normal conditions, it is sufficient to set $N_{step}=1000$. The program keeps track of the total integration error in the positions and the velocities denoted as ε_p and ε_v using

$$\varepsilon_v(t+\Delta t) \approx \varepsilon_v(t) + \frac{|r^{(\mathrm{III})}(t+\Delta t) - r^{(\mathrm{III})}(t)|}{6\Delta t}\Delta t^3$$
$$\varepsilon_p(t+\Delta t) \approx \varepsilon_p(t) + \frac{|r^{(\mathrm{III})}(t+\Delta t) - r^{(\mathrm{III})}(t)|}{24\Delta t}\Delta t^4 + \varepsilon_v(t)\Delta t, \tag{13.4.7}$$

where $r^{(III)}$ represents the same measure for the third time derivative as used in the step-size algorithm. Thus, $r^{(III)} = \langle r^{(III)} \rangle$ when using the first type of algorithm and $r^{(III)} = r^{(III)}{}_{max}$ when using the second type. As the global integration error in the velocities $\varepsilon_v(t)$ is of the second order in Δt, one sees from the second equation of (13.4.7) that the local integration error in the positions is of the third order in Δt (and not of the fourth order as Eq. (13.4.4) suggests). The global integration errors $\varepsilon_v(t)$ and $\varepsilon_p(t)$ are therefore both of the second order in Δt.

In principle, one can use even higher order schemes than the one described by Eq. (13.4.4). Tang (1983) proposed a procedure that includes the evaluation of terms proportional to the fourth time derivative $\mathbf{r}_i^{(IV)}$. Clearly, this permits to take larger time steps Δt while maintaining a certain local integration error. However, at the accuracy level that is normally pursued (typically 1% total integration error in the positions and velocities), it appears to be questionable whether this gain counterbalances the additional cpu time required per step. Furthermore, we note that Tang uses a time-step algorithm in which Δt is taken proportional to the smallest value found for the quantity $[|\mathbf{r}_i^{(II)}|/|\mathbf{r}_i^{(IV)}|]^{1/2}$ (with $i=1, 2, \ldots, N_{sam}$). He claims that the terms associated with $\mathbf{r}_i^{(II)}$ always cause the largest perturbations, while every term in the Taylor expansion is, in general, smaller than its precessor. In our experience, this is not necessarily true for all conditions, and his algorithm is expected to fail occasionally.

Eq. (13.4.1) shows that the presence of a uniform external electrostatic field can simply be taken into account in Eq. (13.4.4) by replacing

$$\mathbf{r}_i^{(II)} \to \mathbf{r}_i^{(II)} + \mathbf{F}_e/m, \qquad (13.4.8)$$

where $\mathbf{F}_e$ is the external field force acting on all particles. In our program, $\mathbf{F}_e$ is always directed along the axial direction and is specified as

$$F_e = \frac{e(V_1 - V_0)}{z_1 - z_0}, \qquad (13.4.9)$$

where V_0 is the beam potential at the initial axial position z_0 (entrance plane) and V_1 is the beam potential at the final axial position z_1 (exit plane). As the field is uniform, the scheme of Eq. (13.4.4) provides an exact calculation of the trajectories, as far as the influence of the external force is concerned. For that reason, the program does not account for the external force in the time step algorithm. Thus, the measure for the acceleration $r^{(II)}$ considered in Eq. (13.4.6) stems from the interaction between the particles only, even in the presence of an external field.

Special care has been taken to handle the calculation of the trajectories at the entrance plane and the exit plane of the beam section. At the start of the ray tracing routine, all particles are projected backwards along their unperturbed trajectories over a time interval T_1. T_1 is here the minimum value of all T_{1i}, with

$$T_{1i} = \frac{z_i - z_0}{v_{zi}}, \qquad i = 1, 2, \ldots, N_{sam}. \tag{13.4.10}$$

After this operation, the foremost particle is located in the entrance plane. The ray tracing is now started. Step by step all particles will pass the entrance plane. In the normal mode of operation (specified by setting $INT = 1$) a particle i, which has passed this plane ($z_0 < z_i < z_1$), experiences the interaction of all other particles, irrespective of their location. The interaction of particle i (with all other particles) continues until it has passed the exit plane ($z_i > z_1$). The ray tracing is stopped when all particles have passed this plane.

In order to clarify this procedure, assume that the ray tracing routine is called again to compute the trajectories from $z = z_1$ to $z = z_2$. The procedure is now repeated. The particles are projected backwards such that all particles are behind the new entrance plane $z = z_1$. Next, they are traced step-wise through the section until all particles have passed the new exit plane $z = z_2$. Instead of calling the ray tracing routine twice, one could have called it only once to perform the ray tracing directly from $z = z_0$ to $z = z_2$. The procedure followed by the program guarantees that both calculations yield the same final coordinates.

A somewhat different procedure is implemented to calculate the trajectories near the source (specified by setting $INT = 2$). In this procedure, two particles only interact when they have *both* passed the entrance plane. As the entrance plane is here assumed to coincide with the emission surface, this seems the most realistic approach. Effectively, the particles are now emitted one by one from the source and start interacting directly after emission. We note that the time-step algorithm always evaluates the perturbation of all particles, whether they are actually taken into account in the ray tracing or not. This way, sudden changes in the time step size Δt are avoided, which might occur otherwise when a particle passes the entrance plane.

In presence of an external field, special care has to be taken to treat the passage of a particle through the entrance plane or exit plane. The program calculates the exact time at which the particle has passed the plane and adjusts the acceleration of the particle experienced during the last time step accordingly. The significance of this procedure becomes clear when one realizes that, for a typical acceleration field of 100 kV/m and a number of steps $N_{step} = 1000$/m, a particles gains on the average an energy of 100 eV per step. Consequently, if one does not account for the exact time of entrance or exit but just starts or stops accelerating in the next time step after this event has taken place, the final energy of a particle can easily be off by 100 eV. This is clearly unacceptable.

In order to avoid the occurrence of large numbers, the coordinates of all particles are scaled at the start of the ray tracing routine, employing the

δ, ν-scaling defined by Eq. (6.13.1). This procedure also leads to some minor reduction of the number of computations in the evaluation of the quantities $r_i^{(II)}$ and $r_i^{(III)}$ from Eqs. (13.4.2) and (13.4.3), since the physical parameter C_0/m does not appear in the scaled equations. Before leaving the ray tracing routine, the coordinates are converted to unscaled quantities, which are used in the other routines of the program.

The total cpu time t_n, used by the numerical ray tracing routine for the calculation of the trajectories of a single sample, consisting of N_{sam} particles through a single section of the system is equal to

$$t_n = n_{step} \frac{1}{2} N_{sam}(N_{sam} - 1) \Delta t_n, \qquad (13.4.11)$$

where n_{step} is the actual number of time steps used in the ray tracing through the section and Δt_n is the cpu time required to evaluated the interaction per pair, per time step. The quantity $N_{sam}(N_{sam}-1)/2$ is equal to the number of pairs in the sample. For the routine implemented in the program, running on an IBM 3083 mainframe computer, it was found that $\Delta t_n \approx 10^{-5}$ s. Thus, for $N_{sam}=100$ and $n_{step}=100$, one obtains $t_n \approx 5$ s.

In order to estimate the total run time of the program, one has to consider that an actual column consists of a number of beam sections. In addition, one usually runs a number of samples in order to obtain sufficient statistics, each sample starting with a different "seed" of initial conditions. Denoting the number of beam sections (or drift sections) as N_{drift} and the number of seeds as N_{seed}, one finds for the total run time T_n required for the numerical ray tracing

$$T_n = N_{seed} N_{drift} \langle n_{step} \rangle \frac{1}{2} N_{sam}(N_{sam} - 1) \Delta t_n, \qquad (13.4.12)$$

in which $\langle n_{step} \rangle$ is the average number of steps per drift section. The quantity $N_{drift} \times \langle n_{step} \rangle$ is equal to the total number of time steps used to calculate the trajectories through the entire column. Taking this number equal to 1000, $N_{seed}=10$ and $N_{sam}=100$, one obtains $T_n \approx 5 \times 10^2$ s ≈ 8 min on an IBM 3083 computer. The cpu time used by the other routines of the program is, in general, negligible compared to the cpu time required for the ray tracing.

The ray tracing can be speeded up by limiting the interaction range of the particles to some value r_{int}. This reduces the number of pairs that has to be considered in the calculation of the interaction force acting on each particle in the sample. The interaction radius r_{int} is, in our program, specified by the input parameter N_{Int}, as $r_{int}=N_{int}\langle s \rangle$, where $\langle s \rangle$ is the average axial separation of the particles. In case r_{int} is large compared to the beam radius, the number of interactions per particle is roughly equal to $2N_{int}$. The total number of interacting pairs in the sample is then equal to $N_{int} \times N_{sam}$, vs

$N_{sam}(N_{sam} - 1)/2$, when the interaction range is not limited. Accordingly the total computation time reduces with a factor $2N_{int}/(N_{sam} - 1)$. In case r_{int} is not large compared to beam radius, the number of interactions considered per particle per step will be smaller than $2N_{int}$, and the computation time reduces even more.

In practice, it is not very convenient to use the feature of interaction range limitation, since one cannot predict beforehand the appropriate value for the parameter N_{int}. Not only does it depend on the beam geometry, but it also differs for the various phenomena of particle interactions. For simplicity one better avoids any restriction of the interaction range and tries to keep the total cpu time within reasonable limits by choosing suitable values for N_{sam} and N_{seed}. However, tor the simulation of high-current and/or low voltage beams, one may prefer a restriction of the interaction range above the choice of an "economical" sample size in order to reduce finite-size effects. We will come back to this subject in Section 13.7.

13.5. Analytical ray tracing

The analytical ray tracing routine (in the program referred to as DRIFT2) is based on a decomposition of the full N-body problem into a sum of two-particle interactions, which are evaluated from analytically derived expressions. In many respects, this method can be considered as the Monte Carlo variant of the *extended two-particle approximation*, which was described in Section 5.3 and utilized in Chapters 6–10. The dynamical aspects of the problem are handled alike in both approaches. However, the systematic integration over the coordinates of a field particle relative to a test particle carried out within the analytical model is in the MC approach replaced by a random selection of initial conditions of a large set of particles. We emphasize that the fundamental assumptions underlying the concept of analytical ray tracing are the same as for the extended two-particle approximation. Accordingly, the reader is referred to Section 5.3 for the basic issues involved and the justification of the approach. Here we will concentrate on the actual realization of the concept within the MC program. The analytical ray tracing routine is written such that it is fully compatible with the ray-tracing routine described in the previous section. The user can easily switch from one method to the other and compare the results.

The reason for implementing the analytical ray tracing routine into the MC program was to verify the premises of the extended two-particle approximation. The practical significance of the routine was first recognized when it was found that, for practical operating conditions, it is as accurate as its numerical counterpart, while it increases the speed of the program with one to two orders of magnitude. It therefore constitutes a very useful enhancement of the MC program. Fast Monte Carlo (FMC) simulations, based on analytical ray tracing, can be carried out on a fast microcomputer, keeping the cpu time within reasonable limits (typically in the order of an hour). A brief description of the method and some examples to verify its validity were reported by Jansen (1987). Here we will outline the method in more detail.

In the analytical ray tracing routine, the final coordinates of the particles (near the exit plane) are directly determined from their initial coordinates (near the entrance plane) using the equations

$$\mathbf{r}_i(t_0 + T) = \tilde{\mathbf{r}}_i(t_0 + T) + \sum_{j \neq i}^{N_{sam}} \Delta \mathbf{r}_{ij}(t_0, t_0 + T)$$

$$\mathbf{v}_i(t_0 + T) = \tilde{\mathbf{v}}_i(t_0 + T) + \sum_{j \neq i}^{N_{sam}} \Delta \mathbf{v}_{ij}(t_0, t_0 + T),$$

(13.5.1)

in which $\tilde{\mathbf{r}}_i(t_0 + T)$ and $\tilde{\mathbf{v}}_i(t_0 + T)$ are the final unperturbed position and velocity of particle i respectively. T denotes the total time of flight and $\Delta \mathbf{r}_{ij}$ and $\Delta \mathbf{v}_{ij}$ are the displacements, in position and velocity respectively, of particle i, due to the interaction with particle j during the flight. In absence of external fields, the unperturbed coordinates are simply given by

$$\begin{aligned}\tilde{\mathbf{r}}_i(t_0 + T) &= \mathbf{r}_i(t_0) + \mathbf{v}_i(t_0)T \\ \tilde{\mathbf{v}}_i(t_0 + T) &= \mathbf{v}_i(t_0).\end{aligned} \quad (V_1 = V_0) \quad (13.5.2)$$

Eq. (13.5.1) is equivalent to Eq. (5.3.12), which is exploited in the extended two-particle approximation. The underlying assumption is that the displacements $\Delta \mathbf{r}_{ij}$ and $\Delta \mathbf{v}_{ij}$ are in general small. As was pointed out in Section 5.3, the method breaks down when the particles are involved in two or more *strong* collisions during their flight. However, such events are exceptional for normal operating conditions. In addition, one should realize that such events mainly affect the tails of the generated energy and trajectory displacement distributions and not the central region. We emphasize that Eq. (13.5.2) refers to a field free drift space. Modifications are required to accommodate external fields, as will be discussed later on in this section.

Eq. (13.5.1) reduce the many particle problem to the calculation of the displacements $\Delta \mathbf{r}_{ij}$ and $\Delta \mathbf{v}_{ij}$. This is a problem of two-particle dynamics, which was considered in Chapter 6. Given the initial relative position $\mathbf{r}_{ij} = \mathbf{r}_i - \mathbf{r}_j$, the initial relative velocity $\mathbf{v}_{ij} = \mathbf{v}_i - \mathbf{v}_j$ and the time of flight T, one has to compute the perturbation of the final coordinates of the particles i and j. As was pointed out in Chapter 6, it is not possible to obtain the exact solution of this problem in an explicit form. However, it is possible to obtain accurate approximations, in an explicit form, for two limiting cases, namely for weak collisions and for nearly complete collisions. If none of these approximations is allowed, the exact solution should be obtained by a numerical procedure.

For convenience of notation, we will from now on denote $\mathbf{r}_{ij}$ as $\mathbf{r}$ and $\mathbf{v}_{ij}$ as $\mathbf{v}$ and take the initial time $t_i = 0$. The coordinates referring to the initial condition are indicated with the subscript 0. The program uses the following scheme to evaluate a two-particle collision:

- The $(\hat{a}, \hat{b})$ coordinate system in the orbital plane is determined from the initial relative position and velocity, $\mathbf{r}_0$ and $\mathbf{v}_0$, using Eq. (6.3.1).
- The parameters $a_0 = (\mathbf{r}_0 \cdot \hat{a})$ and $b = (\mathbf{r}_0 \cdot \hat{b})$ are evaluated from the vectors $\mathbf{r}_0$, $\hat{a}$ and $\hat{b}$. The eccentricity related parameter q_c is determined from Eq. (6.4.16).
- The type of collision is determined and the final displacements in the orbital plane are evaluated, using the appropriate collision dynamics. The program proceeds as follows. The collision is considered as weak if

$$q_c > 10^6 \quad \text{or} \quad a_0 < -50b \quad \text{or} \quad v_0 T - a_0 < -50b \qquad (13.5.3)$$

in accordance with Eq. (6.8.1). When one or more of the inequalities in Eq. (13.5.3) are satisfied, *first order perturbation dynamics* is allowed. The displacements Δa_f, Δb_f, $\Delta \dot{a}_f$ and $\Delta \dot{b}_f$ are then evaluated from Eqs. (6.8.6) and (6.8.5) (notice that in the present notation the subscript i should be replaced by the subscript 0). If Eq. (13.5.3) is not satisfied, the program verifies whether the collision can be considered as nearly complete. This is the case if

$$a_0 > 50b \quad \text{and} \quad v_0 T - a_0 > 50b \quad \text{and} \quad \frac{1}{2} m v_0^2 > 50 \frac{C_0}{r_0} \qquad (13.5.4)$$

in accordance with Eq. (6.7.7). When all inequalities in Eq. (13.5.4) are satisfied, *nearly complete collision dynamics* is allowed, and the displacements Δa_f, Δb_f, $\Delta \dot{a}_f$ and $\Delta \dot{b}_f$ are evaluated from Eqs. (6.7.5) and (6.7.6). Otherwise, the program utilizes a numerical procedure to evaluate these quantities, which is referred to as *full collision dynamics*. The first step of this procedure is the evaluation of the initial value of the polar coordinates r and θ and the sign parameter σ_i, from Eq. (6.3.5). Microscaling is applied, using the prescriptions of Eqs. (6.4.3) and (6.4.4). In the next steps, the program follows the analysis of Section 6.5, which leads, eventually, to the final polar coordinates r_f and θ_f and their derivatives with respect to time $\dot{r}_f$ and $\dot{\theta}_f$; see Eqs. (6.5.8) and (6.5.12). A key step is the evaluation of the final polar coordinate $\rho_f = r_f / d_s$ from Eq. (6.5.7). This is done by numerical inversion, following the scheme of Section 6.6. After removing the microscaling, the displacements Δa_f,

Δb_f, $\Delta \dot{a}_f$ and $\Delta \dot{b}_f$ follow, finally, from Eqs. (6.3.9) and (6.3.7). Fig. 6.4 contains a flow diagram of the essential steps of the calculation in the orbital plane.

- Having evaluated the displacements Δa_f, Δb_f, $\Delta \dot{a}_f$ and $\Delta \dot{b}_f$, the remaining part is to express the results in terms of the displacements of the individual particles in the laboratory system. This is done by means of Eqs. (6.3.8) and (6.3.11). Identifying particle i with the colliding particle (subscript c) and particle j with the test particle (subscript t), these equations yield in the present notation

$$\Delta \mathbf{r}_{ij} = \frac{1}{2}\left(\Delta a_f \hat{a} + \Delta b_f \hat{b}\right), \quad \Delta \mathbf{v}_{ij} = \frac{1}{2}\left(\Delta \dot{a}_f \hat{a} + \Delta \dot{b}_f \hat{b}\right)$$
$$\Delta \mathbf{r}_{ij} = -\Delta \mathbf{r}_{ij}, \quad \Delta \mathbf{v}_{ij} = -\Delta \mathbf{v}_{ij}, \quad (13.5.5)$$

which completes the calculation for a single two-particle interaction. When the interaction has been evaluated for every pair in the sample, the final positions and velocities of the particles follow by means of Eq. (13.5.1).

In the analytical ray tracing routine, the start and the end of the interaction are handled somewhat different than in the numerical ray tracing routine, since one cannot switch the interaction on and off for each particle separately. The routine is organized such that all pairs interact during the same period T, while the absolute time of the start of the interaction may vary per pair. As for the numerical ray tracing, two modes of operation have been incorporated. In the normal mode of operation (specified by setting $INT = 1$), a pair of particles is shifted along the unperturbed trajectories, such that the foremost particle is located in the entrance plane $z = z_0$. The displacements generated by their mutual repulsion is then calculated over the next period T. In the other mode of operation (specified by setting $INT = 2$), the particles are shifted along the unperturbed trajectories, such that the last particle is located in the entrance plane $z = z_0$. This seems more realistic when the entrance plane coincides with the emission surface of the source. In this connection, we should mention that the final results are, in general, insensitive to the choice of the exact initial and final conditions. Finally, we emphasize that the time shifts only concern the calculation of the displacements $\Delta \mathbf{r}_{ij}$ and $\Delta \mathbf{v}_{ij}$ and do not involve the calculation of the unperturbed trajectories. This guarantees that the final coordinates all refer to the same time.

Uniform acceleration fields can be taken into account within the analytical ray-tracing routine. The essential observation is that, in the accelerated frame of reference moving with the beam, the particles can be considered as free, that is not influenced by external fields. The situation is similar to the case of an elevator that is in free-fall in a homogeneous gravitation field. In the elevator system the laws of mechanics can be applied, as if the system is an inertial system in field free space. For the analytical ray tracing routine, this observation implies that the scheme for the calculation of the displacements $\Delta \mathbf{r}_{ij}$ and $\Delta \mathbf{v}_{ij}$ from the initial relative coordinates $\mathbf{r}_{ij}$ and $\mathbf{v}_{ij}$ remains the same in presence of a uniform acceleration field. However, the field impacts the displacements $\Delta \mathbf{r}_{ij}$ and $\Delta \mathbf{v}_{ij}$ through the flight time T and the initial coordinates $\mathbf{r}_{ij}$ and $\mathbf{v}_{ij}$. In addition, one has to modify Eq. (13.5.2) for a correct calculation of the unperturbed trajectories under influence of the external field. We will now consider these aspects in more detail.

The flight time T_i of particle i, with initial axial velocity $v_{zi}(t_0)$, which is accelerated by an external force F_e along the axial direction, is equal to

$$T_i = \frac{m v_{zi}(t_0)}{F_e} \left(\text{sign}\, [v_{zi}(t_0)] \sqrt{1 + \frac{2 F_e (z_1 - z_0)}{m v_{zi}(t_0)^2}} - 1 \right), \qquad (13.5.6)$$

assuming that $2 F_e (z_1 - z_0)/m v_{zi}(t_0)^2 > -1$, which means that the final axial velocity of the particle is expected to be positive. If this is not true, the particle is removed from the sample. The initial velocity $v_{zi}(t_0)$ may here be negative. The force F_e, which is specified by Eq. (13.4.9), can be positive or negative, corresponding to an acceleration or a deceleration of the particle respectively.

The flight time T of the sample as a whole is evaluated from Eq. (13.5.6), taking $v_{zi}(t_0)$ equal to the average axial velocity $\langle v_z \rangle$ of all particles. The final unperturbed coordinates of particle i now follow as

$$\tilde{\mathbf{r}}_i(t_0 + T) = \mathbf{r}_i(t_0) + \mathbf{v}_i(t_0) T + \frac{F_e T_i^2}{2m} + \frac{F_e T_i}{m}\left(T - T_i + \frac{z_i(t_0) - z_0}{v_{zi}(t_0)}\right)$$

$$\tilde{\mathbf{v}}_i(t_0 + T) = \mathbf{v}_i(t_0) + \frac{F_e T_i}{m},$$

$$(13.5.7)$$

which replaces Eq. (13.5.2). The last term in the expression for $\tilde{\mathbf{r}}_i(t_0 + T)$ accounts for the fact that the initial drift velocity of the particle (before reaching the entrance plane) differs by a term $F_e T_i/m$ from its final drift velocity (after passing through the exit plane).

The displacements $\Delta \mathbf{r}_{ij}$ and $\Delta \mathbf{v}_{ij}$ are calculated in the same way as in absence of acceleration, using the appropriate flight time T, in presence of acceleration. With respect to the initial coordinates $\mathbf{r}_{ij}$ and $\mathbf{v}_{ij}$, the program accounts for the fact that the first particle is accelerated during some time Δt, while the last particle has not yet reached the acceleration zone. The time Δt is calculated as

$$\Delta t = \begin{cases} \dfrac{|z_i(t_0) - z_j(t_0)|}{v_{zi}(t_0)} & \text{if } z_j(t_0) \geq z_i(t_0) \\ \dfrac{|z_i(t_0) - z_j(t_0)|}{v_{zj}(t_0)} & \text{if } z_j(t_0) < z_i(t_0). \end{cases} \qquad (13.5.8)$$

The relative position and velocity, $\mathbf{r}_{ij}$ and $\mathbf{v}_{ij}$, are adjusted to account for the acceleration of the first particle during the time Δt.

Finally, it should be recognized that the flight time T_i of particle i through the acceleration zone between z_0 and z_1 is affected by the interaction with the other particles in the sample. The total axial shift Δz_i of particle i due to the interaction with the other particles is computed by

$$\Delta z_i = \sum_{j \neq i}^{N_{sam}} \Delta z_{ij}(t_0, t_0 + T) \tag{13.5.9}$$

in conformity with Eq. (13.5.1). When Δz_i is positive, the actual flight time T_i in the acceleration zone will be shorter than predicted by Eq. (13.5.6) and when Δz_i is negative, the actual flight time T_i will be longer. This effect is compensated by adjusting the flight time in the calculation of the unperturbed trajectories by means of Eq. (13.5.7).

As in the numerical ray tracing routine, the coordinates of the particles are scaled at the start of the ray tracing routine, employing the δ, ν-scaling defined by Eq. (6.13.1). This way, the occurrence of large numbers is avoided. Another advantage is that the parameters C_0/m and the time of flight T do not appear explicitly in the scaled equations, which leads to some simplification of the code. Before terminating the ray tracing routine, the scaling is removed.

The gain in computation speed using analytical ray tracing stems from the ability to determine the final coordinates from the initial coordinates in a single step. The total computation time T_a required for the analytical ray tracing of N_{seed} samples through N_{drift} sections is equal to

$$T_a = N_{seed} N_{drift} \frac{1}{2} N_{sam}(N_{sam} - 1)\langle \Delta t_a \rangle, \tag{13.5.10}$$

where $\langle \Delta t_a \rangle$ is the average cpu time required to evaluate the displacements corresponding to a single two-particle interaction. The actual computation time per two-particle interaction depends on the type of dynamics that is employed. Full collision dynamics requires typically five times more computation time than weak or nearly complete collision dynamics. On an IBM 3083 mainframe computer, it was found that, typically, $\langle \Delta t_a \rangle = 5 \times 10^{-5}$ s. As for the numerical ray tracing, one can increase the computation speed by restricting the interaction range of the particles to some distance r_{int}. For the analytical ray tracing routine, this implies that the interactions of pairs with an impact parameter $b > r_{int}$ are disregarded.

By comparing Eqs. (13.4.12) and (13.5.10), one finds that the ratio of the cpu times required for analytical and numerical ray tracing T_a/T_n is equal to

$$\frac{T_a}{T_n} = \frac{\langle \Delta t_a \rangle}{\langle N_{step} \rangle \Delta t_n} \approx \frac{5}{\langle N_{step} \rangle}. \tag{13.5.11}$$

Thus, the computation speed improvement over the numerical method depends on the average number of steps per section $\langle N_{step} \rangle$ required for the numerical ray tracing. Typically this number varies between 10 and 10^3 depending on the length of the section, the particle density and the specified integration accuracy. Accordingly, the required cpu time reduces with one to two orders of magnitude when switching to analytical ray tracing.

13.6. Simulation of large currents near the source

In Section 13.2, it was pointed out that an accurate simulation of high-current beams operating at a low beam potential requires an extreme large number N_{sam} of particles in the sample. Eqs. (13.4.12) and (13.5.10) show that the total computation time required for the ray tracing increases with the square of N_{sam}. In order to keep the total cpu time within practical limits, N_{sam} should be smaller than typically 10^3 for numerical ray tracing and smaller than 10^4 for analytical ray tracing. When the interaction range is limited to $r_{int} = N_{int}\langle s \rangle$, the maximum N_{sam} can be chosen larger by a factor $N_{sam}/2N_{int}$, as far as cpu constraints are concerned. However, the memory space required for the storage of the coordinates is (at least) $6 \times 8 = 48$ bytes per particle, leading to a practical upper limit for N_{sam} in the order of 10^4.

It is important to observe that one does not need that many particles to obtain acceptable statistics. Therefore, one might as well delete a part of the sample after it has passed through the critical (high current/low voltage) section of the beam. This consideration led us to the introduction of so-called field particles, which are only used to represent the beam and are not stored to memory. The particles in the sample, now referred to as test particles, are traced throughout the system, while the field particles are generated at the start and deleted at the end of a beam section. This concept can most easily be implemented within the analytical ray tracing routine. As the trajectories are calculated in a single time step, analytical ray tracing does not require the storage of intermediate coordinates. This implies that one can generate a field particle, determine its impact on all the test particles and delete its coordinates. The next field particle is then generated, and these steps are repeated, etc. As the field particles are generated one by one, their number can be arbitrarily large without requiring any additional memory space. In this set-up, the interaction between the test particles is ignored. Thus, every test particle is independent from the other test particles and experiences only the interaction with the field particles. The reader should notice that this model comes very close to the analytical theory, which also considers a number of independent test particles experiencing the interaction with the field particles constituting the beam. The main difference with the analytical model, used in Chapters 7–9, is that the motion of the test particles is not confined to the beam axis but resembles the actual properties of the beam.

The gain in computation speed relative to the normal analytical ray tracing routine arises from the fact that every field particle interacts only with a relatively small number of test particles. Let N_{field} be the number of field particles and N_{sam} the number of test particles. The computation time is now proportional to $N_{field} \times N_{sam}$, instead of $N_{sam}(N_{sam}-1)/2$. Taking N_{sam} of the order 10^2, one may choose N_{field} as large as 10^6, while keeping the cpu time within reasonable limits.

The MC program contains a separate analytical ray tracing routine (referred to as DRIFT3) that was developed along these lines of thought. It is especially intended to simulate the beam in the vicinity of the source. The field particles are generated one by one, following the specifications of the source. Each field particle interacts with every test particle in the sample and is deleted afterwards. The set of field particles is centered around the same axial coordinate as the sample of test particles. The final coordinates of the test particles are stored in the usual way and can be traced throughout the remaining part of the column using one of the other ray tracing routines. The analytical ray tracing with field particles has been used to simulate temperature-limited thermionic sources with currents up to $100 \mu A$. In general, the results have to be viewed with some caution, since the deviations from the unperturbed trajectories can become quite large for a significant fraction of the particles, contradicting the premises of the analytical ray tracing method. Clearly, the method is not suited to simulate space charge limited emission in which the trajectories of the individual test particles are strongly affected by the simultaneous action of many field particles.

13.7. Correction of finite-size effects

The qualification "finite-size effect" is used to denote all apparent interaction effects present in the final results of a MC simulation that are related to the inadequate representation of the beam by a sample of particles of finite size. The following finite-size effects are distinguished:
- Underestimation of the interaction effects due to an insufficient sample length L_{sam}. In this case, a particle does not have enough "neighbors" to represents the total beam. To put it differently, the sample length does not exceed the effective interaction range of the particles by a large factor. The effective interaction range is here defined as the maximum distance for which two particles still affect each other strong enough to yield a significant contribution to the macroscopic measure for the considered interaction phenomenon, such as the defocussing distance,

the energy spread and the spatial broadening. The effective interaction range depends on the linear particle density λ and the beam geometry. In addition, it differs for the various interaction phenomena. For instance, the total Boersch effect generated in a beam segment with a narrow crossover stems predominantly from the crossover area, as was discussed in Chapter 7. For the simulation of this effect in this geometry, it is therefore sufficient to take L_{sam} an order of magnitude larger than the crossover radius. However, the trajectory displacement effect generated in the same beam geometry is not necessarily dominated by the contribution of the crossover area. A significant contribution may stem from the diluted parts of the beam, and a proper simulation requires a larger L_{sam} than for the Boersch effect. The space charge effect has an even longer effective interaction range, and a proper simulation requires a larger sample than for the statistical effects. The situation is again different for a nearly cylindrical beam segment in which all parts of the beam contribute equally to the various interaction phenomena, leading to longer effective interaction distances than for a beam segment with a narrow crossover. In all cases the effective interaction range decreases when the average axial separation of the particles $\langle s \rangle$ ($=1/\lambda$) decreases. However, both quantities do not change at the same rate, and the number of "neighbors" having a significant impact increases with the linear density λ. This implies that the minimum required N_{sam} increases with λ but, in general, less than directly proportional. It should be noted that the particles near the edge of the bunch always experience a deficiency of interacting neighbors. N_{sam} should be large enough to render their contribution insignificant relative to the contribution of the particles that have sufficient neighbors.

- Improper estimation of the interaction effects due to the unbalanced space charge force acting on the particles near the edge of the bunch. It may cause an underestimation as well as an overestimation of the interaction effects. The Boersch effect is, in general, the most strongly affected by this kind of finite-size effect. The particles at the front of the bunch experience an artificial acceleration, while the particles in the back are artificially retarded. Accordingly, the generated axial velocity spread (Boersch effect) is overestimated. However, the unbalanced space charge force also causes an increase of the sample length during the flight, which leads to an artificial reduction of the linear particle density and a corresponding reduction of all interaction effects.

It is difficult to formulate general rules for the minimum N_{sam} required for a proper simulation of a certain interaction phenomenon as a function of the beam geometry and the particle density in the beam. To verify the validity of an MC simulation, one has to check whether an increase of N_{sam} affects the outcome of the simulation significantly or not. If so, the sample was too small and one has to perform the simulation again with a larger sample. This aspect of running MC simulations requires a certain skill of the operator. A clear understanding of the mechanisms involved is indispensable for obtaining reliable results without using enormous amounts of cpu time.

Various approaches have been suggested to minimize finite-size effects. Jones et al. (1983) presented a method in which two so-called "ghost" charges travel with the bunch, each at one side, located on the axis. The value of the ghost charge and its distance to the bunch should be chosen such that the space charge force acting on the particles at the edge of the bunch is just balanced. This is the case when one takes the ghost charge equal to the total charge in the bunch and the distance to the edge of the bunch equal to half the sample length L_{sam}. The drawback of the ghost charge approach is that those particles in the bunch that are off-axis experience a lateral interaction force from the ghost charges, which causes an artificial change in the lateral properties of the bunch. Errors are therefore introduced in the simulation of the trajectory displacement effect and space charge defocussing. The remedy would be to consider an extended ghost charge with a lateral dimension that reflects the local width of the beam. However, this seems rather difficult to implement within the ray tracing algorithm.

The approach taken in our program is to introduce a routine (referred to as PROCCO) that processes the final velocities to compensate the artificial acceleration of the particles caused by the unbalanced space charge effect. The routine evaluates the correlation in the final coordinates between the z-velocities and the z-position within the bunch. When specified by the user, the final velocities are adjusted such that the z-dependency of the z-velocities is removed. The major part of the correlation is assumed to be the result of the unbalanced space charge force acting on the particles. However, some correlation is induced by the dispersion of the particles during the time of flight as a consequence of the variation in the initial axial velocities. To put it simply, slow particles have a tendency to run behind, while fast particles have a tendency to run ahead. This source of correlation can be isolated by examining the final coordinates in absence of interaction. When specified by the user, the routine performs this calculation and compensates only for the correlation caused by the interaction between the particles.

The z-dependent velocity correction of the final coordinates removes the direct impact of the second kind of finite size effect on the axial velocity distribution. The main use of this procedure is to prevent an overestimation of the generated energy spread. One can instruct the program to calculate the energy spread both with and without processing of the final coordinates. This provides some insight in the significance of finite-size effects and is helpful in the choice of a proper sample length. It is interesting to study the dependency of the resulting energy spread on the number of particles in the sample N_{sam}. For very small N_{sam}, the energy spread will be underestimated. This can be understood from the fact that the interaction force between the particles in a thin slice is directed perpendicular to the beam axis. With increasing N_{sam} and in absence of correction of the final coordinates, the spread increases to a value that is too large due to the artificial deceleration and acceleration of the particles near the edges of the bunch. With further increasing N_{sam}, their contribution gradually becomes insignificant and the observed energy spread levels off to the correct value. In presence of correction of the final coordinates, the energy spread is expected to increase monotonically with the number of particles in the sample N_{sam} and reaches the correct value for a smaller sample size than without correction. By studying the results both with and without correction, one obtains a lower and an upper limit for the correct energy spread at a relative small N_{sam}.

In extreme cases, the sample can significantly be deformed during the flight due to the unbalanced space charge force. Clearly, the resulting errors in the simulation cannot be removed by processing the final coordinates. Some improvement can be obtained by correcting the velocities during the ray tracing after every time step. A facility is included in the program to perform this task (which can be activated by setting the parameter $IPROC = 1$). As for the correction of the final coordinates, this procedure gives an indication of the error caused by finite size effects and provides a partial correction.

13.8. Data analysis

The source routine, ray tracing routines and optical element routines provide the means to calculate the trajectories of a sample of interacting particles through a user defined column. The output of these routines is a set of $6 \times N_{sam}$ coordinates, specifying the final positions and velocities of the N_{sam} particles. In order to obtain sufficient statistics, the calculation is repeated for a number of seeds N_{seed}, leading to a set of $6 \times N_{seed} \times N_{sam}$ final coordinates. Typically, the total number of particles N_{tot} ($= N_{seed} \times N_{sam}$) is chosen between 1000 and 2000, which corresponds to 6000–12,000 final coordinates. Clearly, some data analysis is required to reduce this information to a limited set of numbers, specifying the characteristic properties of the beam at the target.

Several routines are implemented in the program that serve this purpose. Two routines (referred to as SYMEBR and ASYEBR) analyze the energy broadening of the particles. The first one presupposes that the energy distribution is symmetric and computes the distribution of absolute energy deviations relative to the mean energy of the particles. The second one computes the full energy distribution, which may possibly be asymmetric. The energy distribution generated by the Boersch effect is always symmetric and the first routine (SYMEBR) is appropriate for its analysis. However, the simulation of a thermionic source leads to an asymmetric initial energy distribution, and one needs the second routine (ASYEBR) for an adequate analysis. For both routines, the user can specify whether to inspect the total energy distribution (by setting $ICONS=1$) or the energy distribution associated with the z-component of the velocities (by setting $ICONS=2$).

Three routines (referred to as TBR, RNDTBR and RECTBR) concern the *transverse broadening* of the beam. The first one is a general routine, which considers the displacements from the unperturbed trajectories in a spot of arbitrary shape, while the other two are taylored for the analysis of a round (Gaussian) spot and a rectangular (shaped) spot respectively. The program is organized such that it calculates the trajectories of the particles twice for every sample, one time with the interaction switched off and one time with the interaction switched on (unless specified otherwise by the user). Both the perturbed, and the unperturbed final coordinates of the particles are stored and can be used for further analysis. The general routine (TBR) utilizes both sets of coordinates to evaluate the lateral displacements from the unperturbed trajectories in the target plane. The other routines (RNDTBR and RECTBR) consider the perturbed positions in the target plane only. All routines have a facility to seek the plane of best focus, which is usually located somewhat ahead of the Gaussian image plane, as a result of the defocussing action associated with the space charge effect.

The basic strategy of the analysis is the same for the various routines. A width measure of a distribution that can easily be evaluated is the rms width. The rms spread of a set of N scalar quantities η_i, is given by

$$\langle \Delta \eta^2 \rangle = \frac{1}{N} \sum_{i=1}^{N} \left(\eta_i - \frac{1}{N} \sum_{i=1}^{N} \eta_i \right)^2 = \frac{1}{N} \sum_{i=1}^{N} \eta_i^2 - \left(\frac{1}{N} \sum_{i=1}^{N} \eta_i \right)^2. \quad (13.8.1)$$

The second form permits a straightforward computation of the rms value in a single loop over the index i, which performs the summation of the quantities $\eta_1, \eta_2, \ldots, \eta_N$, as well as the quantities $\eta_1^2, \eta_2^2, \ldots, \eta_N^2$.

Unfortunately, the rms width is, in general, not adequate to characterize the various distributions obtained within the MC program. It is usually dominated by a few large displacements. Therefore, it is rather sensitive to statistical fluctuations. In addition, one should realize that, particularly, the large displacements are susceptible to various kind of errors. One error is the integration error (when using DRIFT1) or the model error (when using DRIFT2) associated with the ray tracing. Another error is related to the modeling of the source. Large displacements are partly generated by pairs of particles that are initially very close to each other. The question arises whether the fully random selection of initial conditions performed by the source routine reflects the actual emission process of a real source. The statistics of near neighbor particles is not known, and the MC simulation may, in this respect, be inappropriate, as will be further discussed in Section 13.9. Summarizing, the rms value is sensitive to statistical errors, ray tracing errors and source modeling errors, and the numbers obtained from Eq. (13.8.1) should be viewed with caution. From the experimental point of view, one should add that predicting a rms value is not very practical either, since it cannot directly be measured.

A reliable width measure, which can straightforwardly be obtained within the MC program, is the smallest Full Width FW_f which contains a certain fraction f of the particles. The program considers the cases $f = 0.1, 0.3, 0.5, 0.7$ and 0.9. The width corresponding to $f = 0.5$ is referred to as the Full Width median value (FW_{50}). For the computation of this type of width measures, the program has to sort the displacements (in energy or position relative to some mean value) on size in ascending order. After sorting, the FW_f value of a *symmetric* distribution can simply be obtained as

$$FW_f = 2 \, |\Delta\eta(f N_{tot})|, \qquad (13.8.2)$$

in which $|\Delta\eta(f N_{tot})|$ denotes the absolute displacement (from the mean value) of the particle with the nearest integer number to $f \times N_{tot}$ in the array of sorted displacements. For a distribution that is possibly *asymmetric*, the program evaluates the set of quantities

$$FW_{f,i} = \Delta\eta(i + f N_{tot}) - \Delta\eta(i), \quad i = 1, 2, \ldots, (1-f)N_{tot} \qquad (13.8.3)$$

and takes FW_f equal to the minimum value.

Sorting algorithms are described in detail by Knuth (1973). A comprehensive discussion is given by Press et al. (1986). The evaluation of Eq. (13.8.2) does not require a completely ordered array of displacements. It only requires that element $f \times N_{tot}$ is on the right position in the array. The MC program exploits a modified "quicksort" routine to perform this task. The routine differs from the normal quicksort algorithm in the respect that it does not order every subfile defined during the sorting process but proceeds with that subfile that contains element number $f \times N_{tot}$ and ignores the others. This procedure provides an extremely fast evaluation of Eq. (13.8.2), even for large N_{tot}. The speed is required to perform the focussing of the spot within acceptable computation time. The focussing is carried out by an iteration-loop, which evaluates Eq. (13.8.2) for various z-values in order to determine the plane for which the width of the spatial distribution becomes minimum. Contrary to Eq. (13.8.2), the evaluation of Eq. (13.8.3) requires a completely ordered array of displacements. Shell's method is used to perform this task. This algorithm is fast enough, since it is only called once by the routine used for the evaluation of an asymmetric energy distribution (ASYEBR).

Width measures other than the rms value and the FW_f value are, in general, more difficult to evaluate. The program has a provision to calculate the *FWHM*, utilizing a least square polynomial fit of the corresponding histogram. A similar method is employed to estimate the edge width of a shaped spot. Histograms are formed by the program, following the specifications of the user. The input required by the program is the number of divisions (or strips) in the histogram and the largest displacement to be included in the histogram. The last value is specified in terms of the fraction of particles to be represented in the histogram or in terms of the ratio between the maximum displacement in the histogram and the Half Width at Half Maximum (*HWHM*) value of the distribution. The program automatically selects a division width that satisfies these specifications. In the case of a shaped spot, one may, alternatively, select the number of divisions in the edge of the spot, instead of the total number of divisions.

A dedicated least square fit procedure has been developed to provide an optimum fit of the histograms. A polynomial fit function is chosen for reasons of flexibility and speed. The general form of the polynomial used by the program is given by

$$y(x) = y_s = \sum_{j=1}^{M} a_j (x - x_s)^{[n_1(M-j)+n_2]/n_3}, \qquad (13.8.4)$$

where n_1, n_2 and n_3 are constants. The user has to specify the number of terms M and the type of polynomial function. The following types are included:

$$\begin{array}{llll}
\text{normal polynomial:} & n_1 = 1, & n_2 = 0, & n_3 = 1 \\
\text{even polynomial:} & n_1 = 2, & n_2 = 0, & n_3 = 1 \\
\text{odd polynomial:} & n_1 = 2, & n_2 = 1, & n_3 = 1 \\
\text{reduced polynomial:} & n_1 = 1, & n_2 = 0, & n_3 = (M-1) \\
\text{reduced odd polynomial:} & n_1 = 2, & n_2 = 1, & n_3 = (2M-1)
\end{array}$$
(13.8.5)

Even polynomials are recommended to fit a symmetric distribution. Odd polynomials, reduced polynomials and reduced odd polynomials are included to fit the edge of a shaped spot. The shifts x_s and y_s are chosen by the program. In most cases the program takes $x_s = y_s = 0$, but for the fit of the edge of a shaped spot, the values for x_s and y_s are chosen such that the origin is centered in the edge. The expectation is that the edge follows an odd function with respect to this origin. An additional feature of the fit algorithm is that one may specify a value $y(x_0)$ of the polynomial at x_0 and a value $y'(x_1)$ of its derivative at x_1, which are imposed on the fit. This feature is included for a shaped spot to obtain a continuous and smooth joint of the fit through the edge and the intensity level at the center of the spot. For the fit of a central distribution, this feature can be used to impose that the fit function becomes zero and/or has zero derivative in the last division of the histogram. A description of the fit algorithm is presented in Appendix 13.B.

The program has a facility to estimate the statistical error in the calculated rms width and FW_{50} width of the energy distribution and the distribution of transverse displacements. It estimates the distribution width for a number of subensembles, selected from the total ensemble of N_{tot} particles. The variance in the results obtained for the individual subensembles is used to estimate the statistical error of the result for the total ensemble. The number of subensembles N_{sub} is taken equal to the nearest integer of $N_{tot}^{1/2}$. This way, the number of subensembles N_{sub} equals, approximately, the number of particles per subensemble $N_p = N_{tot}/N_{sub}$. The minimum total number of particles N_{tot} required to activate this procedure is 100. For each subensemble, the program calculates the corresponding rms width and FW_{50} width value, using equations similar to (13.8.1) and (13.8.2) respectively. The error in the rms value obtained for the total ensemble is now estimated as

$$\sigma_{rms} = \left(\frac{\langle rms^2 \rangle_{se} - \langle rms \rangle_{se}^2}{N_{sub}} \right)^{1/2}, \qquad (13.8.6)$$

where $\langle .. \rangle_{se}$ denotes an average over the results obtained for the individual subensembles. Eq. (13.8.6) exploits that the statistical error in the rms value is expected to decrease with the square root of the number of particles in the ensemble. The program prints out the relative three-sigma error $3\sigma_{rms}/rms$. The same procedure is used to estimate the relative three-sigma error in the FW_{50} width obtained for the total ensemble.

Finally, we should mention that there is some subtlety in the algorithm used to determine the plane of best focus from the distribution of transverse displacements (which is incorporated in TBR). Denoting the perturbed and the unperturbed lateral position at z as $\mathbf{r}_\perp(z)$ and $\tilde{\mathbf{r}}_\perp(z)$ respectively, the program seeks the z-coordinate for which the width of the distribution of displacements $\rho(|\mathbf{r}_\perp(z) - \tilde{\mathbf{r}}_\perp(z_{ref})|)$ becomes minimum. The axial coordinate z_{ref} defines the location of the Gaussian image plane and should be specified by the user. Notice that the algorithm computes the transverse displacement of a particle, as the lateral distance between its perturbed position at z and its unperturbed position in the original Gaussian image plane, which is located at z_{ref}.

13.9. Accuracy limitations of the MC program

The accuracy of a MC simulation is limited by the occurrence of several types of errors. Assuming that trivial errors, such as errors in the program code, errors in the input data and errors due to program abuse by ignorant operators are absent, the following error categories remain:
- *Model errors.* The representation of the beam and the simulation of optical components within the MC program relies on a number of simplifying assumptions. All optical components are assumed to be thin and fully described by their first and third order optical properties and chromatic aberration constants. External fields, other than a uniform acceleration field, are assumed to be absent. Stray fields, mechanical vibrations and electrical charging effects are ignored. Clearly, this list can easily be extended with a number of other obvious simplifications. Less trivial is the model error in the formation of the sample in the vicinity of the cathode surface. The particles are assumed to be randomly distributed over phase space, satisfying the macroscopic properties of the beam. Any correlation in the initial coordinates of the particles is ignored. Various authors have questioned the validity of MC simulations on the grounds that this leads to an unrealistic distribution of the potential energy of the particles. For instance, Rose and Spehr (1983) stated that the MC program allows more than one particle in the same initial position simultaneously, which obviously does not correspond to reality. However, since the source routine exploits a pseudo-random number generator with a cycle of the order 10^9, there is no risk of selecting two identical initial positions as long as the total number of particles is smaller than, say, 10^8. Nevertheless, the initial positions can be very close to each other, leading to the generation of pairs of particles with an extremely large potential energy. For that reason one should disregard large displacements in the data analysis. This is done by evaluating those width measures that do not depend on the exact tails of the distribution, such as the Full Width median value (FW_{50}) and the Full Width at Half Maximum ($FWHM$).
- *Ray tracing errors.* The numerical ray tracing is affected by an integration error that depends on the size of the time step Δt. The local integration error in the positions and velocities is of the third order in the time step Δt, as can be seen from Eq. (13.4.7). Therefore, the total integration error is of the second order in Δt. This implies that an increase of the integration accuracy by a factor A requires a factor $A^{1/2}$ more time steps, leading to a corresponding increase in cpu time. The total integration error is estimated by the program from Eq. (13.4.7), and the user can directly verify whether the accuracy of the ray tracing has been sufficient or not.

The errors generated by the analytical ray tracing are related to the reduction to pair interactions. The calculation of the displacements within the two-particle model can be considered as exact. The reduction to pair interactions is justified as long as the average number of strong interactions per particle experienced during the flight is small. This number is counted by the analytical ray tracing routine and is printed out, which provides an indication of the validity of the Fast Monte Carlo (FMC) approach. In this connection, it should be noted that the errors in the final coordinates due to the occurrence of strong collisions are of a stochastic nature, which has a favorable effect on the overall accuracy of the FMC approach. Furthermore, it was found that the error in the prediction of the statistical properties of the beam (like the width of the energy and trajectory displacement distribution) is significantly smaller than the error in the exact final coordinates of the individual particles. See Jansen (1987) for a more detailed discussion on this topic.

- *Statistical errors.* The accuracy of the estimation of the statistical properties of the beam is related to the total number of particles N_{tot} accumulated in all seeds ($N_{tot} = N_{sam} \times N_{seed}$). The statistical error in the rms value and the FW_{50} value of the energy and trajectory displacement distribution are estimated within the program, which allows a direct verification by the user. As an additional test, one can run the same simulation a number of times using different start values for the random number generator. By observing the variation of the final results, one obtains some idea of the statistical error involved. In order to improve the statistical accuracy by a factor A one has to increase N_{tot} by a factor A^2. This should be done by running more seeds. Accordingly, the cpu time increases with a factor A^2, as can be seen from Eqs. (13.4.12) and (13.5.10).
- *Finite size errors.* The errors related to the finite size of the sample were discussed in Section 13.7. The application of z-dependent velocity correction on the intermediate coordinates or the final coordinates leads to some reduction of these effects and provides insight in their magnitude. In general, one should choose the number of particles in the sample N_{sam} large enough to assure that these effects are insignificant. An increase of N_{sam} by a factor A leads to an increase of the required cpu time by factor A^2, assuming that the interaction range of the particles is not limited.

As far as cpu time is concerned, one sees that the reduction of finite-size errors and statistical errors both lead to a quadratic increase with a certain accuracy improvement factor A, whereas the integration error of the numerical ray tracing shows a square-root dependency on A. Therefore, when using numerical ray tracing, it is a good strategy to choose the nominal number of steps per meter N_{step} large enough to rule out the contribution of integration errors. In order to run the program economically, one should take N_{sam} and N_{seed} such that the finite size error and the statistical error are of the same magnitude.

Clearly, model errors are more difficult to estimate. The major problems occur with the simulation of the beam in the source area, as can be understood from the preceding analysis. Disregarding model errors, the accuracy obtained for the defocussing distance and the FW_{50} values of the energy and spatial distribution is estimated to be better than 10% for normal operating conditions. The accuracy of the calculated $FWHM$ values is in general somewhat worse, which is related to the use of the polynomial fit procedure and the reliability of the estimation of the central height of the distribution. The accuracy of the predictions for the edge width of a shaped spot depends strongly on the number of particles constituting the edge. In general, an accurate estimation of the edge width requires a substantially larger number of particles N_{tot} than the estimation of the FW_{50} or $FWHM$ width of a central distribution.

13.10. MC simulation vs analytical modeling

Our main objective to develop an MC simulation program was to obtain a method to verify the fundamental assumptions underlying the analytical theory. However, the MC program is, in itself, a very useful tool in the design of electron and ion beam columns. A comparison of the pros and cons of the analytical theory and the numerical MC modeling leads to the following conclusions:

- An accurate analytical theory provides insight into the mechanism of the interaction effects and explicitly shows the dependency on the experimental parameters. The MC approach is less suited to gain any physical appreciation for the different phenomena of particle interactions.
- When implemented in a program, the analytical theory can be used for a practically instantaneous evaluation of a column design. It is therefore very suited for the process of design and optimization. This is a clear advantage over the MC simulation, which is, in fact, a rather cumbersome tool for the optimization of a column. Every change in the system parameters requires a new simulation. Performing some kind of optimization, therefore, takes a lot of cpu time (and patience of the operator).
- The analytical theory is less suited to provide accurate predictions for the impact of particle interactions in a complex beam geometry, consisting of many, possibly nonrotational symmetric beam segments. The MC method is for that purpose preferred.
- The combination of an accurate analytical theory and a MC simulation program is very convenient in the design of high-current and/or low-voltage particle beam systems. The analytical theory is

indispensable in the process of optimizing a column. The MC simulation program is most suited to produce an accurate evaluation of the final design and may serve as a backup of the analytical predictions.

13.11. Program organization and examples

In order to demonstrate the organization and performance of the MC program we will, as an example, consider the simulation of the Perkin-Elmer AEBLE 150 shaped beam lithography column. The column is schematically drawn in Fig. 13.1. Most of the input data required for the MC program was obtained from Veneklasen (1985). However, some numbers had to be guessed, and the column dimensions and lens settings used in the simulation may differ slightly from the actual values.

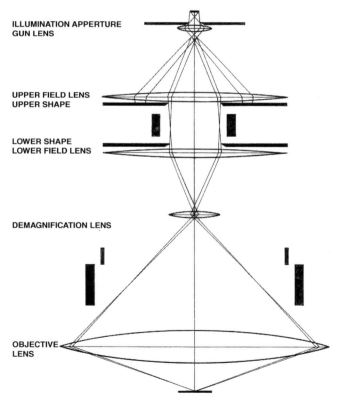

Fig. 13.1 Schematic view of the Perkin-Elmer AEBLE 150 Column.

The MC program requires two input files, the so-called *system* file and the *MC data* file. The *system* file contains the information on the properties of the source and optical components and specifies the column dimensions. It also instructs the program how to perform the ray tracing in the drift sections of the column and how to analyze the final coordinates of the particles. The *MC data* file specifies the mass and charge of the particles in the beam, the emission current, the beam voltage and the initial energy distribution, as well as some global program control parameters.

Fig. 13.2 shows the system file used for the AEBLE 150. All lines with a C or an asterisk (*) in the first column are ignored by the program. The lines starting with a C are used as comment lines; those with an asterisk contain (temporary) deactivated commands. Every command line is separated into a number of fields. The first field contains the name of the routine to be called by the program (SOURCE, DRIFT2, LENS, etc.). The other fields specify the input data required by that routine. For instance, the second field always specifies a z-coordinate. A "−1." as input instructs the program to take over the last value read by the program, and it can be used in fields specifying an axial coordinate or a beam voltage. A more detailed description of the various commands can be found in the reference manual of the program; see Jansen (1988c). The interaction between the particles can be turned on or off, for each drift section separately (by putting a 0 or a 1 in the fifth field following the DRIFT command). In the present example, the interaction is switched off in the sections above the upper shape and switched on for the sections below the upper shape. The specified source does not correspond to the actual source but to the first image of the source crossover, which is located just below the gun lens. The sections above the upper shape serve only to obtain a proper illumination of the upper shape and do not reflect the actual properties of the beam in this part of the column. All system commands can be called more than once except the SOURCE command and the ACCUM command. The ACCUM command instructs the program to accumulate the final coordinates obtained in the individual seeds.

Fig. 13.3 shows the *MC data* file used for a simulation of the AEBLE column for a beam current at the target of $2\mu A$. The first line specifies the particle mass and charge (as the absolute value of the charge in elementary charge units). The second line specifies the beam current and voltage at the source, the initial energy spread, etc. The reader is referred to the reference manual of the program for more detailed information.

The program produces five output files. One file is a general output file, containing the messages of the various routines called by the program and the results of the data analysis. When the STOREC command is included in the *system* file, the program produces a separate output file,

```
C!******!**********!**********!**********!**********!**********!**********!
C
C    AEBLE 150 SHAPED BEAM COLUMN - SYSTEM FILE
C
C!******!**********!**********!**********!**********!**********!**********!
C
C--------ILLUMINATION OF UPPER SHAPE (A1):
C NOTE: SOURCE REPRESENTS FIRST IMAGE OF CROSSOVER (AFTER GUN LENS)
C         ANGLE IS CHOSEN SUCH THAT APERTURE A1 IS JUST COMPLETELY ILLUMINATED
 :SOURCE:  -86.92E-3:    0.     :    0.    :    2    :  8.3E-6  :  8.3E-6  :
 :      :    1      : 1.55E-3   :  3.E3    :
 :DRIFT2:   -1.     : -10.00E-3:   -1.    :   -1.   :    0    :    1     :
C--------UPPER FIELD LENS :
 :LENS  :   -1.     :    0.     :    0.    : 76.92E-3 :
 :DRIFT2:   -1.     :    0.     :   -1.    :   -1.   :    0    :    1     :
C
C--------UPPER SHAPE :
 :APERTR:   -1.     :    0.E0   :  0.E0    : 75.0E-6 : 75.0E-6  :    2     :
C
C--------IN CASE OF NO-SHAPING BY LOWER SHAPE ---------------------------
 :DRIFT2:   -1.     : 60.00E-3  :   -1.    :   -1.   :    1    :    1     :
C
C
C--------IN CASE OF SHAPING BY LOWER SHAPE ------------------------------
*:DRIFT2:   -1.     : 50.00E-3  :   -1.    :   -1.   :    1    :    1     :
C--------LOWER SHAPE
*:APERTR:   -1.     :    0.E0   :  0.E0    : 75.0E-6 : 75.0E-6  :    2     :
*:DRIFT2:   -1.     : 60.00E-3  :   -1.    :   -1.   :    1    :    1     :
C
C
C--------LOWER FIELD LENS :
 :LENS  :   -1.     :    0.E0   :  0.E0    : 76.92E-3 :
 :DRIFT2:   -1.     : 140.00E-3 :   -1.    :   -1.   :    1    :    1     :
C--------DEMAGNIFICATION LENS :
 :LENS  :   -1.     :    0.E0   :  0.E0    : 2.961E-3 :
 :DRIFT2:   -1.     : 297.02E-3 :   -1.    :   -1.   :    1    :    1     :
C--------OBJECTIVE LENS :
 :LENS  :   -1.     :    0.E0   :  0.E0    :39.689E-3 :
 :DRIFT2:   -1.     : 350.49E-3 :   -1.    :   -1.   :    1    :    1     :
C
C--------ACCUMULATION OF PARTICLES :
 :ACCUM :
 :STOREC:
*:READC :
C--------MONTE CARLO STATISTICS :
 :SYMEBR:
 :TBR   :   -1.     :    0      :
 :RECTBR:   -1.     :    0      :
 :TBR   :   -1.     :    1      :
 :RECTBR:   -1.     :    1      :
 :PROCCO:    1      :   10      :    1    :    1    :  0.8E0   :    2     :
 :SYMEBR:
C
 :STOREP:   -1.     :    0      :
 :STOREP:   -1.     :    1      :
```

Fig. 13.2 System file for the MC simulation of the AEBLE 150 Column.

```
C**********************************************************
C
C DATA-INPUT FILE : AEBLEI2.DAT
C
C**********************************************************
C
C THE PARTICLES ARE ELECTRONS:
C:  PM            :  PQ
:   9.10956D-31   :  1.0D0
C
C:  BI            :  BV            :  FWE         :  IDE      :  ICONS
:   4.000D-06     :  2.0D+04       :  0.0D0       :  1        :  1
C:  INTER         :  NSAM          :  NSEED       :  NFIELD   :  NRAND
:   1             :  600           :  5           :  1000     :  123
C:  NSTEP         :  ISTEPA        :  IRLIM       :  NINT     :  IPROC
:   1000          :  1             :  0           :  1000     :  0
C:  NDE           :  NTE           :  NEXPE       :  IFIXE    :  PMAXE
:   40            :  6             :  0           :  3        :  -2.50D0
C:  NDT1          :  NTT1          :  NEXPT1      :  IFIXT1   :  PMAXT1
:   40            :  6             :  0           :  0        :  -2.50D0
C:  NDT2          :  NTT2          :  NEXPT2      :  IFIXT2   :  PMAXT2
*:  40            :  6             :  0           :  0        :  -2.50D0
:   40            :  6             :  1           :  3        :  1.00D0
C:  ISTOREE       :  ISTORET1      :  ISTORET2
:   1             :  1             :  1
C:  IMT1          :  IMT2
:   1             :  1
```

Fig. 13.3 MC data file for the MC simulation of the AEBLE 150 Column.

containing the final positions and velocities of all particles. This file can be read by the program during a next run (using the READC command), which allows for a new analysis of the final coordinates, without repeating the ray tracing. Two separate output files are used to store the final energy distribution (evaluated by the routines SYMBER and ASYEBR) and the final spatial distributions (evaluated by TBR, RNDTBR and RECTBR). Finally, when the STOREP command is issued in the *system* file, the program produces an output file containing the lateral positions of the particles in the plane corresponding to the specified z-coordinate.

A separate plot program (called MCPLOT) is available to read the data from the various output files and plot the energy and spatial distributions, as well as the final positions of the particles. Figs. 13.4 and 13.5 show the plots of the spatial distribution of the particles in the Gaussian image plane, obtained for a target current of 2 and 5 μA respectively. The edge width of the spot is determined by the combined action of the space charge and the trajectory displacement effect. There are no other contributions to the edge width, since all aberration coefficients of all lenses

428   Monte Carlo simulation of particle beams

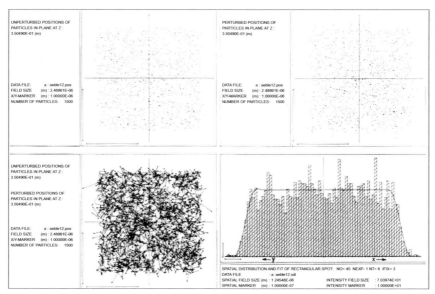

Fig. 13.4 Particle positions in the Gaussian image plane of the AEBLE 150 Column for a 2 μA beam current. The plots at the upper left and the upper right show the unperturbed and the perturbed positions respectively. The plot at the lower left shows the displacements from the unperturbed positions (the squares indicate the perturbed positions). The bars in the lower left corner of these plots is equal to 1 μm. The histogram at the lower right is formed by projecting the perturbed positions to the y-axis (left side of the histogram) and the x-axis (right side of the histogram) and folding the projected positions with respect to the origin. The horizontal bar in the left corner is equal to 0.1 μm. The vertical bar represents 10 particles. The edges are fitted by the MC program using a normal sixth order polynomial function, imposing that the function itself as well as its derivative are connected continuously to the center intensity level of the spot.

are taken equal to zero. Fig. 13.6 compares the intensity distribution in the Gaussian image plane and the plane of best focus for the considered cases of a beam current of 2 and 5 μA at the target. The remaining edge width in the plane of best focus is mainly due to the trajectory displacement effect. Fig. 13.7 shows the energy distribution at the target for the 2 and 5 μA case. As the energy spread at the source was taken equal to zero, the broadening is entirely the result of the interaction between the particles in the part of the beam below the upper shape. More results on the performance of the AEBLE 150, as well as the IBM EL3 shaped beam column, are presented by Jansen (1988a).

Monte Carlo simulation of particle beams   429

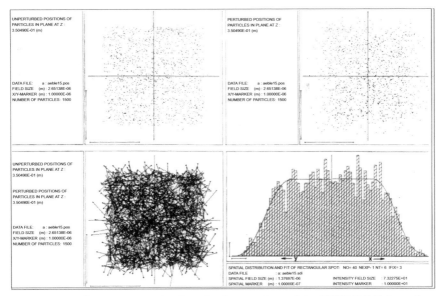

Fig. 13.5 Particle positions in the Gaussian image plane of the AEBLE 150 Column for a 5 μA beam current. See the caption of Fig. 13.4 for further details.

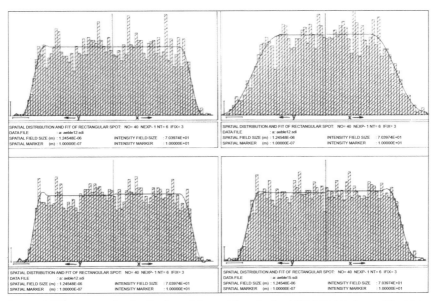

Fig. 13.6 Histograms for the final particle positions in the AEBLE 150 Column. The histograms at the left pertain to a 2 μA beam current. Those at the right to a 5 μA beam current. The upper histograms are formed in the Gaussian image plane and the lower ones in the plane of best focus, which is evaluated by the MC program by seeking the plane for which the edge width becomes minimum. See the caption of Fig. 13.4 for further details.

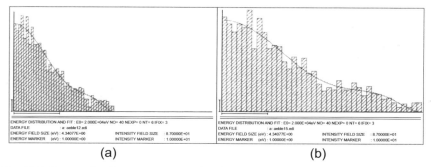

Fig. 13.7 Energy distributions generated in the section of the AEBLE 150 Column below the upper shape for a 2 μA beam current (panel a) and a 5 μA beam current (panel b). The distributions are plotted on the same scale. The horizontal bar indicates a 1 eV energy spread. The vertical bar represents 10 particles. The energy range included in the histograms is chosen by the MC program such that the maximum energy is equal to 2.5 times the Half Width at Half Maximum (*HWHM*) energy width (Thus, the histograms do not contain all particles). The distributions are fitted by the MC program using an even polynomial function of six terms, imposing that the function itself as well as its derivative are zero in the last division of the histogram.

Appendix 13. A. Random number routine

For the generation of pseudo-random numbers, the program employs the so-called linear congruential method, which is described in various textbooks on computer programming. See, for instance, Knuth (1969). Given a number X_i, the next number X_{i+1}, in the sequence is determined using

$$X_{i+1} = (aX_i + c) \bmod m, \qquad (13.A.1)$$

where a, c and m represent some carefully chosen "magic numbers." The constant a is called the multiplier, c the increment and m the modulus. The program uses

$$a = 65539, \quad c = 0, \quad m = 2^{31} - 1 = 2147483647, \qquad (13.A.2)$$

which has a period (or repealing cycle) equal to m. For the start value X_0, one has to choose some number between 1 and m. In order to obtain numbers R_i, which are uniformly distributed between 0 and 1, the program computes

$$R_{i+1} = X_{i+1}/m. \qquad (13.A.3)$$

Double precision reals are used to represent the quantities a, m and R_i, which guarantees that the routine is machine independent. Clearly, this causes some loss in computation speed, but this is irrelevant in our application.

Eqs. (13.A.1)–(13.A.3) constitute the scheme for the generation of a one-dimensional uniform distribution. In addition to this distribution, the source routine requires the simulation of a Gaussian, a Lorentzian and an exponential distribution. Furthermore, these distributions can either be one-dimensional or rotational-symmetric two-dimensional. For the latter, one wishes to simulate the distribution of the radial coordinate. In the remaining part of this appendix, we will outline a method to transform a sequence of random numbers R that are produced by the scheme of Eqs. (13.A.1) and (13.A.3) into either a sequence of random numbers R_1, which follows some one-dimensional distribution $f_1(R_1)$ or into a sequence of random numbers R_2, which follows some two-dimensional distribution $f_2(R_2)$. Let us denote the one-dimensional uniform distribution generated by Eqs. (13.A.1) and (13.A.3) as $f_{1u}(R)$

$$f_{1u}(R) = \begin{cases} 1 & \text{for } 0 \le R \le 1 \\ 0 & \text{for } R < 0 \text{ or } R > 1. \end{cases} \tag{13.A.4}$$

The problem is to generate the following distributions

$$f_{2u}(R_2) = \begin{cases} 1/\pi & \text{for } 0 \le R_2 < 1 \\ 0 & \text{for } R_2 < 0 \text{ or } R_2 > 1. \end{cases} \quad \text{(two-dim. uniform)} \tag{13.A.5}$$

$$f_{1G}(R_1) = \frac{1}{(2\pi)^{1/2}\sigma} e^{-R_1^2/2\sigma^2} \quad \text{(one-dim. Gaussian)} \tag{13.A.6}$$

$$f_{2G}(R_2) = \frac{1}{2\pi\sigma^2} e^{-R_2^2/2\sigma^2} \quad \text{(two-dim. Gaussian)} \tag{13.A.7}$$

$$f_{1L}(R_1) = \frac{1}{\pi A_1 \left[1 + (R_1/A_1)^2\right]} \quad \text{(one-dim. Lorentzian)} \tag{13.A.8}$$

$$f_{2L}(R_1) = \frac{1}{2\pi A_2^2 \left[1 + (R_1/A_2)^2\right]^{3/2}} \quad \text{(two-dim. Lorentzian)} \tag{13.A.9}$$

$$f_{1e}(R_1) = \frac{1}{B} e^{-R_1/B} \quad \text{for } R_1 > 0 \quad \text{(one-dim. exponential)} \tag{13.A.10}$$

For each of the one-dimensional distributions, an auxiliary function $I_1(R_1)$ is defined as

$$I_1(R_1) = \int_0^{R_1} f_1(x) n \, dx, \tag{13.A.11}$$

where $n = 1$ for the uniform and the exponential distribution and $n = 2$ for the other one-dimensional distributions. Similarly, for each of the two-dimensional distributions, an auxiliary function $I_2(R_2)$ is defined as

$$I_2 R_2 = \int_0^{R_2} f_2(x) 2\pi x \, dx. \tag{13.A.12}$$

For the one-dimensional uniform distribution specified by Eq. (13.A.4), one finds from Eq. (13.A.11) that $I_{1u}(R) = R$. In order to transform a random number R into R_1 or into R_2, we will exploit the fact that each distribution is normalized to unity. This implies

$$I_1(R_1) = I_{1u}(R) = R \implies R_1 = I_1^{-1}(R)$$
$$I_2(R_2) = I_{1u}(R) = R \implies R_2 = I_2^{-1}(R),$$
(13.A.13)

where the functions I_1^{-1} and I_2^{-1} denote the inverse functions of I_1 and I_2 respectively. With the Eqs. (13.A.11) and (13.A.12), one obtains from Eqs. (13.A.5)–(13.A.10)

$$I_{2u}^{-1}(R) = \sqrt{R} \qquad \text{(two-dim. uniform)} \qquad (13.A.14)$$
$$I_{1G}^{-1}(R) = \sqrt{2}\sigma \,\text{erf}^{-1}(R) \qquad \text{(one-dim. Gauusian)} \qquad (13.A.15)$$
$$I_{2G}^{-1}(R) = \sigma\sqrt{-2\ln(1-R)} \qquad \text{(two-dim. Gaussian)} \qquad (13.A.16)$$
$$I_{1L}^{-1}(R) = A_1 \tan(\pi R/2) \qquad \text{(one-dim. Lorentzian)} \qquad (13.A.17)$$
$$I_{2L}^{-1}(R) = A_2\sqrt{(1-R)^{-2} - 1} \qquad \text{(two-dim. Lorentzian)} \qquad (13.A.18)$$
$$I_{1e}^{-1}(R) = -B \,\ln(1-R) \qquad \text{(one-dim. exponential)}, \qquad (13.A.19)$$

which solves the problem. Random numbers R_1 or R_2 can be generated from a sequence of uniformly distributed numbers R, using Eq. (13.A.13) in combination with the appropriate function chosen from Eqs. (13.A.14)–(13.A.19).

The transformations outlined above involve only elementary functions, except for the one-dimensional Gaussian distribution. The corresponding Eq. (13.A.15) contains the inverse of the Fresnel error function erf^{-1}, which is defined by the first equation of (4.5.9). In order to avoid the use of an analytical approximation for erf^{-1}, an alternative two-step approach is employed to generate numbers that follow a one-dimensional Gaussian distribution. In the first step, the program selects a numbers R and evaluates R_2 from Eq. (13.A.16), which refers to a two-dimensional Gaussian distribution. In the second step, the program selects another random number R and computes R_1 from R and R_2 as

$$R_1 = R_2 \cos(2\pi R), \qquad (13.A.20)$$

which represents the projection of the radius R_2 on an axis that is rotated over an angle $2\pi R$ with respect to the radius R_2. The numbers R_1 produced this way follow a one-dimensional Gaussian distribution.

In order to standardize the input of the random number routine, the different width measures σ, A and B are expressed in terms of the corresponding Half Width at Half Maximum ($HWHM$) values, using

$$\sigma = HWHM/\sqrt{2\ln 2}$$
$$A_1 = HWHM, \qquad A_2 = \frac{HWHM}{\left(2^{2/3} - 1\right)^{1/2}} \qquad (13.A.21)$$
$$B = HWHM/\ln 2,$$

as follows from Eqs. (13.A.15)–(13.A.19). The input of the random number routine now consists of four parameters: X_i, I_{dis}, I_{dim} and HWHM. X_i is the last random number that was generated by Eq. (13.2.3) or the start value X_0 (in the first call). I_{dis} selects the type of distribution (1=uniform, 2=Gaussian, 3=Lorentzian, 4=exponential). I_{dim} specifies the dimension of the distribution (1 or 2). The routine returns a random number R, which is selected according to the specified distribution.

Appendix 13. B. Polynomial fit algorithm

In this appendix, we will outline a method to fit a set of data points $\{(x_i, y_i) | i = 1, 2, \ldots, N\}$ with the polynomial function of M terms,

$$y(x) = \sum_{j=1}^{M} a_j x^{e_j}, \qquad e_j = \frac{n_1(M-j) + n_2}{n_3}, \qquad (13.B.1)$$

where n_1, n_2 and n_3 are predefined constants and the parameters a_j (where $j = 1, \ldots, M$) are to be determined by the fit procedure. The different sets of values used for n_1, n_2 and n_3 are specified by Eq. (13.8.5). We wish to determine the set of parameters a_j for which the merit function

$$S = \sum_{i=1}^{N} [y_i - y(x_i)]^2 \qquad (13.B.2)$$

has a minimum. This implies

$$\frac{\partial S}{\partial a_j} = 0 \quad \text{for} \quad j = 1, 2, \ldots, M. \qquad (13.B.3)$$

Solving Eq. (13.B.3) constitutes a common least square fit problem. The particular problem we like to investigate here is to solve the least square problem in the subspace of parameters a_j for which the function $y(x)$ fulfills one of the following conditions

(a) $y(x_0) = y_0$ $\qquad\qquad$ $(x_0 \neq 0)$
(b) $y'(x_1) = y_1'$ $\qquad\qquad$ $(x_1 \neq 0)$ $\qquad$ (13.B.4)
(c) $y(x_0) = y_0$ and $y'(x_1) = y_1'$ $\quad (x_0 \neq 0$ and $x_1 \neq 0)$,

where $y'(x)$ is the derivative of $y(x)$. The constants x_0 and x_1 are assumed to be nonzero. We will investigate each of the three cases given by Eq. (13.B.4) separately. The reader may notice that Eq. (13.B.1) differs from Eq. (13.8.4) by the fact that it assumes $x_s = y_s = 0$. This choice simplifies the notation without loss of generality. The shifts x_s and y_s can be incorporated afterwards.

Let us define the auxiliary fit function $y^*(x)$ as

$$y^*(x) = y_0 f_0(x) + y_1 f_1'(x) + \sum_{j=1}^{M-g} a_j^* \left[x^{e_j} - x_0^{e_j} f_0(x) - e_j x_1^{e_j-1} f_1(x) \right]. \quad (13.B.5)$$

Our aim is to find functions $f_0(x)$ and $f_1(x)$ for which $y^*(x)$ fulfills the additional constraint (a), (b) or (c), for any set of parameters a_j^*. The number of terms in the summation is diminished by g, which reflects the reduction of the number of degrees of freedom caused by the additional constraints imposed on the fit. For the cases (a) and (b) $g=1$, while $g=2$ for case (c). Eq. (13.B.5) is organized such that the conditions (a), (b) and (c) are fulfilled for those functions $f_0(x)$ and $f_1(x)$, which show the following properties

$$\begin{array}{ll} \text{(a)} & f_0(x_0) = 1, \quad f_1(x) = 0 \quad \text{for all } x \\ \text{(b)} & f_1'(x_1) = 1, \quad f_0(x) = 0 \quad \text{for all } x \\ \text{(c)} & f_0(x_0) = 1, \quad f_0'(x_1) = 0 \\ & f_1(x_0) = 0, \quad f_1'(x_1) = 1. \end{array} \quad (13.B.6)$$

Thus case (a) only concerns $f_0(x)$, taking $f_1(x) = 0$ for all x; case (b) only concerns $f_1(x)$, taking $f_0(x) = 0$ for all x, while case (c) involves both functions. An additional constraint that should be introduced at this point is that the functions $y^*(x)$ and $y(x)$ are of the same type. This means that for any given set of parameters a_j^* one should be able to find a corresponding set of parameters a_j such that

$$y(x) = y^*(x). \quad (13.B.7)$$

This conditions determines the order of the functions $f_0(x)$ and $f_1(x)$, as will explicitly be shown later. Assuming that one can indeed find functions $f_0(x)$ and $f_1(x)$, which fulfill the conditions of Eqs. (13.B.6) and (13.B.7), the next step would be to solve the set of equations

$$\frac{\partial S}{\partial a_j^*} = 0 \quad \text{for} \quad j = 1, 2, \ldots, M - g, \quad (13.B.8)$$

in which the merit function S is now given by

$$S = \sum_{i=1}^{N} [y_i - y^*(x_i)]^2, \quad (13.B.9)$$

similar to Eq. (13.B.2). By solving Eq. (13.B.8), one obtains a set of parameters a_j^*. Using Eq. (13.B.7), one can express this solution in terms of a set of parameters a_j, and the problem is solved.

Our strategy is now explained and the remaining problem is the choice of the functions $f_0(x)$ and $f_1(x)$. For the separate cases (a), (b) and (c), we found that the following functions $f_0(x)$ and $f_1(x)$ meet the requirements expressed by Eqs. (13.B.6) and (13.B.7)

(a) $f_0(x) = \left(\dfrac{x}{x_0}\right)^p$, $f_1(x) = 0$, $p = \dfrac{n_2}{n_3}$

(b) $f_0(x) = 0$, $f_1(x) = \dfrac{x^q}{qx_1{}^{q-1}}$, $q = \dfrac{n_1 - n_2}{n_3}$

(c) $f_0(x) = \dfrac{qx_1{}^q x^p - px_1{}^p x^q}{qx_1{}^q x_0{}^p - px_1{}^p x_0{}^q}$, $f_1(x) = \dfrac{x_0{}^p x^q - x_0{}^q x^p}{qx_0{}^p x_1{}^{q-1} - px_0{}^q x_1{}^{p-1}}$,

$p = \dfrac{n_2}{n_3}$, $q = \dfrac{n_1 + n_2}{n_3}$.

(13.B.10)

Notice that $p = e_M$, where e_M is defined by Eq. (13.B.1). Thus, p is equal to the order of the lowest order term in $y(x)$. In case (c), we took $q = e_{M-1}$, equal to the order of the next term in $y(x)$. To understand the choice of q in case (b), one has to consider the situations $n_2 = 0$ and $n_2 = 1$ separately. For $n_2 = 0$, we have $q = n_1/n_3 = e_{M-1}$. For $n_2 = 1$, we have $q = (n_1-1)/n_3 = 1/n_3 = e_M$, using that $n_1 = 2$ whenever $n_2 = 1$, as can be seen from Eq. (13.8.5). Thus in both cases, q is chosen equal to the order of the lowest order term in $y(x)$, which is not a constant.

The reader can directly verify that the functions $f_0(x)$ and $f_1(x)$, defined by Eq. (13.B.10), fulfill the conditions of Eq. (13.B.6). One also sees that the functions $y(x)$ and $y^*(x)$ are of the same type, as required by Eq. (13.B.7). One finds that a given set of parameters $a_j{}^*$ is equivalent to a set of parameters a_j, which is for the separate cases (a), (b) and (c) specified by

(a), (c), (b) for $n_2 = 1$: $a_j = a_j{}^*$ for $j = 1, 2, \ldots, M - g$
(b) for $n_2 = 0$: $a_j = a_j{}^*$ for $j = 1, 2, \ldots, M - 2, M$

(a) $a_M = \dfrac{1}{x_0{}^p} \left(y_0 - \displaystyle\sum_{j=1}^{M-1} a_j{}^* x_0{}^{e_j} \right)$

(b) $a_k = \dfrac{1}{qx_1{}^{q-1}} \left(y_1' - \displaystyle\sum_{j=1}^{M-1} a_j{}^* e_j x_1{}^{e_j - 1} \right)$ with $k = \begin{cases} M - 1 & \text{for } n_2 = 0 \\ M & \text{for } n_2 = 1 \end{cases}$

(c) $a_M = \dfrac{1}{qx_1{}^q x_0{}^p - px_1{}^p x_0{}^q} \times \left[y_0 q x_1{}^q - y_1' x_0{}^q x_1 \right.$

$\left. - \displaystyle\sum_{j=1}^{M-2} a_j{}^* \left(x_0{}^{e_j} q x_1{}^q - e_j x_1{}^{e_j} x_0{}^q \right) \right]$

$a_{M-1} = \dfrac{1}{qx_1{}^q x_0{}^p - px_1{}^p x_0{}^q} \times \left[-y_0 p x_1{}^p + y_1' x_0{}^p x_1 \right.$

$\left. - \displaystyle\sum_{j=1}^{M-2} a_j{}^* \left(-x_0{}^{e_j} p x_1{}^p + e_j x_1{}^{e_j} x_0{}^p \right) \right],$

(13.B.11)

which gives the resulting fit function in the form of Eq. (13.B.1).

The set of parameters a_j^* is obtained from Eq. (13.B.8) in the usual way by solving the normal equations

$$0 = \sum_{i=1}^{N} [y_1 - y^*(x_i)] \left[x_i^{e_j} - x_0^{e_j} f_0(x_i) - e_j x_1^{e_j-1} f_1(x_i) \right] \quad (13.B.12)$$

which follow by substitution of Eq. (13.B.9) into Eq. (13.B.8), utilizing Eq. (13.B.5). Alternatively, one may express the normal equation as

$$\sum_{j=1}^{M-g} T_{kj} a_j^* = b_k, \quad (13.B.13)$$

where T is a $(M-g) \times (M-g)$ matrix and b is a vector of length $(M-g)$, with elements

$$T_{kj} = \sum_{i=1}^{N} \left[x_i^{e_k} - x_0^{e_k} f_0(x_i) - e_k x_1^{e_k-1} f_1(x_i) \right] \left[x_i^{e_j} - x_0^{e_j} f_0(x_i) - e_j x_1^{e_j-1} f_1(x_i) \right]$$

$$b_k = \sum_{i=1}^{N} [y_i - y_0 f_0(x_i) - y_1' f_1(x_i,)] \left[x_i^{e_k} - x_0^{e_k} f_0(x_i) - e_k x_1^{e_k-1} f_1(x_i) \right].$$

(13.B.14)

Notice that the elements T_{kj} depend only on x_i and not on y_i, while the elements b_k depend on both x_i and y_i. Eqs. (13.B.13) are solved by means of LU-decomposition and back substitution. For this, the reader is referred to the standard textbooks on numerical programming; see for instance Press et al. (1986). As the matrix T depends only on the set x_i, the decomposition has to be performed only once for a given set x_i and can then be used to solve a_j^* for a number of sets y_i, having the same arguments x_i. In our problem, the quantities x_i are equally spaced with a distance $h = x_{i+1} - x_i$, which is equal to the division width of the histogram. The fit routine uses the scaled arguments x_i/h as input, instead of the arguments x_i themselves. The advantage of this approach is that one knows beforehand that the element T_{11} is, in general, the largest element in the matrix T and the element $T_{(M-g)(M-g)}$ the smallest, since $x_i/h > 1$. This knowledge is exploited to simplify the pivot strategy in the LU-decomposition, which should be incorporated to guarantee an optimum numerical accuracy of the fit.

CHAPTER FOURTEEN

Comparison of analytical theory with Monte Carlo simulations

Contents

14.1. Introduction	437
14.2. General aspects	438
14.3. Voltage and current dependencies for a fixed geometry	441
14.4. Geometry and current dependencies for a fixed beam voltage	445
14.5. Discussion of the results	450

14.1. Introduction

In this chapter, we will compare the analytical theory for the Boersch effect, the trajectory displacement effect and the space charge effect with the results of numerical Monte Carlo simulations. The Monte Carlo technique provides the capability to verify some of the basic assumptions underlying the analytical theory, particularly the reduction of the N-particle dynamical problem to a two-particle problem. It also provides an independent check on the overall correctness of the final equations obtained within the analytical theory.

The analysis of this chapter is restricted to a single segment of an electron beam with a crossover in the middle. The current density distribution and angular distribution in the crossover are assumed to be uniform. The study is based on the results of 276 runs of the Monte Carlo program, covering a wide range of operating conditions. The simulated beam currents varied from 0.01 to 50 μA and the beam potentials from 1 to 50 kV. A number of different beam geometries were considered, taking as extreme cases a nearly cylindrical beam with a beam semi-angle $\alpha_0 = 0.5$ mR and a crossover radius $r_c = 20$ μm and a beam segment with a narrow crossover with $\alpha_0 = 10$ mR and $r_c = 1$ μm. The beam length L was in all simulations equal to 0.1 m.

In general, the Monte Carlo results are in excellent agreement with the predictions of the analytical theory. A good qualitative and quantitative agreement is found for low-current beams as well as for high-current beams, while some quantitative differences occur for intermediate beam currents. These differences are probably caused partly by the inaccuracy of the fit functions employed in the analytical model and partly by the fact that the analytical results refer to the displacements experienced by particles running along the beam axis, while the Monte Carlo program takes the displacements of all particle into account. There also exists some disagreement in the predictions of the space charge defocussing distance for low beam currents, which is due to the incompatibility of the different criteria used in the two models to determine the plane of best focus. However, as the defocussing distances are extremely small for these currents, the differences are irrelevant from an experimental point of view.

The chapter starts with a discussion of the fundamental issues involved. The results of the various Monte Carlo simulations are presented next and are compared with the predictions of the analytical theory. The chapter concludes with a discussion of the results.

14.2. General aspects

In the previous chapter, the technique of numerical Monte Carlo (MC) simulation of charged particle beams was discussed. The Monte Carlo program provides the means to verify some of the basic assumptions underlying our analytical model for Coulomb interactions in particle beams. The principle aspects of the extended two-particle approach used to calculate the Boersch effect and the trajectory displacement effect were outlined in Chapter 5, while the main results of the actual calculations were presented in Chapters 7 and 9 respectively. The analytical model for the space charge effect was outlined in Chapter 11. A summary of the final equations for the Boersch effect, the trajectory displacement effect and the space charge effect will be given in Chapter 16. The final equations of the analytical theory were implemented in a computer program called INTERAC, which provides a convenient tool to evaluate the analytical predictions for a certain set of experimental parameters. For short, we will indicate the predictions of the analytical theory sometimes as the results from INTERAC. In this chapter, the outcome of the INTERAC program is compared to the results of the MC program for the case of a single segment of an electron beam.

The main purpose of the comparison is to verify the accuracy of the results of the analytical theory. It should be recalled that the analytical theory for statistical interactions relies on the following assumptions:

(1) The problem can be modeled by considering the displacements experienced by a set of test particles running along a certain reference trajectory due to the interaction with a set field particles representing

the beam. The results obtained for the central reference trajectory (which is the beam axis) are representative for other (off-axis) trajectories as well.
(2) The initial condition of the beam can be considered monochromatic with respect to the normal energy. In other words, one may assume the initial axial velocity is the same for all particles.
(3) The total displacement of the test particle is equal to the sum of all displacements experienced in the two-particle interactions with the individual field particles, as expressed by Eq. (5.3.12).
(4) The field particles can be considered as statistically independent, as expressed by Eq. (5.3.1).
(5) Magnetic interactions and relativistic effects can be ignored.

Some of these assumptions can be verified within the framework of the analytical model itself. In Chapter 10, we investigated the changes in the final results occurring for off-axis reference trajectories and non-monochromatic beams respectively. According to assumptions (1) and (2), these differences should be minor. With respect to assumption (1), it was found that the results obtained by averaging over all possible reference trajectories are, in general, somewhat smaller than those pertaining to the central reference trajectory. The differences are, in general, less than 20% and for many conditions even less than 10%. The validity of assumption (2) was confirmed by the fact that the width of the initial energy distribution had no significant impact on the final results for practical operating conditions.

Assumption (3), which concerns the dynamical part of the problem, cannot directly be verified within the analytical model. For that purpose, we will use the MC program. As we mentioned in Chapter 13, two types of ray tracing algorithms are implemented in the MC program, referred to as numerical ray tracing (by the routine DRIFT1) and analytical ray tracing (by the routine DRIFT2). The analytical ray tracing algorithm utilizes the same dynamics as the analytical theory. Accordingly, it relies also on assumption (3). Thus, numerical ray tracing is required to verify the validity of assumption (3). The approach taken in this chapter is to perform all MC simulations twice, one time using numerical ray tracing and one time using analytical ray tracing. By comparing the results of both MC simulations, one gets a direct insight in the errors caused by the reduction of the full dynamical N-body problem to pair interactions.

The source routine of the MC program, which generates the initial condition of a bunch of particles, ignores any correlation in the initial coordinates of the particles. This approach is equivalent to assumption (4), underlying the analytical theory. Accordingly, assumption (4) cannot directly be verified with the Monte Carlo program. In this connection, it should be noticed that the MC program could be used to study the correlation in the particle coordinates originating from the mutual Coulomb interaction in the beam. For instance, one could investigate to what extent the interaction in a certain beam segment is affected by the correlation in the initial coordinates arising from the Coulomb interaction in the preceding beam segment. This "source" of correlation is ignored in the analytical theory, which treats every beam segment as independent. However, we did not pursue this problem and restricted the simulations to a single beam segment.

Assumption (5) cannot be verified with our MC program, since the ray tracing algorithms are nonrelativistic and account only for the Coulomb interaction between the particles. However, the implications of this assumption can be studied from first principles, as was shown in Section 10.6.

Summarizing, as far as the fundamental aspects are concerned, the main use of a comparison between analytical theory and MC simulations is to verify assumption (3). One could use MC simulations to verify the validity of assumptions (1) and (2). However, these assumptions were already verified within the analytical model itself, which provides a more elegant method.

Apart from testing some of the premises of the analytical model, the comparison with MC simulations is also useful as an independent check on the overall correctness of the final equations. The MC method is relatively straightforward, and its correct implementation can more easily be verified than the extensive calculations carried out within the analytical model. It is therefore suited as back-up for the analytical predictions.

For the comparison between analytical theory and MC simulations, we selected a large number of different cases, varying over a wide range of experimental conditions. For every case we obtained three predictions for each of the interaction effects: one from the MC program using numerical ray tracing (referred to as DRIFT1), another one from the MC program using analytical ray tracing (referred to as DRIFT2) and the third one from the analytical theory (referred to as INTERAC). Any differences between the results of DRIFT1 and DRIFT2 would imply that assumption (3) is inappropriate for that specific case. Differences between the results of DRIFT2 and INTERAC can have a variety of causes. Firstly, the MC simulation may be off due to the inherent statistical errors or due to the model errors related to the finite size of the particle bunch. These errors were discussed in Section 13.9. Secondly, it should be noticed that the final equations of the theory include functions that are obtained by parameter fitting of numerical data. This introduces some deviations from the exact theoretical predictions. Finally, there are

Comparison of analytical theory with Monte Carlo simulations   441

some fundamental differences between both approaches in the way the lateral displacements are separated into a space charge and a trajectory displacement component. We will return to these subjects in Section 14.5, where the results of the comparison are discussed in some detail.

14.3. Voltage and current dependencies for a fixed geometry

Table 14.1 gives an overview of the cases selected to verify the dependencies of the Coulomb interaction effects on the beam voltage V and the beam current I for a fixed geometry with a beam semi-angle

Table 14.1 Overview of the cases selected to compare the analytical theory with MC simulations, for various beam voltages V and beam currents I (and a fixed geometry). For all cases the beam segment length $L = 0.1$ m and crossover position parameter $S_c = 0.5$. The values chosen for the beam current I are 0.01, 0.02, 0.05, 0.1, 0.2, 0.5, 1, 2, 5, 10, 20 µA.

Experimental parameters				MC param.		Theoretical parameters				
$I(\mu A)$	$V(kV)$	$\alpha_0(mR)$	$r_c(\mu m)$	N_{sam}	N_{seed}	$\bar{\lambda}$	$\bar{r}_c$	K	r_c^*	v_0^*
0.01	1	5	5	100	15	9.58×10^{-5}	174	50	0.163	16.3
...						...				
20.				600	2	0.192				
0.01	2	5	5	100	15	3.39×10^{-5}	347	50	0.206	20.6
...						...				
20.				600	2	6.78×10^{-2}				
0.01	5	5	5	50	30	8.57×10^{-6}	868	50	0.279	27.9
...						...				
20.				600	2	1.71×10^{-2}				
0.01	10	5	5	50	30	3.03×10^{-6}	1736	50	0.351	35.1
...						...				
20.				600	2	6.06×10^{-3}				
0.01	20	5	5	50	30	1.07×10^{-6}	3472	50	0.443	44.3
...						...				
20.				600	30	2.14×10^{-3}				
0.01	50	5	5	50	30	2.71×10^{-7}	8681	50	0.601	60.1
...						...				
20.				600	2	5.42×10^{-4}				

$\alpha_0 = 5\,\text{mR}$ and a crossover radius $r_c = 5\,\mu\text{m}$. The table also gives the selected values for the MC parameters N_{sam} and N_{seed}, which represent the number of particles per sample and the number of seeds respectively. The total number of particles $N_{sam} \times N_{seed}$ is 1500 in the simulations for small beam currents ($\leq 2\,\mu\text{A}$) and 1200 in the simulations for large beam currents ($\geq 5\,\mu\text{A}$). The theoretical parameters $\bar{\lambda}$, $\bar{r}_c$ and K are relevant for the Boersch effect, while the parameters $\bar{\lambda}$, r_c^* and v_0^* apply to the trajectory displacement effect. An overview of the main theoretical parameters can be found in Table 16.1.

The results for the FW_{50} energy spread, the FW_{50} trajectory displacement effect and the space charge defocussing distance Δz are given in Figs. 14.1a–c. In these figures, the data is plotted as function of the beam current I. The different sets of curves correspond to the beam voltages 1, 10 and 50 kV. In Figs. 14.2a–c, the data is plotted as function of the beam potential V, and the different sets of curves correspond to beam currents of 0.01, 0.1, 1 and 10 μA.

Consider the curves for the FW_{50} energy spread presented in Figs. 14.1a and b. The results obtained with DRIFT1 and DRIFT2 are, in general, indistinguishable, except for very high particle densities. This confirms the validity of the reduction to two-particle dynamics (assumption 3). The deviations between MC simulations (DRIFT1 and DRIFT2) and the analytical theory (INTERAC) are larger. The curves in Fig. 14.1a show the transition from the pencil beam regime for small beam currents (I^2-dependency), via the Lorentzian regime (linear I-dependency) and the Holtsmark regime ($I^{2/3}$-dependency) to the Gaussian regime for large beam currents ($I^{1/2}$-dependency). These dependencies are explicitly predicted by Eqs. (7.4.33), (7.5.25), (7.3.40) and (7.3.37) respectively. The corresponding voltage dependencies are predicted to be V^{-1} for the pencil beam regime, $V^{-1/3}$ for the Holtsmark regime, $V^{-1/2}$ for the Lorentzian regime and $V^{-1/4}$ for the Gaussian regime (provided that $\bar{r}_c > 100$; see Eq. (7.3.39)). These dependencies are confirmed by the plots of Fig. 14.2a.

While the agreement by trend is excellent over the full range of operating conditions, one finds for intermediate beam currents somewhat larger values from the analytical theory than from the MC program. A good quantitative agreement is found for small as well as large beam currents. The differences found for intermediate beam currents will be further investigated in Section 14.5.

Similar characteristics are found for the FW_{50} trajectory displacement broadening given in Figs. 14.1b and 14.2b. The curves show the transition from the pencil beam regime for small beam currents (I^3-dependency)

Comparison of analytical theory with Monte Carlo simulations 443

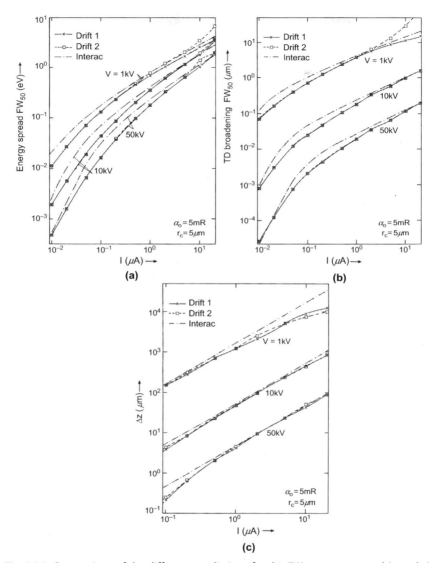

Fig. 14.1 Comparison of the different predictions for the FW_{50} energy spread (panel a), the FW_{50} trajectory displacement broadening (panel b) and the space charge defocussing distance Δz (panel c) produced in a beam segment with $L = 0.1$ m, $\alpha_0 = 5$ mR and $r_c = 5 \mu$m, as function of the beam current I, for different beam voltages V. See Table 14.1 for the detailed input.

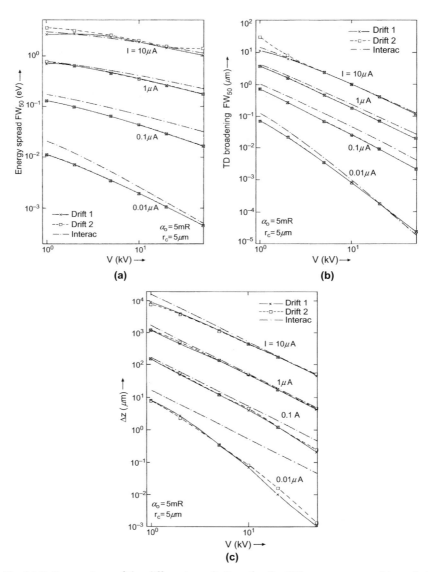

Fig. 14.2 Comparison of the different predictions for the FW_{50} energy spread (panel a), the FW_{50} trajectory displacement broadening (panel b) and the space charge defocussing distance Δz (panel c) produced in a beam segment with $L = 0.1$ m, $\alpha_0 = 5$ mR and $r_c = 5\,\mu$m, as function of the beam voltage V, for different beam currents I. This data is equivalent to that of Fig. 14.1.

via the Holtsmark regime ($I^{2/3}$-dependency) to the Gaussian regime for large beam currents ($I^{1/2}$-dependency). These dependencies are explicitly predicted by Eqs. (9.3.47), (9.3.44) and (9.3.45) respectively. The corresponding voltage dependencies are predicted to be $V^{-5/2}$ for the pencil beam regime, $V^{-4/3}$ for the Holtsmark regime and $V^{-13/12}$ for the Gaussian regime, which is confirmed by the curves of Fig. 14.2b. As for the energy spread, one finds that the analytical theory predicts somewhat larger values than the MC program for intermediate beam currents, while the quantitative agreement is excellent for the full range of operating conditions.

The analytical theory for the space charge effect predicts that the defocussing distance Δz is proportional to the perveance $I/V^{3/2}$, as can be seen from Eq. (11.5.14). This behavior is confirmed by the data plotted in Figs. 14.1c and 14.2c. The figures show a good qualitative and quantitative agreement between theory and MC simulations for most operating conditions. However, for very small and for very large beam currents, the MC simulations predict smaller Δz than the analytical theory. The interpretation of these results is postponed to Section 14.5.

14.4. Geometry and current dependencies for a fixed beam voltage

Table 14.2 gives an overview of the cases selected to verify the dependencies of the Coulomb interaction effects on the beam current I for different geometries and a fixed beam potential V of 10 kV. The beam semi-angle α_0 and crossover radius r_c were chosen such that their product is the same for every case. As in Table 14.1, we included the corresponding values of some theoretical parameters as well as the chosen values for the MC parameters N_{sam} and N_{seed}. Notice that the characteristic beam diameter ratio $K = 2\alpha_0 L/2r_c$ varies between 1.25 and 500 for the different cases.

The results for the FW_{50} energy spread, the FW_{50} trajectory displacement effect and the space charge defocussing distance Δz are given in Figs. 14.3a–c respectively. In these figures, the data is plotted as function of the beam current I. The different sets of curves correspond to the cases ($\alpha_0 = 10$ mR, $r_c = 1\,\mu$m) and ($\alpha_0 = 0.5$ mR, $r_c = 20\,\mu$m), while Fig. 14.3c also includes a set of curves for ($\alpha_0 = 2.5$ mR, $r_c = 4\,\mu$m). In Figs. 14.4a–c, the data is plotted as function of the beam geometry, defined by the values for α_0 and r_c. The different sets of curves correspond to beam currents of 0.01, 0.01, 1 and 10 μA.

Table 14.2 Overview of the cases selected to compare the analytical theory with MC simulations, for various beam geometries (defined by α_0 and r_c) and beam currents I (for a fixed beam potential V). For all cases the beam segment length $L = 0.1$ m and crossover position parameter $S_c = 0.5$. The values chosen for the beam current I are 0.01, 0.02, 0.05, 0.1, 0.2, 0.5, 1, 2, 5, 10, 20 µA.

Experimental parameters				MC param.		Theoretical parameters				
I(µA)	V(kV)	α_0(mR)	r_c(µm)	N_{sam}	N_{seed}	$\bar{\lambda}$	$\bar{r}_c$	K	r_c^*	v_0^*
0.01	10	10	1	50	30	7.58×10^{-7}	1389	500	0.0703	70.3
...						...				
20.				600	2	1.52×10^{-3}				
0.01	10	7.5	1.333	50	30	1.35×10^{-6}	1042	281	0.0937	52.7
...						...				
20.				600	2	2.69×10^{-3}				
0.01	10	5	2	50	30	3.03×10^{-6}	695	125	0.141	35.1
...						...				
20.				600	2	6.06×10^{-3}				
0.01	10	2.5	4	50	30	1.21×10^{-5}	347	31.3	0.281	17.6
...						...				
20.				600	2	2.43×10^{-2}				
0.01	10	1.25	8	50	30	4.85×10^{-5}	174	7.81	0.562	8.79
...						...				
20.				600	2	9.70×10^{-2}				
0.01	10	0.5	20	50	30	3.03×10^{-4}	69.5	1.25	1.41	3.51
...						...				
20.				600	2	0.606				

As for the cases discussed in Section 14.3, the agreement by trend is excellent, while some quantitative differences between the analytical theory and the MC simulations occur for some operating conditions.

Figs. 14.5a and b give the energy distribution obtained by MC simulation for a current of 0.01 and 10 µA respectively in a beam segment with a semi-angle $\alpha_0 = 5$ mR and a crossover radius $r_c = 2$ µm. Both plots give the distribution of the absolute value of the energy displacement with respect to the average energy of 10 keV and contain 90% of the particles. The FW_{50} energy spread is 1.9×10^{-3} eV and 2.5 eV for Figs. 14.5a and b respectively. The distribution of Fig. 14.5a shows distinct non-Gaussian features, as predicted by the analytical model. It has a narrow core and extremely long tails.

Comparison of analytical theory with Monte Carlo simulations

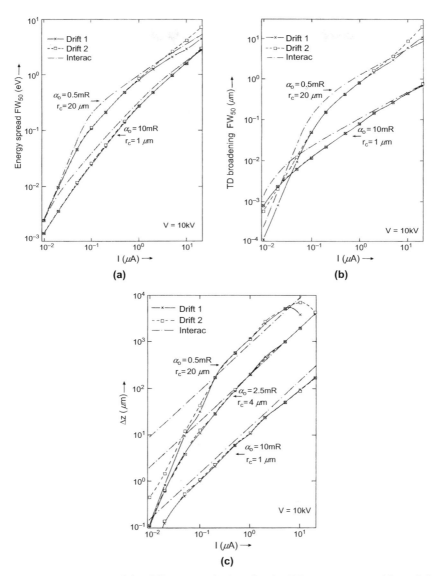

Fig. 14.3 Comparison of the different predictions for the FW_{50} energy spread (panel a), the FW_{50} trajectory displacement broadening (panel b) and the space charge defocussing distance Δz (panel c) produced in a beam segment with $L = 0.1$ m and $V = 10$ kV, as function of the beam current I, for different combinations of α_0 and r_c. See Table 14.2 for the detailed input.

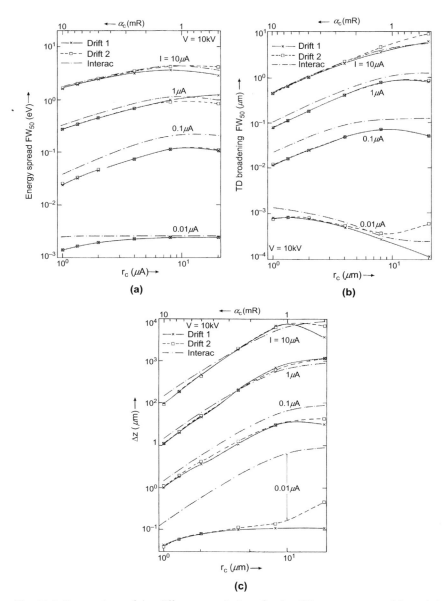

Fig. 14.4 Comparison of the different predictions for the FW_{50} energy spread (panel a), the FW_{50} trajectory displacement broadening (panel b) and the space charge defocussing distance Δz (panel c) produced in a beam segment with $L = 0.1$ m and $V = 10$ kV, as function of the beam geometry represented by α_0 and r_c (which are chosen such that $\alpha_0 \times r_c$ is constant), for different beam currents I. This data is equivalent to that of Fig. 14.3.

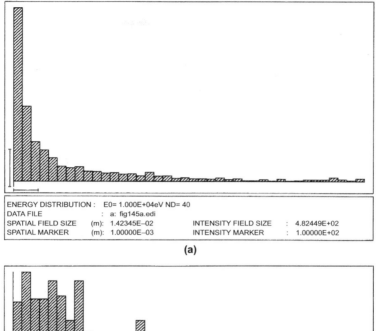
(a)

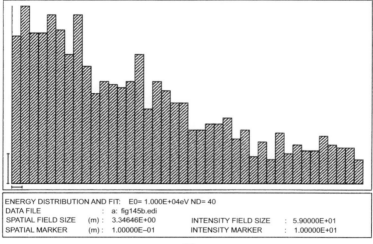
(b)

Fig. 14.5 Energy distribution obtained by MC simulation (using DRIFT1) of a beam segment with $L = 0.1$ m, $\alpha_0 = 5$ mR, $r_c = 2$ μm and $V = 10$ kV, for a beam current of 0.01 μA (panel a) and a beam current of 10 μA (panel b). The horizontal bars in panels a and b indicate an energy width of 1 and 0.1 eV respectively, while the vertical bars represent 100 and 10 particles respectively.

14.5. Discussion of the results

The comparison of the MC results obtained with DRIFT1 and DRIFT2 indicate that the two-particle approach used to solve the N-body dynamical problem is extremely accurate. Only for very large particle densities (occurring for large beam currents and low beam voltages), one finds some minor deviations between the results obtained with the two different ray tracing algorithms. Similar findings were published earlier by Jansen (1987).

The MC results and the predictions of the analytical theory obtained from the INTERAC program are in excellent qualitative agreement for all manifestations of the Coulomb interaction between the particles in the beam. However, for intermediate current densities, the analytical theory predicts somewhat larger values for the generated energy spread and trajectory displacement effect than the MC program. The predictions for the space charge defocussing are, in general, in close agreement, but some significant deviations occur for small defocussing distances (especially in combination with small beam angles), while some differences are also found for very high particle densities.

The presented results for the statistical effects pertain to the Full Width median value (FW_{50}) of the generated energy and trajectory displacement distribution. The MC program also evaluates some other width measures, such as the $FWHM$-value and the FW_{10}, FW_{30}, FW_{70} and FW_{90} values, corresponding to the Full Width of the distribution containing 10%, 30%, 70% and 90% of the particles respectively. The analytical theory gives explicit equations for the FW_{50} and the $FWHM$ value of the distributions. The $FWHM$ value predicted by the MC program shows a larger statistical error than the FW_{50} value. Moreover, the procedure used by the MC program to evaluate the $FWHM$ of a distribution may lead to systematic errors. The program fits a curve through the distribution. From the fit it determines the central height of the distribution first and evaluates the width for which the distribution reaches half this height next. The shape of the energy distribution shown in Fig. 14.5a may elucidate some of the problems that might arise with this procedure. The fraction of the particles present in the central region of this type of non-Gaussian distribution is extremely small. Accordingly, the accuracy of the estimation of the central height will be rather poor, probably leading to an underestimation of the real height. The $FWHM$ value is then overestimated. In order to avoid this kind of statistical and systematic errors, the FW_{50} value is preferred for the comparison.

The comparisons between analytical theory and MC simulations reported earlier by Jansen, Groves and Stickel (1985) and Jansen and

Stickel (1984) pertain to the predictions for the *FWHM* value. Contrary to the results of the present comparison, they found that the analytical theory predicts systematically 45–60% larger values for the trajectory displacement effect than the MC program. Similar results for the energy spread were found in the Lorentzian regime as well as in the Gaussian regime for $\bar{r}_c < 100$. The present study reveals that part of these differences can be resolved by considering the FW_{50} of the distributions instead of the *FWHM* value. It shows that the differences are not systematic over the full range of operating conditions but that the analytical theory predicts slightly larger values than the MC program only for intermediate particle densities.

Although the agreement between the analytical theory and MC simulations is quite satisfactory, one would like to have a better understanding of the quantitative differences occurring for some operating conditions. Let us investigate which of the various sources of errors deteriorating the accuracy of the different models can be held responsible. As the differences in the results are systematic, they cannot be attributed to the *statistical errors* inherent to the MC technique. Model errors of the MC program related to the finite size of the sample (referred to as *finite size errors*) may explain the relatively small values obtained for the defocussing distance Δz at large beam currents. However, part of these differences can also be explained by the fact that the analytical theory for the space charge effect is based on a first order perturbation approximation; see Section 11.4. This approach leads to an over-estimation of the defocussing distance at high current densities, for which the deviations from the unperturbed trajectories can no longer be considered as small. One could isolate the contribution of finite size errors by performing the simulations with a larger sample. However, we did not pursue this problem any further.

The differences in the predictions for the Boersch effect and the trajectory displacement effect occurring for intermediate particle densities cannot be explained by the finite size errors of the MC simulation. Part of these deviations are probably due to the *use of fit functions* within the analytical theory, which leads to some deviation from the exact numerical calculations of the (semi-)analytical model. These deviations can be estimated from Figs. 7.5, 7.10, Fig. 7.14b, 8.7, 9.7 and 9.8 (Notice, however, that these figures pertain to the *FWHM* value instead of the FW_{50} value considered here). Indeed, the deviations are the largest in the transition areas between the different regimes. However, one also sees that these errors are not large enough to fully explain the differences between theory and MC calculations.

As expressed by assumption (1) in Section 14.2, the results of the analytical theory pertain to the displacements experienced by the test particles running along the *central reference trajectory*, which is the beam axis. The

MC program evaluates the displacements of all particles, irrespective of their trajectory. This inconsistency can explain 10–20% of the observed differences for intermediate particle densities, as follows from the analysis of Chapter 10. The Figs. 10.2c, 10.3c, 10.4 and 10.5c show that the results for the "average reference trajectory" are indeed smaller than those for the "central reference trajectory," except for the trajectory displacement effect generated in a pencil beam.

The quantitative differences occurring for intermediate particle densities are most likely due to the combination of the inaccuracy caused by the use of fit functions in the analytical model and the fact that the analytical theory pertains to the central reference trajectory, while the MC program takes all trajectories into account.

With respect to the spatial broadening of the beam, it should be realized that the separation of the total displacements into a space charge component and a trajectory displacement component is handled differently in the analytical theory and the MC model. In the analytical theory, the trajectory displacement effect is isolated by investigating the displacements of the particles running along the central reference trajectory. Due to the rotational symmetry of the beam, the space charge force is zero along this trajectory. (For off-axis reference trajectories the trajectory displacement effect is isolated by ignoring the imaginary part of $p(k)$, which is equivalent to ignoring the odd moments of the two-particle distribution. See Section 5.6). In the MC simulations, the space charge effect and the trajectory displacement effect are separated by seeking the plane of best focus, defined as the plane for which the FW_{50} of the displacement distribution becomes minimum. The remaining broadening is assumed to be entirely the result of the trajectory displacement effect. The incompatibility of the different procedures used to separate statistical and space charge effects may lead to different predictions for both the space charge defocussing distance and the trajectory displacement effect, particularly for small beam currents when the displacements are small compared to the lateral dimensions of the beam. This explains the disagreement in the predictions for the space charge defocussing distance at small beam currents.

In order to get some appreciation for the size of the displacements relative to the beam dimensions, Figs. 14.6a and b show the displacements in the plane of best focus found by MC simulation of a beam segment with $\alpha_0 = 5\,\text{mR}$, $r_c = 2\,\mu\text{m}$ and $V = 10\,\text{kV}$ for a beam current of 0.01 and 10 μA respectively. Fig. 14.6a illustrates that the exact value of the space charge defocussing distance and the trajectory displacement effect is, for such operating conditions, irrelevant from an experimental point of view, since the contribution of the particle interactions to the spot width can be ignored.

Comparison of analytical theory with Monte Carlo simulations 453

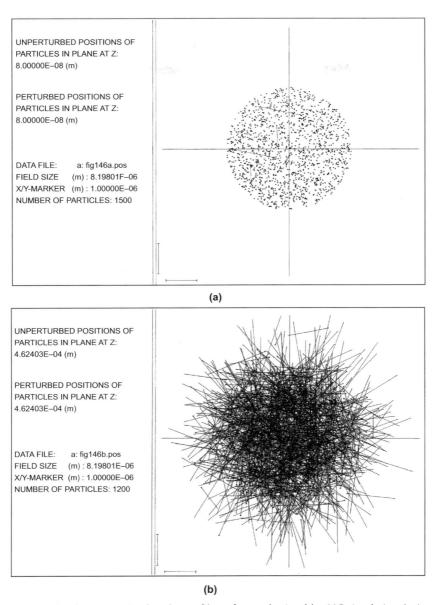

Fig. 14.6 Displacements in the plane of best focus obtained by MC simulation (using DRIFT1) of a beam segment with $L = 0.1$ m, $\alpha_0 = 5$ mR, $r_c = 2$ µm and $V = 10$ kV, for a beam current of 0.01 µA (panel a) and a beam current of 10 µA (panel b). The bars in the panels a and b represent a distance of 1 µm. The perturbed positions are indicated by a square.

Finally, it should be noted that the analytical model indicates that the space charge effect also causes some nonrefocusable blurring if the current density distribution in the cross sections of the beam is non-uniform. In the MC program, this blurring will show up as a contribution to the trajectory displacement effect, which is identified as the remaining blurring after refocussing. This may lead to a larger apparent trajectory displacement effect in the MC simulation than predicted by the analytical theory. However, for the comparison, we considered a beam segment with a uniform angular and spatial distribution in the crossover. The nonrefocusable space charge blurring is for this case negligible.

CHAPTER FIFTEEN

Comparison of recent theories on statistical interactions

Contents

15.1.	Introduction	455
15.2.	Boersch effect	457
	15.2.1 Zimmermann and Knauer	459
	15.2.2 Loeffler	461
	15.2.3 Crewe	462
	15.2.4 Rose and Spehr	463
	15.2.5 De Chambost and Hennion	464
	15.2.6 Sasaki	465
	15.2.7 Massey, Jones and Plummer	466
	15.2.8 Comparison of theories for a beam segment with a crossover	467
15.3.	Statistical angular deflections	467
	15.3.1 Loeffler and Hudgin	471
	15.3.2 Weidenhausen, Spehr and Rose	473
	15.3.3 Massey, Jones and Plummer	474
	15.3.4 Comparison of theories for a beam segment with a crossover	474
15.4.	Trajectory displacement effect	476
	15.4.1 Loeffler and Hudgin	477
	15.4.2 De Chambost	478
	15.4.3 Spehr	479
	15.4.4 Comparison of theories	480
15.5.	Conclusions	482

15.1. Introduction

This chapter gives a quantitative comparison of some recent theories on statistical Coulomb interactions. A qualitative description of the various models was given in Chapter 2, approaching the subject from a historical perspective. In Chapter 5, a systematic overview was presented of the alternative approaches to the N-body dynamical problem and the statistical part of the problem. Some of the models are based on the theoretical concepts developed in the field of plasma physics and stellar dynamics. These concepts were reviewed in Chapter 4.

The comparison of this chapter is restricted to those models that yield their final results in terms of analytical expressions. Numerical models, such as Monte Carlo models and the model proposed by Tang (1987), are therefore disregarded. Furthermore, only those models are considered that treat beams in drift space. This excludes the approaches of Knauer (1981) and Sasaki (1986), dealing with the interaction phenomena occurring in guns with spherical cathodes. In order to cover the results presented by Massey et al. (1981) for a cylindrical beam in a uniform acceleration field, their equations were recalculated for the case that acceleration is absent. The analysis for cylindrical beams is restricted to (nearly) homocentric beams. This excludes cylindrical beams in which the lateral thermal motion is magnetically confined, such as discussed by Zimmermann (1970) and Knauer (1979a). Most theories treat the case of a uniform current density distribution in the cross sections of the beam. For the comparison, we will therefore study this case.

The models developed for the Boersch effect, statistical angular deflections and the trajectory displacement effect will be discussed separately in the successive sections of this chapter. Our approach to these phenomena, which is based on the so-called extended two-particle model, was described in Chapters 7, 8 and 9 respectively. The strategy followed here is to express the final results of the different theories in terms of the scaled quantities employed in our model. This leads to a reduction of the number of parameters involved, which facilitates the comparison.

The extended two-particle model is unique in its ability to predict the full energy, angular and spatial displacement distributions. The model shows that different types of distributions are generated depending on the operating conditions. Due to the variation in the shape of the generated distributions, the choice of the measure to characterize their width is critical. Our model gives explicit expressions for the Full Width at Half Maximum ($FWHM$), the Full Width median (FW_{50}) and the root mean square (rms) value. Most other models evaluate the root mean square (rms) broadening only. More practical width measures are usually obtained from the rms value by assuming that the displacement distribution is Gaussian. However, our model shows that Gaussian behavior occurs only for high particle densities. In the non-Gaussian regimes, the rms value is entirely dominated by the tails of the distribution or in other words by a few large displacements. For those conditions, the rms value provides no information of the central part of the displacement distribution, and a straightforward conversion to more practical width measures as the $FWHM$ and FW_{50} is not possible. Theories predicting a rms value only are therefore essentially incomplete.

Since most theories evaluate a rms value, the comparison focusses on the different predictions for this width measure. The model presented by Loeffler (1969) and Loeffler and Hudgin (1970) requires a separate treatment, since it expresses the final results in terms of the half width median of the displacement distributions. For comparison, these results will be converted to a rms value, employing Loeffler's claim that the distribution is Gaussian. In addition, the results of Loeffler and Hudgin will be compared directly to the median width predicted by our model, which is more appropriate when the displacement distribution is non-Gaussian.

15.2. Boersch effect

We will first review the predictions of our model for the rms energy spread caused by the Boersch effect and next discuss the results of some other theories. As some of the theories make use of the thermodynamic upper limit to the Boersch effect, we will include this limit in the analysis.

With Eqs. (7.3.30), (7.3.32), (7.5.20) and (7.5.37), the rms value of the distribution of axial velocities generated in a *beam segment with a crossover* of arbitrary dimensions can be expressed as (using the d_0, v_0-scaling of Eq. 7.3.2)

$$\langle \Delta \bar{v}_z^2 \rangle^{1/2} = (\pi \bar{\lambda})^{1/2} P_{CE}(\bar{r}_c, K_1, K_2), \tag{15.2.1}$$

where $P_{CE}(\bar{r}_c, K_1, K_2) = [P_{CE}(\bar{r}_c, K_1) + P_{CE}(\bar{r}_c, K_2)]/2$ and

$$P_{CE}(\bar{r}_c, K) = \left[.788 + 6\left(\frac{\bar{r}_c}{K^2}\right)^{2/3} \left[1 + .5\left(\frac{K^2}{\bar{r}_c}\right)^{4/9}\right]^{3/2} \right.$$

$$\left. + \frac{\pi \bar{r}_c}{\{2\ln[.8673(114.6 + \bar{r}_c)]\}^2} \right]^{-1/2}. \tag{15.2.2}$$

The quantities $\bar{\lambda}, \bar{r}_c, K_1$ and K_2 are defined by Eqs. (7.3.22), (7.3.5) and (7.3.1). For large K-values, the function $P_{CE}(\bar{r}_c, K)$ becomes independent of K. For $\bar{r}_c \to 0$, it becomes equal to unity. By removing the scaling in Eq. (15.2.1), using Eq. (7.3.36), one finds for the relative rms energy spread

$$\frac{\langle \Delta E^2 \rangle^{1/2}}{E} = \left(\frac{m}{8\varepsilon_0^2 e}\right)^{1/4} P_{CE}(\bar{r}_c, K_1, K_2) \frac{I^{1/2}}{V^{3/4}}. \tag{15.2.3}$$

If the energy distribution is Gaussian, the FWHM and the rms are related as $FWHM_E = (8\ln 2)^{1/2} \langle \Delta E^2 \rangle^{1/2}$ and Eq. (15.2.3) becomes equivalent to Eq. (7.3.37). (Notice that in the Gaussian regime the functions G_{CE} and P_{CE} are identical.)

In Section 3.8 the thermodynamic upper limit of kinetic energy relaxation was introduced, which is relevant for a beam segment with a narrow crossover. For a uniform angular distribution, the corresponding rms energy spread is given by Eq. (3.8.3). By scaling this expression, one obtains Eq. (7.7.1). Using the form of Eq. (15.2.1), the thermodynamic limit of kinetic energy relaxation can now be expressed in terms of an upper limit $P_{CET}(\bar{\lambda})$ for the function $P(\bar{r}_c, \bar{\lambda}, K_1, K_2)$, which is equal to

$$P_{CET}(\bar{\lambda}) = \frac{1}{(6\pi\bar{\lambda})^{1/2}}. \tag{15.2.4}$$

This upper limit increases with decreasing $\bar{\lambda}$. For $\bar{\lambda} = 1/6\pi = 0.053$, one finds $P_{CET} = 1$, which is equal to the maximum of the function $P_{CE}(\bar{r}_c, K_1, K_2)$, given by Eq. (15.2.2). Thus, according to our model, thermodynamic equilibrium is not reached in a beam with a single crossover for practical particle densities ($\bar{\lambda} < 10^{-2}$).

With Eqs. (7.4.19), (7.4.22) and (7.4.16), the rms value of the distribution of axial velocities generated in a *homocentric cylindrical beam segment* can be expressed as (using the δ, ν-scaling of Eq. 6.13.1)

$$\langle \Delta v_z^{*2} \rangle^{1/2} = 2 [p_2^*(\infty)\lambda^*]^{1/2} P_{PE}(r_0^*), \tag{15.2.5}$$

where $p_2^*(\infty) = 0.1513$ and

$$P_{PE}(r_0^*) = \left(1 + 0.240/r_0^{*8/7}\right)^{-7/8}, \tag{15.2.6}$$

and λ^* is defined by Eq. (7.4.18). For an extended cylindrical beam ($r_0^* \gg 1$), one finds $P_{PE}(r_0^*) = 1$. Removing the scaling in Eq. (15.2.5), using Eq. (7.4.29), yields

$$\frac{\langle \Delta E^2 \rangle^{1/2}}{E} = 0.1414 \frac{e^{1/12} m^{1/4}}{\varepsilon_0^{5/6}} P_{PE}(r_0^*) \frac{I^{1/2} L^{2/3}}{V^{13/12} r_0}. \tag{15.2.7}$$

If the energy distribution is Gaussian, the FWHM and rms are related as $FWHM_E = (8\ln 2)^{1/2} \langle \Delta E^2 \rangle^{1/2}$, and Eq. (15.2.6) leads back to Eq. (7.4.32).

In Section 3.10, the thermodynamic upper limit of potential energy relaxation was discussed, which is relevant for a homocentric cylindrical beam segment. The corresponding rms energy spread is given by Eq. (3.10.11). By scaling, one obtains Eq. (7.7.3). Using the form of Eq. (15.2.5), one can now express this result in terms of an upper limit $P_{PET}(\lambda^*)$ for the function $P_{PE}(r_0^*)$, which is equal to

$$P_{PET}(\lambda^*) = \frac{C_{PET}}{\lambda^{*1/3}}, C_{PET} = \frac{a^{1/3}}{2^{4/3}p_2^*(\infty)^{1/2}} = 0.452. \quad (15.2.8)$$

One finds $P_{PET} = 1$ for $\lambda^* = 0.092$, while P_{PET} increases for smaller λ^*-values. Thus, the upper limit of potential energy relaxation is not reached for practical conditions ($\lambda^* < 10^{-2}$).

To the best of our knowledge, the only other theory treating the case of a homocentric beam segment is that of Massey et al. (1981). We will review their result at the end of this section. For all theories treating a beam segment with a crossover, it was found that the resulting rms energy spread can be expressed in the form of Eq. (15.2.3), provided that the function $P_{CE}(\bar{r}_c, K_1, K_2)$, is generalized to a function $P(\bar{r}_c, \bar{\lambda}, K_1, K_2)$, which may also depend on $\bar{\lambda}$. In order to compare the theories, it is therefore sufficient to compare the corresponding functions $P(\bar{r}_c, \bar{\lambda}, K_1, K_2)$. We will first determine this function for each of the theories and carry out the comparison next. The discussion presented in this section forms an extension to the work presented previously by Jansen et al. (1985).

15.2.1 Zimmermann and Knauer

Zimmermann (1970) and Knauer (1979a,b) attribute the Boersch effect to the relaxation of internal kinetic energy from the longitudinal to the lateral degree of freedom by multiple weak complete two-particle collisions. This approach is closely related to the Fokker-Planck model, used in plasma physics and steller dynamics. In Chapter 4, it was shown that the results presented by Zimmermann and Knauer for a cylindrical beam can directly be obtained from the velocity-diffusion coefficient appearing in the Fokker-Planck equation; see Eq. (4.8.4). The models of Zimmermann and Knauer are less suited to treat noncylindrical beams, in which the current density J and lateral beam temperature $T_\perp$ varies with the axial coordinate. For the case of a beam segment with a crossover, Zimmermann (1970) takes over the analysis given by Loeffler (1969), which will be discussed later on in this section. Knauer's approach for a beam segment with a crossover is to start from the result obtained for a thin cylindrical beam slice and integrate this expression over the axial coordinate z, taking the local current density $J(z)$ and local lateral temperature $T_\perp(z)$ equal to

$$J(z) = J_c \left(\frac{r_c}{r_0(z)}\right)^2, \quad T_\perp(z) = T_{\perp c} \left(\frac{r_c}{r_0(z)}\right)^2, \quad (15.2.9)$$

in which J_c and $T_{\perp c}$ refer to the crossover plane. The first relation is trivial, while the second one was clarified in Chapter 3; see Eq. (3.7.17). In *SI*-units, Knauer's result can be expressed as (Knauer 1979a, Eq. 33)

$$\frac{\langle \Delta E^2 \rangle^{1/2}}{E} = \frac{(em)^{1/4}}{(4\pi)^{3/4}\varepsilon_0} \frac{I^{1/2}}{\alpha_0 r_c^{1/2} V^{5/4}} \left[\ln \Lambda_C \ln(K_1 K_2) - \frac{2}{3} \ln^2 K_1 \right. \quad (15.2.10)$$

$$\left. - \frac{2}{3} \ln^2 K_2 \right]^{1/2},$$

where

$$\Lambda_C = \frac{2^{13/6} \pi^{4/3} \varepsilon_0}{e^{1/2} m^{1/6}} \frac{\alpha_0^2 r_c^{2/3} V^{7/6}}{I^{1/3}}, \quad (15.2.11)$$

Knauer separates the lateral velocities into thermal velocities and crossover velocities. He argues that the impact of both should be combined in a proper way and denotes the total energy broadening as the "composite" Boersch effect. However, one cannot consider thermal and crossover motion as independent quantities. They both originate from the initial thermal motion near the cathode surface and both are affected by the focussing action of the optical components thereafter. One should either express the lateral velocities in terms of geometrical properties (like the beam semi-angle α_0) or in terms of thermodynamical properties (like the beam temperature $T_\perp$). In Eqs. (15.2.10) and (15.2.11), we therefore ignored Knauer's thermal component by taking $T_\perp = 0$.

Knauer truncates the integration when the thermodynamic limit of kinetic energy relaxation is reached. In Eq. (15.2.10) we took the integration limits equal to the physical boundaries of the beam, assuming that the thermodynamic limit is not reached. This assumption should be verified afterwards. Furthermore, one should note that Knauer presupposes a Gaussian angular distribution and defines $\alpha_0 = \langle \alpha^2 \rangle^{1/2}$. For the uniform angular distribution with a cut-off angle α_0 considered in the comparison, one finds $\langle \alpha^2 \rangle^{1/2} = \alpha_0/\sqrt{2}$. Thus, one might argue that Eqs. (15.2.10) and (15.2.11) should be modified by replacing α_0 by $\alpha_0/\sqrt{2}$. However, we prefer to use Knauer's original result and take the angle α_0, appearing in Eqs. (15.2.10) and (15.2.11), equal to the cut-off angle of the uniform angular distribution.

Eqs. (15.2.10) and (15.2.11) can be expressed in the form of Eq. (15.2.3), replacing the function $P_{CE}(\bar{r}_c, K_1, K_2)$, by

$$P_K(\bar{r}_c, \bar{\lambda}, K_1, K_2) = \left(\frac{8}{9\pi \bar{r}_c^2} \right)^{1/4}$$

$$\times \left[\frac{1}{2} \ln\left(\frac{\pi \bar{r}_c^2}{8\bar{\lambda}} \right) \ln(K_1 K_2) - \ln^2 K_1 - \ln^2 K_2 \right]^{1/2}.$$

$$(15.2.12)$$

15.2.2 Loeffler

Loeffler (1969) evaluated the energy spread generated in a beam segment with a narrow crossover by considering the effect of weak interactions on a reference particle moving along the beam axis. Loeffler's statistical treatment of the problem was outlined in Section 5.3; see Eqs. (5.3.14)–(5.3.16). He claims that this procedure leads to a median energy spread rather than a rms energy spread, which is not obvious at all. Moreover, some of the mathematical steps in his derivation are not entirely straightforward. In the approximation for the probability density function of the coordinates of the colliding particles, given by Loeffler's Eq. (19), one would expect that the boundaries are chosen equal to r_0 and $-r_0$ instead of $2r_0$ and $-2r_0$. Furthermore, for the case $8\lambda r_0 > 1$, Loeffler replaces the sum of Nth largest contributions to the total displacement of the reference particle by a "suitable" integral, which can be evaluated analytically. The resulting expression (Loeffler, 1969, Eq. 22) appears to be wrong, as was pointed out by Rose and Spehr (1983). Despite these obscurities, we will use Loeffler's final results in their unmodified form.

Loeffler's result applies to a beam segment of infinite length, in which all collisions can be considered as complete. Loeffler and Hudgin (1970) presented a correction factor, which should be included for beams in which $K = \alpha_0 L/2r_c$ cannot be considered as infinitely large. The resulting equation for the median energy spread generated in a beam segment with a crossover can be expressed as (Loeffler, 1969, Eqs. 22–24; Loeffler and Hudgin, 1970, Eq. 5)

$$\frac{FW_{50,E}}{E} = \frac{e}{2\pi\varepsilon_0} \frac{1}{\alpha_0 r_c V} \Omega(\lambda r_c) \frac{1}{2}[F_L(K_1) + F_L(K_2)], \qquad (15.2.13)$$

where

$$\Omega(\lambda r_c) = \begin{cases} (8\lambda r_c)^{1/2}\left(3 + \ln 2 + 2\ln(8\lambda r_c) + \frac{1}{4}\ln^2(8\lambda r_c)\right)^{1/2} & \text{for } 8\lambda r_c > 1 \\ \frac{\pi}{\sqrt{2}} 8\lambda r_c & \text{for } 8\lambda r_c \ll 1 \end{cases}$$

$$F_L(K) = \left(1 + \frac{11 + 0.4/(\lambda r_c)^2}{K^2}\right)^{-1/2}.$$

(15.2.14)

Loeffler assumed that the resulting energy distribution is Gaussian. In that case, the rms energy spread follows as

$$\langle \Delta E^2 \rangle^{1/2} = FW_{50,E}/2^{3/2}\text{erf}^{-1}(1/2) = 0.74130\,FW_{50,E}. \tag{15.2.15}$$

Eq. (15.2.13) can now be expressed in the form of Eq. (15.2.3), replacing the function $P_{CE}(\bar{r}_c, K_1, K_2)$

$$P_L(\bar{r}_c, \bar{\lambda}, K_1, K_2) = \begin{cases} \dfrac{4.55}{\bar{r}_c^{1/2}}\left[1 + .542\ln(8\bar{\lambda}\bar{r}_c) + .0677\ln^2(8\bar{\lambda}\bar{r}_c)\right]^{1/2} \\ \quad \times \dfrac{1}{2}[F_L(K_1) + F_L(K_2)] \qquad \text{for } 8\bar{\lambda}\bar{r}_c > 1 \\ 14.9\bar{\lambda}^{1/2}\dfrac{1}{2}[F_L(K_1) + F_L(K_2)] \quad \text{for } 8\bar{\lambda}\bar{r}_c \ll 1, \end{cases} \tag{15.2.16}$$

where the function $F_L(K)$ is defined by the second equation of (15.2.14).

Our model shows that the energy distribution generated in an extended beam with a narrow crossover ($K \gg 1$) is Gaussian when $\bar{\lambda} \gg P_{CE}(\bar{r}_c)^2/16\pi$ and Lorentzian when $\bar{\lambda} \ll P_{CE}(\bar{r}_c)^2/16\pi$; see Section 7.3. It is interesting to note that the Gaussian and Lorentzian regime roughly coincide with the regimes distinguished by Loeffler, corresponding to $8\bar{\lambda}\bar{r}_c > 1$ and $8\bar{\lambda}\bar{r}_c \ll 1$ respectively. In the Lorentzian regime, we found for the Full Width median energy spread

$$\frac{FW_{50,E}}{E} = \left(\frac{2m}{\varepsilon_0^2 e}\right)^{1/2} \frac{I}{\alpha_0 V^{3/2}}, \tag{15.2.17}$$

as follows from Eq. (7.3.40), using that for a one-dimensional Lorentzian distribution $FW_{50} = FWHM$. Loeffler's result for $8\lambda r_c \ll 1$, given by Eqs. (15.2.13) and (15.2.14), gives for $K \to \infty$ the same functional dependency, as can be seen by substituting the expression for λ given by Eq. (3.2.4), but is larger by a factor $\sqrt{2}$.

For a pencil beam, our model predicts

$$\frac{FW_{50,E}}{E} = 0.64193\,\frac{m}{\varepsilon_0 e^2}\,\frac{I^2 L}{V^2}, \tag{15.2.18}$$

as follows from Eqs. (7.4.33) and (7.5.36). The same functional dependency follows from Loeffler's result for $8\lambda r_c \ll 1$ and $K \to 0$ (using that $F_L(K) \approx 1.58\lambda r_c K$), but this equation predicts values that are a factor 1.74 larger. The single regime that is not covered by Loeffler's equations for the Boersch effect is the Holtsmark regime, corresponding to a 2/3 power dependency on the beam current. This regime applies to extended beams with small K-values in which weak collisions are dominant.

15.2.3 Crewe

Crewe (1978a) estimated the energy deviation of an on-axis reference particle due to a single complete collision with a particle running with an angle $\alpha = \alpha_0$ and axial separation $b_z = 1/\lambda$ with respect to the reference particle. His result can be expressed as

$$\frac{\Delta E}{E} = \left(\frac{m}{8\varepsilon_0^2 e}\right)^{1/2} \frac{I}{\pi^2 a_0 V^{3/2}} \left[1 + \frac{m}{2^7 \varepsilon_0^2 e \pi^2} \frac{I^2}{a_0^4 V^3}\right]^{-1}. \quad (15.2.19)$$

Since Crewe ignores the statistical nature of the problem, it is not obvious what measure the width ΔE represents. For the comparison, we will take $\langle \Delta E^2 \rangle^{1/2} = \Delta E$. Eq. (15.2.19) can now be expressed in the form of Eq. (15.2.3), replacing the function $P_{CE}(\bar{r}_c, K_1, K_2)$ by

$$P_C(\bar{\lambda}) = \frac{2}{\pi^{3/2}} \frac{\bar{\lambda}^{1/2}}{1 + \bar{\lambda}^2}, \quad (15.2.20)$$

which is independent of $\bar{r}_c, K_1$ and K_2.

15.2.4 Rose and Spehr

Rose and Spehr (1980) introduced the so-called closest encounter approach to evaluate the energy spread generated in a beam segment with a narrow circular or astigmatic crossover. This approach relies on the assumption that the displacement of a reference particle is dominated by the complete collision with its nearest neighbor, which is defined as the particle with the smallest impact parameter with respect to this particle. The statistical procedure followed by Rose and Spehr was outlined in Section 5.3; see Eqs. (5.3.19) and (5.3.20). The final equations of Rose and Spehr do not pertain to a specific reference trajectory but represent an average over all possible reference trajectories in the beam. For that, they followed the method described in Section 10.3.

The scaling used by Rose and Spehr is similar to our d_0, v_0-scaling defined by Eq. (7.3.2). Their results can be converted to our notation, using

$$\begin{aligned} \langle u^2 \rangle &= \langle \Delta \bar{v}_z^2 \rangle = \langle \Delta E^2 \rangle / (2\alpha_0 E)^2 \\ \chi &= \bar{\lambda}/2 \\ \lambda &= 4\bar{r}_c \\ \omega &= \bar{v}/2 \\ \eta &= \bar{r}_\perp / 2\bar{r}_c. \end{aligned} \quad (15.2.21)$$

The equation of Rose and Spehr for the rms energy spread generated in an extended circular crossover with a uniform angular and spatial distribution can directly be expressed in the form of Eq. (15.2.1), provided that the function $P_{CE}(\bar{r}_c, K_1, K_2)$ is replaced by the function $P_{RS}(\bar{r}_c)$, given by Rose and Spehr (1980, Eqs. 38, 40, 22 and 24)

$$P_{RS}(\bar{r}_c)^2 = \frac{8}{\pi} \int_0^2 \bar{v} d\bar{v} F(\bar{v}) \int_0^2 t \, dt \, F(t) \frac{1}{\left[1 + \bar{v}^4 t^2 \bar{r}_c^2 / 4\right]^{1/2}}$$

$$\times K\left(\frac{\bar{v}^2 t \bar{r}_c / 2}{\left[1 + \bar{v}^4 t^2 \bar{r}_c^2 / 4\right]^{1/2}}\right), \quad (15.2.22)$$

where the function K is the complete elliptic integral of the first kind and

$$F(x) = \frac{2}{\pi}\left[\arccos(x/2) - (x/2)\sqrt{1-(x/2)^2}\right]. \tag{15.2.23}$$

This result is identical to our result obtained for an average reference trajectory, as expressed by Eqs. (10.3.13) and (10.3.12), using that $P_{CE}(\bar{r}_c) = [p_2(\bar{r}_c)/\pi]^{1/2}$.

One may wonder why the closest encounter approach yields identical results as our model, in which the total displacement of the test particle is calculated as the sum of the displacements caused by *all* surrounding field particles. The answer can be found in the derivation given by Rose and Spehr (1980) by considering the approximation involved in the transition from Eqs. (34) to (38). They argue that for practical beams, in which $\chi (=\bar{\lambda}/2) \ll 1$, it is allowed to replace the exponential function in Eq. (34) by unity. Doing so, they remove the factor in the probability distribution that suppresses the contribution of the particles that are not the nearest neighbor of the reference particle. By this approximation, their model becomes mathematically identical to ours and therefore yields the same results. The fact that this approximation is allowed implies that the *rms* energy spread generated in a beam segment with a narrow crossover is indeed dominated by collisions between nearest neighbors. Since the impact of more distant neighbors is minor, one may as well include their contribution in the calculation. This is exactly what Rose and Spehr do by ignoring the exponential function in their Eq. (34).

The approximation for the integral of Eq. (15.2.23) given by Rose and Spehr (their Eq. 42b) overestimates the numerical results, and we prefer our approximation given by Eqs. (7.6.11) and (10.3.14). Accordingly, we take

$$P_{RS}(\bar{r}_c) \approx \left(1 + \frac{1.471\pi\bar{r}_c}{4\ln^2(32.42 + 1.276\bar{r}_c)}\right)^{-1/2}, \tag{15.2.24}$$

which is sufficiently accurate for all $\bar{r}_c$ values. The calculation of Rose and Spehr relies on the assumption that the dominant collisions are (nearly) complete, which restricts its applicability to the case $K_1 \gg 1$ and $K_2 \gg 1$. Our calculation also covers the case of small K-values. The function P_{CE}, defined by Eq. (15.2.2), therefore depends also on K_1 and K_2. Notice that, even for $K_1 \gg 1$ and $K_2 \gg 1$, Eqs. (15.2.2) and (15.2.24) differ slightly, since the former pertains to the on-axis reference trajectory while the latter represents an average over all possible reference trajectories.

15.2.5 De Chambost and Hennion

De Chambost and Hennion (1979) estimated the Boersch effect from the mean square fluctuations in the electrostatic force acting on a reference particle. The basics of this type of approach were discussed in Section 5.3; see Eqs. (5.3.21)–(5.3.26). De Chambost and Hennion studied the case of a cylindrical beam in a uniform acceleration zone and the case of a half-crossover in drift space. We will consider the latter. The rms energy spread generated in a full crossover follows in their model by summing the contributions of the two half-crossovers in quadrature. As the two contributions are identical, the rms energy broadening of a full crossover is $\sqrt{2}$ times that of a half-crossover. We checked the derivation given by De Chambost and Hennion and found an error in the arithmetics. Their result is a factor $2^{1/8}$ too large. After correction of this error, one finds (De Chambost and Hennion (1979), Eq. 21 $\times \sqrt{2} \times 2^{-1/8}$)

$$\frac{\langle \Delta E^2 \rangle^{1/2}}{E} = \frac{m^{3/16} e^{1/16}}{18^{3/16} (\pi \varepsilon_0)^{5/6}} \frac{I^{3/8}}{(\alpha_0 r_c)^{1/4} V^{13/16}}. \tag{15.2.25}$$

This result can be expressed in the form of Eq. (15.2.3), replacing the function $P_{CE}(\bar{r}_c, K_1, K_2)$ by

$$P_{CH}(\bar{r}_c, \bar{\lambda}) = \frac{2^{7/8}}{3^{3/8} \sqrt{\pi}} \frac{1}{\bar{\lambda}^{1/8} \bar{r}_c^{1/4}}. \tag{15.2.26}$$

This equation is independent of K_1 and K_2, which are assumed to be large compared to unity.

15.2.6 Sasaki

Sasaki (1984) presented a model for the Boersch effect produced in a beam segment with a crossover that is similar to that of De Chambost and Hennion. However, Sasaki's approach is somewhat more refined in the choice of the cutoff of the interaction force at small interparticle distances and differs in the calculation of the time intervals between successive independent states. His final result can be expressed as (Sasaki, 1984, Eqs. 4.29, 4.30 and 4.27)

$$\frac{\langle \Delta E^2 \rangle^{1/2}}{E} = 0.143 \frac{m^{1/4}}{\varepsilon_0^{1/2} e^{1/4}} \frac{I^{1/2}}{V^{3/4}} G\left[0.881 (\lambda r_c)^{1/3}\right]^{1/2}, \tag{15.2.27}$$

where

$$G(\gamma) = \sum_{k=0}^{N(\gamma)-1} \left[\left(1 - \frac{2k}{3\gamma}\right)^{3/2} - \left(1 - \frac{2(k+1)}{3\gamma}\right)^{3/2} \right] \tag{15.2.28}$$

and $N(\gamma)$ is the integer part of $3\gamma/2$. The plot of the function $G(\gamma)$ given by Sasaki is not consistent with the definition of $G(\gamma)$, reproduced above. It appears that the plot represents the first term only ($k=0$), which is indeed monotonically decreasing with γ. The function of $G(\gamma)$ shown in the plot can be approximated as

$$G(\gamma) \approx \frac{1}{1/3 + \gamma}. \qquad (15.2.29)$$

We will use this approximation in the further analysis. Eqs. (15.2.27) and (15.2.29) can be expressed in the form of Eq. (15.2.3), replacing the function $P_{CE}(\bar{r}_c, K_1, K_2)$ by

$$P_S(\bar{r}_c, \bar{\lambda}) = \frac{0.240}{\left[1/3 + .881\left(\bar{\lambda}\bar{r}_c\right)^{1/3}\right]^{1/2}}. \qquad (15.2.30)$$

This result is independent of K_1 and K_2, which are assumed to be large compared to unity.

15.2.7 Massey, Jones and Plummer

Massey et al. (1981) studied the space charge effect and the statistical interaction effects occurring in the acceleration area of a laser illuminated photo-electron microscope. For the statistical effects, they distinguish two mechanisms, namely potential energy relaxation and velocity (internal kinetic energy) relaxation. For the latter they followed the approach of Knauer (1979a) discussed previously in this section. Here we will investigate their model for potential energy relaxation in a cylindrical beam. As the model of De Chambost and Hennion and the model of Sasaki, the model of Massey et al. starts from the mean square of the fluctuations in the electrostatic force acting on a reference particle. However, their statistical approach differs by the fact that the energy displacements produced in the successive states are added linearly instead of quadratically, as was discussed in Section 5.3; see Eq. (5.3.27). Furthermore, they account for the fact that the smallest inter-particle distance will increase during the flight.

We recalculated their final equation for the energy broadening in a (nearly) homocentric cylindrical beam for the case that acceleration is absent. The final results shows the same functional dependency as the results obtained with acceleration, but it is smaller by a factor 7/10. It can be expressed as (Massey et al., 1981, Eq. $27 \times 7/10$, using Eqs. 22 and 25)

$$\frac{\langle \Delta E^2 \rangle^{1/2}}{E} = 0.09325 \frac{e^{1/12} m^{1/4}}{\varepsilon_0^{5/6}} \frac{I^{1/2} L^{2/3}}{r_0 V^{13/12}}. \qquad (15.2.31)$$

Our result for the rms energy spread generated in a homocentric cylindrical beam is given by Eq. (15.2.7). For an extended beam ($r_0^* \gg 1$), one may take $P_{PE}(r_0^*) \approx 1$. One sees that the resulting equation shows the same functional dependency as Eq. (15.2.31), but it is larger by a factor 1.52.

15.2.8 Comparison of theories for a beam segment with a crossover

We have expressed the energy spread predicted by the various theories for a beam segment with a crossover in terms of the equivalent of the function $P_{CE}(\bar{r}_c, K_1, K_2)$, appearing in Eq. (15.2.3). In order to compare the theories, it is sufficient to compare these functions, which are for the models of Knauer (1979a), Loeffler (1969), Crewe (1978a), Rose and Spehr (1980), De Chambost and Hennion (1979) and Sasaki (1984) given by Eqs. (15.2.12), (15.2.16), (15.2.20), (15.2.24), (15.2.26) and (15.2.30) respectively. Our model is represented by the function $P_{CE}(\bar{r}_c, K_1, K_2)$, given by Eq. (15.2.2). In Figs. 15.1a and b all functions are plotted as functions of $\bar{r}_c$ for the case $K_1 = K_2 = 200$ and $\bar{\lambda} = 10^{-2}$ and 10^{-4} respectively. Fig. 15.1c shows the corresponding results for $K_1 = K_2 = 2$, leaving out those functions that do not depend on K_1 and K_2. In Fig. 15.1A, the thermodynamic upper limit given by Eq. (15.2.4) is also plotted. One sees that it is not reached by any of the curves. The thermodynamic limit goes up for smaller $\bar{\lambda}$-values and exceeds all predictions even further.

The occurrence of Gaussian and Lorentzian type of energy distributions, as predicted by our model, is indicated in Figs. 15.1a and b. For a given $\bar{\lambda}$-value, the transition between both regimes is determined by $\bar{r}_c$ only, as can, for instance, be seen from Fig. 7.6. It should be recalled that the situation is more complicated for small K-values for which the energy distribution can also be Holtsmarkian or of the pencil beam type. For this, the reader is referred to Section 7.5. From Figs. 15.1a and b, one sees that there is some agreement between the theories in the Gaussian regime, while significant differences exist in the Lorentzian regime. The discrepancy in this regime between our result and that of Loeffler is mainly due to the conversion of Loeffler's median width to a rms value by means of Eq. (15.2.15), which requires that the energy distribution is Gaussian. Direct comparison of the predictions for the median energy spreads leads to a better agreement; see Eq. (15.2.17).

15.3. Statistical angular deflections

In Chapter 8 the effect of statistical angular deflections was discussed, which is related to the distribution of transverse velocity displacements $\rho(\Delta v_\perp)$. The calculation is comparable to that of Boersch effect, which is related to the distribution of axial velocity displacements $\rho(\Delta v_z)$. It should be emphasized that the concept of statistical angular deflections is mainly of theoretical significance. In practice, one likes to

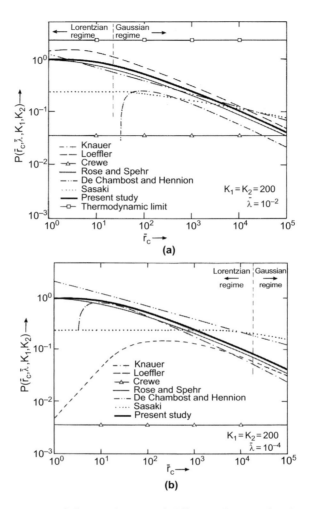

Fig. 15.1 Comparison of the predictions of different theories for the rms energy spread generated in a beam segment with a crossover in the middle. Each theory is represented by a different function $P(\bar{r}_c, \bar{\lambda}, K_1, K_2)$, which is proportional to the rms energy spread; see Eq. (15.2.3). Panels a and b give the results for a narrow crossover ($K_1 = K_2 = 200$) and a scaled linear particle density $\bar{\lambda}$ equal to 10^{-2} and 10^{-4} respectively. Panel c gives the results for a more cylindrical shaped beam segment ($K_1 = K_2 = 2$) and includes only those theories that apply to this case. The thermodynamic limit indicated in panel a pertains to the relaxation of kinetic energy; see Eq. (15.2.4). For smaller $\bar{\lambda}$-values, it exceeds the plotted P-range, and it is therefore not shown in Panel b. The present study reveals that the energy distribution generated in a narrow crossover (panels a and b) is either Gaussian or Lorentzian. These regimes are indicated.

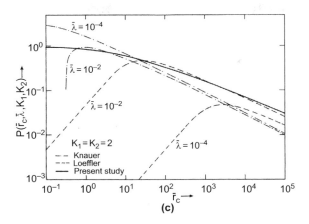

Fig. 15.1—Cont'd

know the trajectory displacements occurring in the final probe, which are only in a few special cases fully determined by angular deflections. The relation between the effect of statistical angular deflections and the trajectory displacement effect was investigated in Section 9.6.

The organization of this section is quite similar to that of the previous section. The predictions of our model for the generated rms angular spread will be reviewed first and the results of some other theories will be considered next. The section ends up with a comparison of the various predictions.

With Eqs. (8.3.18), (8.3.20) and (8.5.27), the rms value of the distribution of transverse velocity displacements generated in a *beam segment with a crossover* of arbitrary dimensions can be expressed as (using the d_0, v_0-scaling of Eq. 7.3.2)

$$\langle \Delta \bar{v}_\perp^2 \rangle^{1/2} = \frac{\langle \Delta \alpha^2 \rangle^{1/2}}{\alpha_0} = (\pi \bar{\lambda})^{1/2} P_{CA}(\bar{r}_c, K_1, K_2), \tag{15.3.1}$$

where $P_{CA}(\bar{r}_c, K_1, K_2) = [P_{CA}(\bar{r}_c, K_1) + P_{CA}(\bar{r}_c, K_2)]/2$ and

$$P_{CA}(\bar{r}_c, K) = \left[.894 + .3 \left(\frac{\bar{r}_c}{K^2} \right)^{2/3} \left[1 + .5 \left(\frac{K^2}{\bar{r}_c} \right)^{4/9} \right]^{3/2} \right.$$

$$\left. + \frac{\pi \bar{r}_c}{\{2 \ln[.8673(114.6 + \bar{r}_c)]\}^2} \right]^{-1/2} \tag{15.3.2}$$

For large K-values, the function $P_{CA}(\bar{r}_c, K)$ becomes independent of K and identical to the function $P_{CE}(\bar{r}_c, K)$, defined by Eq. (15.2.2). For $\bar{r}_c \to 0$, it becomes equal to unity. By removing the scaling in Eq. (15.3.1), using that $\Delta \alpha = \alpha_0 \Delta \bar{v}_\perp$, one finds for the rms angular spread

$$\langle \Delta \alpha^2 \rangle^{1/2} = \frac{1}{2}\left(\frac{m}{8\varepsilon_0^2 e}\right)^{1/4} P_{CA}(\bar{r}_c, K_1, K_2) \frac{I^{1/2}}{V^{3/4}}, \qquad (15.3.3)$$

similar to Eq. (15.2.3). If the angular distribution is Gaussian, the FWHM and rms are related as $FWHM_\alpha = 2(\ln 2)^{1/2} \langle \Delta \alpha^2 \rangle^{1/2}$ and Eq. (15.3.3) becomes equivalent to Eq. (8.3.21).

With Eqs. (8.4.18), (8.4.20) and (8.4.16) the rms value of the distribution of transverse velocity displacements generated in a *homocentric cylindrical beam segment* can be expressed as (using the δ, v-scaling of Eq. 6.13.1)

$$\langle \Delta v^*{}_\perp^2 \rangle^{1/2} = 2\left[p_2^*(\infty)\lambda^*\right]^{1/2} P_{PA}(r_0^*), \qquad (15.3.4)$$

where $p_2^*(\infty) = 0.3026$ and

$$P_{PA}(r_0^*) = \left(1 + .185/r_0^{*8/7}\right)^{-7/4} \qquad (15.3.5)$$

and λ^* is defined by Eq. (7.4.18). For an extended cylindrical beam $(r_0^* \gg 1)$, one finds $P_{PA}(r_0^*) = 1$. By removing the scaling in Eq. (15.3.4), using Eq. (8.4.27), one finds for the rms angular spread

$$\langle \Delta \alpha^2 \rangle^{1/2} = 0.1000 \frac{e^{1/12} m^{1/4}}{\varepsilon_0^{5/6}} P_{PA}(r_0^*) \frac{I^{1/2} L^{2/3}}{V^{13/12} r_0}. \qquad (15.3.6)$$

If the angular distribution is Gaussian, the FWHM follows as $FWHM = 2(\ln 2)^{1/2} \langle \Delta \alpha^2 \rangle^{1/2}$, in agreement with Eq. (8.4.30).

The rms angular spread and the rms energy spread generated in an extended cylindrical beam in which the interaction effects are not limited by the beam boundaries are related by

$$\langle \Delta \alpha^2 \rangle^{1/2} = \frac{\langle \Delta E^2 \rangle^{1/2}}{\sqrt{2} E}, \qquad (15.3.7)$$

as can directly be verified from Eqs. (15.3.6) and (15.2.7), taking $r_0^* \gg 1$. From Eqs. (15.3.3) and (15.2.3), one sees that a similar relation holds true for the interaction effects occurring in a beam segment with a narrow crossover (thus K_1 and $K_2 \gg 1$), which differs, however, by a factor $\sqrt{2}$.

For the theories treating a beam segment with a crossover, it was found that the resulting rms angular spread can be expressed in the form of Eq. (15.3.3), provided that the function $P_{CA}(\bar{r}_c, K_1, K_2)$ is generalized to a function $P(\bar{r}_c, \bar{\lambda}, K_1, K_2)$, which may also depend on $\bar{\lambda}$. We will first determine this function for each of the theories. The comparison between the theories will he carried out next by comparing these functions.

15.3.1 Loeffler and Hudgin

Loeffler and Hudgin (1970) extended the model of Loeffler (1969) to include the impact of the lateral components of the Coulomb force, acting on a reference particle that runs along the axis in a beam segment with a narrow crossover. First order perturbation dynamics is used to evaluate the impact of a single weak two-particle collision analytically. They separate the lateral force exerted by a colliding particle on the reference particle in a component in the direction of the relative transverse velocity ($\mathbf{F}_{\perp 1}$) and a component perpendicular to this direction ($\mathbf{F}_{\perp 2}$); see Fig. 2.1. They state that the latter is responsible for an angular deflection of the reference particle $\Delta \alpha$, while the former causes a lateral shift Δr only. Their reasoning is that this component changes of sign when the particles pass the point of closest approach, causing a cancellation of the angular deflections picked up in the first and the second half of the flight.

One may wonder to what image plane the shift Δr, evaluated by Loeffler and Hudgin, refers. As it is not accompanied by a net angular shift, the size of the corresponding virtual shift will be the same for any image plane. However, the force component causing the angular deflection $\Delta \alpha$ will also produce a spatial shift Δr. The corresponding virtual shift will only be negligible for an image plane that coincides with the crossover plane, as can be understood from the symmetry of the trajectory with respect to the crossover. Accordingly, we conclude that the trajectory displacement effect predicted by Loeffler and Hudgin pertains to the virtual broadening observed in the crossover plane, while the virtual spatial broadening will be larger for other planes.

It should also be noticed that the force component producing the angular deflection will be zero when the colliding electron follows a meridian trajectory. The effect of angular deflections is therefore expected to vanish in a homocentric beam segment ($r_c = 0$). Finally, it should be emphasized that the reasoning of Loeffler and Hudgin is only valid for weak collisions and requires that the crossover is narrow and located somewhere near the middle of the beam segment.

Loeffler and Hudgin solved the statistical part of the problem by numerical means but presented their final results in an analytical form. For the Full Width median of the distributions of angular changes $\Delta \alpha$ and trajectory displacements Δr, they give respectively (Loeffler and Hudgin, 1970, Eqs. 2 and 3 after correction of a misprint)

$$FW_{50\alpha} = \frac{e}{4\pi\varepsilon_0} \frac{1}{\alpha_0 r_c V} \Lambda(\lambda r_c) \qquad (15.3.8)$$

$$FW_{50r} = \frac{e}{4\pi\varepsilon_0} \frac{1}{\alpha_0^2 V} \Gamma(\lambda \alpha_0 L). \qquad (15.3.9)$$

From the plot of the functions Λ and Γ, one finds

$$\Lambda(x) \approx \begin{cases} 5.0 x^{2/3} & \text{for } x \gg 1 \\ 13 x^2 & \text{for } x \ll 1 \end{cases} \qquad (15.3.10)$$

$$\Gamma(x) \approx \begin{cases} 2.8 x^{5/6} & \text{for } x \ll 1 \\ 10 x^3 & \text{for } x \ll 1, \end{cases} \qquad (15.3.11)$$

which should be compared to the function $\Omega(x)$, given by the first equation of (15.2.14).

In this section, we will consider the result of Loeffler and Hudgin for the angular displacements, while the result for the trajectory displacements will be discussed in the next section. By naively assuming that the distribution of angular displacements is a (two-dimensional) Gaussian distribution, one obtains the following relation:

$$FW_{50,\alpha} = FWHM_\alpha = 2(\ln 2)^{1/2} \langle \Delta \alpha^2 \rangle^{1/2}. \qquad (15.3.12)$$

Eqs. (15.3.8) and (15.3.10) can now be expressed in the form of Eq. (15.3.3), replacing the function $P_{CA}(\bar{r}_c, K_1, K_2)$

$$P_{LH}(\bar{r}_c, \bar{\lambda}) = \begin{cases} \dfrac{5.0}{(\pi \ln 2)^{1/2}} \dfrac{\bar{\lambda}^{1/6}}{\bar{r}_c^{1/3}} & \text{for } \bar{\lambda}\bar{r}_c \gg 1 \\ \dfrac{13}{(\pi \ln 2)^{1/2}} \bar{\lambda}^{3/2} \bar{r}_c & \text{for } \bar{\lambda}\bar{r}_c \ll 1. \end{cases} \qquad (15.3.13)$$

This equation is independent of K_1 and K_2, which are assumed to be large.

As for the Boersch effect, our model indicates that distribution becomes non-Gaussian for low and moderate particle densities. For moderate $\bar{\lambda}$-values and K_1 and $K_2 \gg 1$ one finds from Eqs. (8.3.24) and (8.5.43) for the Full Width median angular displacement generated in a beam segment with a crossover

$$FW_{50,\alpha} = .8793 \frac{m}{\varepsilon_0^2 e}(1 + .02069\bar{r}_c) \frac{I^2}{\alpha_0^3 V^3}$$

$$= \begin{cases} .4572 \dfrac{m}{\varepsilon_0 e^2} \dfrac{I^2 r_c}{\alpha_0 V^2} & \text{for } \bar{r}_c \gg 100 \\ .8793 \dfrac{m}{\varepsilon_0^2 e} \dfrac{I^2}{\alpha_0^3 V^3} & \text{for } \bar{r}_c \ll 100, \end{cases} \qquad (15.3.14)$$

using the approximation for $f_\infty(\bar{r}_c)$ given by Eq. (8.3.10) and the definition of $\bar{r}_c$ given by Eq. (7.3.5). The functional dependency of Eq. (15.3.14) occurs when weak complete collisions are dominant and the corresponding regime is called the "weak complete collision regime." The result of Loeffler and Hudgin, expressed by Eqs. (15.3.8) and (15.3.10), leads, for $\lambda r_c \ll 1$, to the same functional dependency as Eq. (15.3.14), for $\bar{r}_c \gg 100$, but is larger by a factor 1.13. We conclude that their result for $\lambda r_c \ll 1$ is not generally valid but predicts accurate results for the weak complete collision regime when $\bar{r}_c \gg 100$. It should be noticed that for very small particle densities, the angular displacement distribution will change to the pencil beam type, leading to the functional dependency given by Eq. (8.5.35). This regime is not covered by the results of Loeffler and Hudgin.

Let us now investigate the result of Loeffler and Hudgin given for $\lambda r_c \gg 1$ From Eqs. (15.3.8) and (15.3.10), one obtains by using the definition of λ given by Eq. (3.2.4)

$$FW_{50,\alpha} = 0.32 \frac{m^{1/3}}{\varepsilon_0} \frac{I^{2/3}}{V^{4/3}\alpha_0 r_c^{1/3}} = 0.40 \frac{m^{1/3}}{\varepsilon_0} \frac{I^{2/3}K^{1/3}}{V^{4/3}\alpha_0^{4/3}L^{1/3}}, \quad (15.3.15)$$

where $K = \alpha_0 L/2r_c$. The 2/3 power dependency on the beam current suggest that this regime coincides with the Holtsmark regime distinguished in our model. This regime occurs for an extended beam in which the effect is dominated by weak incomplete interactions. From Eqs. (8.5.31), (8.5.20) and (8.5.43), one finds for the Holtsmark regime

$$FW_{50,\alpha} = 0.4810 \frac{m^{1/3}}{\varepsilon_0} \left| \frac{1}{(1/K+2S_c)^{1/3}} - \frac{1}{[1/K+2(1-S_c)^{1/3}]} \right| \frac{I^{2/3}}{V^{4/3}\alpha_0^{4/3}L^{1/3}},$$

$$(15.3.16)$$

in which one should take $S_c = 0$ or $S_c = 1$ when $K \to 0$. For $K \to \infty$, the $FW_{50,\alpha}$ value predicted by Eq. (15.3.16) vanishes for a crossover in the middle ($S_c = 1/2$), contrary to the result given by Loeffler and Hudgin. For a half-crossover ($S_c = 0$ or $S_c = 1$) and $K \gg 1$, one finds the same functional dependency as predicted by Loeffler and Hudgin. Finally, one should realize that for very large particle densities, the angular displacement distribution will become Gaussian, leading to the functional dependency given by Eq. (8.3.21). This regime is not covered by the results of Loeffler and Hudgin.

15.3.2 Weidenhausen, Spehr and Rose

Weidenhausen et al. (1985) employed the "closest encounter approach" of Rose and Spehr (1980) to evaluate the mean square angular spread generated in a beam segment with a narrow circular or astigmatic crossover. The angular and current density distribution in the crossover are assumed to be Gaussian. Using the relations of Eq. (15.2.21), their final result can directly be expressed in the form of Eq. (15.3.1), provided that the function $P_{CA}(\bar{r}_c, K_1, K_2)$ is replaced by Weidenhausen et al. (1985, Eqs. 20, 22 and 23)

$$P_{WSR}(\bar{r}_c) = \begin{cases} (4/\pi - 1)^{1/2} & \text{for } \bar{r}_c \to 0 \\ \dfrac{(4/\pi - 1)^{1/2}}{2\bar{r}_c^{1/2}} \left(32.4 - 33.6 \ln \bar{r}_c + 16.5 \ln^2 \bar{r}_c\right)^{1/2} & \text{for } \bar{r}_c > 25. \end{cases}$$

(15.3.17)

This result applies to the case $\bar{\lambda} \ll 1$ and $\bar{\lambda}\bar{r}_c < 1/2$. The numerical data plotted by Weidenhausen et al. shows that the function $P_{WSR}(\bar{r}_c)$ has a maximum for $\bar{r}_c \approx 5$, which exceeds the value for $\bar{r}_c = 0$ by about 30%.

15.3.3 Massey, Jones and Plummer

Massey et al. (1981) evaluated the impact of the fluctuations in the transverse component of the electrostatic force, acting on a reference particle that runs through an extended (nearly) homocentric beam segment in a uniform acceleration zone. Using this model we evaluated the rms angular spread for the case that acceleration is absent. We found that the resulting expression for the angular spread $\langle \Delta \alpha^2 \rangle^{1/2}$ can directly be obtained from the energy spread $\langle \Delta E^2 \rangle^{1/2}$ predicted by Eq. (15.2.31) by employing Eq. (15.3.7). As the same relation holds true for the corresponding equations derived within our model we end up with the same conclusions as reached for the Boersch effect, namely that the predictions of both models show the same functional dependency for the case of an extended homocentric cylindrical beam but differ by a factor 1.52.

15.3.4 Comparison of theories for a beam segment with a crossover

The theories of Loeffler and Hudgin (1970) and Weidenhausen et al. (1985) give a prediction for the angular spread generated in a beam segment with a crossover. We have expressed their results in terms of the equivalent of the function $P_{CA}(\bar{r}_c, K_1, K_2)$ appearing in Eq. (15.3.3), see Eqs. (15.3.13) and (15.3.17) respectively. Our model is represented by the function defined by Eq. (15.3.2). Contrary to the other models, it is also valid for beams with small K-values. The comparison will be restricted to the case K_1 and $K_2 \gg 1$.

In Fig. 15.2, the functions $P(\bar{r}_c, \bar{\lambda}, K_1, K_2)$ representing the different theories are plotted as functions of $\bar{r}_c$ for the case $K_1 = K_2 - 200$. The

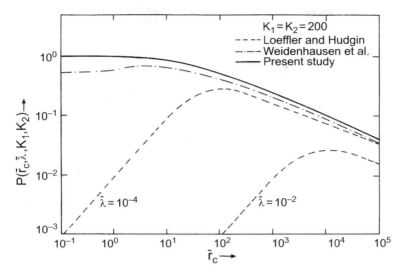

Fig. 15.2 Comparison of the predictions of different theories for the rms angular spread generated in a beam segment with a crossover in the middle. Each theory is represented by a different function $P(\bar{r}_c, \bar{\lambda}, K_1, K_2)$ which is proportional to the rms angular spread, see Eq. (15.3.3). All theories, apart from the present one, assume a narrow crossover and the comparison is therefore restricted to the case $K_1 = K_2 = 200$. The result of Loeffler and Hudgin is the only one that depends on the scaled linear particle density $\bar{\lambda}$. The values $\bar{\lambda} = 10^{-2}$ and $\bar{\lambda} = 10^{-4}$ are considered.

function representing the theory of Loeffler and Hudgin is the only one that depends on $\bar{\lambda}$. It is plotted for the cases $\bar{\lambda} = 10^{-2}$ and 10^{-4}. As for the different predictions for the rms energy spread, which were compared in the previous section, one finds some agreement between the different theories for large values of $\bar{r}_c$. For these $\bar{r}_c$-values, the angular displacement distribution is of the Gaussian type.

The substantial disagreement for smaller $\bar{r}_c$-values between our result and that of Loeffler and Hudgin is partly due to the conversion of their result from a median width to a rms value by means of Eq. (15.3.12), which requires the angular displacement distribution to be Gaussian. As the distribution is non-Gaussian for small $\bar{r}_c$-values, the direct comparison of the predictions for the median angular spreads is more appropriate and shows a better agreement; see Eq. (15.3.14).

The result of Weidenhausen et al., based on the closest encounter approach, is in reasonable agreement with our prediction for most $\bar{r}_c$-values but leads to a 0.52 times smaller value for $\bar{r}_c \to 0$. Due to the approximations made within this model for practical particle densities

$(\bar{\lambda} \ll 1)$, it is mathematically equivalent to our model, as was pointed out in the previous section. The differences between both results mainly arise from the fact that our result pertains to an on axis reference trajectory and assumes a uniform angular and spatial distribution in the crossover, while that of Weidenhausen et al. represents an average over all possible reference trajectories and assumes a Gaussian angular and spatial distribution in the crossover. The difference between both results found for $\bar{r}_c \to 0$ is due to the different treatment of strong scattering events, as was pointed out by Rose (1988). In the calculation of Weidenhausen et al., a strong collision with a scattering angle $\chi > \pi/2$ is replaced by a collision with scattering angle $\pi - \chi$ (see Fig. 6.3). Their argument is that one cannot differentiate between both events, due to the indistinguishability of the particles. It should be added that this effect is mainly of academic interest, since practical conditions correspond to $\bar{r}_c \gg 1$ as can be seen from Fig. 7.6.

In experimental systems, one is mainly concerned with the spatial broadening of the final probe and the angular spread has to be translated into a spatial broadening in some reference plane, which is imaged on the target. The angular spread should therefore be multiplied with some distance, which will be proportional to the beam segment length L. In Section 9.2, it was pointed out that this procedure implies that one cannot use complete collision dynamics to evaluate the angular spread, since this dynamics neglects the terms proportional to $1/L$ (or equivalently $1/T$). See the discussion of Eq. (9.2.5). This is apparently not recognized by Weidenhausen et al. (1985) who claim that their result for the rms angular spread (which is based on complete collisions) can be used to evaluate the spatial blurring in some image plane that is located at some distance from the crossover.

15.4. Trajectory displacement effect

In Chapter 9, the extended two-particle model was employed to describe the trajectory displacement effect, which is related to the distribution of virtual displacements $\rho(\Delta r)$, observed in a certain reference (or image) plane. With Eqs. (9.3.28), (9.3.24), (9.3.25) and (9.5.10), the rms value of the distribution of virtual displacements generated in a *beam segment with a crossover* of arbitrary dimensions can be expressed as (using the δ, ν-scaling of Eq. 6.13.1)

$$\langle \Delta r^{*2} \rangle^{1/2} = 2 \left[p_2^*(\infty, 1/2, 1/2) \bar{\lambda} \right]^{1/2} P_{CT}(v_0^*, r_c^*, S_c, S_i), \qquad (15.4.1)$$

where $p_2^*(\infty, 1/2, 1/2) = 0.273$ and

$P_{CT}(v_0^*, r_c^*, S_c, S_i)$

$$\approx \left(\frac{1 + 0.682(S_c - 0.5) - 0.739(S_c - 0.5)^2}{\left(1 + 1.40/v_0^{*8/7}\right)^{7/2} (1 + 0.30 r_c^*)^{1/2}} + 113 v_0^{*1/2}(S_c - S_i)^2 \right)^{1/2}.$$

(15.4.2)

By removing the scaling in Eq. (15.4.1) using Eq. (6.13.1) one obtains

$$\langle \Delta r^2 \rangle^{1/2} = 9.50 \times 10^{-2} \frac{e^{1/12} m^{1/4}}{\varepsilon_0^{5/6}} P_{CT}(v_0^*, r_c^*, S_c, S_i) \frac{I^{1/2} L^{2/3}}{\alpha_0 V^{13/12}}. \qquad (15.4.3)$$

If the trajectory displacement distribution is Gaussian, the FWHM and rms value are related as $FWHM_r = 2(\ln 2)^{1/2} \langle \Delta r^2 \rangle^{1/2}$, and Eq. (15.4.3) becomes equivalent to Eq. (9.3.45). It should be noted that the Gaussian regime occurs only for relatively large particle densities ($\bar{\lambda} > 0.05$). The practical use of the rms value to characterize the trajectory displacement effect is, therefore, even further restricted than the use of the rms energy spread and the rms angular broadening.

Similar to the approach taken in the previous sections, our strategy is to express the results of the different theories for the trajectory displacement effect in the form of Eq. (15.4.3), replacing the function $P_{CT}(v_0^*, r_c^*, S_c, S_i)$ by a generalized version, which will be different for each theory. The comparison of the theories will be carried out at the end of the section by comparing these functions.

15.4.1 Loeffler and Hudgin

In the previous section, we reproduced the results of Loeffler and Hudgin (1970) for the trajectory displacement generated in a beam segment with a narrow crossover, see Eqs. (15.3.9) and (15.3.11) and (15.3.11). By naively assuming that the trajectory displacement distribution is a (two-dimensional) Gaussian distribution, the rms value can be calculated from the FW_{50} value as $\langle \Delta r^2 \rangle^{1/2} = FW_{50,r}/2(\ln 2)^{1/2}$, similar to Eq. (15.3.12). The results of Loeffler and Hudgin can now be expressed in the form of Eq. (15.4.3), replacing the function $P_{CT}(v_0^*, r_c^*, S_c, S_i)$ by

$$P_{LH}(v_0^*, \bar{\lambda}) = \begin{cases} 2.6 \, v_0^{*1/2} \bar{\lambda}^{1/3} & \text{for } 4\bar{\lambda} v_0^{*3} \gg 1 \\ 1.8 \times 10^2 \bar{\lambda}^{5/2} v_0^{*7} & \text{for } 4\bar{\lambda} v_0^{*3} \ll 1. \end{cases} \qquad (15.4.4)$$

This equation pertains to the case $S_c = S_i \approx 1/2$ and requires a narrow crossover.

Our model shows that the trajectory displacement distribution will be non-Gaussian for most operating conditions. Accordingly, it is more appropriate to compare the $FW_{50,r}$-value predicted by Loeffler and

Hudgin directly to our results obtained for this width measure. Loeffler and Hudgin give for $\lambda a_0 L \ll 1$

$$FW_{50,r} = 0.28 \frac{m^{3/2}}{\varepsilon_0 e^{7/2}} \frac{I^3 a_0 L^3}{V^{5/2}}, \qquad (15.4.5)$$

as follows from Eqs. (15.3.9) and (15.3.11). Our model predicts a similar functional dependency for the pencil beam regime. From Eq. (9.3.47), (9.5.6) and (8.5.43), one finds for this regime in case $S_c = S_i$

$$FW_{50,r} = 0.03622 \frac{m^{3/2}}{\varepsilon_0 e^{7/2}} \left| 4 - 12S_c(1 - S_c) + \frac{3 - 6S_c(1 - S_c)}{K} \right| \frac{I^3 a_0 L^3}{V^{5/2}}. \qquad (15.4.6)$$

For $K \to 0$, one should take $S_c = 0$ or $S_c = 1$. For $K \to \infty$ and $S_c = 1/2$, we obtain the same functional dependency as Loeffler and Hudgin, but their result is larger by a factor 7.8.

For $\lambda a_0 L \gg 1$, one obtains from Eqs. (15.3.9) and (15.3.11)

$$FW_{50,r} = 0.17 \frac{m^{5/12}}{\varepsilon_0 e^{1/4}} \frac{I^{5/6} L^{5/6}}{V^{17/12} a_0^{7/6}}. \qquad (15.4.7)$$

This functional dependency is not confirmed by any of the regimes distinguished within our model. According to our model, the trajectory displacement distribution is Holtsmarkian for moderate particle densities and Gaussian for large particle densities, leading to the functional dependencies of Eqs. (9.3.44) and (9.3.45) respectively. These regimes are not covered by the results of Loeffler and Hudgin.

15.4.2 De Chambost

De Chambost (1982) relates the trajectory displacement effect generated in a beam segment with a crossover to the energy spread generated in such a beam segment. For the energy spread he utilizes the result given by De Chambost and Hennion (1979). Following this approach but using the corrected Eq. (15.2.25), one finds

$$\begin{aligned}\langle \Delta r^2 \rangle^{1/2} &= \frac{\langle \Delta E^2 \rangle^{1/2} L}{2E} F_C(K, S_c) \\ &= \frac{m^{3/16} e^{1/16}}{18^{3/16} 2 (\pi \varepsilon_0)^{5/8}} \frac{I^{3/8} L}{(a_0 r_c)^{1/4} V^{13/16}} F_C(K, S_c), \end{aligned} \qquad (15.4.8)$$

where

$$F_C(K, S_c) = \left(S_c^2 h(2S_c K)^2 + (1 - S_c)^2 h[2(1 - S_c)K]^2 \right)^{1/2}$$

$$h(x) = \frac{x^{1/2}}{(x-1)^{3/2}} \left(x - \frac{1}{x} - 2 \ln x \right)^{1/2}. \qquad (15.4.9)$$

Eq. (15.4.8) can be expressed in the form of Eq. (15.4.3), replacing the function $P_{CT}(v_0^*, r_c^*, S_c, S_i)$ by

$$P_C(v_0^*, r_c^*, S_c, \bar{\lambda}) = 0.822 \frac{v_0^{*1/2}}{\bar{\lambda}^{1/8} r_c^{*1/4}} F_C(v_0^*/2r_c^*, S_c), \qquad (15.4.10)$$

using that $K = v_0^*/2r_c^*$. This equation pertains to the case $S_c = S_i$ and requires that $K \gg 1$.

15.4.3 Spehr

Spehr (1985a) relates the trajectory displacement effect to a transfer of internal kinetic energy from the longitudinal to the lateral degree of freedom by weak complete collisions. According to Spehr, this transfer occurs in the dilute parts of the beam, where the longitudinal beam temperature exceeds the transverse beam temperature. In this model, the trajectory displacement effect stems mainly from the initial energy spread. The magnitude of the initial energy spread, which plays no role in the other models, is therefore an essential parameter in Spehr's model.

For the broadening of an initially homocentric crossover ($r_c = 0$) Spehr gives (Spehr, 1985a, Eqs. 19–21, 17)

$$\Delta r = \left(\frac{em}{2^9 \pi^3 \varepsilon_0^4}\right)^{1/4} \frac{I^{1/2} L^{1/2}}{V^{5/4} \alpha_0} \frac{E^{1/2}}{\langle \Delta E^2 \rangle^{1/4}} \left[3.22 + \ln^2\left(\frac{4\pi\varepsilon_0}{7.98e} V \alpha_0 L \frac{\langle \Delta E^2 \rangle}{E^2}\right)\right]^{1/2}, \qquad (15.4.11)$$

provided that

$$\frac{\langle \Delta E^2 \rangle}{E^2} > 100 \frac{C_0}{\alpha_0 LE} \quad \text{and} \quad \bar{\lambda} < \frac{\langle \Delta E^2 \rangle}{24 \alpha_0 E^2}. \qquad (15.4.12)$$

By assuming that we may identify Δr with an rms value, this result can be expressed in the form of Eq. (15.4.3), replacing the function $P_{CT}(v_0^*, r_c^*, S_c, S_i)$ by

$$P_S(v_0^*, \langle \Delta v_z^{*2} \rangle) = \frac{0.901}{\langle \Delta v_z^{*2} \rangle^{1/4}} [3.22 + \ln^2(1.00 v_0^* \langle \Delta v_z^{*2} \rangle)]^{1/2}. \qquad (15.4.13)$$

The conditions of Eq. (15.4.12) can be expressed as

$$\langle \Delta v_z^{*2} \rangle > 12.5/v_0^* \quad \text{and} \quad \bar{\lambda} < \alpha_0 \langle \Delta v_z^{*2} \rangle / 6 v_0^{*2}. \qquad (15.4.14)$$

Spehr's result pertains to the virtual broadening of an originally homocentric crossover ($r_c^* = 0$, $S_i - S_c$). It is independent of the location of the crossover, which is here specified by S_c. The initial energy spread is represented in Eq. (15.4.13) by the parameter $\langle \Delta v_z^{*2} \rangle^{1/2}$. The relation with the experimental parameters can be obtained from Eq. (10.4.5). For practical beams, the value of $\langle \Delta v_z^{*2} \rangle^{1/2}$ is typically in the range 0.1–1. When the first constraint of Eq. (15.4.14) is not fulfilled, it is not permitted to use the approximating result of Eq. (15.4.13). Instead, one has to evaluate the integral appearing in Spehr's model by numerical means (Spehr, 1985a, Eq. 20). The outcome will be smaller than predicted by Eq. (15.4.13) and leads to a zero broadening for $\langle \Delta v_z^{*2} \rangle^{1/2} \to 0$.

15.4.4 Comparison of theories

The theories of Loeffler and Hudgin (1970), De Chambost (1982) and Spehr (1985a) give a prediction for the trajectory displacement effect generated in a beam segment with a crossover. We have expressed their results in terms of the equivalent of the function $P_{CT}(v_0^*, r_c^*, S_c, S_i)$, appearing in Eq. (15.4.3), see Eqs. (15.4.4), (15.4.10) and (15.4.13) respectively. Our model is represented by the function, which is defined by Eq. (15.4.2). This function is valid for the full range of operating conditions represented by the parameters v_0^*, r_c^*, S_c and S_i. The other models apply to large values of $K = v_0^*/2r_c^*$ (narrow crossover) and $S_i = S_c$ (image plane in the crossover). The result of Loeffler and Hudgin requires, in addition, that the crossover is located somewhere near the middle of the beam segment ($S_c \approx 1/2$).

In Fig. 15.3a, the functions $P\left(v_0^*, r_c^*, S_c, S_i, \bar{\lambda}, \langle \Delta v_z^{*2} \rangle^{1/2}\right)$ representing the different theories are plotted as functions of v_0^* for the case $S_c = S_i = 1/2$ and $r_c^* = 0.02$. Fig. 15.3b gives the corresponding results for $r_c^* = 2$. The function representing Spehr's model is not included in this figure, since its use is restricted to a nearly homocentric crossover ($r_c^* \ll 1$). For the parameter $\langle \Delta v_z^{*2} \rangle^{1/2}$ appearing in Spehr's model, we considered the values 1 and 10, which correspond to a normal and an extremely large initial energy spread respectively. The functions representing the theory of Loeffler and Hudgin and the theory of De Chambost depend on $\bar{\lambda}$. They are plotted for the cases $\bar{\lambda} = 10^{-2}$ and 10^{-4}.

Figs. 15.3a and b show that there exists a significant disagreement between the predictions of the various theories. It should be noticed that for the theory of Loeffler and Hudgin it is more appropriate to compare the predictions for the median width of the trajectory displacement distribution. This leads to a better agreement with our theory for some conditions, as was pointed out previously in this section.

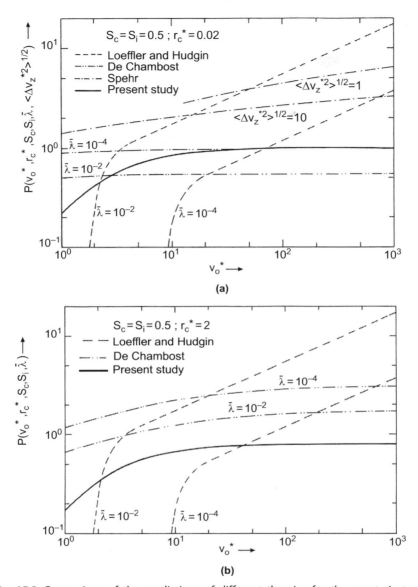

Fig. 15.3 Comparison of the predictions of different theories for the rms trajectory displacement broadening generated in a beam segment with a crossover in the middle ($S_c = 1/2$). Each theory is represented by a different function $P(v_0^*, r_c^*, S_c, S_i, \bar{\lambda}, \langle \Delta v_z^{*2} \rangle^{1/2})$, which is proportional to the rms spread of the virtual displacements Δr in the image plane; see Eq. (15.4.3). The results pertain to the virtual broadening observed in the crossover plane ($S_i = S_c$). Panels a and b apply to a small crossover ($r_c^* = 0.02$) and a large crossover ($r_c^* = 2$) respectively. Spehr's result pertains to a (nearly) homocentric crossover and is therefore not included in panel b. The model of Loeffler and Hudgin and the model of De Chambost depend on the scaled linear particle density $\bar{\lambda}$. The values $\bar{\lambda} = 10^{-2}$ and $\bar{\lambda} = 10^{-4}$ are considered. Spehr's result depends on the initial energy spread, which is represented by the parameter $\langle \Delta v_z^{*2} \rangle^{1/2}$. The cases of a normal energy spread ($\langle \Delta v_z^{*2} \rangle^{1/2} = 1$) and an extreme large energy spread ($\langle \Delta v_z^{*2} \rangle^{1/2} = 10$) are considered.

15.5. Conclusions

A number of recent theories for the different manifestations of the statistical Coulomb interactions in a beam in drift space have been compared with each other and with the predictions of the extended two-particle model. Most theories handle the case of beam segment with a crossover, which is usually assumed to be narrow ($K \gg 1$). Massey et al. (1981) considered a (nearly) homocentric beam segment. The Boersch effect is the most studied manifestation of statistical interactions, for which seven other theories were considered. For the effect of statistical angular deflections, we studied three other theories. The same number of theories were investigated for the trajectory displacement effect. Apart from the present study, only two models cover all manifestations of statistical interactions. One is that of Loeffler (1969) and Loeffler and Hudgin (1970) and the other is that of the Darmstadt group, presented by Rose and Spehr (1980, 1983), Weidenhausen et al. (1985) and Spehr (1985a,b). Most theories evaluate the rms width of the displacement distribution. Loeffler and Hudgin expresses their final result in terms of a median width.

The models of Zimmermann (1970) and Knauer (1979a,b) for the Boersch effect are based on the concepts developed in the field of plasma physics and stellar dynamics. In this type of approach, the effect is attributed to a large number of independent weak collisions. Accordingly, the generated energy distribution is assumed to be Gaussian. Although not explicitly stated by the authors, the typical duration of a collision is assumed to be short compared to the time scale at which the macroscopic properties of a bunch of particles in the beam change. Furthermore, the lateral dimensions of such a bunch are assumed to be large enough to treat it as an extended three-dimensional system, ignoring the influence of the beam boundaries.

In actual beams, the particle motion cannot be represented as a random thermal motion superimposed on a smooth systematic motion, since the typical collision duration is of the same order as the total flight time through the beam segment. Furthermore, the number of collisions is not necessarily large, and the contribution of strong interactions cannot always be ignored. The validity of the models of Zimmermann and Knauer is typically restricted to beams with a relatively narrow crossover (K of the order 10^2) and a high particle density. The models do not apply to beams of low particle density and/or nearly cylindrical beams (for which the displacement of a particle is determined by only a few collisions that are predominantly incomplete) nor to beams with an extremely small crossover (for which the influence of the beam boundary cannot be ignored).

The approach of Loeffler (1969) and Loeffler and Hudgin (1970) is based on a more realistic representation of the particle motion in the beam. Their collision dynamics is based on first order perturbation theory. Both weak complete collisions and weak incomplete collisions are, therefore, evaluated in a proper way. However, this type of dynamics is not suited for strong collisions, which restricts the approach to beams of moderate and low particle density. By evaluating a median, instead of a rms value, their results are suited to characterize the width of the non-Gaussian distributions generated in beams of low and moderate particle densities, which distinguishes this model from all others. The comparison with the median values predicted by our model shows that the results of Loeffler and Hudgin are not generally valid, as they suggest, but they do provide reasonable accurate predictions for some operating conditions. See Eqs. (15.2.17), (15.2.18), (15.3.14) and (15.4.5). Unfortunately, the discussion presented by Loeffler and Hudgin is extremely brief and does not give any physical explanation for the different predicted parameter dependencies. It appears that the relation between the parameter dependency of the median width and the type of displacement distribution involved was not fully understood by these authors. With respect to the comparison of the rms values in Figs. 15.1–15.3, it should be noted that the conversion from a median to an rms value is not correct for the non-Gaussian regimes. The direct comparison of the median values is for these regimes preferred.

The model presented by De Chambost and Hennion (1979) and De Chambost (1982) and the model of Sasaki (1984) are not based on a reduction to two-particle collisions but start from the mean square of the fluctuations in the electrostatic force in the beam. The time development of the distribution of field particles, surrounding a test particle, is represented as a succession of independent states. The electrostatic interaction force experienced by the test particle varies randomly from state to state. This approach ignores any correlation in the particle coordinates in successive independent states, which leads to a rather artificial truncation of the interaction for small interparticle distances. Furthermore, the size of the time intervals between successive independent states represents a critical parameter in these models, but its choice seems rather arbitrary. Accordingly, one cannot expect that these approaches predict accurate absolute values, which is indeed confirmed by the comparison. Nevertheless, there is some agreement by trend with the results of our model for extended beams with a narrow crossover.

The model of Massey et al. (1981) for the rms energy spread and rms angular spread, generated by potential energy relaxation in a nearly homocentric cylindrical beam, also starts from the mean square fluctuations in the electrostatic force in the beam, but accounts for the correlation between the successive states. Their results are in exact agreement by trend with our results but smaller by a factor of approximately 2/3. It should be noticed that the practical use of the rms width is restricted to the Gaussian regime, which occurs in (nearly) homocentric cylindrical beams only for rather high particle densities ($\lambda^* \gtrsim 1$).

Crewe (1978a) evaluated the energy change caused by a single complete collision, taking some representative initial condition. This approach entirely ignores the statistical nature of the problem. Figs. 15.1a and b show that Crewe's equation strongly underestimates the Boersch effect generated in a beam segment with a crossover. The predicted trend is not correct either.

The so-called closest encounter approach, used by Rose and Spehr (1980, 1983) to evaluate the rms energy spread and by Weidenhausen et al. (1985) to evaluate the rms angular spread, both generated in a beam segment with a narrow crossover, is, for practical particle densities ($\bar{\lambda} \ll 1$), in close agreement with the results of our model. However, this agreement is not surprising, as is suggested by Weidenhausen et al. (1985), but a direct consequence of the approximations made within the closest encounter approach. For practical particle densities, they do not exclusively consider the closest encounter but also take the contribution of the more distant neighbors into account. This makes their model mathematically equivalent to ours. The remaining differences are mainly due to the fact that our final equations refer to the on-axis reference trajectory, while their results represent an average over all possible reference trajectories, following the method outlined in Section 10.3. It should be noticed that the use of complete collision dynamics restricts the applicability of their results to beams with a narrow crossover ($K \gg 1$). Our model uses a more refined collision dynamics, which extends its applicability to beams of arbitrary dimensions.

We emphasize once more that the rms value provides only practical predictions in the Gaussian regime. The rms value shows for all conditions a square root dependency on the beam current. Other width measures, such as the *FWHM* and the FW_{50}, show different current dependencies for the different regimes distinguished within our model. This is apparently not recognized by Rose and Spehr (1983) who stated that the linear current dependency of the energy spread predicted by Loeffler for small current densities ($\lambda r_c \ll 1$) is incorrect.

The model of Spehr (1985a) relates the trajectory displacement effect generated in a beam segment with a narrow crossover to the energy distribution of the particles entering the beam segment. His result, therefore, vanishes for initially monochromatic beams. For large initial energy spreads, his predictions become insensitive to the exact value of the energy spread, but overestimate the spatial broadening. See Fig. 15.3a. Spehr's physical explanation of the probe broadening due to the trajectory displacement effect is not confirmed by our calculations. Our results for non-monochromatic beams, reported in Section 10.4, show that the trajectory displacement effect is, for practical operating conditions, almost independent of the initial energy spread.

We conclude this section with some remarks concerning the fundamental aspects of the dynamical part of the problem. In our model, the total displacement of the test particle is calculated as the sum of all displacements caused by the individual field particles. This reduction of the N-particle problem to a two-particle problem is allowed, provided that the test particle is not involved in two or more *strong* collisions simultaneously or successively in the same beam segment. Such events are very exceptional for practical operating conditions. The Monte Carlo simulations reported in Chapter 14 demonstrate that this two-particle approach is extremely accurate, even for relatively high particle densities.

The validity of the reduction to pair interactions is not generally recognized. For instance, Rose and Spehr (1983) motivate their closest encounter approach by the argument that an approximation that neglects all correlation between collisions (taking place simultaneously or successively) cannot be expected to yield results whose range of validity exceeds that of a single scattering model. Similar statements are made by Sasaki (1984, 1986) and Tang (1987). Indeed, if all two-particle interactions were strong, it would be erroneous to ignore the correlation between them. However, in practical beams, most collisions are weak, and the total displacement of a particle can, in good approximation, be calculated as the sum of all contributions of the individual two-particle encounters, based on the argument that forces are additive. The inaccuracy of some theories based on two-particle collisions is due to the fact that the two-particle dynamical problem is oversimplified. In order to obtain results that are valid for arbitrary beam geometries, it is essential to solve the two-particle problem without making any assumptions on the strength or the completeness of the collision.

CHAPTER SIXTEEN

Summary for the one-minute designer

Contents

16.1.	Introduction	487
16.2.	Physical aspects	488
16.3.	Parameter dependencies	499
16.4.	Equations for the Boersch effect	507
16.5.	Equations for the trajectory displacement effect	513
16.6.	Equations for the space charge effect	516
16.7.	Addition of the effects generated in individual beam segments	518

16.1. Introduction

This chapter summarizes the results of this monograph that arc relevant for the designer of charged particle beam systems. The physical aspects of our model for Coulomb interactions in particles beams are reviewed with the objective of providing insight in the essential steps of the calculation and the basic assumptions involved. The extended two-particle approach used for the modeling of the Boersch effect and the trajectory displacement effect (statistical interactions) can predict the detailed shape of the generated energy and trajectory displacement distribution. Different regimes of operating conditions are found, leading to different types of distributions. Gaussian behavior occurs at relatively high particle densities. Each of the various types of non-Gaussian distributions, arising at low and moderate particle densities, can be related to a different type of collision, which is dominant for those conditions.

A number of measures are used to characterize the width of the distributions. The root mean square (rms) is only suited for a Gaussian distribution. Therefore, we prefer to use the Full Width at Half Maximum ($FWHM$) and the Full Width median (FW_{50}) value. In order to evaluate the performance of a shaped beam lithography system, the trajectory displacement broadening is also expressed in terms of the corresponding d_{1288} and d_{2575} edge width. The relation between the different width measures

Advances in Imaging and Electron Physics, Volume 230
ISSN 1076-5670
https://doi.org/10.1016/B978-0-443-29784-7.00016-9

depends on the type of distribution involved. The theory provides explicit equations for the $FWHM$, the FW_{50}, the d_{1288} and the d_{2575} values.

The dependency of the width of the distributions on the experimental parameters differs for the various regimes. The different dependencies can be understood on the basis of some elementary physical arguments. A more detailed analysis is required to obtain expressions that are also quantitatively correct. The first two sections of this chapter focus on the qualitative analysis. The next three sections review the detailed analytical prescriptions obtained for the Boersch effect, the trajectory displacement effect and the space charge effect. These results pertain to a single beam segment in drift space. The method of adding the contributions of the different segments, constituting the beam, is outlined in the last section.

The equations resulting from the analytical model are particularly suited for the implementation in a computer program. Such a program has been developed and is called INTERAC. It allows for an interactive evaluation of the impact of Coulomb interactions on the performance of a column. Due to the speed of this approach, it provides a powerful tool for the optimization of a certain column design.

16.2. Physical aspects

The model schematizes the beam as a succession of beam segments separated by thin optical components. The impact of particle interactions on the properties of the beam is studied for a single beam segment. The basic expectation of this approach is that the correlation between the contributions of the various segments can be ignored. Fig. 3.1 shows an individual rotational symmetric beam segment in drift space. Its macroscopic condition is determined by the experimental parameters: the beam current I, beam voltage V, length L, crossover radius r_c, beam semi-angle α_0, crossover position parameter $S_c = L_1/L$, image plane position parameter $S_i = L_i/L$ and initial energy spread ΔE_0. L_1 is the distance between entrance plane and crossover plane and L_i the distance between entrance plane and image plane. The image plane is optically conjugated to the target plane of the column. The (virtual) spatial broadening of the beam observed in this plane can, therefore, directly be related to the broadening of the probe at the target. In most applications, the image plane coincides with the crossover plane ($S_i = S_c$). However, this is not true for all beam segments in a shaped beam lithography column that employs Koehler illumination. The current density distribution and angular distribution in the crossover are, in general, assumed to be uniform, but Gaussian distributions have been considered, too.

Given the spatial and velocity distribution of the particles at the entrance of the beam segment, the final distribution has to be computed, taking the mutual Coulomb repulsion into account. Consider a system of N identical particles of charge e and mass m. The particle density is assumed to be low enough to treat the "gas" as a classic system, neglecting quantum mechanical effects. Given the positions $\mathbf{r}_j(t)$ of all particles ($j = 1, 2, \ldots, N$) at time t, the total Coulomb force acting on a particle i can be evaluated as

$$\mathbf{F}_i(t) = C_0 \sum_{j \neq i}^{N} \frac{\mathbf{r}_i(t) - \mathbf{r}_j(t)}{|\mathbf{r}_i(t) - \mathbf{r}_j(t)|^3}, \qquad (16.2.1)$$

in which $C_0 = e^2/4\pi\varepsilon_0$. Magnetic interactions and relativistic effects are ignored.

In order to compute the trajectories of the individual particles, one has to solve Newton's equations of motion

$$\frac{d^2 \mathbf{r}_i(t)}{dt^2} = \frac{\mathbf{F}_i(t)}{m} \qquad (i = 1, 2, \ldots, N), \qquad (16.2.2)$$

in which the interaction force $\mathbf{F}_i(t)$ is given by Eq. (16.2.1). Notice that the evaluation of $\mathbf{F}_i(t)$ requires the knowledge of the trajectories of all particles up to time t. The coupling of Eqs. (16.2.1) and (16.2.2) constitutes a many particle problem that seems impossible to solve by analytical means for $N > 2$. Even for $N = 2$, one cannot obtain the particle trajectories $\mathbf{r}_1(t)$ and $\mathbf{r}_2(t)$ in an explicit form, as will be discussed later on.

A detailed calculation of all particle trajectories is not required for all manifestations of particle interactions. Consider a set of test particles running successively along a specified trajectory in the beam, called the reference trajectory. Each of the test particles experiences the Coulomb repulsion of the particles by which it is surrounded, referred to as field particles. Due to the statistical nature of the beam, the configuration of field particles will vary randomly for the different test particles. As a result, the displacements of the test particles arriving at the exit plane of the segment will be statistically distributed. It can be demonstrated that the average displacement of the test particles is related to the average electrostatic force experienced along the reference trajectory. This force is strictly deterministic. Using a continuum-model, it can be represented as the force generated by the smoothed-out distribution of charge, referred to as the space charge force. On the other hand, the widths of the distributions of displacements in position and velocity experienced by the test particles are directly related to the discrete nature of the beam. They are determined by the stochastic fluctuations in the charge density rather than the average (smoothed-out) charge density. The distribution of displacements of the test particles relative to their average shift represent the so-called statistical Coulomb interactions. For the calculation of statistical interactions, one has to consider the individual particle trajectories, whereas for the calculation of the space charge effect it is sufficient to consider the smoothed-out distribution of charge.

The lateral component of the space charge force, acting on a test particle in a rotational symmetric beam with a uniform current density distribution is proportional to its radial distance to the beam axis. This implies that the impact of the space charge effect on the lateral properties of the beam corresponds to the action of a negative lens, causing a defocussing of the final probe as well as some (de)magnification. These effects can be compensated by proper lens adjustment and do not affect the resolution of the system. However, for a non-uniform current density distribution, the space charge lens will also introduce aberrations, leading to a nonrefocusable blurring of the probe. In general, the impact of the space charge effect can be described by specifying the first and third order optical properties of the corresponding space charge lens. These properties have been calculated in first order perturbation approximation, assuming that particle density is low enough to assure that the perturbations are small.

It should be emphasized that the model for the space charge effect (as well as the model for statistical interactions) applies to beams of low or moderate density. It is not suited to describe the beam spreading observed in high intensity beams, for instance, those used in oscilloscope tubes. It does not cover the space charge effects that may occur in the vicinity of the cathode of a thermionic gun either.

Our model for statistical interactions, called the extended two-particle model, is based on a decomposition of the full N-particle problem into a sum of two-particle effects. For convenience of notation, we will denote the total displacement of particle i in position and velocity, Δr_i and Δv_i, in general as $\Delta \eta_i$. The Boersch effect corresponds to a broadening of the axial velocities. Thus, for this effect, one should consider $\Delta \eta = \Delta v_z$. The trajectory displacement effect corresponds to random (virtual) shifts observed in the image plane, thus $\Delta \eta = \Delta r_\perp$. For some conditions, the shift $\Delta r_\perp$ is entirely determined by the change in transverse velocity of the test particle. In that case, it is sufficient to consider $\Delta \eta = \Delta v_\perp$. The reduction of the N-particle dynamical problem used within the model can be expressed as

$$\Delta \eta_i = \sum_{j \neq i}^{N} \Delta \eta_{ij}, \quad (16.2.3)$$

in which $\Delta \eta_{ij}$ is the displacement of particle i due to the two-particle interaction with particle j. The validity of Eq. (16.2.3) will be discussed later on in this section.

The calculation of $\Delta \eta_{i\,j}$ constitutes a two-particle problem, which is defined by Eqs. (16.2.1) and (16.2.2), taking $N = 2$. One would like to have an explicit analytical expression for the displacement $\Delta\boldsymbol{\eta}_{ij}$ terms of the initial coordinates $\mathbf{r}_i$, $\mathbf{r}_j$, $\mathbf{v}_i$, $\mathbf{v}_j$ and the time of flight T. However, in general, such a solution cannot be obtained. The classic solution of the Kepler problem specifies the relative trajectory of the particles in the orbital plane, which is, in polar coordinates, represented by a function $r(\theta)$. Our problem requires a more detailed analysis, which takes the time of flight T through the beam segment into account. For that, one needs the coordinates of the particles explicitly as function of time t. In terms of the polar coordinates in the orbital plane, one requires the functions $r(t)$, $\theta(t)$ and their derivatives with respect to time, $\dot{r}(t)$ and $\dot{\theta}(t)$. As it turns out one can obtain the time t explicitly as function of the relative distance r, thus $t = t(r)$, but the function involved cannot be inverted. This prohibits an exact explicit analytical calculation of the particle trajectories as function of time t.

A good approximation of the exact final coordinates of the particles involved in a two-particle collision can be obtained for two types of collisions. One is a *complete collision* (or *nearly complete collision*), in which the particles effectively come from infinity and recede to infinity. For this type of collision, it is sufficient to consider the asymptotic values of the polar coordinates. The other type is a *weak collision*, in which the deviations from the unperturbed trajectories are small. First order perturbation dynamics can be used to describe such a collision. In this approach, one approximates the actual interaction force by the force that would occur when the particles would follow their unperturbed trajectories. The remaining collisions, those that are both strong and incomplete, cannot be handled analytically. For these collisions, one has to perform the inversion of the function $t(r)$ by numerical means. This procedure is referred to as *full collision dynamics*.

We note that the calculation of the spatial displacements of the particles $\Delta\mathbf{r}_1$ and $\Delta\mathbf{r}_2$ caused by a nearly complete collision is more involved than calculation of the corresponding velocity displacements $\Delta\mathbf{v}_1$ and $\Delta\mathbf{v}_2$. The displacements $\Delta\mathbf{v}_1$ and $\Delta\mathbf{v}_2$ caused by a complete collision are finite, despite the fact that the corresponding time of flight T is infinitely large. However, the results for $\Delta\mathbf{r}_1$ and $\Delta\mathbf{r}_2$ diverge when T is taken infinitely large. One has to account for the fact that the collision is nearly, but not entirely, complete. In other words, the initial and final state of the collision may not be identified with the asymptotic conditions. The type of dynamics that takes this into account is referred to as *nearly complete collision* dynamics.

Let us return to the general N-particle problem. The validity of the decomposition of Eq. (16.2.3) depends on the strength of the two-particle interactions. The decomposition is entirely justified when all interactions are weak. First order perturbation dynamics is then appropriate to calculate all displacements $\Delta\boldsymbol{\eta}_{ij}$, using the interaction force that would occur when the particles would follow their unperturbed trajectories. Since forces are additive, the decomposition of Eq. (16.2.3) is then justified. Eq. (16.2.3) also provides accurate results when the total displacement of a particle is dominated by a single strong interaction. The particle will experience a significant deviation from its unperturbed trajectory, which should be calculated using nearly complete collision dynamics or full collision dynamics. As the deviation from the unperturbed trajectory is large, the calculation of the weak interactions with the other particles will be inaccurate. However, as the total displacement of the particle is dominated by that single strong interaction, the contributions of the weak interactions can be ignored, and Eq. (16.2.3) is still valid. Clearly, the decomposition of Eq. (16.2.3) breaks down when the particles are involved in two or more strong collisions simultaneously or successively. However, such events are rare for the particle densities encountered in systems used for lithography and microscopy applications.

Eq. (16.2.3) reduces the dynamical part of the many-particle problem to a proper analysis of the two-particle problem. The statistical part of the problem is to determine the probability that a test particle is surrounded by a certain configuration of field particles and, given this probability distribution, to compute the corresponding distribution of displacements of the test particle $\rho(\Delta\eta)$. Let us denote the set of unperturbed coordinates of field particle i relative to the test particle as ξ_i. A certain configuration of field particles relative to the test particle is then specified by the set $\xi_1, \xi_2, ..., \xi_{N-1}$. In order to determine the probability $P_N(\xi_1, \xi_2, ..., \xi_{N-1})$ of the configuration of field particles $\xi_1, \xi_2, ..., \xi_{N-1}$, we assume that the field particles are statistically independent,

$$P_N(\xi_1, \xi_2, ..., \xi_{N-1}) = \sum_{i=1}^{N-1} P_2(\xi_i), \quad (16.2.4)$$

in which $P_2(\xi)$ is the probability that a field particle has coordinates ξ relative to the test particle. Eq. (16.2.4) ignores any correlation in the coordinates of the field particles. Such correlations might arise from the emission process at the cathode or the Coulomb interaction in the beam segments preceding the one considered. Ignoring the last "source" of correlation is equivalent to the assumption that Debeye screening is absent. This is indeed the case for practical operating conditions, as can be verified by comparing the relaxation time involved to the time of flight through the column.

The model concentrates on the displacements experienced along the central (on-axis) reference trajectory in a rotational symmetric beam. The space charge force is zero on this trajectory and the resulting displacements are purely statistical which facilitates the analysis. In addition, it is assumed that the beam is initially monochromatic with respect to the normal energy, which means that all particles initially have the same axial velocity. This assumption is reasonable, since the velocity spread of particles at emission is strongly reduced during the succeeding acceleration. As a result of these two simplifications, the number of independent components of the vector ξ is only four, whereas the general case would require six components to describe the position and velocity of a field particle relative to the test particle. The validity of these assumptions has been verified within the framework of the model by performing the full calculation for some specific beam geometries. It was found that, average broadening for all possible reference trajectories is, in general, somewhat smaller than the broadening obtained for the central reference trajectory. In general, the differences are within 20% and for many cases even within 10%. Furthermore, it was found that the impact of the initial energy distribution on the magnitude of the statistical effects can be ignored for practical operating conditions.

Eq. (16.2.4) enables us to express the distribution of field particles relative to the test particle in terms of the experimental parameters, which were specified previously. The problem is now to compute the distribution of displacements of the test particle $\rho(\Delta\eta)$ from the distribution of field particles, employing the solution of the dynamical problem expressed by Eq. (16.2.3). This problem can be solved using a procedure that is well known from statistical mechanics (sometimes referred to as Markoff's method—for instance, see Chandrasekhar, 1943). This method permits a calculation of the detailed shape of the full displacement distribution. It was found that the type of distribution depends on the beam geometry as well as the particle density. For large particle densities, the total deviation of the test particle is built up by a large number of independent collisions. The corresponding distribution is Gaussian, as can be understood from the central limit theorem. For lower particle densities, the distribution becomes non-Gaussian, showing more pronounced tails. It was also found that different types of displacement distributions are generated in extended beams and pencil beams. A beam is called extended (or three-dimensional) if its transverse dimensions are large compared to the average axial separation of the particles. A pencil beam represents the opposite case. In such a beam all particles are nearly on a row.

Without going into a detailed discussion, we wish to clarify the occurrence of the different types of non-Gaussian displacement distributions and indicate their relation with the beam geometry and the particle density. As an example, let us consider an extended monochromatic cylindrical beam. Assume that the beam is homocentric, which means that the particles run along parallel trajectories. In the frame of reference moving with the beam, the particles are initially at rest. Assume that the particle density is low enough to assure that the deviations from the unperturbed trajectories are small. Thus, first order perturbation dynamics can be used to compute the displacements. The displacements in transverse and axial velocity now follow as

$$\Delta \mathbf{v}_\perp = \frac{\mathbf{F}_\perp T}{m}, \qquad \Delta v_z = \frac{F_z T}{m}, \qquad (16.2.5)$$

where T is the time of flight. Clearly, if first order perturbation dynamics holds true, the distribution of displacements is directly related to the distribution of the fluctuating component of the interaction force that would occur in the unperturbed beam. For the specific case of particles that are at rest relative to each other, as in our example, this distribution was first calculated by Holtsmark (1919). He considered the problem of determining the probability $\rho(\mathbf{F})$ that a given electric field strength $\mathbf{F}$ acts at a point in a gas of density n_d, composed of randomly distributed ions. His result can be expressed as

$$\rho(\mathbf{F}) = \frac{1}{2\pi^2 F} \int_0^\infty k\, dk\, \sin(Fk) e^{-A_{3/2} k^{3/2}}, \qquad (16.2.6)$$

in which $A_{3/2}$ depends on the particle mass m, particle charge e and the density n_d

$$A_{3/2} = \frac{4}{15} (2\pi C_0)^{3/2} n_d, \qquad (16.2.7)$$

where $C_0 = e^2/4\pi\varepsilon_0$. It can be demonstrated that the distribution of the longitudinal component of the interaction force F_z is given by the one-dimensional variant of the Holtsmark distribution

$$\rho(F_z) = \frac{1}{\pi} \int_0^\infty dk\, \cos(kF_z) e^{-A_{3/2} k^{3/2}}, \qquad (16.2.8)$$

while the distribution of the lateral component $\mathbf{F}_\perp$ is given by

$$\rho(\mathbf{F}_\perp) = \frac{1}{2\pi} \int_0^\infty k\, dk\, J_0(kF_\perp) e^{-A_{3/2} k^{3/2}}, \qquad (16.2.9)$$

in which J_0 is the Bessel function of the first kind and zero order. Notice that the argument of the exponential function is the same in Eqs. (16.2.6), (16.2.8) and (16.2.9). The 3/2 power dependency on the integration variable k is characteristic for the Holtsmark distribution. For the beam of our example, it follows from Eq. (16.2.5) that the generated distributions of transverse and axial velocities, $\rho(\Delta \mathbf{v}_\perp)$ and $\rho(\Delta v_z)$, are of the same type as the distributions $\rho(\mathbf{F}_\perp)$ and $\rho(F_z)$ respectively.

A more general analysis shows that the distributions $\rho(\Delta \mathbf{v}_\perp)$ and $\rho(\Delta v_z)$ are Holtsmarkian for any beam geometry, provided that the beam is extended and provided that the perturbations of the particles are small. Other types of distributions will be generated when these conditions are not fulfilled. However, the form of Eqs. (16.2.8) and (16.2.9) can be generalized to include those distributions as well. Denoting the displacements in position or velocity again as $\Delta\eta$, one can express the one-dimensional distribution of axial displacements, in general, as

$$\rho(\Delta\eta) = \frac{1}{\pi} \int_0^\infty dk \cos(k\Delta\eta) e^{-A_\gamma k^\gamma} \qquad (16.2.10)$$

and the two-dimensional distribution of transverse displacements as

$$\rho(\Delta\eta) = \frac{1}{2\pi} \int_0^\infty k\,dk\, J_0(k\Delta\eta) e^{-A_\gamma k^\gamma}. \qquad (16.2.11)$$

The numerical constant γ is characteristic for the type of distribution involved, while the parameter A_γ specifies the dependency on the experimental parameters. The different types of distributions, encountered in the calculation of statistical effects, can be summarized as

$\gamma = 2$ — Gaussian distribution
$\gamma = 3/2$ — Holtsmark distribution
$\gamma = 1$ — Lorentzian distribution $\qquad (16.2.12)$
$\gamma = 1/2$ — distribution of F_z in a pencil beam
$\gamma = 1/3$ — distribution of $F_\perp$ in a pencil beam.

The shape of the corresponding one- and two-dimensional distributions are depicted in Figs 5.3 and 5.4 respectively. These figures clearly show that a small γ-value leads to a distribution with a narrow core and long tails.

The representation of Eqs. (16.2.10) and (16.2.11) still contains some degree of approximation. The extended two-particle approach, which is based on a strict derivation of the distribution $\rho(\Delta\eta)$ starting from the assumptions expressed by Eqs. (16.2.3) and (16.2.4), shows that the argument of the exponential function appearing in Eqs. (16.2.10) and (16.2.11) should, in general, be expressed as a function $\lambda p(k)$, where λ is the linear particle density. Thus, Eqs. (16.2.10) and (16.2.11) rely on the approximation

$$\lambda p(k) \approx A_\gamma k^\gamma. \qquad (16.2.13)$$

The function $p(k)$ is quadratic in k for small k-values, but with increasing k, its power dependency becomes less strong. For extremely large k-values, it will ultimately increase proportional to $k^{1/2}$ or $k^{1/3}$. The value of the linear particle density λ determines what part of the $p(k)$ function dominates the displacement distribution $\rho(\Delta\eta)$. Eq. (16.2.13) simplifies the evaluation of the distribution $\rho(\Delta\eta)$ by assuming that a single k-regime dominates the entire distribution. However, this approximation is only justified for a certain range of $\Delta\eta$-values. The large $\Delta\eta$-behavior of $\rho(\Delta\eta)$ is always dominated by the small k-behavior of $p(k)$, which is quadratic in k. This implies that the tails of any type of distribution $p(\Delta\eta)$ ultimately fall off exponentially, as for a Gaussian distribution.

Most of the regimes manifest in the $p(k)$-transform can directly be related to different types of collisions involved. For instance, the 3/2 power dependency on k corresponding to the Holtsmark regime stems from weak incomplete collisions in an extended beam, as can be understood from the previous discussion. The same type of collisions occurring in a pencil beam lead to a 1/2 power dependency on k for the distribution of axial displacements and a 1/3 power dependency for the distribution of transverse displacements. Weak nearly complete collisions, which do only take place in an extended beam, lead to a linear dependency on k for the distribution of axial displacements and a 1/2 power dependency for the distribution of transverse displacements. The quadratic k-behavior for small k is *not* connected to a single type of collision. All types of collisions contribute to this part of the $p(k)$ transform, but strong collisions are usually dominant. We emphasize that the function $p(k)$ is entirely determined by the geometry of the beam segment. The linear particle density λ enters the model in the evaluation of the distribution $p(\Delta\eta)$ from the quantity $\lambda p(k)$. The value of λ determines what part of the $p(k)$-transform dominates $p(\Delta\eta)$, or in other words, which type of collision contributes the most.

The distinction of the various regimes covered by different types of collisions is connected to the distinction of different interaction mechanisms. The axial velocity spread (Boersch effect) generated in a narrow crossover is sometimes attributed to the conversion of internal kinetic energy (kinetic energy in the frame of reference moving with the beam) from the lateral degrees of freedom to the longitudinal degree of freedom. The acceleration of the beam causes a reduction of the axial internal energy spread but leaves the lateral spread unaffected. Acceleration, therefore, generates a non-equilibrium condition. Relaxation will occur towards an isotropic distribution of internal kinetic energy. This mechanism is referred to as *relaxation of kinetic energy*. It leads to an increase of the internal energy spread in the axial direction. The maximum energy spread is generated when the relaxation is complete. This condition defines the upper (thermodynamic) limit of kinetic energy relaxation.

It should be noted that the concept of relaxation of kinetic energy is unable to explain the Boersch effect generated in a homocentric cylindrical beam segment, since the transverse velocity spread is zero in such a beam. The mechanism that occurs here is *relaxation of potential energy*. At the entrance of the beam segment, the particles are more or less randomly distributed over the beam volume. Each pair of particles has a certain mutual potential energy, depending on their initial separation. During the flight, part of this potential energy will be converted into kinetic energy, leading to the generation of microscopic random velocity components. This process should be separated from the expansion of the beam volume under influence of the space charge force, which corresponds to the conversion of potential energy into a macroscopic systematic motion. The upper (thermodynamic) limit of potential energy relaxation is reached when the microscopic density distribution obeys Maxwell-Boltzmann statistics. In this description the kinetic energy of the particles is represented by a beam temperature.

Our model implicitly accounts for the different mechanisms through the detailed treatment of the two-particle problem. The potential energy of the particles involved in a nearly complete collision is effectively zero in both the initial and the final state. The main effect of such a collision is a directional change of the particles and a transition of kinetic energy from one particle to the other. Thus, when nearly complete collisions are dominant, the mechanism involved is relaxation of kinetic energy. Conversion of potential energy into kinetic energy (and vice versa) occurs in incomplete collisions. When weak incomplete collisions are dominant, the mechanism involved is relaxation of potential energy.

The width of the distributions specified by Eqs. (16.2.10) and (16.2.11) depends on the parameter A_γ. The Full Width at Half Maximum (*FWHM*) and the Full Width median (*FW*$_{50}$) can be expressed as

$$FWHM = A_\gamma^{1/\gamma} FWHM_\gamma, \quad FW_{50} = A_\gamma^{1/\gamma} FW_{50\gamma}, \qquad (16.2.14)$$

in which *FWHM*$_\gamma$ and *FW*$_{50\gamma}$ are numerical constants, which depend on the dimension n of the distribution as well as the value of γ. Table 5.1 gives the values of *FWHM*$_\gamma$ and *FW*$_{50\gamma}$ for $n = 1, 2$ and 3 and the γ-values listed in Eq. (16.2.12). We note that the distributions of Eqs. (16.2.10) and (16.2.11) have no finite second or higher moment for $\gamma < 2$, due to their long tails. However, if one calculates the distribution $\rho(\Delta\eta)$ from the full $p(k)$ transform, the tails will ultimately fall of exponentially, as was mentioned previously. This leads to a finite value of the second and higher order moments. In this context it should be recognized that the root mean square (rms) is not a practical width measure for the non-Gaussian distributions generated by statistical interactions. It is entirely determined by the tails, which show a different dependency on the experimental parameters as the bulk part of the distribution.

The parameter γ is a key parameter in the theory. It specifies the type of distribution that is generated. Eq. (16.2.14) shows that it also determines the dependency of the *FWHM* and *FW*$_{50}$ width measures on the experimental parameters, which are represented by the quantity A_γ. For instance A_γ is directly proportional to the beam current I. Thus, if one knows γ, one can directly say that the *FWHM* and *FW*$_{50}$ value of the displacement distribution $\rho(\Delta\eta)$ will be proportional to $I^{1/\gamma}$. The other way around, if one knows that the *FWHM* or *FW*$_{50}$ increases as I^p, one may conclude that $\gamma = 1/p$. We emphasize that the rms width is dominated by the tails of

$\rho(\Delta\eta)$, which ultimately follow a Gaussian type of behavior ($\gamma = 2$). Accordingly, the rms width is for all conditions proportional to $I^{1/2}$.

The parameter γ also determines how the effects generated in the individual segments constituting the beam should be added. Assuming that the displacements experienced by the test particle in the successive beam segments are not correlated and that all distributions are of the same type (same γ), one finds for the FWHM of total displacement distribution $\rho(\Delta\eta)$ generated in the entire beam

$$FWHM_T = \left(\sum_{j=1}^{N_b} FWHM_j^\gamma \right)^{1/\gamma}, \qquad (16.2.15)$$

where $FWHM_j$ is the FWHM of $\rho(\Delta\eta)$ generated in segment number j and N_b is the total number of beam segments. The same equation holds true for the median width FW_{50}, while rms values should always be added quadratically ($\gamma = 2$). When the distributions are not all of the same type (same γ), one should, in principle, add the corresponding $\lambda p(k)$ functions and determine the FWHM from the resulting total $\lambda p(k)$ function. However, in practice, it is more convenient to define some kind of effective γ-value for which Eq. (16.2.15) holds approximately. This approach will be discussed in some detail in Section 16.7.

Eq. (16.2.15) presupposes that the displacements generated in the individual beam segments are uncorrelated. In the opposite case that the contributions are entirely correlated, their widths should be added linearly. This situation is relevant for the so-called *slice method*. In this approach, one subdivides a beam segment into thin cylindrical slices and calculates the total FWHM as a linear sum of the contributions of the individual slices,

$$FWHM = \sum_{s=1}^{N_s} FWHM_P(I, V, r_0, \Delta z_s) \approx \int_{z_0}^{z_1} dz \left(\frac{FWHM_P(I, V, r_0, \Delta z)}{\Delta z} \right), \qquad (16.2.16)$$

where $FWHM_P(I, V, r_0, \Delta z)$ is the FWHM of the distribution generated in a cylindrical slice (with Parallel rays) of length Δz, radius r_0, carrying a current I at a potential V. This expression is expected to be linear in Δz. Accordingly, the integrand $FWHM_P(I, V, r_0, \Delta z)/\Delta z$ will only depend on z through the beam radius r_0. The quantities z_0 and z_1 are the axial coordinates of the entrance and exit plane of the beam segment respectively; thus $L = z_1 - z_0$. The validity of Eq. (16.2.16) requires the particle density in the beam be low enough to guarantee that the displacements from the unperturbed trajectories are small. In that case, the displacements of the test particle experienced in the individual slices will indeed be entirely correlated, since they are generated by the same complexion of neighbor field particles. The slice method will be employed in the next section to derive the parameter dependencies in a beam segment with a crossover of low particle density, from the results obtained for a cylindrical beam segment.

16.3. Parameter dependencies

The extended two-particle model has the capability to predict the detailed shape of the full displacement distribution $\rho(\Delta\eta)$ generated by the statistical interactions and includes a calculation of different width measures of $\rho(\Delta\eta)$. In the previous section, it was pointed out that different types of distributions may occur, depending on the beam geometry and the particle density in the beam. The dependency on the experimental parameters of the characteristic width measures, such as the $FWHM$ and the FW_{50} value, will be different in the various regimes. Prior to the presentation of the detailed equations we like to show that the dependency on the experimental parameters can be predicted on the basis of some elementary physical arguments. We will first consider a beam of low particle density in which the deviations from the unperturbed trajectories are small. In terms of "collisions," this implies that weak incomplete collisions are dominant. Beams of moderate particle density in which the contribution of weak complete collisions as well as strong collisions may become significant will be considered next.

In order to specify the qualifications "low," "moderate" and "high" particle density, let us first introduce the dimensionless quantities $\bar{\lambda}$ and λ^*. For a beam segment with a crossover, the particle density can best be expressed in terms of the scaled linear particle density $\bar{\lambda}$, defined as

$$\bar{\lambda} = \frac{m^{1/2}}{2^{7/2}\pi\varepsilon_0 e^{1/2}} \frac{I}{\alpha_0^2 V^{3/2}}. \tag{16.3.1}$$

For a cylindrical beam segment ($\alpha_0 = 0$), one cannot use $\bar{\lambda}$. Instead, we use the scaled linear particle density λ^*, defined as

$$\lambda^* = \frac{m^{1/2}}{2^{7/2}\pi\varepsilon_0 e^{1/2}} \frac{IL^2}{r_0^2 V^{3/2}}. \tag{16.3.2}$$

As a rough indication, low particle densities correspond to values for $\bar{\lambda}$ and λ^* below 10^{-4}, moderate particle densities correspond to values between 10^{-4} and 10^{-2}, and high particle densities to values above 10^{-2}.

The Boersch effect corresponds to a broadening of the axial velocity distribution. It is usually expressed in terms of the generated energy spread. An energy displacement ΔE is related to an axial velocity displacement Δv_z as

$$\frac{\Delta E}{E} \approx \frac{m v_z \Delta v_z}{\frac{1}{2} m v_z^2} = 2\frac{\Delta v_z}{v_z}, \qquad (16.3.3)$$

provided that $\Delta v_z \ll v_z$. The trajectory displacement effect corresponds to the generation of random displacements in position and velocity in the lateral direction of the beam. An angular displacement $\Delta \alpha$ is related to a transverse velocity displacement $\Delta v_\perp$ as

$$\Delta \alpha \approx \frac{\Delta v_\perp}{v_z}, \qquad (16.3.4)$$

provided that $\Delta v_\perp \ll v_z$. The corresponding virtual spatial displacement Δr in the image plane (which is optically conjugated to the target plane) is given by

$$\Delta r \approx \Delta \alpha |z - z_i|, \qquad (16.3.5)$$

in which $|z - z_i|$ is the distance between the location where the deflection $\Delta \alpha$ occurs and the image plane.

Consider a *homocentric cylindrical beam segment* with radius r_0. Given a force $\mathbf{F} = (F_\perp, F_z)$ acting on the test particle, the corresponding velocity displacements $\Delta v_\perp$ and Δv_z follow, in first order perturbation approximation, from Eq. (16.2.5). The time of flight $T = L/v_z$. We emphasize that Eq. (16.2.5) presupposes that particles are initially at rest relative to each other, while the deviations from the unperturbed trajectories are small. The force component F_z will scale with $C_0 d_z/d^3$, where d and d_z denote the average distance and the average axial distance between neighbor particles respectively, and $C_0 = e^2/4\pi\varepsilon_0$. Employing Eq. (16.3.3), we therefore expect that

$$\frac{\Delta E}{E} \sim \frac{C_0}{m} \frac{L d_z}{v_z^2 d^3}. \qquad (16.3.6)$$

For an *extended beam*, both d and d_z will scale with $n_d^{-1/3}$, where n_d denotes the particle density, which is given by

$$n_d = \frac{I}{\pi r_0^2} \sqrt{\frac{m}{2e^3 V}}. \qquad (16.3.7)$$

Taking $d \sim n_d^{-1/3}$ and $d_z \sim n_d^{-1/3}$, Eq. (16.3.6) yields

$$\frac{\Delta E}{E} \sim \frac{m^{1/3}}{\varepsilon_0} \frac{I^{2/3} L}{V^{4/3} r_0^{4/3}}, \qquad (16.3.8)$$

using that $v_z = (2eV/m)^{1/2}$. Eq. (16.3.8) specifies the dependency on the experimental parameters of the energy spread generated in an extended homocentric cylindrical beam segment in which the deviations from the unperturbed trajectories are small. From the dependency on the beam current I, one may conclude that $\gamma = 3/2$. Thus, the distribution $\rho(\Delta E)$ is of the Holtsmark type, as specified by Eq. (16.2.12). The reader may verify that

Eq. (16.3.8) can also be obtained from the Holtsmark distribution $\rho(F_z)$ given by Eq. (16.2.8), employing Eqs. (16.2.14), (16.2.7), (16.2.5) and (16.3.3).

In a *pencil beam*, both d and d_z will scale with $1/\lambda$, where the linear particle density λ is given by

$$\lambda = I\sqrt{\frac{m}{2e^3 V}}. \tag{16.3.9}$$

Taking $d \sim 1/\lambda$ and $d_z \sim 1/\lambda$, Eq. (16.3.6) yields

$$\frac{\Delta E}{E} \sim \frac{m}{\varepsilon_0 e^2} \frac{I^2 L}{V^2}, \tag{16.3.10}$$

using that $v_z = (2\,eV/m)^{1/2}$. Eq. (16.3.10) specifies the dependency on the experimental parameters of the energy spread generated in a pencil beam in which the deviations from the unperturbed trajectories are small. Notice that Eq. (16.3.10) does not depend on the radial dimensions of the beam, which are assumed to be small compared to the average axial separation of the particles.

For the angular displacement $\Delta \alpha$, one finds from Eqs. (16.3.4) and (16.2.5)

$$\Delta \alpha \sim \frac{C_0}{m} \frac{L d_r}{v_z^2 d^3}, \tag{16.3.11}$$

similar to Eq. (16.3.6). The quantity d_r is the average radial distance between neighbor particles. For an *extended beam*, both d and d_r will scale with $n_d^{-1/3}$, where n_d is the particle density given by Eq. (16.3.7). Accordingly, one finds

$$\Delta \alpha \sim \frac{m^{1/3}}{\varepsilon_0} \frac{I^{2/3} L}{V^{4/3} r_0^{4/3}}, \tag{16.3.12}$$

using that $v_z = (2\,eV/m)^{1/2}$. Eq. (16.3.12) is similar to Eq. (16.3.8) and applies to the same conditions. The similarities between the results obtained for $\Delta E/E$ and $\Delta \alpha$ can be understood from the fact that the particles are initially at rest in the frame of reference moving with the beam. Random velocity components are generated during the flight, due to the relaxation of potential energy. The generated velocity distribution will be rotational symmetric (in the frame of reference moving with the beam) provided that the beam is extended.

Differences between the results for $\Delta E/E$ and $\Delta \alpha$ are expected to occur in pencil beams, in which the distribution of field particles around a test particle is nonrotational symmetric. In a *pencil beam*, the distance d scales with $1/\lambda$, where λ is the linear particle density, given by Eq. (16.3.9). However, the radial distance d_r is expected to scale with the beam radius $r_0 (r_0 \ll 1/\lambda)$. Substitution into Eq. (16.3.11) yields

$$\Delta \alpha \sim \frac{m^{3/2}}{\varepsilon_0 e^{7/2}} \frac{I^3 L r_0}{V^{5/2}}, \tag{16.3.13}$$

which specifies the dependency on the experimental parameters of the angular deflections generated in a cylindrical pencil beam in which the deviations from the unperturbed trajectories are small. In the extreme case that all particles are on a line ($r_0 = 0$), there is no transverse component of the interaction force and $\Delta \alpha$ becomes zero. From the dependency on the beam current I, one may conclude that $\gamma = 1/3$, which defines the type of two-dimensional distribution involved.

In order to express the angular displacement $\Delta \alpha$ into a spatial displacement Δr, one should realize that a cylindrical beam segment is usually succeeded by a lens, which focuses the beam in its back focal plane. Let f be the focal distance of the lens. The spatial displacement in the back focal plane of the lens Δr caused by an angular displacement $\Delta \alpha$ is equal to

$$\Delta r \approx f \Delta \alpha. \tag{16.3.14}$$

The parameter dependency of the trajectory displacement effect generated in an extended cylindrical beam follows by substitution of Eq. (16.3.12), while the result for a pencil beam follows with Eq. (16.3.13).

Now, consider a *beam segment with a narrow crossover* of radius r_c, with a beam semi-angle α_0. We will employ the slice method of Eq. (16.2.16) to determine the parameter dependencies for $\Delta E/E$ and $\Delta \alpha$ from the results obtained for a cylindrical beam segment. The characteristic beam geometry quantities K, K_1 and K_2 are defined as

$$K = \frac{\alpha_0 L}{2r_c}, \quad K_1 = 2KS_c, \quad K_2 = 2K(1 - S_c), \tag{16.3.15}$$

where $S_c = L_1/L$ is the crossover position parameter. The restriction that the crossover is narrow implies that $K \gg 1$. We will assume that both $K_1 \gg 1$ and $K_2 \gg 1$, which is justified for $K \gg 1$ when the crossover is located somewhere near the middle of the beam segment ($S_c \approx 1/2$). Let us define the axial coordinate of the crossover as $z = 0$. The beam radius at position z is then approximately given by

$$r_0(z) \approx r_c + \alpha_0 |z|. \tag{16.3.16}$$

For the energy spread generated in an *extended beam*, one should start from Eq. (16.3.8), where r_0 is now given by Eq. (16.3.16). Substitution into Eq. (16.2.16) yields

$$\frac{\Delta E}{E} \sim \frac{m^{1/3}}{\varepsilon_0} \frac{I^{2/3}}{V^{4/3}} \int_{z_0}^{z_1} \frac{dz}{[r_c + \alpha_0 |z|]^{4/3}} \sim \frac{m^{1/3}}{\varepsilon_0} \frac{I^{2/3}}{V^{4/3} r_c^{1/3} \alpha_0}, \tag{16.3.17}$$

using that $K_1 \gg 1$. This equation specifies the dependency on the experimental parameters of the energy spread generated by *weak incomplete collisions* in an extended beam segment with a narrow crossover. The energy spread generated in a *pencil beam* does not depend on r_0, as is expressed by Eq. (16.3.10). Accordingly, this equation applies also to a pencil beam with a narrow crossover.

The virtual spatial displacement Δr in the crossover ($S_i = S_c$) of an *extended* beam follows from Eq. (16.2.16), substituting Eqs. (16.3.5), (16.3.12) and (16.3.16)

$$\Delta r \sim \frac{m^{1/3}}{\varepsilon_0} \frac{I^{2/3}}{V^{4/3}} \int_{z_0}^{z_1} \frac{|z|\, dz}{[r_c + \alpha_0 |z|]^{4/3}}$$

$$\sim \frac{m^{1/3}}{\varepsilon_0} \frac{I^{2/3} L^{2/3}}{V^{4/3} \alpha_0^{4/3}} |S_c^{2/3} + (1 - S_c)^{2/3}| \quad (S_c = S_i),$$

(16.3.18)

using that $K_1 \gg 1$ and $K_2 \gg 1$. This equation applies to the same conditions as Eq. (16.3.17). Similarly, one finds with Eq. (16.3.13) for a *pencil beam*

$$\Delta r \sim \frac{m^{3/2}}{\varepsilon_0 e^{7/2}} \frac{I^3 L^3 \alpha_0}{V^{5/2}} |3S_c(1 - S_c) - 1| \quad (S_c = S_i) \tag{16.3.19}$$

using that $K_1 \gg 1$ and $K_2 \gg 1$. This equation specifies the dependency of the trajectory displacement effect generated by weak incomplete collisions in a pencil beam with a narrow crossover.

So far, we restricted the analysis to beams of low particle density, in which the interaction effects are dominated by weak incomplete collisions. With increasing particle density, other types of collisions will become significant. *Weak complete collisions* can occur in beams with a narrow crossover in which the lateral dimensions of the beam in the entrance plane ($\sim \alpha_0 L S_c$) and exit plane ($\sim \alpha_0 L [1 - S_c]$) are large compared to the average axial separation of the particles $1/\lambda$. These conditions are typical for moderate particle densities. With further increasing particle density the contribution of *strong complete* and *strong incomplete collisions* becomes significant.

The velocity shifts of an on-axis test particle involved in a *complete collision* are given by

$$\Delta v_z \cong \frac{m\alpha^3 v_z^3 b_z}{2C_0 \left[1 + \left(b_z^2 + r_\perp^2 \sin^2\Phi\right)(m\alpha^2 v_z^2/2C_0)^2\right]} \tag{16.3.20}$$

$$\Delta v_\perp \cong \frac{\alpha v_z \left[1 + \left(m\alpha^3 v_z^3 r_\perp \sin(\Phi)/2C_0\right)^2\right]^{1/2}}{\left[1 + \left(b_z^2 + r_\perp^2 \sin^2\Phi\right)(m\alpha^2 v_z^2/2C_0)^2\right]}, \tag{16.3.21}$$

in which the coordinates $\xi = (\alpha, b_z, r_\perp, \Phi)$ specify unperturbed coordinates of the field particle relative to those of the test particle; see Fig. 5.1. Eqs. (16.3.20) and (16.3.21) follow from the angle of deflection caused by a complete collision (which is the angle between the asymptotes of the relative trajectory in the orbital plane) and the transformation to coordinates in the laboratory system.

For a *weak complete collision*, one may neglect the term 1 between the brackets in the denominator of Eqs. (16.3.20) and (16.3.21). In addition we assume that $r_\perp \ll b_z$, which is justified for a narrow crossover. This gives

$$\Delta v_z \cong \frac{2C_0}{m\alpha v_z b_z}, \quad \Delta v_\perp \cong \frac{4C_0^2}{m^2\alpha^3 v_z^3 b_z^2}. \qquad (16.3.22)$$

The distance b_z will scale with $1/\lambda$ and the angle α will scale with the beam semi-angle α_0. Accordingly, one finds, with Eq. (16.3.3), for the parameter dependency of the energy spread generated by weak complete collisions in a beam segment with a narrow crossover

$$\frac{\Delta E}{E} \sim \frac{m^{1/2}}{\varepsilon_0 e^{1/2}} \frac{I}{\alpha_0 V^{3/2}}, \qquad (16.3.23)$$

using that $v_z = (2eV/m)^{1/2}$. Similarly, it follows from Eqs. (16.3.4) and (16.3.22) that

$$\Delta\alpha \sim \frac{m}{\varepsilon_0^2 e} \frac{I^2}{\alpha_0^3 V^3}, \qquad (16.3.24)$$

which refers to the same conditions as Eq. (16.3.23). The corresponding virtual shift in the image plane follows by multiplication with the distance between crossover and image plane

$$\Delta r \sim \frac{m}{\varepsilon_0^2 e} \frac{I^2 L}{\alpha_0^3 V^3} |S_c - S_i|. \qquad (16.3.25)$$

This equation expresses that the virtual shift generated by a weak complete collision is zero when crossover and image plane coincide ($S_c = S_i$). For that condition, the spatial and angular shift cancel out.

With further increasing particle densities, the contribution of *strong collisions* will become significant, and the deviations from the unperturbed trajectories can no longer be ignored. The total displacement of each test particle will be built up by a number of uncorrelated interactions leading to a Gaussian type of displacement distribution $\rho(\Delta\eta)$. In general, this distribution is not dominated by a single type of collision. In order to determine the dependency on the experimental parameters, one has to carry out the statistical calculation of the rms value, which now provides a proper width measure. For a narrow crossover ($K \gg 1$), one may assume that the main contribution to the distribution of axial and transverse velocity displacements comes from (nearly) complete collisions and employ Eqs. (16.3.20) and (16.3.21). For an on-axis test particle and uniform spatial and angular distribution, the calculation of the mean square velocity displacement $\langle \Delta v^2 \rangle$ consists of the evaluation of the following integral,

$$\langle \Delta v^2 \rangle = \lambda \int_0^{v_0} \frac{2v\,dv}{v_0^2} \int_0^{2\pi} \frac{d\Phi}{2\pi} \int_0^{r_c} \frac{2r_\perp\,dr_\perp}{r_c^2} \int_{-S_c L}^{(1-S_c)L} db_z [\Delta v(v,\Phi,r_\perp,b_z)]^2, \qquad (16.3.26)$$

Summary for the one-minute designer

where $\Delta v(v, \Phi, r_\perp, b_z)$ denotes the expression given by either Eq. (16.3.20) or Eq. (16.3.21) for the calculation of the quantities $\langle \Delta v_z^2 \rangle$ and $\langle \Delta v_\perp^2 \rangle$ respectively. In both cases, all integrals can be solved analytically but one. Using Eqs. (16.3.3) and (16.3.4), the results can be expressed as

$$\frac{\langle \Delta E^2 \rangle^{1/2}}{E} = 2\langle \Delta \alpha^2 \rangle^{1/2} = \left(\frac{m}{8\varepsilon_0^2 e}\right)^{1/4} P_C(\bar{r}_c) \frac{I^{1/2}}{V^{3/4}}, \qquad (16.3.27)$$

where the scaled crossover radius $\bar{r}_c$ and the function P_C are given by

$$\bar{r}_c = \frac{8\pi\varepsilon_0}{e} \alpha_0^2 V r_c \qquad (16.3.28)$$

$$P_C(\bar{r}_c) = \left[\frac{8}{\pi \bar{r}_c} \int_0^1 dy \frac{\sqrt{1-y^2}}{y} \sinh^{-1}\left(\frac{1}{2}\bar{r}_c y\right)\right]^{1/2} \qquad (16.3.29)$$

A good approximation of $P_C(\bar{r}_c)$ is given by

$$P_{Ca}(\bar{r}_c) = \left(\frac{1}{1 + \pi\bar{r}_c/\{2 \ln [.8673(114.6 + \bar{r}_c)]\}^2}\right)^{1/2}. \qquad (16.3.30)$$

Eq. (16.3.27) shows a square root dependency on the beam current I. Thus $\gamma = 1/2$, confirming that the corresponding distribution is Gaussian.

The dependency on the other experimental parameters depends on the value of the scaled crossover radius $\bar{r}_c$. For $\bar{r}_c \to 0$, one finds from Eq. (16.3.29) or Eq. (16.3.30) that $P_C(\bar{r}_c) = 1$. The corresponding relative energy spread $\langle \Delta E^2 \rangle^{1/2}/E$ and angular spread $\langle \Delta \alpha^2 \rangle^{1/2}$ are proportional to the square root of the perveance $I/V^{3/2}$, as can be seen from Eq. (16.3.27). For $\bar{r}_c \to \infty$, one finds from Eqs. (16.3.27), (16.3.28) and (16.3.30)

$$\frac{\langle \Delta E^2 \rangle^{1/2}}{E} \approx .616 \frac{(me)^{1/4}}{\varepsilon_0} \left[1 + .217 \ln\left(1 + \frac{\bar{r}_c}{114.6}\right)\right] \frac{I^{1/2}}{V^{5/4} r_c^{1/2} \alpha_0} \qquad (16.3.31)$$

The same result applies to $2\langle \Delta \alpha^2 \rangle^{1/2}$, as indicated by Eq. (16.3.27). Eq. (16.3.31) provides an accurate approximation for $\bar{r}_c > 100$.

It should be emphasized that Eq. (16.3.27) relies on the assumption that complete collisions are dominant. For high particle densities, this will be true if $K_1 \gg 1$ and $K_2 \gg 1$, which requires $K \gg 1$ and $S_c \approx 1/2$. The analysis for smaller K values is more complex, since one has to account for *strong incomplete collisions*, which cannot be handled analytically, as was discussed in the previous section. Accordingly, the integral of Eq. (16.3.26) has to be performed numerically, employing the numerical solution of the two-particle problem. This was done for various K and $\bar{r}_c$ values. The result can again be expressed by Eq. (16.3.27), provided that the function $P_C(\bar{r}_c)$ is replaced by a function $P_C(\bar{r}_c, K_1, K_2)$, which depends also on K_1 and K_2. An analytical approximation for this function was obtained by fitting the numerical data. Different functions $P_C(\bar{r}_c, K_1, K_2)$ are required to describe the small K-behavior of $\langle \Delta E^2 \rangle^{1/2}$ and $\langle \Delta \alpha^2 \rangle^{1/2}$.

The trajectory displacement distribution $\rho(\Delta r)$ also becomes Gaussian for high particle densities. In this regime, the contribution of strong collisions is significant, leading to substantial deviations from the unperturbed trajectories. Contrary to the analysis of the velocity shifts $\Delta v_\perp$ and Δv_z, one cannot use complete collision dynamics to calculate the shift Δr caused by a two-particle collision, as was discussed in the previous section. One has to account for the fact that the interaction time T is finite. Unfortunately, the complexity of the resulting equation for Δr prohibits an exact analytical evaluation of the mean square displacement $\langle \Delta r^2 \rangle$. The corresponding integral, which is similar to Eq. (16.3.26), has to be performed numerically. At this point it is advantageous to scale the equations in order to reduce the number of independent parameters. In addition to the quantities S_c and S_i and the scaled linear particle density $\bar{\lambda}$, the resulting equation depends on the scaled crossover radius r_c^* and scaled transverse velocity v_0^*, which are defined as

$$r_c^* = \left(\frac{2\pi\varepsilon_0}{e} \right)^{1/3} \frac{V^{1/3} r_c}{L^{2/3}} \qquad (16.3.32)$$

$$v_0^* = \left(\frac{2\pi\varepsilon_0}{e} \right)^{1/3} V^{1/3} L^{1/3} \alpha_0. \qquad (16.3.33)$$

Notice that the scaling quantities differ from the ones used for the Boersch effect ($\bar{r}_c$ and K). However, in both cases, the quantity $\bar{\lambda}$ is used as the measure for the particle density. For the case $S_c = S_i = 1/2$, $r_c^* = 0$ and $v_0^* \to \infty$, the rms value of $\rho(\Delta r)$ is given by

$$\langle \Delta r^2 \rangle^{1/2} = .0950 \frac{e^{1/12} m^{1/4}}{\varepsilon_0^{5/6}} \frac{I^{1/2} L^{2/3}}{V^{13/12} \alpha_0}. \qquad (16.3.34)$$

In general, the width of the distribution increases strongly when $S_i \neq S_c$. When the crossover plane and the image plane do coincide ($S_c = S_i$), the rms value shows a slight increase with S_c ($0 < S_c < 1$). The dependency on the scaled crossover radius r_c^* is very weak and can practically be ignored for $r_c^* < 1$. The rms is independent of v_0^* for large v_0^*-values (typically $v_0^* > 100$) and decreases with v_0^* for small v_0^*.

This completes our qualitative analysis. In the next sections, the detailed equations are reviewed, which yield accurate quantitative predictions for the various manifestations of Coulomb particle-particle interactions in a rotational symmetric beam segment in drift space. The form of each of these equations is derived from the analytical analysis performed for the extreme cases that a single regime is dominant. Some of the constants are determined by parameter fitting of the numerical data obtained for the general case, using a suitable fit function. The final equations explicitly show the functional dependency corresponding to the regime that is dominant for most practical operating conditions. The dependencies of the other regimes are not explicitly represented but can be retrieved from the asymptotic behavior of the fit functions. These functions are expressed in terms of dimensionless parameters. Table 16.1 gives an overview of the scale measures and the dimensionless parameters used in the model. The various functional dependencies and physical constants appearing in the final equations for the Boersch effect, the effect of statistical angular deflections and the trajectory displacement effect are summarized by Tables 16.2a, 16.2b, and 16.2c respectively.

16.4. Equations for the Boersch effect

The theory for the Boersch effect is presented in Chapter 7. Here, we will restrict ourselves to a review of the essential results. For the case of a homocentric cylindrical beam segment, we will add an equation for the calculation of the FW_{50} value, which follows straightforwardly from the analysis of Section 7.4.

The Full Width (FWHM or FW_{50}) of the normal energy distribution generated in a *beam segment with a crossover* is given by

$$\frac{FW_E}{E} = G_{FW} C_{CGE} \frac{1}{2} [G_{CE}(\bar{\lambda}, \bar{r}_c, K_1) + G_{CE}(\bar{\lambda}, \bar{r}_c, K_2)] \sqrt{\frac{I}{V^{3/2}}}, \quad (16.4.1)$$

where the constant C_{CGE} is equal to

$$C_{CGE} = 1.4002 \frac{m^{1/4}}{\varepsilon_0^{1/2} e^{1/4}}. \quad (16.4.2)$$

In the case of electrons, one finds $C_{CGE} = 726.62$ in *SI*-units. The function $G_{CE}(\bar{\lambda}, \bar{r}_c, K)$ represents

$$G_{CE}(\bar{\lambda}, \bar{r}_c, K) = \left[\frac{1}{P_{CE}(\bar{r}_c, K)^4} + \frac{A\bar{r}_c^{4/3}}{\bar{\lambda}^{2/3} H(K)^4} + \frac{B}{D\bar{\lambda}^2} + \frac{C}{\bar{\lambda}^6 \bar{r}_c^4 K^4} \right]^{-1/4}, \quad (16.4.3)$$

where the functions $P_{CE}(\bar{r}_c)$ and $H(K)$ are given by

$$P_{CE}(\bar{r}_c) = \left[0.788 + 0.6 \left(\frac{\bar{r}_c}{K^2}\right)^{2/3} \left[1 + \frac{1}{2}\left(\frac{K^2}{\bar{r}_c}\right)^{4/9}\right]^{3/2} \right.$$

$$\left. + \frac{\pi F \bar{r}_c}{\{2 \ln [.8673(E + F\bar{r}_c)]\}^2} \right]^{-1/2} \quad (16.4.4)$$

Table 16.1 Overview of the scale measures and scaled parameters used in the extended two-particle model for Coulomb interactions in particle beams.

		Unit	Definition	Functional dependency	Value physical const. for electrons (SI)
Scale measures	d_0	m	$\dfrac{e^2}{4\pi\varepsilon_0 m v_0^2}$	$\dfrac{e}{8\pi\varepsilon_0}\dfrac{1}{\alpha_0^2 V}$	7.1999×10^{-10}
	v_0	$\dfrac{m}{s}$	$\alpha_0 v_z$	$\left(\dfrac{2e}{m}\right)^{1/2} \alpha_0 V^{1/2}$	5.9309×10^{5}
	δ	m	$\left(\dfrac{e^2 L^2}{\pi\varepsilon_0 m v_z^2}\right)^{1/3}$	$\left(\dfrac{e}{2\pi\varepsilon_0}\right)^{1/3}\dfrac{L^{2/3}}{V^{1/3}}$	1.4228×10^{-3}
	ν	$\dfrac{m}{s}$	$\left(\dfrac{e^2 v_z}{\pi\varepsilon_0 mL}\right)^{1/3}$	$\dfrac{2^{1/6} e^{5/6}}{(\pi\varepsilon_0)^{1/3} m^{1/2}}\dfrac{V^{1/6}}{L^{1/3}}$	8.4382×10^{2}
Density measures	λ	$\dfrac{1}{m}$	$\dfrac{I}{e v_z}$	$\left(\dfrac{m}{2e^3}\right)^{1/2}\dfrac{I}{V^{1/2}}$	1.0524×10^{13}
	$\bar{\lambda}$	–	λd_0	$\dfrac{m^{1/2}}{2^{7/2}\pi\varepsilon_0 e^{1/2}}\dfrac{I}{\alpha_0^2 V^{3/2}}$	7.5768×10^{3}
	λ^*	–	$\lambda\dfrac{\delta^3}{4 r_0^2}$	$\dfrac{m^{1/2}}{2^{7/2}\pi\varepsilon_0 e^{1/2}}\dfrac{IL^2}{r_0^2 V^{3/2}}$	7.5768×10^{3}
	χ_c	–	$\lambda \alpha_0 L$	$\left(\dfrac{m}{2e^3}\right)^{1/2}\dfrac{I\alpha_0 L}{V^{1/2}}$	1.0524×10^{13}
	χ_p	–	λr_0	$\left(\dfrac{m}{2e^3}\right)^{1/2}\dfrac{I r_0}{V^{1/2}}$	1.0524×10^{13}
Main parameters	$\bar{r}_c$	–	$\dfrac{r_c}{d_0}$	$\dfrac{8\pi\varepsilon_0}{e}\alpha_0^2 r_c V$	1.3889×10^{9}
	K	–	$\dfrac{\alpha_0 L}{2 r_c}$	$\dfrac{\alpha_0 L}{2 r_c}$	–
	r_c^*	–	$\dfrac{r_c}{\delta}$	$\left(\dfrac{2\pi\varepsilon_0}{e}\right)^{1/3}\dfrac{r_c V^{1/3}}{L^{2/3}}$	7.0286×10^{2}
	v_0^*	–	$\dfrac{v_0}{\nu}$	$\left(\dfrac{2\pi\varepsilon_0}{e}\right)^{1/3}\alpha_0 V^{1/3} L^{1/3}$	7.0286×10^{2}

Related parameters

$S_c = \dfrac{L_1}{L}$ $S_i = \dfrac{L_i}{L}$

$K_1 = 2 S_c K$ $K_2 = 2(1 - S_c) K$

Table 16.2a Overview of the results for the relative energy spread $FWHM_E/E$ in the different regimes for a beam segment with a crossover and a homocentric cylindrical beam segment. The equations for the cylindrical beam apply to a uniform current density distribution. Those for a beam with a crossover apply to a uniform spatial and a uniform angular distribution in the crossover and $K_1 = K_2 = K$ (crossover in the middle). Effective measures for Gaussian density distributions are presented in Section 7.6. The results for the beam segment with a crossover can be generalized for $K_1 \neq K_2$, using the approach of Eq. (7.5.37).

$\dfrac{FWHM_E}{E}$	Beam segment with a crossover			Homocentric cylindrical beam		
	Code + equation	Factor + value for electrons (SI)	Funct. dependency + add. equations	Code + equation	Factor + value for electrons (SI)	Funct. dependency + add. equations
G	CGE (7.3.37)	$\left(\dfrac{8(\ln 2)^2 m}{\varepsilon_0^2 e}\right)^{1/4}$ 726.62	$P_{CE}(\bar{r}_c, K)\dfrac{I^{1/2}}{V^{3/4}}$ (7.5.21), (7.5.20)	PGE (7.4.32)	$0.3330\dfrac{e^{1/2}m^{1/4}}{\varepsilon_0^{5/6}}$ 0.4537	$P_{PE}(r_0^*)\dfrac{I^{1/2}L^{2/3}}{r_0 V^{13/12}}$ (7.4.22), (7.4.16)
H	CHE (7.5.25)	$1.3233\dfrac{m^{1/3}}{\varepsilon_0}$ 14.488	$H(K)\dfrac{I^{2/3}}{r_c^{1/3}\alpha_0 V^{4/3}}$ (7.5.23)	PHE (7.4.31)	$0.22056\dfrac{m^{1/3}}{\varepsilon_0}$ 2.4147	$\dfrac{I^{2/3}L}{r_0^{4/3}V^{4/3}}$
L	CLE (7.3.40)	$\left(\dfrac{2m}{\varepsilon_0^2 e}\right)^{1/2}$ 3.8085×10^5	$\dfrac{I}{\alpha_0 V^{3/2}}$	Not applicable		
P	Equal to result for cylindrical beam			PPE (7.4.33)	$0.11178\dfrac{m}{\varepsilon_0 e^2}$ 4.4800×10^{17}	$\dfrac{I^2 L}{V^2}$

Table 16.2b Overview of the results for the angular spread $FWHM_\alpha$ in the different regimes for a beam segment with a crossover and a homocentric cylindrical beam segment. The equations apply to the same conditions as described in the caption of Table 16.2a. Effective measures for Gaussian density distributions are presented in Section 8.7.

		Beam segment with a crossover			Homocentric cylindrical beam	
$FWHM_\alpha$	Code+equation	Factor+value for electrons (SI)	Funct. dependency + add. equations	Code+equation	Factor+value for electrons (SI)	Funct. dependency + add. equations
G	CGA (8.3.21)	$\left(\dfrac{(\ln 2)^2 m}{8\varepsilon_0^2 e}\right)^{1/4}$ 256.90	$P_{CA}(\bar{r}_c, K)\dfrac{I^{1/2}}{V^{3/4}}$ (8.5.28), (8.5.27)	PGA (8.4.30)	$0.1665\dfrac{e^{1/12}m^{1/4}}{\varepsilon_0^{5/6}}$ 0.2269	$P_{PA}(r_0*)\dfrac{I^{1/2}L^{2/3}}{r_0 V^{13/12}}$ (8.4.20), (8.4.16)
H	CHA (8.5.31)	$0.19233\dfrac{m^{1/3}}{\varepsilon_0}$ 2.1057	$S_{HA}\dfrac{I^{2/3}}{\alpha_0^{4/3}V^{4/3}L^{1/3}}$ (8.5.20)	PHA (8.4.29)	$0.10177\dfrac{m^{1/3}}{\varepsilon_0}$ 1.1142	$\dfrac{I^{2/3}L}{r_0^{4/3}V^{4/3}}$
W	CWA (8.3.24)	$0.033329\dfrac{m}{\varepsilon_0^2 e}$ 2.4172×10^9	$f_\infty(\bar{r}_c)^2\dfrac{I^2}{\alpha_0^3 V^3}$ (8.3.10)	Not applicable		
P	CPA (8.5.35)	$4.1531\dfrac{10^{-4}m^{3/2}}{\varepsilon_0 e^{7/2}}$ 2.4772×10^{28}	$S_{PA}\dfrac{I^3\alpha_0 L^2}{V^{5/2}}$ (8.5.21)	PPA (8.4.32)	$8.3061\dfrac{10^{-4}m^{3/2}}{\varepsilon_0 e^{7/2}}$ 4.9544×10^{28}	$\dfrac{I^3 r_0 L}{V^{5/2}}$

Summary for the one-minute designer

Table 16.2c Overview of the results for the *FWHM* trajectory displacement effect in the different regimes for a beam segment with a crossover. The equations apply to the case of a uniform spatial and a uniform angular distribution in the crossover. Effective measures for Gaussian density distributions are presented in Section 9.7. For a cylindrical beam the spatial broadening in the backfocal plane of a lens, which is located at the end of the segment, is equal to $FWHM_\alpha \times f$, where *f* is the focal length of the lens. The reader is referred to Table 16.2b for the functional dependency of the angular spread $FWHM_\alpha$ generated in a homocentric cylindrical beam.

		Beam segment with a crossover	
$FWHM_r$	Code + equation	Factor + value for electrons (SI)	Funct. dependency + add. equations
G	CGT (9.3.45)	$0.158 \dfrac{e^{1/2} m^{1/4}}{\varepsilon_0^{5/6}}$ 0.215	$P_{CT} \dfrac{I^{1/2} L^{2/3}}{\alpha_0 V^{13/12}}$ (9.3.25), (9.5.10)
H	CUT (9.3.44)	$0.17205 \dfrac{m^{1/3}}{\varepsilon_0}$ 1.8837	$S_{HT} \dfrac{I^{2/3} L^{2/3}}{\alpha_0^{4/3} V^{4/3}}$ (9.5.3)
W	CWA (9.3.46)	$0.033329 \dfrac{m}{\varepsilon_0^2 e}$ 2.4172×10^9	$\lvert S_c - S_i \rvert f_\infty^2 \dfrac{I^2 L}{\alpha_0^3 V^3}$ (8.3.10)
P	CPT (9.3.47)	$6.2918 \dfrac{10^{-5} m^{3/2}}{\varepsilon_0 e^{7/2}}$ 4.1287×10^{27}	$S_{PT} \dfrac{I^3 \alpha_0 L^3}{V^{5/2}}$ (9.5.6)

$$H(K) = \dfrac{1}{\left[1 + 9K^{-2} + 2K^{-1/3}\right]^{1/2}}. \qquad (16.4.5)$$

In order to obtain a *FWHM* value, the constant G_{FW} in Eq. (16.4.1) and the constants *A*, *B* and *C* in Eq. (16.4.3) should be taken equal to

$$G_{FW} = 1, \quad A = 1.575 \times 10^{-3}, \quad B = 7.606 \times 10^{-4}, \quad C = .3045 \quad (FWHM), \qquad (16.4.6)$$

while for a FW_{50}, one should use

$$G_{FW} = .57288, \quad A = 8.246 \times 10^{-4}, \quad B = 8.192 \times 10^{-5},$$
$$C = 3.015 \times 10^{-5} \quad (FW_{50}). \qquad (16.4.7)$$

The constant *D* in Eq. (16.4.3) and the constants *E* and *F* in Eq. (16.4.4) are, for a uniform angular and spatial distribution, given by

$$D = 1, \quad E = 114.6, \quad F = 1 \qquad (16.4.8)$$

Gaussian angular and spatial distributions can in reasonable approximation be represented by the same set of equations provided that one replaces α_0 and r_c by the effective values α_{eff} and r_{eff}, defined as

$$\alpha_{\mathit{eff}} \cong 1.6\sigma_\alpha, \qquad r_{\mathit{eff}} \cong 1.6\sigma_r, \qquad (16.4.9)$$

where $\sigma_\alpha^2 = \langle \alpha^2 \rangle/2$ and $\sigma_r^2 = \langle r^2 \rangle/2$. A more accurate approach is to take $\alpha_0 = \sqrt{2}\sigma_\alpha$ and $r_c = \sqrt{2}\sigma_r$ and use the following modified values for the constants D, E and F,

$$\begin{aligned}
D_{gg} &= .6169, & E_{gg} &= 40.74, & F_{gg} &= 1.315 \\
D_{gu} &= .6169, & E_{gu} &= 49.19, & F_{gu} &= 1.138 \\
D_{ug} &= 1, & E_{ug} &= 100.0, & F_{ug} &= 1.160 \\
(D_{uu} &= 1, & E_{uu} &= 114.6, & F_{uu} &= 1),
\end{aligned} \qquad (16.4.10)$$

in which the first subscript indicates the type of angular distribution and the second subscript the type of spatial distribution (u = uniform, g = Gaussian).

In order to obtain an effective γ-value, which characterizes the type of distribution, one should evaluate

$$\gamma_{\mathit{eff}} = \frac{G_{CE}(\bar{\lambda}, \bar{r}_c, K_1)\gamma(\bar{\lambda}, \bar{r}_c, K_1) + G_{CE}(\bar{\lambda}, \bar{r}_c, K_2)\gamma(\bar{\lambda}, \bar{r}_c, K_2)}{G_{CE}(\bar{\lambda}, \bar{r}_c, K_1) + G_{CE}(\bar{\lambda}, \bar{r}_c, K_2)}, \qquad (16.4.11)$$

where

$$\gamma(\bar{\lambda}, \bar{r}_c, K) = G_{CE}(\bar{\lambda}, \bar{r}_c, K)$$

$$\times \left[\frac{2^4}{P_{CE}(\bar{r}_c, K)^4} + \frac{3^4}{2^4} \frac{A\bar{r}_c^{4/3}}{\bar{\lambda}^{2/3} H(K)^4} + \frac{B}{D\bar{\lambda}^2} + \frac{1}{2^4} \frac{C}{\bar{\lambda}^6 \bar{r}_c^4 K^4} \right]^{1/4} \qquad (16.4.12)$$

and the function $G_{CE}(\bar{\lambda}, \bar{r}_c, K)$ is given by Eq. (16.4.3). For the constants A, B and C, we use the values for the FW_{50} value, specified by Eq. (16.4.7). The constant D depends on the type of distribution, as specified by Eq. (16.4.10). The outcome of Eq. (16.4.11) is a value for γ_{eff} between 1/2 and 2. The parameter γ_{eff} serves as input for the algorithm used for the addition of the results obtained for the individual beam segments, which is the subject of Section 16.7.

The Full Width ($FWHM$ or FW_{50}) of the normal energy distribution generated in a *homocentric cylindrical beam segment* is given by

$$\frac{FW_E}{E} = H_{FW} C_{PHE} H_{PE}(\lambda^*, r_0^*) \frac{I^{2/3} L}{V^{4/3} r_0^{4/3}}, \qquad (16.4.13)$$

where the constant C_{PHE} is equal to

$$C_{PHE} = 0.22056 \frac{m^{1/3}}{\varepsilon_0}. \tag{16.4.14}$$

For electrons, one finds $C_{PHE} = 2.4147$ in SI-units. The function $H_{PE}(\lambda^*, r_0^*)$ represents

$$H_{PE}(\lambda^*, r_0^*) = \left[\left(1 + \frac{A\lambda^*}{P_{PE}(r_0^*)^6} \right)^{1/4} + \frac{B}{\chi_p^2} \right]^{-2/3}, \tag{16.4.15}$$

where χ_p is the pencil beam factor for a cylindrical beam ($\chi_p = 4\lambda^* r_0^{*3}$) and the function $P_{PE}(r_0^*)$ is given by

$$P_{PE}(r_0^*) = \left(1 + 0.240/r_0^{*8/7} \right)^{-7/8}. \tag{16.4.16}$$

In order to obtain a FWHM value, the constant H_{PW} in Eq. (16.4.13) and the constants A and B in Eq. (16.4.15) should be taken equal to

$$H_{FW} = 1, \quad A = 2.997, \quad B = 1.386, \quad (FWHM), \tag{16.4.17}$$

while for a FW_{50}, one should use

$$H_{FW} = .67347, \quad A = 7.911, \quad B = 5.566 \times 10^{-2} \quad (FW_{50}). \tag{16.4.18}$$

A Gaussian current density distribution leads to the same set of equations, provided that one takes $r_0 = \sqrt{2}\sigma_r$, with $\sigma_r^2 = \langle r^2 \rangle/2$.

In order to obtain an effective γ-value that characterizes the type of distribution, one should evaluate

$$\gamma_{eff} = H_{PE}(\lambda^*, r_0^*) \left[\left(\frac{3^6}{2^6} + \frac{2^6 A\lambda^*}{P_{PE}(r_0^*)^6} \right)^{1/4} + \frac{1}{2^{3/2}} \frac{B}{\chi_p^2} \right]^{2/3}, \tag{16.4.19}$$

where the function $H_{PE}(\lambda^*, r_0^*)$ is given by Eq. (16.4.15). For the constants A and B, we use the values for the FW_{50} value, specified by Eq. (16.4.18). The outcome of Eq. (16.4.19) is a value for γ_{eff} between 1/2 and 2.

16.5. Equations for the trajectory displacement effect

The theory for the trajectory displacement effect is presented in Chapter 9. Chapter 8 covers the related effect of statistical angular deflections. Here we will restrict ourselves to a review of the essential results and will add some equations for the calculation of the d_{1288} and d_{2575} edge width of shaped spots, utilizing the analysis of Chapter 12. For the case of a homocentric cylindrical beam, we will add an equation for the FW_{50} value, which follows straightforwardly from the analysis of Sections 8.4 and 9.4.

The Full Width (*FWHM*, FW_{50}, d_{1288} or d_{2575}) of the trajectory displacement distribution generated in a *beam segment with a crossover* is given by

$$FW_r = H_{FW} C_{CHT} H_{CT}(\bar{\lambda}, v_0^*, r_c^*, S_c, S_i) \frac{I^{2/3} L^{2/3}}{V^{4/3} \alpha_0^{4/3}}, \qquad (16.5.1)$$

where the constant C_{CHT} is equal to

$$C_{CHT} = 0.17205 \frac{m^{1/3}}{\varepsilon_0}. \qquad (16.5.2)$$

For electrons $C_{CHT} = 1.8837$ in *SI*-units. The function $H_{CT}(\bar{\lambda}, v_0^*, r_c^*, S_c, S_i)$ is given by

$$H_{CT}(\bar{\lambda}, v_0^*, r_c^*, S_c, S_i) = H_{CT1} + |S_c - S_i| H_{CT2}, \qquad (16.5.3)$$

where the functions H_{CT1} and H_{CT2} are equal to

$$H_{CT1} = \left[\left(\frac{1}{S_{HT}(S_c, S_i, K)^6} + \frac{A\bar{\lambda}}{P_{CT1}(v_0^*, r_c^*, S_c)^6} \right)^{1/7} + \frac{B}{v_0^{*6} \bar{\lambda}^2 S_{PT}(S_c, S_i, K)^{6/7}} \right]^{-7/6}$$

$$H_{CT2} = \left(\frac{C\bar{\lambda}^{1/2}}{v_0^{*3/4}} + \frac{D}{(E + F v_0^{*2} r_c^*)^3 v_0^{*3} \bar{\lambda}^4} \right)^{-1/3}.$$

$$(16.5.4)$$

The functions $P_{CT1}(v_0^*, r_c^*, S_c)$, $S_{HT}(S_c, S_i, K)$ and $S_{PT}(S_c, S_i, K)$ represent

$$P_{CT1}(v_0^*, r_c^*, S_c) = \frac{\left[1 + 0.682(S_c - 0.5) - 0.739(S_c - 0.5)^2\right]^{1/2}}{\left(1 + 1.40/v_0^{*8/7}\right)^{7/4} \left(1 + 0.30 r_c^*\right)^{1/2}} \qquad (16.5.5)$$

$$S_{HT} S_c, S_i K = \left| \frac{3S_c - 2S_i + 3/2K}{1/K + 2S_c^{1/3}} + \frac{31 - S_c - 21 - S_i + 3/2K}{[1/K + 21 - S_c]^{1/3}} - \frac{3}{K^{2/3}} \right|$$

$$(16.5.6)$$

$$S_{PT}(S_c, S_i, K) = G \left| 4 - 6(S_c + S_i) + 12 S_c S_i + H \frac{3 - 6 S_c^2 - 6 S_i (1 - 2 S_c)}{K} \right|.$$

$$(16.5.7)$$

In order to obtain a *FWHM* value, the constant H_{FW} in Eq. (16.5.1) and the constants *A*, *B*, *C* and *D* in Eq. (16.5.4) should be taken equal to

$$H_{FW} = 1, \quad A = 58.88, \quad B = 25.42, \quad C = .00638,$$
$$D = 5.415 \times 10^{-4} \quad (FWHM), \qquad (16.5.8)$$

while for a FW_{50}, d_{1288} or d_{1275} value, one should use respectively

$$H_{FW} = 1.2505, \quad A = 225.2, \quad B = .1439, \quad C = .0125,$$
$$D = 5.768 \times 10^{-8} \quad (FW_{50})$$
$$H_{FW} = 1.3836, \quad A = 418.2, \quad B = .02458, \quad C = .0170,$$
$$D = 1.758 \times 10^{-9} \quad (d_{1288}) \quad (16.5.9)$$
$$H_{FW} = .72980, \quad A = 251.7, \quad B = .1576, \quad C = .0132,$$
$$D = 7.248 \times 10^{-8} \quad (d_{2575}).$$

The constants E and F in the expression for H_{CT2}, given by Eq. (16.5.4), and the constants G and H in Eq. (16.5.7) should, for a uniform angular and spatial distribution, be taken equal to

$$E = 1, \quad F = .08275, \quad G = 1, \quad H = 1. \quad (16.5.10)$$

Gaussian angular and spatial distributions can best be represented by taking $\alpha_0 = \sqrt{2}\sigma_\alpha$ and $r_c = \sqrt{2}\sigma_r$ (with $\sigma_x^2 = \langle \alpha^2 \rangle/2$ and $\sigma_r^2 = \langle r^2 \rangle/2$) and using the following modified values for the constants E, F, G and H,

$$\begin{aligned}
E_{gg} &= .7854, & F_{gg} &= .08971 & G_{gg} &= 1.268, & H_{gg} &= 1 \\
E_{gu} &= .7854, & F_{gu} &= .06990, & G_{gu} &= 1.268, & H_{gu} &= .7887 \\
E_{ug} &= 1, & F_{ug} &= .1062, & G_{ug} &= 1, & H_{ug} &= 1.268 \\
(E_{uu} &= 1, & F_{uu} &= .08275, & G_{uu} &= 1, & H_{uu} &= 1),
\end{aligned} \quad (16.5.11)$$

in which the first subscript indicates the type of angular distribution and the second subscript the type of spatial distribution (u = uniform, g = Gaussian).

In order to obtain an effective γ-value, which characterizes the type of distribution, one should evaluate

$$\gamma_{eff} = H_{CT1} \left[\left(\frac{(3/2)^6}{S_{HT}(S_c, S_i, K)^6} + \frac{2^6 A \bar{\lambda}}{P_{CT1}(v_0^*, r_c^*, S_c)^6} \right)^{1/7} \right.$$
$$\left. + \frac{B/3^{6/7}}{v_0^{*6} \bar{\lambda}^2 S_{PT}(S_c, S_i, K)^{6/7}} \right]^{7/6}, \quad (16.5.12)$$

where the function H_{CT1} is given by the first equation of (16.5.4). For the constants A and B, we use the values for the FW_{50} value specified by the first line of Eq. (16.5.9). The outcome of Eq. (16.5.12) is a value for γ_{eff} between 1/3 and 2.

The angular deflections generated in a *homocentric cylindrical beam segment* produce a trajectory displacement distribution in the back focal plane of the lens succeeding the beam segment, which has a Full Width (FWHM, FW_{50}, d_{1288} or d_{2575}) given by

$$FW_r = FW_\alpha f = H_{FW} C_{PHA} H_{PA}(\lambda^*, r_0^*) \frac{I^{2/3} Lf}{V^{4/3} r_0^{4/3}}, \quad (16.5.13)$$

where f is the focal distance of the lens and the constant C_{PHA} is equal to

$$C_{PHA} = 0.10177 \frac{m^{1/3}}{\varepsilon_0}, \tag{16.5.14}$$

For electrons, one finds $C_{PHA} = 1.1142$ in SI-units. The function $H_{PA}(\lambda^*, r_0^*)$ represents

$$H_{PA}(\lambda^*, r_0^*) = \left[\left(1 + \frac{A\lambda^*}{P_{PA}(r_0^*)^6}\right)^{1/7} + \frac{B}{C\chi_p^2} \right]^{-7/6}, \tag{16.5.15}$$

where χ_p is the pencil beam factor for a cylindrical beam ($\chi_p = 4\lambda^* r_0^{*3}$) and the function $P_{PA}(r_0^*)$ is given by

$$P_{PA}(r_0^*) = \left(1 + 0.185/r_0^{*8/7}\right)^{-7/4}. \tag{16.5.16}$$

In order to obtain a $FWHM$ value, the constant H_{FW} in Eq. (16.5.13) and the constants A and B in Eq. (16.5.15) should be taken equal to

$$H_{FW} = 1, \quad A = 1.851, \quad B = 30.82 \quad (FWHM), \tag{16.5.17}$$

while for a FW_{50}, d_{1288} or d_{1275} value, one should use respectively

$$\begin{aligned} H_{FW} &= 1.2505, & A &= 7.078, & B &= .1745 & (FW_{50}) \\ H_{FW} &= 1.3836, & A &= 13.15, & B &= 2.980 \times 10^{-2} & (d_{1288}) \\ H_{FW} &= .72980, & A &= 7.914, & B &= .1910 & (d_{2575}). \end{aligned} \tag{16.5.18}$$

For a uniform current density distribution $C = 1$. For a Gaussian current density distribution, one should take $C = 1.226$ and $r_0 = \sqrt{2}\sigma_r$, with $\sigma_r^2 = \langle r^2 \rangle / 2$

In order to obtain an effective γ-value, which characterizes the type of distribution, one should evaluate

$$\gamma_{eff} = H_{PA}(\lambda^*, r_0^*) \left[\left(\frac{3^6}{2^6} + \frac{2^6 A\lambda^*}{P_{PA}(r_0^*)^6}\right)^{1/7} + \frac{1}{3^{6/7}} \frac{B}{\chi_p^2} \right]^{7/6}, \tag{16.5.19}$$

where the function $H_{PA}(\lambda^*, r_0^*)$ is given by Eq. (16.5.15). For the constants A and B, we use the values for the FW_{50} value, specified by the first line of Eq. (16.5.18). The outcome of Eq. (16.5.19) is a value for γ_{eff} between 1/3 and 2.

16.6. Equations for the space charge effect

The theory of the space charge effect occurring in beams of low particle density is presented in Chapter 11. We will review the essential

Summary for the one-minute designer

results here. For a *beam segment with a crossover*, the defocussing distance Δz_f and magnification M of the crossover due to the space charge effect are given by

$$\Delta z_f = C_{SC} \frac{1}{2K}(K_1 Z(K_1) + K_2 Z(K_2)) \frac{IL}{\alpha_0^2 V^{3/2}} \qquad (16.6.1)$$

$$M = 1 + C_{SC}[R_1(K_1) - R_1(K_2)] \frac{I}{\alpha_0^2 V^{3/2}}, \qquad (16.6.2)$$

where the constant C_{SC} is equal to

$$C_{SC} = \frac{m^{1/2}}{2^{5/2} \pi \varepsilon_0 e^{1/2}}. \qquad (16.6.3)$$

For electrons, one finds $C_{SC} = 1.5154 \times 10^4$ in *SI*-units. The functions $Z(K)$ and $R_1(K)$ are given by

$$Z(K) = 1 + \frac{1}{1+K} - \frac{2}{K}\ln(1+K) \qquad (16.6.4)$$

$$R_1(K) = \ln(1+K) - \frac{K}{1+K}. \qquad (16.6.5)$$

The functions $R_1(K)$ and $Z(K)$ are plotted in Figs. 11.2 and 11.3 respectively. Eqs. (16.6.1)–(16.6.5) are valid for uniform as well as Gaussian angular and spatial distributions. For the latter, one should use $\alpha_0 = \sqrt{2}\sigma_\alpha$ and $r_c = \sqrt{2}\sigma_r$, where $\sigma_\alpha^2 = \langle \alpha^2 \rangle /2$ and $\sigma_r^2 = \langle r^2 \rangle /2$.

For non-uniform distributions, the space charge effect will also introduce aberrations. When the current density distribution is Gaussian in every cross section of the beam, the virtual transverse displacement in the crossover $\Delta r_\perp$ caused by the third order aberration of the space charge lens is, for a trajectory with unperturbed coordinates a and $r_\perp$, equal to

$$\frac{\Delta r_\perp}{r_c} = \frac{1}{2}C_{SC}\left[D\left(\frac{r_\perp}{r_c}\right)^3 + F\left(\frac{r_\perp}{r_c}\right)^2 \frac{\alpha}{\alpha_0} + C\left(\frac{\alpha}{\alpha_0}\right)^2 \frac{r_\perp}{r_c} + S\left(\frac{\alpha}{\alpha_0}\right)^3\right]\frac{I}{\alpha_0^2 V^{3/2}}, \qquad (16.6.6)$$

where the functions D, F, C and S are defined as

$$D = D_1(K_1) - D_1(K_2), \quad D_1(K) = \frac{3K^2 + K^3}{12(1+K)^3} \qquad (16.6.7)$$

$$F = F_1(K_1) + F_1(K_2), \quad F_1(K) = \frac{K^3}{2(1+K)^3} \qquad (16.6.8)$$

$$C = C_1(K_1) - C_1(K_2),$$

$$C_1(K) = \frac{3}{2}\ln(1+K) - \frac{6K + 15K^2 + 11K^3}{4(1+K)^3} \qquad (16.6.9)$$

$$S = S_1(K_1) + S_1(K_2),$$
$$S_1(K) = \frac{1}{2}K - 2\ln(1+K) + \frac{9K + 21K^2 + 13K^3}{6(1+K)^3}. \qquad (16.6.10)$$

The functions $D_1(K)$, $F_1(K)$, $C_1(K)$ and $S_1(K)$ are plotted in Fig. 11.4. For $K \gg 1$, the major contribution stems from spherical aberration, described by the term proportional to $S\alpha^3$. The corresponding coefficient of spherical aberration C_s can be expressed as

$$C_s = C_{SC} \frac{1}{4K}(S_1(K_1) + S_1(K_2)) \frac{IL}{\alpha_0^4 V^{3/2}}, \qquad (16.6.11)$$

where the constant C_{SC} is given by Eq. (16.6.3).

The space charge effect generated in a *homocentric cylindrical beam segment* leads to a defocussing Δz_f of the crossover formed by the succeeding lens, which is equal to

$$\Delta z_f = C_{SC} \frac{ILf^2}{r_0^2 V^{3/2}}, \qquad (16.6.12)$$

where f is the focal distance of the lens and the constant C_{SC} is given by Eq. (16.6.3). This equation applies to a uniform as well as a Gaussian spatial distribution. For the latter, one should use $r_c = \sqrt{2}\sigma_r$, where $\sigma_r^2 = \langle r^2 \rangle / 2$.

Since the beam is homocentric, there is only one third order aberration term. The corresponding aberration in the crossover formed by the succeeding lens is proportional to $C_s \alpha^3$, where α is the angle of the unperturbed trajectory at the crossover. For a Gaussian current density distribution in the cylindrical beam, the coefficient of spherical aberration C_s is equal to

$$C_s = C_{SC} \frac{ILf^4}{4r_0^2 V^{3/2}}, \qquad (16.6.13)$$

where the constant C_{SC} is given by Eq. (16.6.3).

16.7. Addition of the effects generated in individual beam segments

Eq. (16.2.15) specifies how to add the FWHM of the distributions of displacements $\rho(\Delta E)$ and $\rho(\Delta r)$ generated in the individual beam segments constituting the beam. Denoting, in general, the FWHM, FW_{50}, d_{1288} or d_{2575} value of the distribution generated in beam segment j ($j = 1, 2, \ldots, N_b$) as FW_j, one can express the summation rule as

$$FW_T = \left(\sum_{j=1}^{N_b} FW_j^\gamma\right)^{1/\gamma}. \qquad (16.7.1)$$

It presupposes that the individual displacement distributions are uncorrelated and all of the same type, that is, described by the same parameter γ. The corresponding one- and two-dimensional distributions are specified by Eqs. (16.2.10) and (16.2.11) respectively.

In Section 16.2, it was pointed out that the representation of Eqs. (16.2.10) and (16.2.11) is only approximate. The actual distributions cannot fully be characterized by a single γ-value. For all conditions, the distribution will ultimately fall off exponentially for large displacements. Such a fall off corresponds to $\gamma = 2$. Accordingly, the core and the tails of the displacement distribution are often determined by different γ-values. The effective parameter γ_{eff}, which is for the different distributions defined by Eqs. (16.4.11), (16.4.19), (16.5.12) and (16.5.19), roughly corresponds to the γ-value of the distribution at the boundary of the 50% volume. For the summation rule of Eq. (16.7.1), we now take $\gamma = \gamma_{eff}$ for all which measures.

There is still another problem in the application of Eq. (16.7.1). It presupposes that all beam segments give the same γ_{eff}-value. However, it should be anticipated that this is usually not the case. A way to handle this problem is to define a total effective value $\gamma_{eff,T}$ as a weighted average of the individual γ_{eff}-values

$$\gamma_{eff,T} = \frac{\sum_{j=1}^{N_b} FW_{50,j}\gamma_{eff,j}}{\sum_{j=1}^{N_b} FW_{50,j}}, \qquad (16.7.2)$$

where $FW_{50,j}$ and $\gamma_{eff,j}$ are the Full Width median and the effective γ-value for beam segment j respectively. The summation can now be performed with Eq. (16.7.1), taking $\gamma = \gamma_{eff,T}$. A somewhat more sophisticated approach is to subdivide the full γ-range into a number of intervals and perform the summation in two steps. In the first step, one adds those FW_j with a $\gamma_{eff,j}$ that are in the same γ-interval, using Eq. (16.7.1), with γ equal to the midpoint of the γ-interval. In the second step, the resulting FW values for each γ-interval are used to compute $\gamma_{eff,T}$ from Eq. (16.7.2) and summed next by means of Eq. (16.7.1), taking $\gamma = \gamma_{eff,T}$.

For the trajectory displacement effect as well as the space charge effect, one has to take the transverse magnification M_j of the image plane in segment j to the target into account. The FW_j trajectory displacement values, used in Eqs. (16.7.1) and (16.7.2), should all refer to the target plane. Thus, the quantities FW_j are assumed to include a factor M_j. The space charge defocussing distances Δz_j should be added as

$$\Delta z_T = \sum_{j=1}^{N_b} M_j^2 \Delta z_j, \qquad (16.7.3)$$

using that the axial magnification is the square of the transverse magnification. The third order space charge aberrations $\Delta r_{\perp,j}$, given by Eq. (16.6.6), should be multiplied with M_j, as for the trajectory displacement effect. When adding the coefficients of spherical aberration $C_{s,j}$, one should account for the (de)magnification of the beam semi-angle ($\alpha_{0,j} = M_j \alpha_{0,T}$). This gives

$$C_{sT} = \sum_{j=1}^{N_b} M_j^4 C_{s,j}, \qquad (16.7.4)$$

similar to the summation of the spherical aberration of the electrostatic and magnetic lenses in the column.

References

Aarseth, S.J., 1972. Gravitational N-Body Problem. In: Lecar, M. (Ed.), Proceedings of IAU Colloquium, vol. 10. D. Reidel Publishing Company, Dordrecht, Holland, p. 373.
Abramowitz, M., Stegun, S., 1965. Handbook of Mathematical Functions. Dover Publications, Inc., New York, NY.
Allee, D.R., Pehoushek, J.D., Pease, R.F.W., 1988. J. Vac. Sci. Technol. B 6, 1989.
Alles, D.S., Thomson, M.G.R., 1987. VLSI Electronics Microstructure Science. vol. 16 Academic Press, New York, p. 57.
Andersen, W.H.J., 1967. Br. J. Appl. Phys. 18, 1573.
Andersen, W.H.J., Mol, A., 1968. Proc. Eur. Reg. Conf. Electron Microsc. 4th, 339.
Andretta, M., Marini, M., Zanarini, G., 1986. IEEE Trans. Electron Dev. ED-33, 1084.
Ash, E.A., Gabor, D., 1954. Proc. R. Soc. A228, 477.
Barth, J., 1988. Private communications.
Beck, A.H., 1973. Int. J. Electron. 36, 121.
Bell, A.E., Swanson, L.W., 1979. Phys. Rev. B 19, 3353.
Boersch, H., 1954. Z. Phys. 139, 115.
Bohm, D., Pines, D., 1951. Phys. Rev. 82, 625.
Brody, I., 1987. Int. J. Electron. 62, 1.
Brody, I., et al., 1981. IEEE Trans. Electron Dev. ED-28, 1422.
Broers, A.N., 1981a. J. Electrochem. Soc. Solid. State. Sci. Technol. 128, 166.
Broers, A.N., 1981b. IEEE Trans. ED-28, 1268.
Busch, H., 1926. Z. Phys. 81, 974.
Chandrasekhar, S., 1941. Astrophys. J. 94, 511.
Chandrasekhar, S., 1942. Principles of Stellar Dynamics. University of Chicago Press.
Chandrasekhar, S., 1943. Rev. Mod. Phys. 15, 1.
Chandrasekhar, S., von Neumann, J., 1942. Astrophys. J. 95, 489.
Chandrasekhar, S., von Neumann, J., 1943. Astrophys. J. 97, 1.
Chang, T.H.P., Wilson, A.D., Speth, A.J., Ting, C.H., 1976. Proc. Electron Ion Sic. Technol. 7th Int. Conf., p. 392.
Chapman, S., Cowling, T.G., 1970. The Mathematical Theory of Non-Uniform Gases. Cambridge University Press, p. 359.
Chu, H.C., Munro, E., 1982. Optik 61, 121.
Contopoulos, G., 1972. Gravitational N-Body Problem. In: Lecar, M. (Ed.), Proceedings of IAU Colloquium, vol. 10. D. Reidel Publishing Company, Dordrecht-Holland, p. 169.
Cottrell, G.A., 1981. Rev. Sci. Instrum. 52, 1174.
Crewe, A.V., 1978a. Optik 50, 205.
Crewe, A.V., 1978b. Optik 52, 337.
Crewe, A.V., 1987. Ultramicroscopy 23, 159.
Cummings, K.D., Harriot, L.R., Chi, G.C., Ostermayer, F.W., 1986. SPIE 632, 93.
Cutler, C.C., Hines, M.E., 1955. Proc. I.R.E. 43, 307.
Davis, D.E., Gillespie, S.J., Silverman, S.L., Stickel, W., 1983. J. Vac. Sci. Technol. B 1, 1003.
Dayan, P.S., Jones, G.A.C., 1981. J. Vac. Sci. Technol. 19, 1094.
De Chambost, E., 1982. Optik 62, 189.

De Chambost, E., Hennion, C., 1979. Optik 55, 357.
De Chambost, E., Frichet, A., Chartier, M., Ta The, H., Trotel, J., 1986. J. Vac. Sci. Technol. B 4, 78.
Debeye, P., Hückel, E., 1923. Z. Phys. 24, 185.
Degenhardt, V., Koops, H., 1982. Optik 61, 395.
Dietrich, W., 1958. Z. Phys. 152, 306.
Ditchfield, R.W., Whelan, M.J., 1977. Optik 48, 163.
Doran, S., Perkins, M., Stickel, W., 1975. J. Vac. Sci. Technol. 12, 1174.
El-Kareh, A.B., El-Kareh, J.C.J., 1970. Electron Beams, Lenses and Optics. Academic Press, New York, p. 250.
El-Kareh, A.B., Smither, M.A., 1979. J. Appl. Phys. 50, 5596.
Eng, G., 1985. J. Appl. Phys. 58, 4365.
Epstein, B., 1958. Compt. Rend. 246, 586.
Erdélyi, A., Magnus, W., Oberhettinger, F., Tricomi, F.G., 1954. Tables of Integral Transforms (California Institute of Technology. Bateman manuscript project). McGraw-Hill Book Company Inc., New York.
Fack, H., 1955. Physikalischer Verh. 6, 6.
Feller, J., 1966. An Introduction to Probability Theory and Its Applications. vol. 2, J. Wiley & Sons Inc., New York, p. 165.
Fischer, M., 1970. J. Appl. Phys. 41, 3615.
Fowler, R.D., Gibson, G.E., 1934. Phys. Rev. 46, 1075.
Franzen, W., Porter, J.H., 1975. Advances in Electronics and Electron Physics. In: Marton, L. (Ed.), vol. 39. Academic Press, New York, p. 73.
Fujinami, M., Shimazu, N., Hosokawa, T., Shibayama, A., 1987. J. Vac. Sci. Technol. B 5, 61.
Furukawa, Y., 1986. J. Vac. Sci. Technol. A 4, 1908.
Gadzuk, W., Plummer, E.W., 1973. Rev. Mod. Phys. 45, 487.
Gaukler, K.H., Speidel, R., Vorster, F., 1975. Optik 42, 391.
Gesley, M.A., Swanson, L.W., 1984. J. Phys. C9 (45), 167.
Gilbert, H., 1972. Gravitational N-Body Problem. In: Lecar, M. (Ed.), Proceedings of IAU Colloquium, vol. 10. D. Reidel Publishing Company, Dordrecht-Holland, p. 5.
Glaser, W., 1952. Grundlagen der Elektronenoptik. Springer-Verlag, Vienna, p. 66.
Goldstein, 1980. Classical Mechanics, second ed. Addison-Wesley Publishing Company, Reading, MA.
Goto, E., Soma, T., 1977. Optik 48, 255.
Goto, E., Soma, T., Idesawa, M., 1978. J. Vac. Sci. Technol. 15, 883.
Gradshteyn, I.S., Ryzhik, I.M., 1980. Table of Integrals Series and Products. Academic Press, New York.
Grivet, P., 1965. Electron Optics. Pergamon Press, Oxford, p. 277.
Grivet, P., 1971. Electron Optics, second ed. Pergamon Press, Oxford.
Groves, T., 1981a. J. Vac. Sci. Technol. 19, 110.
Groves, T., 1981b. J. Vac. Sci. Technol. 19, 1106.
Groves, T., 1984. Private communications.
Groves, T., Hammond, D.L., Kuo, H., 1979. J. Vac. Sci. Technol. 16, 1680.
Gryzinski, M., 1964a. Phys. Rev. 138, A305.
Gryzinski, M., 1964b. Phys. Rev. 138, A322.
Gryzinski, M., 1964c. Phys. Rev. 138, A336.
Haberstroh, G., 1956. Z. Phys. 145, 20.
Haef, A.V., 1939. Proc. Inst. Radio Eng. 27, 586.
Hamaguchi, S., Kai, J., Yasuda, H., 1988. J. Vac. Sci. Technol. B 6, 204.

Hamish, H., Loeffler, K.H., Kaiser, H.J., 1964. Proc. Eur. Reg. Conf. Electron Microsc. 3, 11.
Hanson, G.R., Siegel, B.M., 1979. J. Vac. Sci. Technol. 16, 1875.
Hanszen, K.J., Lauer, R., 1967. Z. Naturforsch. 22a, 238.
Hanszen, K.J., Lauer, R., 1969. Z. Naturforsch. 24a, 214.
Hartl, W.A.M., 1966. Z. Phys. 191, 33.
Hartwig, D., Ulmer, K., 1963a. Z. Angew. Phys. 15, 309.
Hartwig, D., Ulmer, K., 1963b. Z. Phys. 173, 294.
Hassel Shearer, M., Takemura, H., Isobe, M., Goto, N., Tanaka, K., Miyauchi, S., 1986. J. Vac. Sci. Technol. B 4, 64.
Hauke, R., 1977. Ph.D. Thesis, University of Tübingen.
Heinrich, H., Essig, M., Geiger, J., 1977. J. Appl. Phys. 12, 197.
Herriot, D.R., 1982. J. Vac. Sci. Technol. 20, 781.
Hertz, P., 1909. Math. Ann. 67, 387.
Hines, M.E., 1951. J. Appl. Phys. 22, 1385.
Hoeberechts, A.M.E., van Gorkum, G.G.P., 1986. J. Vac. Sci. Technol. B 4, 105.
Holtsmark, J., 1919. Ann. Phys. 58, 38.
Hosokawa, T., Morita, H., 1983. J. Vac. Sci. Technol. B 1, 1293.
Hutter, R., 1967. Focussing of Charged Particles. In: Septier, A. (Ed.), vol. 2. Academic Press, New York, p. 3.
Ichimaru, S., 1973. Basic Principles of Plasma Physics. W. A. Benjamin Inc., London.
Ichinokawa, T., 1968. *Jpn.* J. Appl. Phys. 7, 799.
Ichinokawa, T., 1969. *Jpn.* J. Appl. Phys. 8, 137.
Ishitani, T., Umemura, K., Hosoki, S., Takayama, S., Tamura, H., 1984. J. Vac. Sci. Technol. A 2, 1365.
Ishitani, T., Kawanami, Y., Shukuri, S., 1987a. Jpn. J. Appl. Phys. 26, 1777.
Ishitani, T., Umemura, K., Kawanami, Y., 1987b. J. Appl. Phys. 61, 748.
Ishitani, T., Kawanami, Y., Oshinishi, T., Umemura, K., 1987c. Appl. Phys. A. 44, 233.
Ishitani, T., Umemura, K., Aida, T., 1987d. J. Vac. Sci. Technol. A 5, 2907.
Jansen, G.H., 1987. J. Vac. Sci. Technol. B 5, 146.
Jansen, G.H., 1988a. J. Vac. Sci. Technol. B 5, 146.
Jansen, G.H., 1988b. J. Vac. Sci. Technol. B 6, 1977.
Jansen, G.H., 1988c. Reference Manual MC-Simulation Program. Delft University of Technology.
Jansen, G.H., 1988d. Reference Manual INTERAC Program. Delft University of Technology.
Jansen, G.H., Stickel, W., 1984. In: Heuberger, A., Beneking, H. (Eds.), Proceedings of the Microcircuits Engineering Conference, p. 167.
Jansen, G.H., van Leeuwen, J.M.J., van der Mast, K.D., 1983. In: Ahmed, H., Cleaver, J.R.A., Jones, G.A.C. (Eds.), Proceedings of the Microcircuits Engineering Conference, p. 99.
Jansen, G.H., Groves, T.R., Stickel, W., 1985. J. Vac. Sci. Technol. B 3, 190.
Jones, G.A.C., Rao, V.R.M., Sun, H.T., Ahmed, H., 1983. J. Vac. Sci. Technol. B 1, 1298.
Jones, G.A.C., Sargent, P.M., Norris, T.S., Ahmed, H., 1985. J. Vac. Sci. Technol. B 3, 124.
Kasper, E., 1982. Advances in Optical and Electron Microscopy. In: Barer, R., Cosslett, V.E. (Eds.), vol. 8. Academic Press, New York, p. 207.
Katsuhiro, K., Ozasa, S., Komoda, T., 1983. J. Vac. Sci. Technol. B 1, 1303.
Kawanami, Y., Ishitani, T., Umemura, K., Shukuri, S., 1987. J. Vac. Sci. Technol. B 5, 1364.

Kelly, J., Groves, T., Kuo, H.P., 1981. J. Vac. Sci. Technol. 19, 936.
King, H.J., et al., 1985. J. Vac. Sci. Technol. B 3, 106.
Kirstein, P.T., Kino, G.S., Walters, W.E., 1967 (digitized 2007). SpaceCharge Flow. McGraw-Hill, New York.
Kittel, C., 1958. Elementary Statistical Mechanics. J. Wiley & Sons Inc., New York.
Klemperer, O., 1953. Electron Optics. Cambridge University Press.
Klemperer, O., Barnett, M.E., 1971. Electron Optics, third ed. Cambridge University Press.
Knauer, W., 1979a. Optik 54, 211.
Knauer, W., 1979b. J. Vac. Sci. Technol. 16, 1676.
Knauer, W., 1981. Optik 59, 335.
Knuth, D.E., 1969. The Art of Computer Programming. Addison-Wesley Publishing Company, Reading, MA, p. 2.
Knuth, D. E., (1973). "The Art of Computer Programming," Addison-Wesley Publishing Company., Reading, MA (1973), 3.
Komuro, M., Kanayama, T., Hiroshima, H., Tanoue, H., 1983. Appl. Phys. Lett. 42, 908.
Krohn, V.E., Ringo, G.R., 1975. Appl. Phys. Lett. 27, 249.
Kuroda, K., 1984. J. Vac. Sci. Technol. A 2, 68.
Langmuir, D.B., 1937. Proc. Inst. Radio Eng. 25, 977.
Lauer, R., 1982. Advances in Optical and Electron Microscopy. In: Barer, R., Cosslett, V.E. (Eds.), 8. Academic Press, New York, p. 137.
Lawson, J.D., 1977. The Physics of Charged-Particle Beams. Clarendon Press, Oxford.
Lejeune, C., Aubert, J., 1980. Applied Charged Particle Optics: Part A. In: Septier, A. (Ed.), Academic Press, New York, p. 159.
Lenz, F., 1958. Proc. Int. Conf. Electron Microsc. 4, 39.
Lindsay, P.A., 1960. Adv. Electron. Electron Phys. 13, 181.
Livesay, W.R., Greeneich, J.S., Wolfe, J.E., Felker, R.J., 1983. Solid State Technol. 9 (83), 137.
Loeffler, K.H., 1963. Physikalischer Verh. 3, 104.
Loeffler, K.H., 1964. Ph.D. Thesis, University of Berlin.
Loeffler, K.H., 1969. Z. Angew. Phys. 27, 145.
Loeffler, K.H., Hudgin, R.H., 1970. Proc. Int. Conf. Electron Microsc. 7, 67.
Mair, G.L.R., Mulvey, T., 1985. In: van der Mast, K.D., Radelaar, S. (Eds.), Proceedings of the Microcircuits Engineering Conference, vol. 3, p. 133.
Massey, G.A., Jones, M.D., Plummer, B.P., 1981. J. Appl. Phys. 52, 3780.
Mayer, H.P., 1985. Appl. Phys. Lett. 47, 1247.
McClelland, J.J., Ratliff, J.M., Fink, M., 1981. J. Appl. Phys. 52, 7039.
McGregor-Morris, J.T., Mines, R., 1925. J. Inst. Elect. Eng. 63, 1065.
Meisburger, W.D., 1983. Private communications.
Melngailis, J., 1987. J. Vac. Sci. Technol. B 5, 469.
Meyer, W.E., 1958. Optik 15, 398.
Mihran, T.G., 1966. J. Appl. Phys. 38, 159.
Miller, M.H., Dow, W.G., 1961. J. Appl. Phys. 32, 274.
Moore, R.D., Caccoma, G.A., Pfeiffer, H.C., Weber, E.V., Woodard, O.C., 1981. J. Vac. Sci. Technol. 19, 950.
Morita, H., Hosokawa, T., Fujinama, M., 1985. In: van der Mast, K.D., Radelaar, S. (Eds.), Proceedings of the Microcircuits Engineering Conference, vol. 3, p. 53.
Moriya, S., Komatsu, K., Harada, K., Kitayama, T., 1983. J. Vac. Sci. Technol. B 1, 990.
Mory, C., 1985. Ph.D. Thesis, University of Orsay.
Mory, C., Colliex, C., Cowley, J.M., 1987. Ultramicroscopy 21, 171.

Mott-Smith, H.M., 1953. J. Appl. Phys. 24, 249.
Munro, E., 1987. Nucl. Instrum. Meth. Phys. Res. A258, 443.
Murata, K., Kyser, D.F., 1987. Advances in Electronics and Electron Physics. vol. 69 Academic Press, New York, p. 175.
Nagy, G.A., Szilagyi, M., 1974. Introduction to the Theory of Space Charge Optics. Macmillan Press Ltd, London.
Namkung, W., Chojnacki, E.P., 1986. Rev. Sci. Instrum. 57, 341.
Narum, D.H., Pease, R.F.W., 1986. J. Vac. Sci. Technol. B 4, 154.
Newton, L., 1687. Philosophiae Naturalis Principia Mathematica, In: Koyre, A., Cohen, I.B. (Eds.), Reprint of the third edition (1726). Harvard University Press, p. 1972.
Norris, T.S., Jones, G.A.C., Ahmed, H., 1987. In: Castagné, C., Perrocheau, J. (Eds.), Proceedings of the Microcircuits Engineering Conference, vol. 6, p. 99.
Nottingham, W.B., 1956. Handbuck der Physik. In: Flugge, S. (Ed.), vol. XXI. Springer-Verlag, Berlin, p. 1.
Orloff, J., 1985. J. Microsc. 140, 303.
Parzen, P., Goldstein, L., 1951. J. Appl. Phys. 22, 398.
Petric, P., Woodard, O., 1983. Solid State Technol. 9 (83), 154.
Pfeiffer, H.C., 1971. Rec. Symp. Electron Ion Laser Beam Technol. 11, 239.
Pfeiffer, H.C., 1972. Proc. Scanning Electron Microsc. 5, 113.
Pfeiffer, H.C., 1979. Recent advances in electron beam lithography for the high volume production of VLSI devices. IEEE Trans. Electron Dev. ED-26, 663.
Pfeiffer, H.C., 1984. Solid State Technol. 9 (84), 223.
Pfeiffer, H.C., Loeffler, K.H., 1970. Proc. Int. Conf. Electron Microsc. 17, 63.
Pierce, J.R., 1954. Theory and Design of Electron Beams. D. van Nostrand Company Inc., New York.
Pines, D., Bohm, D., 1952. Phys. Rev. 85, 338.
Pipes, L.A., Harvill, L.R., 1983. Applied Mathematics for Engineers and Physicists, ninth printing. McGraw-Hill, New York.
Piwczyk, B.P., Williams, A.E., 1983. Solid State Technol. 9 (83), 145.
Press, W.H., Flannery, B.P., Teukolsky, S.A., Vetterling, W.T., 1986. Numerical Recipes. Cambridge University Press, Cambridge.
Prewett, P.D., Kellogg, E.M., 1985. Nuclear Instrum. Meth. Phys. Res. B 6, 135.
Rehmet, M., 1965. Ph.D. Thesis, University of Berlin.
Reimer, L., 1984. Transmission Electron Microscopy. Springer-Verlag, Berlin.
Rose, H., 1988. Private communications.
Rose, H., Spehr, R., 1980. Optik 57, 339.
Rose, H., Spehr, R., 1983. Advances in Electronics and Electron Physics. vol. 13C Academic Press, New York, p. 475.
Saitou, N., Ozasa, S., Komoda, T., Tatsuno, G., Uno, Y., 1981. J. Vac. Sci. Technol. 19, 1087.
Saitou, N., Okumura, M., Matsuoka, G., Matsuzaka, T., Komoda, T., Sakitani, Y., 1985. J. Vac. Sci. Technol. B 3, 98.
Saitou, N., Hosoki, S., Okumura, M., Matsuzaka, T., Matsuoka, G., Ohyama, M., 1986a. In: Lehmann, W.W., Bleiker, Ch. (Eds.), Proceedings of the Microcircuits Engineering Conference, vol. 5, p. 123.
Saitou, N., Okazaki, S., Murai, F., Ozasa, S., Konishi, T., 1986b. J. Vac. Sci. Technol. B 4, 265.
Saitou, N., Okazaki, S., Nakamura, K., 1987. Solid State Technol. 11 (87), 65.
Sasaki, T., 1979. Conference on VLSI: Architecture, Design, Fabrication. California Institute of Technology.
Sasaki, T., 1982. J. Vac. Sci. Technol. 21, 695.
Sasaki, T., 1984. J. Vac. Sci. Technol. A 2, 1352.

Sasaki, T., 1986. J. Vac. Sci. Technol. B 4, 135.
Schiske, P., 1961. Physikalischer Verh. 12, 143.
Schiske, P., 1962. Proc. Int. Conf. Electron Microsc. 5, KK-9.
Schwartz, J.W., 1957. RCA Rev. 18, 3.
Seliger, R.L., Ward, J.W., Wang, V., Kubena, R.L., 1978. Appl. Phys. Lett. 34, 310.
Septier, A., 1967. Focussing of Charged Particles. Academic Press, Inc., New York. I, II.
Shao, Z., Crewe, A.V., 1987. Ultramicroscopy 23, 169.
Simpson, J.A., Kuyatt, C.E., 1966. J. Appl. Phys. 37, 3805.
Sivukhin, D.V., 1966. Reviews of Plasma Physics. In: Leontovich, M.A. (Ed.), vol. 4. Consultants Bureau Enterprises, Inc., New York, p. 93.
Smith, L.P., Hartman, P.L., 1940. J. Appl. Phys. 11, 220.
Spehr, R., 1985a. Optik 70, 109.
Spehr, R., 1985b. In: van der Mast, K.D., Radelaar, S. (Eds.), Proceedings of the Microcircuits Engineering Conference, vol. 3, p. 61.
Spehr, R., Rose, H., 1979. Annu. EMSA Meet. 37, 570.
Speidel, R., (1965/1966). Optik 23, 125.
Speidel, R., Gaukler, K.H., 1968. Z. Phys. 208, 419.
Speidel, R., Kurz, D., Gaukler, K.H., 1979. Optik 54, 257.
Speidel, R., Brauchle, P., Kramer, B., Schwab, U., 1985. Optik 71, 167.
Spitzer, L., 1962. Physics of Fully Ionized Gases. J. Wiley & Sons Inc., New York.
Steffen, K.G., 1965. High Energy Beam Optics. J. Wiley & Sons Inc., New York.
Stickel, W., Langner, G.O., 1983. J. Vac. Sci. Technol. B 1, 1007.
Stickel, W., Pfeiffer, H.C., 1978. Proceedings of the Symposium on Electron and Ion Beam Science and Technology., p. 32.
Stumpff, K., 1959. Himmelsmechanik Band I. VEB Deutscher Verlag der Wissenschaften, Berlin.
Swanson, L.W., 1975. J. Vac. Sci. Technol. 12, 1228.
Swanson, L.W., Schwind, G.A., Bell, A.E., Brady, J.E., 1979. J. Vac. Sci. Technol. 16, 1864.
Swanson, L.W., Schwind, G.A., Bell, A.E., 1980. J. Appl. Phys. 51, 3453.
Szilagyi, M., 1988. Electron and Ion Optics. Plenum Press, New York.
Takaoka, A., Ura, K., 1986. Optik 74, 71.
Takaoka, A., Sato, K., Ura, K., 1986. Proc. Int. Conf. Electron Microsc., Kyoto XI, 269.
Tang, T.T., 1983. Optik 64, 237.
Tang, T.T., 1987. Optik 76, 38.
Thompson, B.J., Headrick, L.B., 1940. Proc. Inst. Radio Eng. 28, 319.
Troyon, M., 1976. Optik 46, 439.
Troyon, M., 1987. J. Microsc. Spectrosc. Electr. 12, 431.
Troyon, M., 1988. J. Microsc. Spectrosc. Electr. 13, 49.
Troyon, M., Zinzindohoue, P., 1986. Proc. Int. Conf. Electron Microsc., Kyoto 11, 273.
Trubnikov, B.A., 1965. Reviews of Plasma Physics. In: Leontovich, M.A. (Ed.), vol. 1. Consultants Bureau Enterprises, Inc., New York, p. 105.
Tscharnuter, W., 1972. Gravitational N-Body Problem. Proceedings of IAU Colloquium, In: Lecar, M. (Ed.), vol. 10. D. Reidel Publishing Company, Dordrecht-Holland, p. 10.
Tuggle, D.W., Li, J.Z., Swanson, L.W., 1985. J. Microsc. 140, 293.
Tuggle, D.W., Swanson, L.W., Gesley, M.A., 1986. J. Vac. Sci. Technol. B 4, 131.
Ulmer, K., Zimmermann, B., 1964. Z. Phys. 182, 194.
Umemura, K., Ishitani, T., Tamura, H., 1986. Jpn. J. Appl. Phys. 25, L885.

van den Broek, M.H.L.M., 1984. Optik 67, 69.
van den Broek, M.H.L.M., 1986a. J. Appl. Phys. 59, 3923.
van den Broek, M.H.L.M., 1986b. Ph.D. Thesis, Delft University of Technology..
van der Mast, K.D., Jansen, G.H., 1987. In: Lehmann, W.W., Bleiker, Ch. (Eds.), Proceedings of the Microcircuits Engineering Conference, vol. 5, p. 93.
van der Mast, K.D., Jansen, G.H., Barth, J.E., 1985. In: van der Mast, K.D., Radelaar, S. (Eds.), Proceedings of the Microcircuits Engineering Conference, vol. 3, p. 43.
van Gorkum, G.G.P., Hoeberechts, A.M.E., 1986. J. Vac. Sci. Technol. B 4, 108.
van Haeringen, H., 1985. Charged Particle Interactions. Coulomb Press Leyden, Leiden, The Netherlands.
van Kampen, N.G., 1981. Stochastic Processes in Physics and Chemistry. North-Holland Publishing Company, Amsterdam.
van Leeuwen, J.M.J., Jansen, G.H., 1983. Optik 65, 179.
Varnell, G.L., Spicer, D.F., Hebley, J., Robbins, R., Carpenter, C., Malone, M., 1979. J. Vac. Sci. Technol. 16, 1787.
Veith, W., 1955. Z. Angew. Phys. 152, 306.
Venables, J.A., Cox, G., 1987. Ultramicroscopy 21, 33.
Venables, J.A., Janssen, A.P., 1980. Ultramicroscopy 5, 297.
Veneklasen, L.H., 1985. J. Vac. Sci. Technol. B 3, 185.
von Borries, D., Dosse, J., 1938. Arch. Electrotech. 32, 221.
Ward, J.W., 1984. J. Vac. Sci. Technol. B 3, 207.
Ward, J.W., Utlaut, M.W., Kubena, R.L., 1987. J. Vac. Sci. Technol. B 5, 169.
Ward, J.W., Kubena, R.L., Utlaut, M.W., 1988. J. Vac. Sci. Technol. B 6, 2090.
Watson, E.E., 1927. Philos. Mag. 3, 849.
Watson, K.W., 1955. Phys. Rev. 102, 12.
Weidenhausen, A., Spehr, R., Rose, H., 1985. Optik 69, 126.
Wendt, G., 1942. Z. Phys. 119, 423.
Wendt, G., 1943. Z. Phys. 120, 720.
Wendt, G., 1948. Ann. Phys. 2, 256.
Wilson, A.D., 1983. Proc. IEEE 71, 575.
Wolf, E.D., 1983. Proc. IEEE 71, 589.
Wolf, D.A., 1985a. J. Appl. Phys. 58, 3692.
Wolf, D.A., 1985b. J. Appl. Phys. 58, 3697.
Yau, Y.W., Groves, T.R., Pease, R.F.W., 1983. J. Vac. Sci. Technol. B 1, 1141.
Zimmermann, B., 1968. Ph.D. Thesis, University Karlsruhe.
Zimmermann, B., 1969. Rec. Symp. Electron Ion Laser Beam Technol. 10, 297.
Zimmermann, B., 1970. Advances in Electronics and Electron Physics. vol. 29 Academic Press, New York, p. 257.
Zinzindohoue, P., 1986. Optik 74, 131.
Zinzindohoue, P., Troyon, M., 1986. Proc. Int. Conf. Electron Microsc., Kyoto 11, 271.
Zvorykin, V.K., Morton, G.A., Ramberg, E.G., Hillier, J., Vance, A.W., 1961. Electron Optics and the Electron Microscope, fifth ed. J. Wiley & Sons Inc., New York. sect. 16.9.

Index

Note: Page numbers followed by "*f*" indicate figures, and "*t*" indicate tables.

A

Acceleration coefficient κ, 322
Addition of effects per beam segment.
 See Beam segment
Analytical ray tracing, 406–412
Angular deflections, statistical, 221–266
 in beam with crossover
  narrow, 225–231, 503–504
  of arbitrary dimensions, 240–261
 comparison of theories for, 474–476
 definition of, 5, 222
 distribution of, 229
 general aspects of, 222–225
 in homocentric cylindrical beam, 231–240, 501–502
 table of resulting equations for, 510*t*
Aperture, MC routine for, 400
ASYEBR, MC routine, 416–417
Average
 displacement/shift of test particle, 331, 341–342
 interaction force, 62–63, 332, 336
 reference trajectory, 302, 327–328
Axial velocity
 in terms of experimental parameters, 30
 spread
  generation of (*see* Boersch effect)
  reduction of, 46

B

BBGKY hierarchy, 36
Beam geometry quantities K, K_1, and K_2, 177, 502, 508*t*
Beam parameters. *See* Experimental parameters
Beam segment
 addition of effects per, 127–131, 518–520
 classification of, 31–32
 with crossover
  of arbitrary dimensions
   angular deflections in, 240–261
   Boersch effect in, 198–212, 507–513
   space charge effect in, 516–518
   trajectory displacement effect in, 288–292, 513–516
  narrow
   angular deflections in, 225–231, 504
   Boersch effect in, 177–189, 502, 504
   space charge effect in, 350–351
   trajectory displacement effect in, 272–285, 502, 504
 cylindrical, homocentric, 31
  angular deflections in, 231–240, 502
  Boersch effect in, 189–198, 500, 512–513
  space charge effect in, 352–355, 518
  trajectory displacement effect in, 286–288, 502, 515–516
Beam temperature, 41–49
 definition of, 42
 longitudinal, in terms of experimental parameters, 45–46
 transverse, in terms of experimental parameters, 48
Binary interaction approximation, 81
Boersch effect, 173–219
 in beam with crossover
  of arbitrary dimensions, 198–212, 507–513
  narrow, 177–189, 506
 comparison of theories for, 457–467
 definition of, 5, 174–175
 experimental data of, 12–13
 general aspects of, 174–176
 historical notes on, 3–6

Boersch effect (*Continued*)
  in homocentric cylindrical beam, 189–198, 500, 507
  resulting equations for, 507–513
    table of, 509*t*
  using Fokker–Planck approach, 85
Boltzmann
  constant, 41
  transport equation, 36
Brightness
  differential, 40
  normalized, 40, 48

C

Cathode temperature, 45, 396
Central collision, 150–151, 155
Central limit theorem, 68–69, 79–80
Center of mass system
  angular momentum in, 138
  coordinates in, 136–137
  kinetic energy in, 137
Chromatic aberration
  constant of, 371–372, 398
  MC routine for, 398, 400
  spot-width measures for, 370–384
Classification
  of beams, 31–32
  of collisions, 133–134, 491
  of interaction phenomena, 3–6
  of particle densities, 1–2, 499
Closest encounter approximation, 101–102, 463, 473–474, 484
Collective interaction, 4–5, 10–11
Colliding particle. *See* Field particles
Collision dynamics. *See* Two-particle, dynamics
Comparison
  of axial and lateral velocity displacements, 239–240
  of beams with and without crossover trajectory displacement effect in, 286–288
  of MC simulation and analytical theory, 421–422, 437–454
  of theories on statistical interactions, 99–106, 455–485

Complete collision
  conditions for, 150, 199
  definition of, 70–71, 133–134, 491
  dynamics of, 70–71, 141–144
  energy shift by, 177
Congruent flow, 338–339
Conjugated planes, optically, 38–40, 269
Continuity equation, 34
Continuum condition, 337
Coordinate
  system in orbital plane, 138–141
  transformation
    from orbital plane to laboratory system, 140
    from off-axis to on-axis reference system, 307
Coulomb
  force
    definition of, 61, 488–489
    displacements by components of, 20–21
    fluctuating and average component of, 61
    in MC sample, 401
    in terms of charge density distribution, 61, 333
  interaction phenomena
    definition of, 4, 487
    final equations for, 507–513
    physical aspects of, 488–499
  logarithm, 73, 75–80, 82
  scaling, 166–167
Cpu time
  for analytical ray tracing, 406–407
  for numerical ray tracing, 405, 411
Crossover. *See also* Beam segment
  position parameter, definition of, 28–30, 488, 508*t*
  radius r_c, definition of, 29*f*
  scaled
    $\bar{r}_c$, 178, 504–505, 508*t*
    r_c^*, 506, 508*t*
Cumulant
  of displacement distribution, 109
  generating function, 111
Current density distribution
  at emission, 397
  of Gaussian spot, 358*f*

general form, 334
of shaped spot, 362f
of spot limited by
   chromatic aberration, 370–384, 375f
   space charge defocussing, 388
   spherical aberration, 385
   trajectory displacement effect, 365

D

Data analysis, in MC program, 416–420
Debeye screening, 51–55, 97, 492
   length, 52, 73, 82
De Broglie wavelength, 32
Deflection angle
   for complete collision, 70–71, 70f, 143
   of deflector, 400
Deflector, MC routine for, 400
Defocussing distance. *See* Space charge, lens
Delta function, 106–107, 182
Differential cross section, 143–144
Diffusion
   approximation, 66–69
   equation, 67–68
   tensor
     calculation of, 69–74, 84
     definition of, 66
   in velocity space, 67, 79–80
Displacement
   distribution
     for axial velocity
       in beam with crossover of arbitrary dimensions, 206
       narrow, 236
       in homocentric cylindrical beam, 194
     definition of, 106, 330–331
     calculation method for, 106–109
     for lateral position
       in beam with crossover, 280f
       in homocentric cylindrical beam, 286
     for lateral velocity
       in beam with crossover of arbitrary dimensions, 252–253
       narrow, 230f
       in homocentric cylindrical beam, 183–184
     moments and cumulants of, 109–112

representation in k-domain, 120–127
shape of, 125f, 495
tails of, 126–127
width measures of, 125f, 365–370
   (*see also* Full width)
spatial, in reference plane, 163–166, 267, 502–503
vector $\Delta\eta$, 94–95, 135–136
in velocity
   longitudinal component, 160–162, 504–505
   as used in plasma physics, 71–72
   transverse component, 162–163, 223, 225, 504–505
$d_{p,1-p}$ width
   in MC program, 421–423
   of spot limited by
     chromatic aberration, 381, 383f
     space charge defocussing, 389f
     spherical aberration, 387f
     trajectory displacement effect, 369f
DRIFT1, MC routine, 401, 439–442, 450
DRIFT2, MC routine, 401, 439–442, 450
DRIFT3, MC routine, 413
Dynamical
   friction, coefficient of
     calculation of, 69–74, 84
     definition of, 66
     part of calculation, in analytical theory
       definition of, 93, 95
       in terms of interaction force, 98, 330–331
   variables, 135–136
d_{1288} and d_{2575} width, due to trajectory displacement effect
   resulting equations, 514–516

E

Eccentricity, 142
Edge-width. *See* $d_{p,1-p}$ width
Effective
   beam length (*see* Space charge, lens)
   defocussing distance (*see* Space charge, lens)
   parameter γ (*see* Shape parameter γ)
   width measure
     for Gaussian distribution, 216–218, 296

Effective (*Continued*)
  for truncated Gaussian distribution, 390–392
Electron beam lithography, 1
  variable shaped spot system, 28, 269, 361–362, 424
Emittance
  invariance of, 37–38
  plots of, 39f
Energy distribution
  and axial velocity distribution, 46, 187, 197, 313, 371, 499–500
  at emission, 44
  generated by Coulomb interactions (*see* Boersch effect)
  in MC calculation, 395, 415–416, 430f, 449f
Ensemble average, 33–34, 330–331
Entrance and exit plane, for ray tracing, 403, 409
Equation of motion
  Hamiltonian, 33
  Newtonian, 489
Experimental parameters
  definition of, 28–31, 29f, 488
  dependency of interaction phenomena on, 499–507
Exponential distribution, 431
Extended beam
  definition of, 32, 493
  interaction force in, 117–118
Extended two-particle model
  basic assumptions of, 299–300, 438, 490
  fundamentals of, 89–90, 103–104, 106
  in 1, 2 and 3 dimensions, 113–115
External field, uniform axial
  within analytical theory, 320–324
  within MC program, 403, 409–410

F

Fast Monte Carlo simulation, 406–407
Field emission, operating conditions for, 187
Field particles
  correlations in coordinates of, 96–97, 492
  definition of, 3–4, 93, 489–490
  within MC program, 412–413
  unperturbed trajectories of, 94, 134–135, 301–302, 301f
  unperturbed configuration of, 94, 492
Finite-size effects, 413–416, 422, 451
First order perturbation
  approximation, for reduction of N-particle problem, 99–101
  dynamics, 76, 153–156, 491
  conditions for, 153, 408
  theory for space charge effect, 341–343
First order properties, of space charge lens, 343–346
Fit algorithm, for MC program, 419–420, 433–436
Fit functions, within analytical theory
  error introduced by, 451
  for Full Width at Half Maximum, 186–187, 196–197, 211, 238, 256, 283–284, 290, 507, 509–511t, 511–515
  for Full Width median, 212, 284
  for small k-behaviour, 183, 185, 229, 235, 254, 337, 339
Fluctuation-dissipation theorem, example of, 69
Fluctuating interaction force
  distribution of, 86, 115–120
  model for, 63–67, 104–106
  separation from average force, 61–62, 79–80
  time scale of, 63–64, 82
Focused ion beam system, 3
Fokker–Planck
  approach, 63–67
  discussion of, 79–81
  validity of, 81–85
  equation, 66–67
Full collision dynamics, 408–409, 491
Full Width at Half Maximum (*FWHM*)
  of angular displacements
    in beam with crossover
      of arbitrary dimensions, 259–260
      narrow, 230
    in homocentric cylindrical beam, 238–240
    table of results for, 510t
  of axial velocity distribution
    in beam with crossover
      of arbitrary dimensions, 206, 207–209f, 209–212

Index

narrow, 187
in homocentric cylindrical beam, 196
of displacement distribution (general), 121–126, 125f, 126t
for different shapes (γ-values), 124–126, 125f, 497
of energy distribution
in beam with crossover
of arbitrary dimensions, 211–212, 507
narrow, 187–188
in homocentric cylindrical beam, 196–197, 512–513
table of results for, 509t
of lateral velocity distribution
in beam with crossover
narrow, 229–230
in homocentric cylindrical beam, 236–238
in MC program, 419, 450–451
of trajectory displacement distribution
in beam with crossover, 282–284, 290–291, 507
for different shapes (γ-values), 124–126, 125f, 370
in homocentric cylindrical beam, 286, 515–516
table of results for, 511t
Full Width median (FW_{50})
of axial velocity distribution, 212
definition of, 121, 361
of displacement distribution (general), 121–126, 125f, 126t
for different shapes (γ-values), 124–126, 125f, 497
of energy distribution, 507, 512–513
of lateral velocity distribution, 260–261
of trajectory displacement distribution, 284, 514–516
FW_f spot-width
definition of, 361
of energy distribution, 370
in MC program, 418, 450
of spot limited by
chromatic aberration, 381, 382f
space charge defocussing, 389
spherical aberration, 386
trajectory displacement effect, 368

G

Gaussian
angular and spatial distributions, 96, 213–218, 262–266, 296–297, 372, 431–432
displacement distribution, 120–121
regime
for axial velocity distribution
in beam with crossover of arbitrary dimensions, 206
narrow, 184
in homocentric cylindrical beam, 195–196
for lateral velocity distribution
in beam with crossover of arbitrary dimensions, 253
narrow, 230
in homocentric cylindrical beam, 237
for trajectory displacement distribution, 282
truncated distribution, 371–372, 390–392
Gauss's theorem, 333
Geometrical aberration. *See* Third order aberration
Geometrical variables ξ, 135, 157, 301–302, 307–308, 324
Ghost charge, 415

H

Half-complete collision, 158
Hamilton
equations of motion, 33
formalism, 32–35
function, 33
Helmholtz-Lagrange, law of, 40
Homocentric beam. *See* Beam segment
Holtsmark distribution
definition of, 12–13, 86, 87f, 493–494
relation with energy distribution, 88
representation in k-domain, 122
1-dimensional form, 119, 196
2-dimensional form, 118, 237–238
3-dimensional form, 115–116
Holtsmark regime
for axial velocity distribution
in beam with crossover, 206
in homocentric cylindrical beam, 196

Holtsmark regime (*Continued*)
  for lateral velocity distribution
    in beam with crossover, 254
    in homocentric cylindrical beam, 237–238
  for trajectory displacement distribution, 282

I

Image plane, 10, 28–30, 37–38, 499–500
  position parameter, definition of, 28–30, 488
Impact parameter
  definition of, 14, 101, 138
  distribution of, 16, 102, 463
Input files, for MC program, 425
INTERAC program, 438, 440–442, 450
Interaction
  range, 405–406, 413
  mechanism, 496
  phenomena (*see* Coulomb)
Internal kinetic energy
  definition of, 13, 42
  relaxation of (*see* Relaxation)

J

Jump moments, 65–66, 69, 73–74
  definition of, 65–66
  results for, 73–74

K

Kepler
  problem, 133–134, 490–491
  trajectory-equation, 142
Kinetic theory of gases, relation with methods of, 61
Knife-edge scans, 358–361. *See also* $d_{p,1-p}$ width

L

Lambert's (cosine) law, 44–45
Laminar flow
  condition for, 340
  definition of, 10–11, 337
  space charge model based on, 338–341
Langmuir's equation, 48
Lateral displacement. *See* Displacement
Lens, MC routine for, 398–401.
      See also Space charge, lens
Linear particle density. *See* Particle, density

Liouville's theorem
  consequences of, 37–41
  definition of, 34, 36
  derivation of, 34
Longitudinal velocity. *See* Axial velocity
Lorentz
  distribution, 121, 184, 431–432
  transformation, 325
Lorentzian regime, 184, 210

M

Magnetic interaction, 32
Markoff
  method, 89–90, 493
  process, 65
Maxwell-Boltzmann distribution, 41, 55
  half, 44, 395
MC data file, for MC program, 425, 427f
MCPLOT, MC plot program, 427–428
Mean square field fluctuation
  approximation, 104–106, 466
Meridian ray, 37–38, 341
Micro scales, 141, 157
Microscopic particle distribution, 51–52, 333
Model error, in MC program, 421
Moment
  of displacement distribution, 110
  generating function, 110–111
Monochromatic beam, 30, 92–93.
      See also Nonmonochromatic beams
Monte Carlo (MC) simulation of particle beams
  introduction to, 393–395
  historical notes on, 25–26
  program organization, 424–429
  versus analytical theory, 423–424, 437–454
    for geometry and current dependency, 445–449
    for voltage and current dependency, 441–445
Minimum beam cross section, for laminar flow, 339

N

N-body problem, reduction to 2-body problem, 81, 490–492
Nearest neighbour distribution, 102, 463–464

Nearly complete collision, dynamics of, 148–153, 167–172, 270, 491
  conditions for, 150, 198, 408–409
Nonmonochromatic beams, 312–320, 315–320f
Numerical ray tracing, 401–406
  integration error in, 402–403, 421–423

O

Object plane, 9–10, 38–40
Off/on-axis. See Reference trajectory
Optical elements, MC routines for, 398–401
Optically conjugated planes, 38–40
Orbital plane
  coordinate representation in, 138–141
  orientation of, 71f, 134f
Oscilloscope tube, 280
Output files, for MC program, 425–428

P

Particle
  density
    classification of, 2, 499
    linear, λ
      definition of, 30, 508t
      relativistic correction on, 326
      scaled
      k, 21
      λ, 183–184, 499, 508t
      λ^*, 194, 499, 508t
    three-dimensional, 30
  distribution function $f(r, v, t)$, 35, 85
  optics, brief description of, 9–10
Pencil beam
  definition of, 32, 493
  factor
    for beam with crossover χ_c, 210–211
    for cylindrical beam χ_p, 195
  interaction force in, 118–120
  regime
    for axial velocity distribution
      in beam with crossover, 210–211
      in homocentric cylindrical beam, 196
    for lateral velocity distribution
      in beam with crossover, 259
      in homocentric cylindrical beam, 238
    for trajectory displacement distribution, 283

Perihelion
  angle, 142
  time of passage, 144
Perveance, 187
Phase space
  of axial coordinates, 37
  of lateral coordinates, 37–38
  6-dimensional, 35
  6N-dimensional, 33
Plane of best focus, 24–25, 418–419, 452
Plasma frequency, 53, 78, 82
Plasma physics, relation with methods of, 11, 62–63, 84, 89
Poisson equation, 51–52
Polynomial fit function. See Fit algorithm
Potential energy of beam
  definition of, 55
  relaxation of (see Relaxation)
PROCCO, MC routine, 415
$p(k)$-transform
  approximation for
    for Boersch effect in narrow crossover, 183f
    by power law, 120–121, 365, 495
  for axial velocity displacements
    in beam with crossover
      of arbitrary dimensions, 198–212
      narrow, 180–183, 213
    in homocentric cylindrical beam, 189–198
  definition of, 107
  large k-behaviour of, 181, 215–216, 227–228, 263–264, 310
  for lateral velocity displacements
    in beam with crossover
      of arbitrary dimensions, 240–261
      narrow, 227–229, 263
    in homocentric cylindrical beam, 231–240
  real and imaginary part of, 112
  small k-behaviour of, 120–121, 181, 193, 205, 215, 227–228, 235, 252, 263, 278, 310
  for trajectory displacements, in beam with crossover, 276–278, 279f, 280

Q

Quadrupole, MC routine for, 400
Quantum mechanical effects, importance of, 32

R

Random number generator, 430–433
Ray-equation, 9–10
　incorporating space charge force, 336
　paraxial, 9–10
Ray tracing, in MC program,
　　401–406
READC, MC routine, 425–427
RECTBR, MC routine, 417
Reference particle. *See* Test particle
Reference plane. *See* Image plane
Reference trajectory
　average, 302
　definition of, 3–4, 93–94, 489–490
　off-axis, 112–113, 300–312, 452
　　approximating approach for, 307–312,
　　　311–312*f*
　　exact approach for, 300–307, 301*f*,
　　　304–306*f*
Regimes, 174, 222, 268, 496. *See also* and
　　Weak collision; Gaussian;
　　Holtsmark; Lorentzian; Pencil
　　beam
Relativistic
　beam potential, 325
　effects, 32, 324–327
Relaxation
　of internal kinetic energy, 13, 22, 50, 89,
　　110–111, 218, 496
　of potential energy, 50, 55–57, 89,
　　110–111, 219, 496–497
RNDTBR, MC routine, 417
Root mean square width
　of angular displacements, 470
　of axial velocity distribution
　　in beam with crossover, 184–185,
　　　207–209*f*
　　in homocentric cylindrical beam, 193
　definition of, 95, 184–185
　of energy distribution, 458
　of lateral velocity distribution
　　in beam with crossover, 229,
　　　255–256*f*
　in MC program, 417–418
　shortcomings of, 456, 484
　of trajectory displacement distribution,
　　280, 292, 477
Rotational error of lens, 399
Rutherford's scattering law, 143–144

S

Sample of particles
　definition of, 395
　length, 397, 413
　minimum size, 414–415
Scale
　measures
　　for d_0, v_0-scaling, 177
　　for micro-scaling, 141, 157
　　table of, 508*t*
　　for δ, v-scaling, 166
　transformation, between δ, v- and d_0,
　　　v_0-scaling, 201
Scanning electron microscope, 3
Seed, of initial conditions, 25, 405, 416, 422
Shape parameter γ, 495, 498, 519
　effective, 512–513, 515–516, 519
　total effective, 519
Shift. *See* Displacement
Slice method
　application of
　　in external acceleration field, 320–321
　　for *FWHM*
　　　of angular deflection distribution,
　　　　261–262
　　　of energy distribution, 210
　　　of trajectory displacement
　　　　distribution, 289–291
　definition of, 131–132, 498–499
Smoothed out distribution of charge, 4,
　　61–62, 334, 489–490
Sorting algorithm, 418–419
Source
　MC routine for, 395–398, 412–413
　velocity distribution at, 44, 395
Space charge
　effect, 329–356, 342*f*, 344–345*f*, 348*f*, 353*f*
　　in beam with laminar flow, 338–341
　　in beam with narrow crossover,
　　　350–351, 516–518
　definition of, 4, 330, 489–490
　effective beam length for, 355–356
　general aspects of, 330–338
　historical notes on, 23–25
　in homocentric cylindrical beam,
　　352–355, 353*f*, 518
　resulting equations for, 516–518
　spot-width measures for, 370–384
　and trajectory displacement effect, 452

Index

lens, 335
  defocussing by
    definition of, 5–6, 24–25, 452, 490
    results for, 345, 349, 354, 516–520
  first order optical properties of, 341–343
    addition of, 350
    magnification by, 346, 351, 516–517
    nonrefocusable broadening due to, 5–6, 24–25, 346–350, 490
    spherical aberration of, 348, 351, 354, 518
      addition of, 518
      disk of least confusion due to, 349, 351
      and effective defocussing, 349
    third order optical properties of, 346–351
Spatial shift. *See* Displacement
Spherical aberration
  constant, 384–385
  by space charge (*see* Space charge, lens)
  spot-width measures for, 384–387
Statistical
  angular deflections (*see* Angular deflections)
  effects, definition of, 4, 489–490
  error, in MC program, 420, 422
  part of calculation, in analytical theory
    for angular deflections, 224–225
    for Boersch effect, 175–176
    definition of, 93–95
    for trajectory displacement effect, 272
Stellar dynamics, relation with methods of, 35, 62–63, 78–79
Step function, definition of, 44
Stochastic
  probe broadening, 22
  ray deflections, 21–22
STOREC, MC routine, 425–427
STOREP, MC routine, 425–427
Subensemble, in MC program, 420
SYMEBR, MC routine, 416–417
System file, for MC program, 425–427, 427*f*

T

TBR, MC routine, 417
Temperature. *See* Beam temperature
Test particle, 3–4, 93, 489–490
Thermal velocities, 460

Thermionic emission
  operating conditions for, 188*f*
  velocity distribution for, 44, 396
Thermodynamical, description of the beam, 43–44
Thermodynamic limits
  for relaxation of kinetic energy, 50, 218, 458, 496
  for relaxation of potential energy, 50, 55–57, 218–219, 458–459, 496–497
Thin-lens approximation, 398
Third order aberration
  by space charge (*see* Space charge, lens)
  of thin-lens, 398–399
  of thin-quadrupole, 400
Time of flight
  in absence of acceleration, 30
  in presence of acceleration, 322, 409–410
  relativistic correction on, 325–326
Time step algorithm, 402–403, 421
Trajectory displacement effect, 267–297
  in beam with crossover
    of arbitrary dimensions, 288–292, 513–516
    homocentric, 272–285
  comparison of theories for, 476–481
  definition of, 5, 268
  experimental data of, 19
  general aspects of, 268–272
  historical notes on, 19–23, 20*f*
  in homocentric cylindrical beam, 286–288, 502, 515–516, 518
  relation with angular deflections, 269–270, 293–296
  resulting equations for, 513–516
  table of, 511*t*
  spot-width measures for, 365–370
Transition probability function, 65, 85
Two-particle
  distribution function
    for axial velocity displacements
      in beam with crossover of arbitrary dimensions, 198–212
      narrow, 177–189
      in homocentric cylindrical beam, 189–198

Two-particle (*Continued*)
  definition of, 108
  for lateral velocity displacements
    in beam with crossover of arbitrary dimensions, 240–261
    narrow, 240–261
    in homocentric cylindrical beam, 231–240
  for off-axis reference trajectories, 302, 307–308
  for trajectory displacements
    in beam with crossover, 272, 273f, 277
  dynamics, 133–172, 491
    calculation scheme for, 147f, 408–409
    of collision with zero initial relative velocity, 157–160
    of complete collision, 70–71, 141–144
    in Fast Monte Carlo program, 406–409
    of nearly complete collision, 148–153, 167–172
    numerical approach to, 147–148
    of weak collision, 76, 76f, 153–156

U

Universal beam spreading curve, 339–340
Unperturbed trajectory
  within analytical model (*see* Field particles)
  within MC model, 407, 410, 417

V

Velocity
  axial (*see* Axial velocity)
  scaled, transverse, v_0^*, 201, 506, 508t
  shift in (*see* Displacement)
Virtual spatial displacement. *See* Displacement, spatial
Vlasov equation, for single component, nonrelativistic, 62–63

W

Weak
  collision (*see* First order perturbation, dynamics)
  complete collision regime
    for lateral velocity distribution, 230–231, 254
    for trajectory displacement distribution, 282
Width measures, 124–126, 125f, 357–392, 358–360f, 362–363f, 367f, 369f, 372f, 374–375f, 377f, 379–380f, 382–383f, 387f, 389f, 391f. *See also* $d_{p,1-p}$ width; Full Width at Half Maximum; Full Width median; FW_f width; Root mean square width

Z

Z-dependent velocity correction, 415–416

Printed and bound by CPI Group (UK) Ltd, Croydon, CR0 4YY
09/12/2024
01802749-0001